Parasite Genomics Protocols

METHODS IN MOLECULAR BIOLOGY™

John M. Walker, SERIES EDITOR

289. **Epidermal Cells,** *Methods and Applications,* edited by *Kursad Turksen,* 2004
288. **Oligonucleotide Synthesis,** *Methods and Applications,* edited by *Piet Herdewijn,* 2004
287. **Epigenetics Protocols,** edited by *Trygve O. Tollefsbol,* 2004
286. **Transgenic Plants:** *Methods and Protocols,* edited by *Leandro Peña,* 2004
285. **Cell Cycle Control and Dysregulation Protocols:** *Cyclins, Cyclin-Dependent Kinases, and Other Factors,* edited by *Antonio Giordano and Gaetano Romano,* 2004
284. **Signal Transduction Protocols,** *Second Edition,* edited by *Robert C. Dickson and Michael D. Mendenhall,* 2004
283. **Bioconjugation Protocols,** edited by *Christof M. Niemeyer,* 2004
282. **Apoptosis Methods and Protocols,** edited by *Hugh J. M. Brady,* 2004
281. **Checkpoint Controls and Cancer, Volume 2:** *Activation and Regulation Protocols,* edited by *Axel H. Schönthal,* 2004
280. **Checkpoint Controls and Cancer, Volume 1:** *Reviews and Model Systems,* edited by *Axel H. Schönthal,* 2004
279. **Nitric Oxide Protocols,** *Second Edition,* edited by *Aviv Hassid,* 2004
278. **Protein NMR Techniques,** *Second Edition,* edited by *A. Kristina Downing,* 2004
277. **Trinucleotide Repeat Protocols,** edited by *Yoshinori Kohwi,* 2004
276. **Capillary Electrophoresis of Proteins and Peptides,** edited by *Mark A. Strege and Avinash L. Lagu,* 2004
275. **Chemoinformatics,** edited by *Jürgen Bajorath,* 2004
274. **Photosynthesis Research Protocols,** edited by *Robert Carpentier,* 2004
273. **Platelets and Megakaryocytes, Volume 2:** *Perspectives and Techniques,* edited by *Jonathan M. Gibbins and Martyn P. Mahaut-Smith,* 2004
272. **Platelets and Megakaryocytes, Volume 1:** *Functional Assays,* edited by *Jonathan M. Gibbins and Martyn P. Mahaut-Smith,* 2004
271. **B Cell Protocols,** edited by *Hua Gu and Klaus Rajewsky,* 2004
270. **Parasite Genomics Protocols,** edited by *Sara E. Melville,* 2004
269. **Vaccina Virus and Poxvirology:** *Methods and Protocols,* edited by *Stuart N. Isaacs,* 2004
268. **Public Health Microbiology:** *Methods and Protocols,* edited by *John F. T. Spencer and Alicia L. Ragout de Spencer,* 2004

267. **Recombinant Gene Expression:** *Reviews and Protocols, Second Edition,* edited by *Paulina Balbas and Argelia Johnson,* 2004
266. **Genomics, Proteomics, and Clinical Bacteriology:** *Methods and Reviews,* edited by *Neil Woodford and Alan Johnson,* 2004
265. **RNA Interference, Editing, and Modification:** *Methods and Protocols,* edited by *Jonatha M. Gott,* 2004
264. **Protein Arrays:** *Methods and Protocols,* edited by *Eric Fung,* 2004
263. **Flow Cytometry,** *Second Edition,* edited by *Teresa S. Hawley and Robert G. Hawley,* 2004
262. **Genetic Recombination Protocols,** edited by *Alan S. Waldman,* 2004
261. **Protein–Protein Interactions:** *Methods and Applications,* edited by *Haian Fu,* 2004
260. **Mobile Genetic Elements:** *Protocols and Genomic Applications,* edited by *Wolfgang J. Miller and Pierre Capy,* 2004
259. **Receptor Signal Transduction Protocols,** *Second Edition,* edited by *Gary B. Willars and R. A. John Challiss,* 2004
258. **Gene Expression Profiling:** *Methods and Protocols,* edited by *Richard A. Shimkets,* 2004
257. **mRNA Processing and Metabolism:** *Methods and Protocols,* edited by *Daniel R. Schoenberg,* 2004
256. **Bacterial Artifical Chromosomes, Volume 2:** *Functional Studies,* edited by *Shaying Zhao and Marvin Stodolsky,* 2004
255. **Bacterial Artifical Chromosomes, Volume 1:** *Library Construction, Physical Mapping, and Sequencing,* edited by *Shaying Zhao and Marvin Stodolsky,* 2004
254. **Germ Cell Protocols, Volume 2:** *Molecular Embryo Analysis, Live Imaging, Transgenesis, and Cloning,* edited by *Heide Schatten,* 2004
253. **Germ Cell Protocols, Volume 1:** *Sperm and Oocyte Analysis,* edited by *Heide Schatten,* 2004
252. **Ribozymes and siRNA Protocols,** *Second Edition,* edited by *Mouldy Sioud,* 2004
251. **HPLC of Peptides and Proteins:** *Methods and Protocols,* edited by *Marie-Isabel Aguilar,* 2004
250. **MAP Kinase Signaling Protocols,** edited by *Rony Seger,* 2004
249. **Cytokine Protocols,** edited by *Marc De Ley,* 2004
248. **Antibody Engineering:** *Methods and Protocols,* edited by *Benny K. C. Lo,* 2004
247. **Drosophila Cytogenetics Protocols,** edited by *Daryl S. Henderson,* 2004
246. **Gene Delivery to Mammalian Cells: Volume 2:** *Viral Gene Transfer Techniques,* edited by *William C. Heiser,* 2004

METHODS IN MOLECULAR BIOLOGY™

Parasite Genomics Protocols

Edited by

Sara E. Melville

*Department of Pathology,
University of Cambridge, Cambridge, UK*

HUMANA PRESS ✳ TOTOWA, NEW JERSEY

© 2004 Humana Press Inc.
999 Riverview Drive, Suite 208
Totowa, New Jersey 07512

www.humanapress.com

All rights reserved. No part of this book may be reproduced, stored in a retrieval system, or transmitted in any form or by any means, electronic, mechanical, photocopying, microfilming, recording, or otherwise without written permission from the Publisher. Methods in Molecular Biology™ is a trademark of The Humana Press Inc.

All papers, comments, opinions, conclusions, or recommendations are those of the author(s), and do not necessarily reflect the views of the publisher.

This publication is printed on acid-free paper. ∞
ANSI Z39.48-1984 (American Standards Institute)
Permanence of Paper for Printed Library Materials.
Cover illustration: Background: Bloodstream-form Trypanosoma brucei parasite, magnified at 10,000 X. Courtesy of Dr. Laurence Tetley, University of Glasgow. Foreground: Figure 4A in chapter 21, *FISH Mapping for Helminth Genomes,* by Hirohisa Hirai and Yuriko Hirai.

Production Editor: Angela L. Burkey.
Cover design by Patricia F. Cleary.

For additional copies, pricing for bulk purchases, and/or information about other Humana titles, contact Humana at the above address or at any of the following numbers: Tel.: 973-256-1699; Fax: 973-256-8341; E-mail: humana@humanapr.com; or visit our Website: www.humanapress.com

Photocopy Authorization Policy:
Authorization to photocopy items for internal or personal use, or the internal or personal use of specific clients, is granted by Humana Press Inc., provided that the base fee of US $25.00 copy is paid directly to the Copyright Clearance Center at 222 Rosewood Drive, Danvers, MA 01923. For those organizations that have been granted a photocopy license from the CCC, a separate system of payment has been arranged and is acceptable to Humana Press Inc. The fee code for users of the Transactional Reporting Service is: [1-58829-062-X /04 $25.00].
e-ISBN: 1-59259-793-9

Printed in the United States of America. 10 9 8 7 6 5 4 3 2 1

Library of Congress Cataloging-in-Publication Data

Parasite genomics protocols / edited by Sara E. Melville.
  p. ; cm. -- (Methods in molecular biology ; v. 270)
Includes bibliographical references and index.
 ISBN 1-58829-062-X (alk. paper)
 1. Parasites--Molecular genetics--Methodology.
 [DNLM: 1. Genomics--methods. 2. Parasites--genetics. 3. Genetic Techniques. QU 58.5 P223 2004] I. Melville, Sara E. II. Series:
Methods in molecular biology (Clifton, N.J.) ; v. 270.
 QL757.P25 2004
 616.9'6--dc22
                               2003025343

Preface

Parasitic diseases remain a major health problem throughout the world, for both humans and animals. For many of us, our technologically advanced lifestyle has decreased the prevalence and transmission of parasitic diseases, but for the majority of the world's population, they are ever present in homes, domestic animals, food, or the environment. The study of parasites and parasitic disease has a long and distinguished history. In some cases, it has been driven by the great importance of the presence of the parasite to the community, for example, those that affect our livestock. In other cases, it is clear that applied research has suffered for lack of funding because the parasite affects people with few resources, such as the rural poor in resource-poor countries. These instances include the so-called "neglected diseases," as defined by the World Health Organization (WHO).

Parasites have complicated life cycles, and a thorough understanding of the unique characteristics of a particular parasite species is vital in attempts to avoid, prevent, or cure infection or to alleviate symptoms. Of course, the biological characteristics that each parasite has developed to aid survival and transmission, to avoid destruction by the immune system, and to adapt to a changing environment are of lasting fascination to basic biologists as well. The elegance of these biological systems has ensured that the study of protozoan and metazoan parasites also remains an active field of research in countries where the diseases are not a threat to the population.

Over the last decade, we have seen the increasing application of genomics technology to the study of parasites and parasitic disease. Malaria genomics was first, and we can thank the timely intervention of the World Bank/United Nations Development Program/WHO Special Program in Tropical Diseases Research for the fact that several parasites causing neglected diseases also entered the genome sequencing programs at a relatively early stage. The nuclear genomes of the protozoan parasites malaria, African and South American trypanosomes, and *Leishmania* are complete or close to completion. Comparative genomics programs involving shotgun sequencing of genomes of related species are already underway. Several other organisms have benefited from substantial expressed sequence tag programs, such as the protozoans *Toxoplasma* and *Eimeria* and the metazoan parasites *Schistosoma* and *Filaria*, prior to entering a genome-sequencing program.

However, relatively few of the world's parasites will have the luxury of a genome sequencing program. Also, the field of parasitology encompasses a vast range of organisms, single-celled and multicellular, extracellular and intracellular, that offer different advantages and disadvantages for genome analysis. For this reason, it is not simple to compile a volume of methods that are widely applicable. For example, extraction of DNA or RNA from extracellular parasites that may be cultured in vitro or in vivo is usually relatively facile, but extraction of pure parasite nucleic acids is more difficult when working with intracellular parasites. Where neither ex vivo culture nor in vivo growth in laboratory rodents is possible, the use of many of the techniques described in *Parasite Genomics Protocols* is severely limited. Many of the protozoan parasites have no or few introns, such that genome sequencing is particularly efficient for gene discovery. However, the very high (80%) adenine-thymine content of the *Plasmodium falciparum* genome and the astonishing polymorphism of the *Trypanosoma cruzi* genome both required the development of new methods for sequencing and algorithms for analysis. Transformation of organisms with foreign DNA is considered a vital part of efficient functional genomic analysis, yet this is only possible in few species.

Parasite Genomics Protocols begins with three reviews from two major sequencing centers to set in context the genome data that are available. Different sequencing approaches result in different types of datasets that can be confusing for the user. Annotation is an often misunderstood area, and it is vital to understand its strengths and its limitations to avoid over- or underinterpretation of the data presented. The multiplicity of databases can also prove frustrating if the user is not able to appreciate the basis of the different decisions taken by developers. Chapters 1–3 aim to help researchers avoid the obvious pitfalls and to gain access to these resources with minimum frustration.

The remaining chapters describe methods that may be loosely grouped under one or more umbrellas headed genomics, functional genomics, or postgenomics. Some of these protocols have been enabled or facilitated by the availability of DNA sequence and/or by the biological resources created by the sequencing programs, such as microarray analysis. Some techniques were developed without regard to the availability of whole-genome sequence data, but are now seen as major players in the functional analysis of novel genes identified in sequencing programs, such as RNA interference, gene knockout, mutagenesis, and others. Included are protocols that are also applicable to organisms for which limited sequence data are available, such as rapid amplification of cDNA ends (RACE), subtraction libraries, amplified fragment length polymorphism (AFLP) analysis and others. We can see from the various methods presented that genome sequencing, annotation, and the development of user-friendly databases are essentially enabling technologies, allowing researchers to progress more rapidly in their genetic research while opening new avenues for global analysis.

I would like to thank all the authors of *Parasite Genomics Protocols* for the care they took in preparing their detailed protocols and for their suggestions of additional chapters. I would also like to thank John Walker, the series editor, for his help with editing. I especially thank Vanessa Toone for her time and care during the critical phase of preparation for publishing. I am very aware that this is but a snapshot of the relevant research being carried out in laboratories across the world and, indeed, that new applications are already on the horizon. However, we have aimed to provide sufficient details and tips to allow interested researchers to assess which of these techniques may be applied and developed for the study of their parasite of interest.

The full scientific potential of the availability of genome sequence and associated resources is not yet realized, neither in actuality nor in our imagination. For those of us who will remember the before and after, it will be exciting to reflect in years to come on their contribution to rapid progress in the study of the molecular biology and biochemistry of these fascinating organisms in the 21st century.

Sara Melville

Contents

Preface v
Contributors xi

1 Sequencing Strategies for Parasite Genomes
   Daniella Bartholomeu and Najib M. El-Sayed 1

2 Annotation of Parasite Genomes
   Matthew Berriman and Midori Harris 17

3 Parasite Genome Databases and Web-Based Resources
   Christiane Hertz-Fowler and Neil Hall 45

4 Expressed Sequence Tags: *Medium-Throughput Protocols*
   *Claire Whitton, Jennifer Daub, Marian Thompson,
   and Mark Blaxter* 75

5 Expressed Sequence Tags: *Analysis and Annotation*
   John Parkinson and Mark Blaxter 93

6 Positive Selection Scanning of Parasite DNA Sequences
   Winston A. Hide and Raphael D. Isokpehi 127

7 RACE and RAGE Cloning in Parasitic Microbial Eukaryotes
   Bryony A. P. Williams and Robert P. Hirt 151

8 Amplified (Restriction) Fragment Length Polymorphism
   (AFLP) Analysis
   Daniel K. Masiga and C. Michael R. Turner 173

9 Minisatellites and MVR-PCR for the Individual Identification
   of Parasite Isolates
   Annette MacLeod 187

10 Analysis of Differentially Expressed Parasite Genes and Proteins
   Using Transcriptomics and Proteomics
   Daniel C. Gare 203

11 Gene Expression Studies Using Self-Fabricated Parasite cDNA
   Microarrays
   Karl F. Hoffmann and Jennifer M. Fitzpatrick 219

12 DNA Content Analysis on Microarrays
   Upinder Singh, Preetam H. Shah, and Ryan C. MacFarlane 237

13 Typing Single-Nucleotide Polymorphisms in *Toxoplasma gondii*
   by Allele-Specific Primer Extension and Microarray Detection
   *Chunlei Su, Christian Hott, Bernard H. Brownstein,
   and L. David Sibley* 249

14 Transfection of the Human Malaria Parasite *Plasmodium falciparum*
    Brendan S. Crabb, Melanie Rug, Tim-Wolf Gilberger, Jennifer K. Thompson, Tony Triglia, Alexander G. Maier, and Alan F. Cowman 263

15 A PCR-Based Method for Gene Deletion and Protein Tagging in *Trypanosoma brucei*
    George K. Arhin, Shuiyuan Shen, Elisabetta Ullu, and Christian Tschudi 277

16 Analysis of Gene Function in *Trypanosoma brucei* Using RNA Interference
    Appolinaire Djikeng, Shuiyuan Shen, Christian Tschudi, and Elisabetta Ullu 287

17 In Vitro Shuttle Mutagenesis Using Engineered Mariner Transposons
    Kelly A. Robinson, Sophie Goyard, and Stephen M. Beverley 299

18 Random Mutagenesis Strategies for Construction of Large and Diverse Clone Libraries of Mutated DNA Fragments
    Sudsanguan Chusacultanachai and Yongyuth Yuthavong 319

19 Separation, Digestion, and Cloning of Intact Parasite Chromosomes Embedded in Agarose
    Vanessa Leech, Michael A. Quail, and Sara E. Melville 335

20 Chromosome Fragmentation in *Leishmania*
    Pascal Dubessay, Christine Blaineau, Patrick Bastien, and Michel Pagès 353

21 FISH Mapping for Helminth Genomes
    Hirohisa Hirai and Yuriko Hirai 379

22 Fiber-FISH: Fluorescence *In Situ* Hybridization on Stretched DNA
    Klaus Ersfeld 395

23 Yeast Two-Hybrid Assay for Studying Protein–Protein Interactions
    Ahmed Osman 403

24 From Genomes to Vaccines for Leishmaniasis
    Carmel B. Stober 423

Index 439

Contributors

GEORGE K. ARHIN • *Department of Internal Medicine, Yale University School of Medicine, New Haven, CT*
DANIELLA BARTHOLOMEU • *Department of Parasite Genomics, The Institute for Genomic Research, Rockville, MD*
PATRICK BASTIEN • *CNRS UMR5093 "Génome et Biologie Moléculaire des Protozoaires Parasites," Laboratoire de Parasitologie-Mycologie, Faculté de Médecine, Montpellier, France*
MATTHEW BERRIMAN • *Pathogen Sequencing Unit, Wellcome Trust Sanger Institute, Hinxton, Cambridgeshire, UK*
STEPHEN M. BEVERLEY • *Department of Molecular Microbiology, Washington University Medical School, St. Louis, MO*
CHRISTINE BLAINEAU • *CNRS UMR5093 "Génome et Biologie Moléculaire des Protozoaires Parasites," Laboratoire de Parasitologie-Mycologie, Faculté de Médecine, Montpellier, France*
MARK BLAXTER • *Institute of Cell, Animal, and Population Biology, Ashworth Laboratories, University of Edinburgh, Edinburgh, UK*
BERNARD H. BROWNSTEIN • *Department of Genetics, Washington University School of Medicine, St. Louis, MO*
SUDSANGUAN CHUSACULTANACHAI • *National Center for Genetic Engineering and Biotechnology, National Science and Technology Development Agency, Science Park, Pathumthani, Thailand*
ALAN F. COWMAN • *Division of Infection and Immunity, The Walter and Eliza Hall Institute of Medical Research, Victoria, Australia*
BRENDAN S. CRABB • *Division of Infection and Immunity, The Walter and Eliza Hall Institute of Medical Research, Victoria, Australia*
JENNIFER DAUB • *Institute of Cell, Animal, and Population Biology, Ashworth Laboratories, University of Edinburgh, Edinburgh, UK*
APPOLINAIRE DJIKENG • *Department of Internal Medicine, Yale University School of Medicine, New Haven, CT*
PASCAL DUBESSAY • *CNRS UMR5093 "Génome et Biologie Moléculaire des Protozoaires Parasites," Laboratoire de Parasitologie-Mycologie, Faculté de Médecine, Montpellier, France*
NAJIB M. EL-SAYED • *Department for Parasite Genomics, The Institute for Genomic Research, Rockville, MD*
KLAUS ERSFELD • *Department of Biological Sciences, University of Hull, Hull, UK*

JENNIFER M. FITZPATRICK • *Department of Pathology, University of Cambridge, Cambridge, UK*
DANIEL C. GARE • *Department of Pathology, University of Cambridge, Cambridge, UK*
TIM-WOLF GILBERGER • *Division of Infection and Immunity, The Walter and Eliza Hall Institute of Medical Research, Victoria, Australia*
SOPHIE GOYARD • *Immunophysiologie et Parasitisme Intracellulaire, Institut Pasteur, Paris, France*
NEIL HALL • *Informatics, The Institute for Genomic Research, Rockville, MD*
MIDORI HARRIS • *European Bioinformatics Institute, Hinxton, Cambridgeshire, UK*
CHRISTIANE HERTZ-FOWLER • *Pathogen Sequencing Unit, Wellcome Trust Sanger Institute, Hinxton, Cambridgeshire, UK*
WINSTON A. HIDE • *Pathogen Bioinformatics Unit, South African National Bioinformatics Institute, University of the Western Cape, Bellville, South Africa*
HIROHISA HIRAI • *Center for Human Evolutionary Modeling Research; Department of Cellular and Molecular Biology, Primate Research Institute, Kyoto University, Inuyama, Aichi, Japan*
YURIKO HIRAI • *Center for Human Evolutionary Modeling Research; Department of Cellular and Molecular Biology, Primate Research Institute, Kyoto University, Inuyama, Aichi, Japan*
ROBERT P. HIRT • *Department of Zoology, The Natural History Museum, London, UK*
KARL F. HOFFMANN • *Department of Pathology, University of Cambridge, Cambridge, UK*
CHRISTIAN HOTT • *Department of Genetics, Washington University School of Medicine, St. Louis, MO*
RAPHAEL D. ISOKPEHI • *Pathogen Bioinformatics Unit, South African National Bioinformatics Institute, University of the Western Cape, Bellville, South Africa*
VANESSA LEECH • *Department of Pathology, University of Cambridge, Cambridge, UK*
RYAN C. MACFARLANE • *Departments of Internal Medicine and Microbiology, Stanford University, Stanford, CA*
ANNETTE MACLEOD • *Wellcome Centre for Molecular Parasitology, Anderson College, University of Glasgow, Glasgow, UK*
ALEXANDER G. MAIER • *Division of Infection and Immunity, The Walter and Eliza Hall Institute of Medical Research, Victoria, Australia*

Contributors

DANIEL K. MASIGA • *Molecular Biology and Biotechnology Unit, International Centre of Insect Physiology and Ecology (ICIPE), Nairobi, Kenya*

SARA E. MELVILLE • *Department of Pathology, University of Cambridge, Cambridge, UK*

AHMED OSMAN • *Department of Microbiology, School of Medicine and Biomedical Sciences, SUNY at Buffalo, Buffalo, NY*

MICHEL PAGÈS • *CNRS UMR5093 "Génome et Biologie Moléculaire des Protozoaires Parasites," Laboratoire de Parasitologie-Mycologie, Faculté de Médecine, Montpellier, France*

JOHN PARKINSON • *Institute of Cell, Animal, and Population Biology, Ashworth Laboratories, University of Edinburgh, Edinburgh, UK*

MICHAEL A. QUAIL • *Pathogen Sequencing Unit, Wellcome Trust Sanger Institute, Hinxton, Cambridgeshire, UK*

KELLY A. ROBINSON • *Department of Molecular Microbiology, Washington University Medical School, St. Louis, MO*

MELANIE RUG • *Division of Infection and Immunity, The Walter and Eliza Hall Institute of Medical Research, Victoria, Australia*

PREETAM H. SHAH • *Departments of Internal Medicine and Microbiology, Stanford University, Stanford, CA*

SHUIYUAN SHEN • *Department of Internal Medicine, Yale University School of Medicine, New Haven, CT*

L. DAVID SIBLEY • *Department of Molecular Microbiology, Washington University School of Medicine, St. Louis, MO*

UPINDER SINGH • *Departments of Internal Medicine and Microbiology, Stanford University, Stanford, CA*

CARMEL B. STOBER • *Department of Medicine, Cambridge Institute for Medical Research, University of Cambridge, Cambridge, UK*

CHUNLEI SU • *Department of Molecular Microbiology, Washington University School of Medicine, St. Louis, MO*

JENNIFER K. THOMPSON • *Division of Infection and Immunity, The Walter and Eliza Hall Institute of Medical Research, Victoria, Australia*

MARIAN THOMPSON • *Institute of Cell, Animal, and Population Biology, Ashworth Laboratories, University of Edinburgh, Edinburgh, UK*

TONY TRIGLIA • *Division of Infection and Immunity, The Walter and Eliza Hall Institute of Medical Research, Victoria, Australia*

CHRISTIAN TSCHUDI • *Departments of Epidemiology & Public Health and Internal Medicine, Yale University School of Medicine, New Haven, CT*

C. MICHAEL R. TURNER • *Division of Infection and Immunity, Institute of Biomedical and Life Sciences (IBLS), Glasgow University, Glasgow, UK*

ELISABETTA ULLU • *Departments of Internal Medicine and Cell Biology, Yale University School of Medicine, New Haven, CT*

CLAIRE WHITTON • *Institute of Cell, Animal, and Population Biology, Ashworth Laboratories, University of Edinburgh, Edinburgh, UK*

BRYONY A. P. WILLIAMS • *Department of Botany, University of British Columbia, Vancouver, Canada*

YONGYUTH YUTHAVONG • *National Center for Genetic Engineering and Biotechnology, National Science and Technology Development Agency, Science Park, Pathumthani, Thailand*

1

Sequencing Strategies for Parasite Genomes

Daniella Bartholomeu and Najib M. El-Sayed

Summary

Recent advances in the field of sequencing have enabled the determination of the complete nucleotide sequence of a large number of complex genomes. The complete genome sequence of the parasite *Plasmodium falciparum* has been published recently, and many other parasite genome initiatives are underway. Parasite genomes vary in size, nucleotide composition, polymorphism level, content, and distribution of repetitive elements. These genomic features affect the performance of sequencing strategies. As a consequence, each of the ongoing parasite genome projects has adopted distinct sequencing approaches. The degree of completeness and accuracy desired as well as available funds should be considered carefully when choosing the most appropriate sequencing strategy.

Key Words: BAC ends database; clone-by-clone strategy; fingerprinting; map-as-you-go; minimal tiling path; optical mapping; parasite genomes; physical mapping; sequence assembly; sequence-ready map; sequencing strategies; whole-chromosome shotgun; whole-genome shotgun.

1. Introduction

Improvements in the technologies and strategies used for DNA sequence determination have enabled the sequencing of a growing number of complex genomes in an affordable, faster, and more accurate manner. Parasite genomes vary in size, nucleotide composition, polymorphism level, content, and distribution of repetitive elements. These genomic features affect the performance of sequencing strategies. This chapter discusses the principal strategies adopted by ongoing parasite genome projects and the factors that affect the choice of an optimal method.

2. Clone-by-Clone Strategy

The clone-by-clone strategy is one of the most commonly used approaches for the sequencing of complex genomes and has been adopted by the *Trypanosoma brucei* (*1*) and *Leishmania major* genome projects (*2*).

The initial step of the clone-by-clone strategy consists of cutting the DNA into large fragments of 40–300 kb and cloning them into a suitable vector. Overlapping clones are identified, and their relative order along the chromosome is determined. Each clone is then fragmented into smaller pieces (1–3 kb), which are cloned into plasmid vectors. The subclones are end-sequenced and the reads are assembled to generate the sequence of the large insert clone.

The conventional clone-by-clone approach requires *a priori* physical mapping for each region targeted for sequencing. The physical mapping consists of identifying and ordering overlapping fragments that together span a region of interest. For example, if the objective is to sequence a chromosome, a map of ordered clones is generally identified before the initiation of the sequencing phase. Typically, three vector systems are used to construct physical maps: yeast artificial chromosome (YAC), bacterial artificial chromosome (BAC), and cosmid vectors. Inserts up to 1 megabase (Mb) in size can be cloned in YACs, up to 200 kb in BACs, and up to 40 kb in cosmid vectors. The choice of the cloning system will depend primarily on the genome size. In general, a bigger genome requires a larger average insert size to minimize the number of clones necessary to cover one genome equivalent. However, other aspects, such as clone stability, should be taken into account. For example, BAC clones are more stable than YACs and cosmids, where the occurrence of deletions, chimeric inserts, and DNA rearrangements is common. In addition, BAC clones are an optimal substrate for shotgun sequencing because of their low vector-to-insert ratio, whereas the contamination with yeast DNA during the subcloning of short fragments of YAC inserts is a common occurrence. On the other hand, highly repetitive sequences seem to be underrepresented in bacterial systems, and yeast is able to maintain AT-rich DNA more effectively than *Escherichia coli (3)*. Therefore, besides the genome size, intrinsic genome and cloning vector features should be evaluated and, often, more than one cloning system is chosen.

The most common strategy used to construct physical maps involves a combination of fingerprinting and marker hybridization. Briefly, restriction digestion is performed on a large number of clones, and computer algorithms are used to compare the resulting patterns (fingerprints), assemble the clones into contigs, and infer the extent of clone overlap. The contigs can then be assigned to a chromosome by the hybridization of clone-derived markers to pulse-field gel (PFG) blots (*see* Chapter 19). Alternatively, cloned DNA markers (such as expressed sequence tags [ESTs, *see* Chapter 4] or sequenced tagged sites [STSs], perhaps resulting from a genome survey sequence project [*see* Chapter 3]) previously assigned to a unique chromosome can be used to screen a high-density filter containing the gridded library.

The contig map formed by a set of ordered clones that confer around 10-fold coverage to a genomic region is known as a sequence-ready map. Once this map

Sequencing Strategies

is constructed, the next step is to select a minimally overlapping path of clones. This collection of clones, which together provides complete coverage across a genomic region with minimal overlap, is called a minimal tiling path. Next, the selected clones are submitted to shotgun sequencing.

The physical mapping step is an extremely laborious and time-consuming process. To circumvent these limitations, an alternative clone-by-clone strategy that eliminates the need for any prior mapping was proposed (4). This strategy, known as map-as-you-go, uses BAC libraries as the main sequencing substrate and can be divided into two main steps. Briefly, the first step consists of the generation of paired end-sequence markers achieved through the sequencing of both ends of a large number of BAC clones. These sequences are used to construct a BAC ends database. In a second step, BAC clones containing sequences whose chromosomal localizations were previously assigned, known as seed BACs, are selected from the BAC ends database or identified by hybridization to high-density filters containing all clones from the BAC library. Once the seed BAC is sequenced to contiguity, this sequence is checked against the BAC ends database to identify overlapping clones, which are then fingerprinted. Two BAC clones showing internal consistency among the fingerprints and minimal overlap at either end of the seed BAC are selected for the next round of sequencing. This strategy, illustrated in **Fig. 1**, has been used successfully in various sequencing projects, including the *T. brucei* genome project at The Institute for Genomic Research (TIGR) (1), as well as the *Arabidopsis thaliana* genome (5). Below, each step is discussed in further detail.

2.1. Generation of Paired End-Sequence Markers

Because the map-as-you-go strategy does not require prior physical mapping data, the BAC ends database is the main source for markers used to select clones for sequencing. Assuming a random BAC library, the end sequences provide markers distributed uniformly throughout the genome. The marker density depends basically on the number of end sequences generated, the library insert size, and the genome size. Once a BAC is completely sequenced, the BAC ends database is used to identify clones that minimally overlap this BAC sequence by virtue of their end sequence. The fingerprints of a selection of the overlapping BAC clones are compared to identify clones containing artifacts or inconsistencies for elimination as a sequencing substrate. These discrepancies can be generated by incorrect mapping resulting from the presence of repeats at the end of the BAC (*see* below) or anomalies (deletions, insert or vector DNA rearrangements, presence of chimeric inserts) and will result in a restriction pattern that differs from the one observed in other overlapping clones. Two BAC clones that minimally overlap the 5'- and 3'-ends of the sequenced BAC are selected for sequencing. In addition to providing paired end-sequence markers,

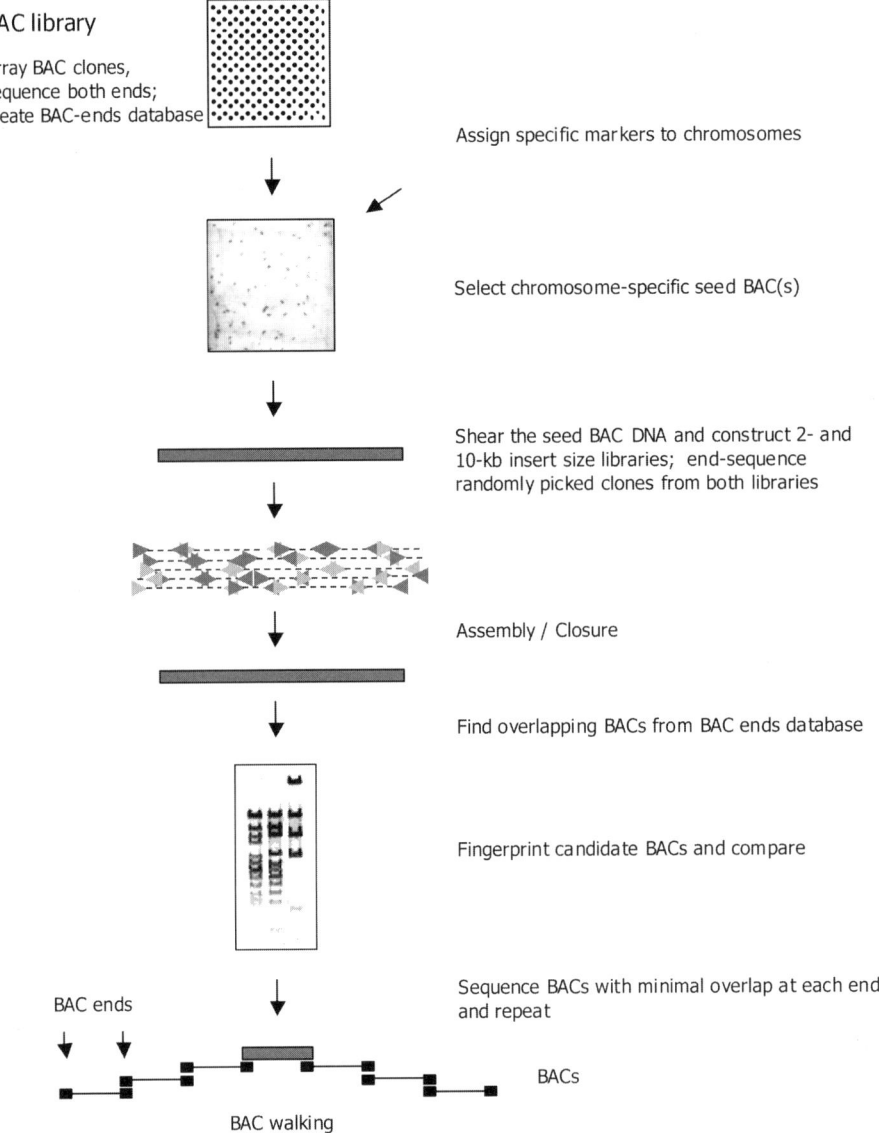

Fig. 1. Schematic representation of the map-as-you-go sequencing approach.

the generation of a searchable database of discontinuous single-pass sequences enhances gene discovery during the initial phase of a genome project.

The first step toward the generation of the BAC ends database is the construction of a high-quality BAC library. There are important factors that need to be considered in this step.

Sequencing Strategies

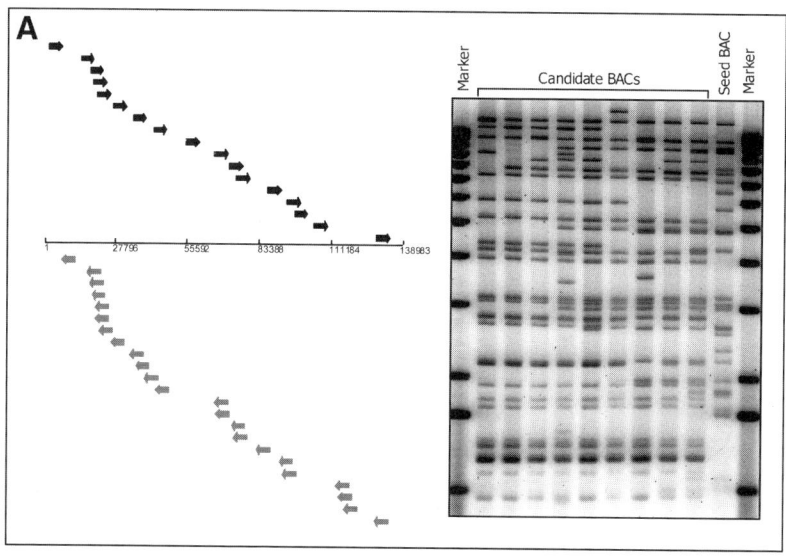

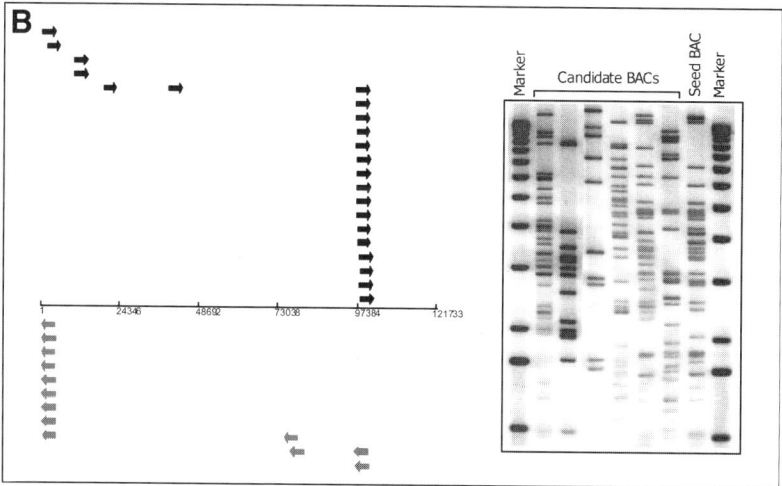

Fig. 2. BAC walking. The left segment in panels **A** and **B** shows a graphical representation of the results from a search of a query sequence (typically a finished BAC) against a BAC ends database. The location and orientation of the matches are indicated. The right panels show the fingerprinting of candidate BACs. (**A**) Successful BAC extension. Note that the matches are distributed uniformly along the BAC sequence (left panel). The fingerprinting reveals that the BAC clones display similar restriction patterns, indicating they are anchored in the same genomic region. (**B**) Problematic BAC extension. The BAC ends search reveals areas of deep coverage, indicating the presence of repeats (left panel). The fingerprinting reveals pattern inconsistency among the clones, indicating that they represent different areas of the genome (right panel).

First, the genomic DNA should be partially digested with at least two different restriction enzymes. This strategy results in library segments with different genomic representations, avoiding the overrepresentation of regions in which one of the selected enzymes cuts frequently. In addition, the construction of another library obtained from randomly large insert sheared DNA is also highly recommended. The end sequencing of randomly sheared DNA clones eliminates the bias inherent to restriction fragments and results in the generation of markers that are not anchored on a restriction pattern. Furthermore, it permits the identification of telomere-proximal sequences (*see* below).

Second, if the genome is highly repetitive, it is important to select enzymes whose restriction sites are not frequent in the repeat elements. BAC end sequences anchored in repetitive sequences can pose serious limitations to contig extension (*see* **Fig. 2**).

Third, because the main objective of the BAC ends database is to generate markers distributed evenly throughout the genome, another important point is to evaluate the randomness of the library. It is known that GC content as well as the nature and distribution of repeats can generate biased libraries (*6*), resulting in areas with low depth of coverage. One way to assess the randomness of a BAC library is to hybridize single-copy genes with high-density filters containing the gridded library. The number of positive signals for each gene should concur with the library coverage. In addition, the BAC ends database can be searched against repetitive sequences. The number of matches obtained should agree with the number of copies for each query sequence and the genome coverage.

Fourth, the total number of desired clones in the library depends primarily on the size and repetitive content of the genome. Typically, for lower eukaryotes, more than 10-fold genome coverage is desired. In the case of the parasite *T. brucei*, which contains a low number of repeats and has a haploid nonminichromosomal genome size of 28 Mb, the RPCI-93 BAC library (http://www.chori.org/bacpac/tbrucei93.htm) consists of 18,000 clones. With an average insert size of 150 kb, this library represents around 90-fold haploid genome equivalents. The end sequences of 5000 clones represent about 26-fold clone coverage of the genome and provides, on average, a marker of 500–600 bp every 2700 bp. Such high marker density is quite useful for the construction of a high-resolution sequence-ready map. Genomes with a large content of repetitive sequences require a higher marker density.

2.2. Generation of Continuous Sequences and Chromosome Sequencing

The first step toward the sequencing of an entire chromosome or genome using the map-as-you-go strategy is the identification of sequence markers, such as STSs or ESTs, that are specific for each chromosome. The chromosome-

Sequencing Strategies

specific markers are identified by hybridizing ESTs and STSs sequences to PFGE blots. These markers are then used to search against the BAC ends database and to screen a high-density filter containing the gridded BAC library. The selected BAC clones are analyzed by fingerprinting for consistency, and the selected seed BAC is submitted to shotgun sequencing. Generally, more than one seed BAC is selected per chromosome.

2.2.1. Sequencing a BAC Clone to Contiguity

To obtain the complete sequence of the selected BAC, its DNA is randomly fragmented into smaller pieces and subcloned into plasmid vectors. Clones from this shotgun library are then picked at random, end-sequenced, and assembled into contigs automatically using computer algorithms. Discontinuities, or gaps, between the contigs are filled using various strategies. These steps are described here.

2.2.1.1. CONSTRUCTION OF A SHOTGUN LIBRARY

The first step toward the construction of a shotgun library is DNA fragmentation, which is a critical factor for the success and efficiency of shotgun sequencing. Physical shearing methods, such as nebulization, sonication, or hydrodynamic shearing are preferred over enzymatic digestion because they generate random fragments, eliminating the bias inherent to a restriction pattern (*see* Chapter 19). Whereas nebulization and sonication methods are more accessible to many laboratories, hydrodynamic shearing requires more expensive equipment but produces fragments with a narrower size range *(7)*. For all methods, shearing conditions can be adjusted to maximize the generation of fragments of the desired size. Mechanical fragmentation methods generate DNA molecules containing 5'- and 3'-overhang ends with variable length. To clone these fragments, their ends must be converted into blunt ends. Various enzymatic procedures can be used for this purpose. Single-strand exonucleases, such as mung bean nuclease, digest specifically single-strand regions generating blunt ends. Alternatively, the DNA can be treated with T4 DNA polymerase, which removes 3'-overhangs and has a polymerase activity that fills in 5'-overhangs. To generate libraries with a narrow insert size range and to remove the enzymes of the end-repair reaction, the blunt-end fragments are size-selected by electrophoresis (*see* Chapter 19). Next, adaptor sequences are added to the end of the fragments, which are then cloned into the chosen plasmid vector.

2.2.1.2. SHOTGUN SEQUENCING AND ASSEMBLY

Sequencing both ends of a large number of clones selected randomly from the shotgun library provides the substrate for the assembly process. The initial assembly is carried out when there is a sufficient number of reads. The Lander-Waterman model *(8)* is used to calculate the number of random clones that must be end-sequenced to obtain 6–8 times sequence coverage and a minimum number of gaps. For instance, assuming a BAC insert of 150 kb and an average length of sequence read of 500 nucleotides, the sequencing of both ends of 1050 clones (2100 reads) will result in approx sevenfold coverage. In theory, at this level of sequence coverage, only about two 150-bp gaps of totally unsequenced DNA are predicted. However, in practice, the numbers of actual gaps are frequently greater than simulation. In general, most BACs have fewer than 10 gaps following random sequencing to ×7 coverage.

The assembly process has become highly accurate as a result of the development of base-calling programs. Phred, the preferred base-calling program *(9, 10)*, reads the chromatograms generated by the automated sequencers, attributes a base for each identified peak (base call), and assigns a quality value (Phred value) for each base sequenced, in which the quality value is associated with the probability that a base is called incorrectly. One error per 1000 sequenced bases is acceptable for most sequencing projects. The base-calling programs generate two output files: the sequence and the quality value files. After the removal of vector sequence, assembler programs use both sequence and quality value files to generate consensus sequences by picking the highest-quality read at each position in a contig.

2.2.1.3. Finishing Phase: Filling the Gaps

The finishing phase consists of resequencing regions with low base quality, ensuring that every base is covered by reads from at least two different subclones and sequenced on both strands, filling gaps, and correcting misassemblies. Problems can be generated by the presence of repetitive sequences, unclonable regions, high GC content, palindromic regions, and homopolymeric areas. The use of alternative sequencing chemistries (dye primer instead of dye terminator, dimethyl sulfoxide, 7-deaza-dGTP analog nucleotide, etc.) can solve high GC content, palindromic regions, and homopolymeric areas. Generally, misassemblies resulting from the presence of repeats and gaps are the most difficult problems to resolve during the finishing phase.

There are two types of gaps: sequence and physical gaps. Sequence gaps are those for which a template is available and are usually easier to solve. Frequently, the use of primers pointing outward from the ends of contigs in sequencing reactions on the appropriate template is sufficient to close a sequence gap. On the other hand, physical gaps span regions that are not covered by an existing clone and result in contigs whose order with respect to each other is unknown. Various approaches are used during the closure phase. They include primer walking, transposon-mediated sequencing, and combinatorial and multiplex polymerase chain reaction (PCR), among others. The extensive collection of end sequences should mean that a very small number of physical gaps will remain in each BAC after the shotgun phase. However, it should be emphasized that the finishing process is a laborious and time-consuming phase because a wide variety of problems can appear for which a high level of technical expertise is required. Each BAC has its own set of sequence problems that should be addressed on a case-by-case basis to design the most appropriate strategy to close the BAC.

2.3. BAC Walking

To generate BAC contigs, the sequence of a closed BAC is searched against the BAC ends database to identify overlapping clones. Alternatively, overlapping clones can be identified through the screening of high-density filters containing the gridded BAC library, using as probes DNA fragments amplified from the end of the BAC. The candidate clones for extension are fingerprinted, and a new BAC clone showing internal consistency among the fingerprints and minimal overlap at the end of the closed BAC is then selected for sequencing. **Figure 2A** represents fingerprinting data leading to a successful BAC extension. In this case, the overlapping clones are distributed uniformly along the closed BAC and the fingerprinting analysis reveals excellent consistency among the candidate clones. All of them display similar restriction patterns, suggesting that the clones are located in the same genomic region. In this case, any BAC clone mapped close to the end of the sequenced BAC can be selected for sequencing. However, there are other examples in which the extension is not a straightforward process. Difficulties arise when the search results in few numbers of candidates for extension; this occurs more frequently when the contig reaches the end of a chromosome. Telomeres and subtelomeres are very dynamic regions enriched with repetitive sequences and, therefore, can undergo frequent rearrangements. Because of their repetitive nature, a scarcity of restriction sites, and the presence of single-strand 3'-OH overhangs, telomeres are usually rearranged, missing, or underrepresented in genomic libraries. To sequence these areas, specific approaches using telomeric adaptors have been utilized to construct telomeric libraries *(11)*. In addition, we have recently observed that telomere sequences can be well represented in large insert sheared libraries, as in the case of the fosmid segment of CHORI-105 (http://bacpac.chori.org/tcruzi105.htm), a library constructed in Dr. de Jong's laboratory for the *Trypanosoma cruzi* genome project. Another adverse situation is when the 5'- or 3'-end of a sequenced BAC is anchored in a repetitive region. In this case, the search against the BAC ends database results in high-depth coverage at the end of the closed BAC, and the fingerprints reveal discrepancy among the candidates, pointing to

BACs that are scattered throughout the genome (**Fig. 2B**). The best way to deal with this problem is to move inward until a region devoid of repeats is identified and select overlapping BAC clones anchored in that region for fingerprint analysis.

2.4. Validation

Because no known cloning system is free of artifacts and mapping errors can occur, the collinearity of the assembled sequence and the genome must be checked. The assembly validation can be performed at the BAC or chromosomal level. At the BAC level, there are at least three methods to identify misassemblies. First, the most accurate method is to compare the *in silico* restriction digest of the assembled sequence with the fingerprinting profile obtained prior to sequencing. Any observed discrepancy suggests sequence misassembly or rearrangement of the BAC insert before sequencing. Second, if the closed BAC contains sequence-based markers, such as ESTs or STSs, the assembled BAC can be checked for the presence of such sequences. Third, any discontinuities and/or deletions within the coding sequences can represent areas that might require further scrutiny. These suspicious regions can be PCR-amplified from the original BAC, or from other BAC clones that span the same region, and resequenced.

At the chromosomal level, the optical mapping technique can be used to ensure collinearity of the BAC contig and the chromosome sequence *(12)*. Briefly, the genomic DNA is sheared into very large fragments, typically 0.4–12 Mb in size, elongated on a silanized glass surface, and digested *in situ* by a restriction enzyme. Because the ends of digested fragments "shrink," cleavage sites are visible as small gaps between fragments, which retain their original order. To visualize the fragments, the DNA is stained with an intercalating dye and inspected under fluorescence microscopy. Digital images are recorded, and the fragments are sized by measuring the distance between the gaps and by quantifying the integrated fluorescence intensity. Maps of single molecules are then constructed. The optical map of an entire chromosome cannot be obtained from a single molecule because large molecules break during the procedure, so the experiment should be repeated several times to generate a complete chromosome dataset. The many overlapping restriction site patterns generated are used to assemble maps of individual molecules into contigs, and a consensus map for the whole chromosome is obtained. The assembled chromosome or BAC sequence is then digested *in silico* with the same enzyme used to generate the optical mapping data, and the fragment sizes obtained from the two approaches are optimally aligned. As in any other assembly method, the consensus optical map is generated computationally. Here, the best way to avoid misassembly is to obtain a high-resolution optical map by using at least two restriction enzymes.

3. Shotgun Strategies

Until recently, the clone-by-clone approach was the most commonly used strategy for sequencing parasite genomes. The *Leishmania major* Friedlin genome project used cosmid and BAC clones as the main substrate for sequencing (http://www.sanger.ac.uk/Projects/L_major/) and the *T. brucei* chromosomes II–VIII were sequenced at TIGR using the map-as-you-go approach (http://www.tigr.org/tdb/e2k1/tba1/). However, with the improvement of assembly algorithms, the "assembly unit" of the eukaryotic genome projects has increased from large insert clones (clone-by-clone approach) to chromosomes (whole-chromosome shotgun, [WCS]), to entire genomes (whole-genome shotgun, [WGS]).

3.1. WCS Strategy

The first step of the WCS strategy involves the construction of chromosome-specific libraries (*see* Chapter 19). Briefly, the chromosomal bands are gel-purified from pulsed-field gels, extracted by agarase digestion, sheared into 1–2 kb fragments, and cloned into plasmid vectors. Randomly picked clones are then end-sequenced and the sequences assembled into contigs. The gaps are closed using targeted closure procedures. STSs, microsatellite markers, and optical mapping data are used, among other techniques, to order the contigs on the chromosome and to validate the genome assembly.

Variation in size between homologous chromosomes and comigration of nonhomologous chromosomes in PFGs are common occurrences in parasite genomes *(13–16)*. Therefore, construction of chromosome-specific libraries may represent a challenge and often results in contamination with clones derived from other chromosomes. This hampers the assembly and closure processes and requires a larger number of reads to produce the necessary sequence coverage. In general, more than one PFGE condition is used in an attempt to resolve all chromosomes pairs *(14–16)*. Despite such efforts, *Plasmodium falciparum* chromosomes 6, 7, and 8 could not be successfully resolved using PFGE and were sequenced as a mixed pool *(17)*. Also, the generation of such chromosome-enriched libraries can be unfeasible for genomes composed of large chromosomes. For example, the smallest chromosomes of helminth parasites, such as *Schistosoma* and *Brugia*, are too large to be separated by PFGE (*see* Chapter 21), and, therefore, sufficient DNA from an individual chromosome cannot be isolated *(18)*.

The *Plasmodium falciparum* genome project at the Wellcome Trust Sanger Institute (http://www.sanger.ac.uk/Projects/P_falciparum/) adopted a modified WCS strategy, which has also been used to sequence the *T. brucei* chromosomes I, IX, X, and XI (http://www.sanger.ac.uk/Projects/T_brucei/). This modified WCS strategy incorporates elements from both clone-by-clone and WCS approaches.

In addition to the sequencing of chromosome-specific libraries, low-coverage shotgun sequences from a minimal tiling set of overlapping YAC (*P. falciparum*) or P1 and BAC (*T. brucei*) clones on the chromosomes are also obtained. This approach is used to screen out the reads that do not belong to the target chromosome but derive from cross-contamination with other chromosomes during the library construction. In addition, it facilitates the assembly of the whole chromosome by grouping the chromosome library sequences into smaller subsets deriving from individual clones, reducing the complexity of the assembly and finishing processes.

3.2. WGS Strategy

The WGS strategy consists of shearing the genomic DNA into smaller fragments, which are size-selected in agarose gels. Multiple libraries are constructed, and a very large number of clones are randomly selected and sequenced at both ends. The reads are assembled into contigs, and mate-pair information is used to order them (a mate-pair is a pair of sequences derived from each end of the same clone). A set of contigs that are ordered, oriented, and positioned with respect to each other by mate-pair reads are known as a scaffold. Scaffolds are the main product of a WGS strategy and can be assigned to chromosomes using chromosome-specific markers. Linkage groups and optical maps, among other techniques, can be used to order the contigs and validate the genome assembly. The WGS approach does not require any previous mapping effort before the sequencing phase and has been the method of choice for many parasite projects. Here are some examples of parasite projects using the WGS strategy:

- *Trypanosoma cruzi* (http://www.tigr.org/tdb/e2k1/tca1/)
- *Toxoplasma gondii* (http://www.tigr.org/tdb/e2k1/tga1/ and http://www.sanger.ac.uk/Projects/T_gondii/)
- *Plasmodium vivax* (http://www.tigr.org/tdb/e2k1/pva1/)
- *Plasmodium yoelii yoelii* (http://www.tigr.org/tdb/e2k1/pya1/)
- *Cryptosporidium parvum* (http://www.parvum.mic.vcu.edu/)
- *Theileria parva* (http://www.tigr.org/tdb/e2k1/tpa1/)
- *Theileria annulata* (http://www.sanger.ac.uk/Projects/T_annulata/)
- *Giardia lamblia* (http://jbpc.mbl.edu/Giardia-HTML/index2.html)
- *Entamoeba histolytica* (http://www.tigr.org/tdb/e2k1/eha1/ and http://www.sanger.ac.uk/Projects/E_histolytica/)
- *Schistosoma mansoni* (http://www.tigr.org/tdb/e2k1/sma1/ and http://www.sanger.ac.uk/Projects/S_mansoni/)
- *Brugia malayi* (http://www.tigr.org/tdb/e2k1/bma1/)

An important feature of this strategy is the use of libraries of distinct insert sizes as sequencing substrate, because each class has a distinct role during the assembly. Typically, small (2–3 kbp), intermediate (10–15 kbp), and large (e.g.,

50 and 100 kbp) insert size libraries are constructed. The end sequences from small and intermediate insert size libraries provide the bulk of the sequence coverage, whereas end sequences from large insert clones are essential for ordering of the contig groups and independent verification of overall genome structure.

The assembly process in the WGS strategy is more difficult than in other sequencing strategies, in which the problem is confined to large insert clones (clone-by-clone) or chromosomes (WCS). It requires a robust assembler that is capable of handling a very large number of sequence reads. Recently, the adoption of the WGS strategy for sequencing complex genomes, such as *Drosophila melanogaster* *(19)* and the human genome by Celera Genomics *(20)*, has greatly accelerated the improvement of software devoted to the assembly process. These projects resulted in the development of the Celera Assembler, a program that has been used at TIGR to assemble many parasites' genomes. Other WGS assembly programs include Arachne *(21)*, Jazz *(22)*, RePS *(23)*, and Phusion *(24)*.

Because of the complexity of the assembly, the quality of the input sequences must be much higher than that of other sequencing strategies *(19)*. First, regions of low-quality sequence need to be removed much more aggressively. This reduces the chance of false overlaps when comparing millions of reads (in the case of parasite genomes) in the initial phase of the assembly process. Second, most of the reads must be in mate pairs. Typically, more than 75% of reads in pairs is acceptable. Paired end sequences greatly improve the efficiency of the sequence assembly by providing spacing and orientation information, which are key elements in the WGS strategy. Third, variance in the insert length in a library should be less than 10%. This may represent a challenge, especially when constructing large insert size libraries, because large DNA fragments cannot be resolved accurately in ordinary agarose gels. After library construction, the insert size of a large number of clones needs to be checked on agarose gels and the average and the variance of insert sizes estimated. These values can be validated by matching the reads against a previously sequenced contig, and, therefore, a more accurate measure of the distance between mate pairs can be determined.

In general, repeats are the main challenge for assembly programs. The basic principle of the assembly process is to build contigs based on sequence similarity between reads *(25)*. However, overlapping reads can share a common repetitive element and may originate from distinct regions of the genome. A well-designed assembler should distinguish true overlaps from overlaps induced by repeats quickly and efficiently. The details of how each assembler deals with repeats differ from program to program and are beyond the scope of this chapter. However, the key to resolving repeats resides in the use of mate-pair information. A clone that spans a repeat and whose insert size is longer than the repetitive element is helpful if one of its ends is anchored on the repeat. This is because its mate pair would be located in a neighboring nonrepetitive region, therefore linking the repeat to a unique and correctly assembled region.

Polymorphism is another factor that may complicate a WGS assembly. Assembler algorithms need to distinguish ambiguities resulting from polymorphism from those originating from sequencing errors. This is possible because base-calling errors frequently are associated with low-quality values, and they do not tend to be confirmed by other reads. Highly polymorphic genomes, however, require a much higher level of sequence coverage to ensure these inferences are reliable. In homologous regions displaying high level of polymorphism, two contigs can be generated, each one corresponding to one haplotype. Because assembly programs generally are conservative in building scaffolds, they prevent two contigs from covering the same region. Recently, a scaffolder program able to deal with this problem was developed at TIGR *(26)*. The haplotype structure of a genome is clearly evident in the output of the program Bambus, whereby homologous polymorphic regions are represented as bubbles (each path corresponding to one haplotype) that are then rejoined in similar homologous regions. In addition to the ability to deal with polymorphic genomes, Bambus is capable of using linking information other than mate pairs to build scaffolds. This includes sequence from a reference genome, physical markers, and gene synteny data. More important, Bambus works in a hierarchical fashion by incorporating the highest-quality data in the initial phase and then progressively incorporating lower-quality data without disturbing the initial scaffolds. Bambus is freely available under an open-source license from http://www.tigr.org/software/bambus.

4. Choosing a Sequencing Strategy

Each genome project has its own goals and limitations. A balance among various considerations should guide the choice for the most appropriate sequencing strategy. These typically include the desired final product (level of completeness), available funds, and limitations associated with intrinsic genome characteristics. Clone-by-clone and WCS approaches result in a high level of sequence accuracy and completeness, but the data are not generated very rapidly. In contrast, WGS allows the fast generation of a large amount of discontinuous sequence, which yields tremendous insights into genome content in the early phases of a project and allows assessment of polymorphism, identification of new classes of repeats, and comparative analyses with related organisms. DNA polymorphism data are usually not generated when using a clone-by-clone approach, because only one variant of each genomic region is sequenced (e.g., if a genome is diploid). Regions of particular biological interest can be identified and sequenced faster using the map-as-you-go strategy when compared with the classical clone-by-clone strategy that requires extensive physical mapping efforts prior to the sequencing phase. On the other hand, the WGS strategy can be used to study the genome in a global and comprehensive fashion.

However, all available information related with the genome composition (content and distribution of repeats, polymorphism level, GC content) needs to be examined carefully. For example, the fact that the construction of high-quality large insert libraries of A+T-rich DNA in the *E. coli* system is a difficult task was the main reason for which the *P. falciparum* genome sequencing consortium adopted the WCS strategy *(27)*. One disadvantage of this approach is an inevitable cross-contamination with DNA from other chromosomes that occurs during the preparation of PFG-purified chromosome samples. Clone-by-clone strategies provide modularity. This has several implications. First, the assembly process is local (at the clone level); therefore, problems related with this step, such as repeats, are also local. Misassemblies are more easily and rapidly detected when a clone-by-clone approach is used, because the assembly validation can be performed more accurately at the individual clone level. Second, annotation can be performed incrementally (and in a distributed manner if needed) on individual clones as they are finished. In contrast, WCS and WGS approaches yield annotated sequences only in the late phases of the project (*see* Chapter 2). Third, the modularity provided by clone-by-clone and WCS approaches greatly facilitates the establishment of sequencing and annotation consortia, because the clones and chromosomes can be distributed among the sequencing centers. In the case of the WGS strategy, raw sequences can be generated in different laboratories, but the assembly and closure processes need to be centralized, or at least highly coordinated.

References

1. El-Sayed, N. M., Hegde, P., Quackenbush, J., et al. (2000) The African trypanosome genome. *Int. J. Parasitol.* **30,** 329–345.
2. Myler, P. J., Audleman, L., deVos, T., et al. (1999) *Leishmania major* Friedlin chromosome 1 has an unusual distribution of protein-coding genes. *Proc. Natl. Acad. Sci. USA* **96,** 2902–2906.
3. Gardner, M. J. (2001) A status report in the sequencing and annotation of the *P. falciparum* genome. *Mol. Biochem. Parasitol.* **118,** 133–138.
4. Venter, J. C., Smith, H. O., and Hood, L. (1996) A new strategy for genome sequencing. *Nature* **381,** 364–366.
5. The Arabidopsis Genome Initiative. (2000) Analysis of the genome sequence of the flowering plant *Arabidopsis thaliana. Nature* **408,** 796–815.
6. Green, P. (1997) Against a whole-genome shotgun. *Genome Res.* **7,** 410–417.
7. Thorstenson, Y. R., Hunicke-Smith, S. P., Oefner, P. J., et al. (1998) An automated hydrodynamic process for controlled, unbiased DNA shearing. *Genome Res.* **8,** 848–855.

8. Lander, E. S. and Waterman, M. S. (1988) Genomic mapping by fingerprinting random clones: a mathematical analysis. *Genomics* **2**, 231–239.
9. Ewing, B., Hillier, L., Wendl, M. C., et al. (1998) Base-calling of automated sequencer traces using phred. I. Accuracy assessment. *Genome Res.* **8**, 175–185.
10. Ewing, B. and Green, P. (1998) Base-calling of automated sequencer traces using phred. II. Error probabilities. *Genome Res.* **8**, 186–194.
11. Chiurillo, M. A., Santos, M. R., Franco Da Silveira, J., et al. (2002) An improved general approach for cloning and characterizing telomeres: the protozoan parasite *Trypanosoma cruzi* as model organism. *Gene* **294**, 197–204.
12. Aston, C., Mishra, B., and Schwartz, D. C. (1999) Optical mapping and its potential for large-scale sequencing projects. *Trends Biotechnol.* **17**, 297–302.
13. Santos, M. R., Cano, M. I., Schijman, A., et al. (1997) The *Trypanosoma cruzi* Genome Project: nuclear karyotype and gene mapping of clone CL Brener. *Mem. Inst. Oswaldo Cruz* **92**, 821–828.
14. Ivens, A. C., Lewis, S. M., Bagherzadeh, A., et al. (1998) A physical map of the *Leishmania major* Friedlin genome. *Genome Res.* **8**, 135–145.
15. Melville, S. E., Leech, V., Navarro, M., et al. (2000) The molecular karyotype of the megabase chromosomes of *Trypanosoma brucei* stock 427. *Mol. Biochem. Parasitol.* **111**, 261–273.
16. Carucci, D. J., Horrocks, P., and Gardner, M. J. (2002) Purification of chromosomes from *Plasmodium falciparum*. *Meth. Mol. Med.* **72**, 235–240.
17. Hall, N., Pain, A., Berriman, M., et al. (2002) Sequence of *Plasmodium falciparum* chromosomes 1, 3–9, and 13. *Nature* **419**, 527–531.
18. Johnston, D. A., Blaxter, M. L., Degrave, W. M., et al. (1999) Genomics and the biology of parasites. *BioEssays* **21**, 131–147.
19. Adams, M. D., Celniker, S. E., Holt, R. A., et al. (2000) The genome sequence of *Drosophila melanogaster*. *Science* **287**, 2185–2195.
20. Venter, J. C., Adams, M. D., Myers, E. W., et al. (2001) The sequence of the human genome. *Science* **291**, 1304–1351.
21. Batzoglou, S., Jaffe, D. B., Stanley, K., et al. (2002) Arachne: a whole-genome shotgun assembler. *Genome Res.* **12**, 177–189.
22. Aparicio, S., Chapman, J., Stupka, E., et al. (2002) Whole-genome shotgun assembly and analysis of the genome of *Fugu rubripes*. *Science* **297**, 1301–1310.
23. Wang, J., Wong, G. K., Ni, P., et al. (2002) RePS: a sequence assembler that masks exact repeats identified from the shotgun data. *Genome Res.* **12**, 824–831.
24. Mullikin, J. C. and Ning, Z. (2003) The Phusion Assembler. *Genome Res.* **13**, 81–90.
25. Pop, M., Salzberg, S. L., and Shumway, M. (2002) Genome sequence assembly: algorithms and issues. *IEEE Computer* **35**, 47–54.
26. Pop, M., Kosack, D. S., and Salzberg, S. L. (2004) Hierarchical scaffolding with Bambus. *Genome Res.* **14**, 149–159.
27. Gardner, M. J., Hall, N., Fung, E., et al. (2002) Genome sequence of the human malaria parasite *Plasmodium falciparum*. *Nature* **419**, 498–511.

2

Annotation of Parasite Genomes

Matthew Berriman and Midori Harris

Summary

Genome annotation is the application of useful biological descriptions to sequence data. Different levels of time and effort can be invested to produce correspondingly different depths of annotation depending on what methods are employed. Researchers using genome data should, therefore, understand how annotations are generated to assess their validity correctly and to determine what level of inferences can be made accordingly. Thorough annotation requires a large range of procedures, most of which involve manual reviews of all available evidence. First, gene structures often are computed algorithmically and edited based on in-depth analyses of the underlying sequence data. Second, functional predictions draw on data from various sources. Finally, the use of structured and controlled descriptions, such as those provided by gene ontology, can be used so that final descriptions are not only consistent and unambiguous, but capable of being used in further downstream analyses such as cross-species comparisons.

Key Words: Algorithm; annotation; BLAST; classification; domain; EC number; FASTA; gene finding; gene ontology; gene prediction; genome; hidden Markov models; metabolism; ortholog; paralog; pseudogene; RNA genes; sequence similarity.

1. Introduction

Useful biological descriptions increase the utility of a genome sequence. Annotating genomes with features such as predicted genes or possible regulatory elements provides a framework for researchers to interrogate data and can accelerate hypothesis-driven research. Good annotation should provide a shortcut for researchers, allowing them to narrow their searches to the genes or sequences that interest them. Furthermore, big picture analyses can be performed based on the roles of sequences, particularly genes, if they are accurately described. For instance, researchers can reconstruct whole pathways *in silico* when all of the annotation is used together.

The depth of annotation of different genome projects varies, often reflecting the project's status. Noncontiguous sequence data are difficult to annotate thoroughly. Gene predictions may change as contigs (individual contiguous sequences) are merged or split as sequence assembly progresses, or the sequence itself may change during the course of the project. Thorough manual reviews are required to ensure quality, but these take time. For projects such as whole-genome shotgun "skims," in which the number of sequencing reads that cover each base is lower, or for data released during the course of an ongoing project, often some kind of automated analysis is employed. These methods provide a first analysis of the content of a genome but are error-prone and must be used with caution; the error rate is much lower with manual annotation, and often the detail provided is greater. The choice between manual and automatic annotation consequently is made by balancing the conflicting needs for speed and accuracy. An appreciation of the methods involved in genome annotation is essential before using and interpreting these data.

This chapter aims to illustrate how parasite genomes are annotated. Ideally, annotation includes an accurate description of the coordinates, properties, and biological role of all genes and other features of interest within a genome. Although this is a moving target, numerous key approaches are involved throughout the annotation process and will be described throughout this chapter. First, the genes are predicted using a combination of specialized gene-finding tools and similarity searching. Features other than genes may also be annotated, such as repetitive sequences and transposable elements, but these are often project-specific and, therefore, beyond the scope of this chapter. After predicting the position of all genes in a genome, numerous methods can be used to help ascribe functions to them. For convenience, gene prediction and ascribing function are often considered as discrete steps, but in reality, these steps are performed concurrently and feed back upon one another. For example, similarity searches are used to great effect to predict the function of genes but are also an essential part of gene prediction, in which they help to define boundaries for coding sequences.

2. Gene Prediction

Genes are the focus of most annotation, and when a genome project is finished, its annotated genes often are presented like the solution to a puzzle. This is an oversimplification—gene prediction is an inexact science. Although the accuracy of sequence data may be extremely high, gene predictions can be variable, and their accuracy is harder to determine. Many factors influence this accuracy: larger genomes may not be finished (i.e., contiguated) to the same degree as smaller genomes, resulting in different parts of a gene being found on unconnected contigs. Unusual genome composition can make some gene predictions

Annotation of Parasite Genomes 19

easier by exaggerating differences between coding and noncoding sequences, or it can make others more difficult by reducing the performance of prediction algorithms. In addition, many organisms employ splicing to remove introns from transcripts. This adds a further layer of complexity when predicting coding sequences and will be discussed later. The true accuracy of gene predictions can only really be determined over time as more evidence accumulates or gene expression is confirmed by further experimentation in the laboratory.

Often, references to genes are made incorrectly. Gene finding should often more correctly be described as coding sequence prediction, the translation of which into an amino sequence provides clues to the gene's function. In addition to its coding sequence, a eukaryotic gene also includes 5'- and 3'-untranslated regions. Therefore, precise annotation of a gene would require the correct identification of its transcriptional start and stop sites, but it is not usually feasible to annotate these *ab initio*. One exception is when complementary DNA (cDNA) sequence data are available; this will be discussed later. For the purposes of most annotation, the coding sequence does contain most of the information about what a gene does. It is normally assumed that a predicted coding sequence indicates a fully functional gene is present, albeit without the exact coordinates being known. Hence, "coding sequence" and "gene" often are used interchangeably, although the precise coordinates of each will differ. The coordinates of the gene, including its untranslated regions, will in fact extend beyond the coordinates of a predicted coding sequence.

2.1. Open Reading Frames and Coding Sequences

In its simplest form, gene prediction can involve searching for potentially transcribed open reading frames (ORFs). However, finding ORFs in itself cannot be regarded as gene prediction. An ORF is a length of DNA that contains a contiguous set of amino acid-encoding codons that begins with a start codon (usually ATG in eukaryotes) and ends with a stop codon. The number of ORFs within a genome is dependent on, among other things, its G+C content; the three stop codons are AT-rich and, therefore, rare in G+C-rich genomes. This was illustrated in a recent comparison *(1)* of the number of ORFs in the genomes of two prokaryotes: *Campylobacter jejuni*, which has a G+C content of approx 31%, and *Streptomyces coelicolor*, which has a G+C content of approx 72%. Prokaryotes have a fairly constant coding capacity of 1 gene every 1.1 kb, but if all ORFs greater than 100 codons are counted, *S. coelicolor* has 30,595 ORFs in 8.7 Mb (3.53 per kb), whereas *C. jejuni* has only 1783 ORFs in 1.6 Mb (1.1 per kb). Clearly, the presence of an ORF is not a guarantee that a sequence is transcribed. Therefore, a major challenge of annotation is to distinguish real coding ORFs from a background of noncoding ORFs. Where RNA *cis* splicing occurs, the problem is further compounded. Here, coding sequences are fragmented, with coding ORFs interrupted by intron sequences. Exons still require contiguous sequences of in-frame amino acid-encoding codons but are not punctuated by stop codons; thus, distinguishing real exons from all the available possibilities is complicated. Identification of full-length coding sequences *de novo* from putative exon sequences often is extremely time-consuming. The key is to evaluate numerous other signals based on the properties of coding sequences. This is most effectively done using a combination of computer algorithms to make coding sequence predictions followed by a manual review of available evidence using specialized sequence visualization and annotation software. One such software tool is Artemis *(2,3)*, which is freely available (http://www.sanger.ac.uk/Software/Artemis/) and compatible with most computer operating systems.

2.2. Gene-Finding Algorithms

A common principle behind gene-finding tools is deciding whether a given sequence has similar properties to those of known exons within the same organism. Known coding sequences are provided as a training set, and the algorithm looks for other sequences within the genome that share similar properties. When very few genes have been characterized from an organism, the training set may be too small; ideally, 100 or more genes are preferred. In these cases, different approaches can be taken to prepare training sets. First, comparing the genomic sequence to cDNA sequences can reveal coding sequences. Second, longer ORFs from a genome often are assumed more likely to correspond to real coding sequences.

Most popular gene-finding tools use some kind of statistical model. Hidden Markov models (HMMs) are the most commonly employed and have been reviewed elsewhere *(4,5)*. Briefly, HMMs are statistical models generated from sequences that calculate probabilities of a sequence being in a particular state (intragenic, intron, exon, etc.) based on the state of the preceding sequence. The detail of how each program builds its gene models differs from program to program and is beyond the scope of this chapter. However, several signals must be recognized and combined to make predictions that must then obey a set of rules. In the program GlimmerM *(6)*, for example, the context around splice sites is important. To build a model of a gene, it looks at a 16-base region around donor and a 29-base region around splice acceptor sites. Because each gene-finding program uses a different set of rules to model what a gene looks like, different programs will frequently come up with different predictions based on identical input sequences. **Figure 1** illustrates how the outputs of several gene-finding programs, Phat *(7)*, GlimmerM, and Gene Finder (P. Green, unpublished), differ quite markedly for a region of chromosome 5 in the human malaria parasite *Plasmodium falciparum*. In annotating gene PFE0480c (**Fig. 1A**, boxed), the

Annotation of Parasite Genomes

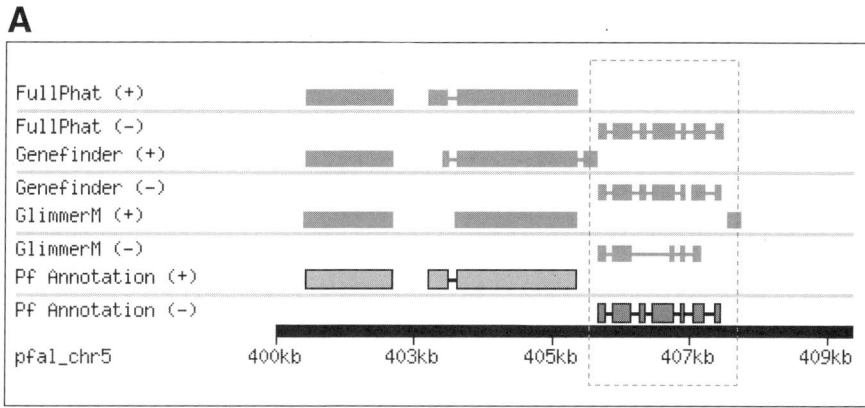

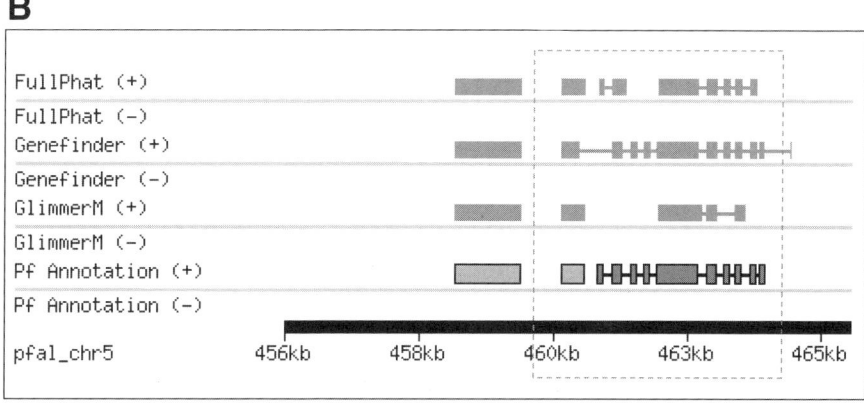

Fig. 1. Genes are predicted in the *P. falciparum* genome sequence using the output from several gene-finding algorithms. Panels **A** and **B** are screen shots from the *P. falciparum* genome database PlasmoDB (http://www.plasmodb.org) showing the output from the PHAT, Genefinder, and GlimmerM algorithms and the corresponding human-annotated gene prediction ("Pf annotation"). Forward (+) and reverse (−) DNA strands are shown above and below each gray line.

three gene-finding programs broadly agree that a coding sequence is present. After close manual inspection, an annotator has decided that the prediction by Phat is the most convincing. In another region (**Fig. 1B**, boxed), integrating the data is much harder: Phat, Genefinder, and GlimmerM predict a different number of coding sequences. In the absence of manual annotation, the output from any of these algorithms would give an incorrect gene model. In such cases, a careful manual review is required to construct a gene model based on aspects of all three predictions, using any other available evidence to guide the decision.

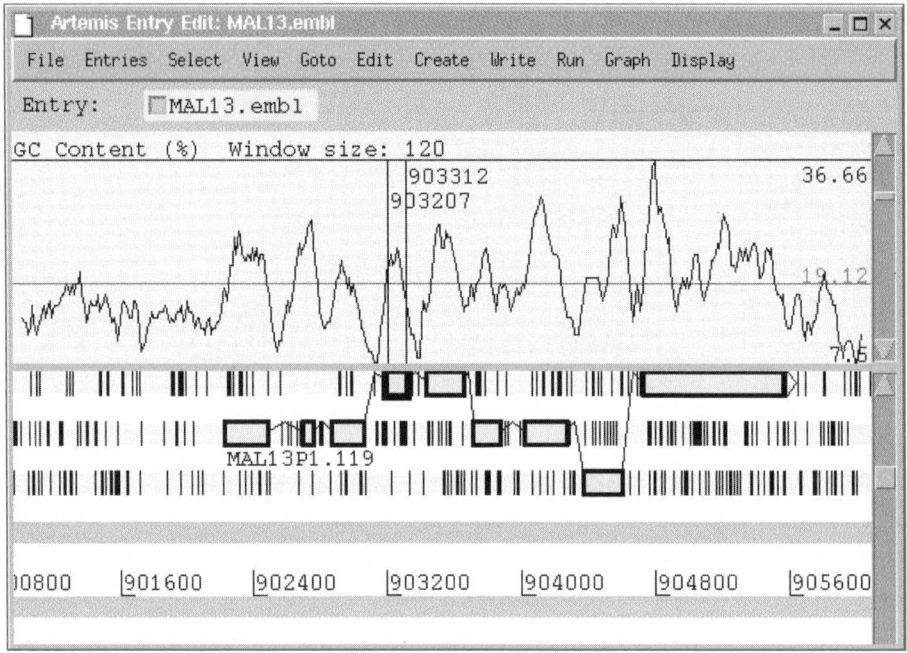

Fig. 2. Peaks in G+C content often indicate the presence of exons in *P. falciparum*. A coding sequence from chromosome 13 of *P. falciparum* is shown in the Artemis sequence visualization tool. The sequence is translated in all three reading frames with top codons indicated by vertical bars. Above the sequence, a graph of G+C content is shown, with peaks corresponding to exons in the gene.

2.3. Using Base Composition as Evidence

Gene-finding algorithms are never perfect. Unusual genes, particularly those that differ greatly from training sets, will always be poorly detected. Therefore, it can be important to use clues based on the properties of DNA to refine the output of gene-finding software. Sometimes, careful investigators can even use the DNA properties alone to produce their own *ab initio* gene predictions that were not detected by any software. A good example is the *P. falciparum* genome, in which annotation was greatly aided by considering G+C content. The average G+C content of this parasite is unusually low at 19.4% *(8)*, but it is the considerable difference in base composition between exons (23.7% G+C) and introns/intergenic regions (approx 13.5% G+C) that can be put to good use. In **Fig. 2**, a putative coding sequence from *P. falciparum* is shown in Artemis. Large spaces between stop codons and peaks in G+C content provide clues to the existence of exons. A reading frame-specific view of G+C content such as the GC Frame Plot within Artemis (an implementation of the program FramePlot

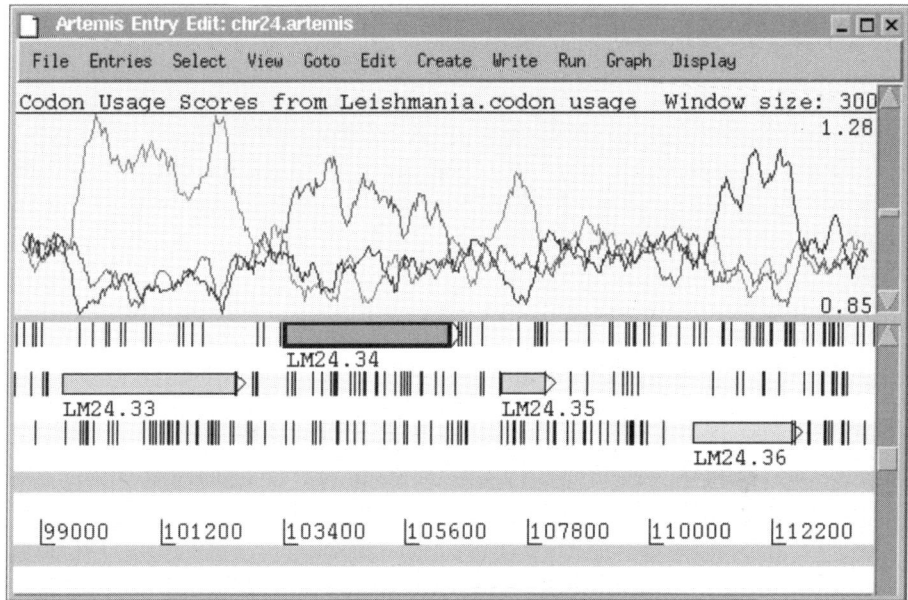

Fig. 3. Codon usage helps identify genes in *Leishmania major*. Several putative coding sequences from *L. major* are shown in the sequence visualization tool Artemis. Each coding sequence coincides with a reading frame-specific peak in a graph show the correlation with *L. major* codon usage. (Data courtesy of A. Ivens and C. S. Peacock.)

[9]) can also be used to show any strong bias toward a G or C at the third base, or "wobble" position, of codons. Third-base G+C content is calculated within a moving window and plotted. Three separate plots are produced by starting the window at position one, two, or three relative to the start of a sequence. Peaks reveal the positions of increased third-base G+C content in each of the possible reading frames. Looking for codons based on biased base composition can be taken one step further by considering codon usage. There are characteristic patterns in the usage of individual codons for different species (http://www.kazusa.or.jp/codon/), i.e., codons for the same amino acid are used with different frequencies. **Figure 3** shows how codon usage is used to help predict coding sequences in *Leishmania major*. Here, coding sequences often are indicated by ORFs that coincide with peaks in reading frame-specific codon usage.

None of the preceding examples of properties represents a universal solution for gene finding. The sequence properties that can be exploited best usually are discovered empirically after assessing the usefulness of several lines of evidence by trial and error.

2.4. Refining Gene Models

At an early stage of annotation, comparisons of genomic sequence to cDNA or peptide sequences can enable preliminary gene predictions to be refined. In particular, complex exon structures can be resolved; because the sequence of the cDNA is reverse-transcribed from messenger RNA, it contains no intron sequences. Therefore, comparing its sequence to that of genomic DNA highlights the position of exons, including the hard-to-predict untranslated regions. Obtaining a good-quality library of full-length cDNA sequences is laborious and time-consuming, so an alternative approach normally is employed. A library of partial cDNA sequences is prepared, and high-throughput sequencing is performed to generate a large number of lower-quality partial sequences known as expressed sequence tags (ESTs). Though they rarely contain full-length cDNA sequences, ESTs can still be invaluable for increasing the accuracy of predicted exon coordinates. In **Fig. 4**, a search of chromosome 13 against a database of ESTs provides a "hit" that is used to confirm the boundaries of two coding sequences in *P. falciparum*. An EST matches the sequence of the last four exons of one coding sequence (PFM1140w) and another EST matches four exons from PFM1145c on the opposite strand. Similar comparisons can be made using peptide sequences from proteomics experiments. Here, the peptide data correspond only to coding sequences and, therefore, are useful for refining exon coordinates. This is illustrated in **Fig. 5**, which shows an *ab initio* gene prediction on chromosome 10 of *P. falciparum*. The initial predicted gene (PF10_0107) had only three exons; however, a comparison with a proteomic fragment provides evidence for a fourth exon. The exact coordinates of the new exon are obtained by extending the fragment in either direction to the next available splice sites.

2.5. Pseudogenes

Genes that are not translated into products are collectively referred to as pseudogenes. Their inactivation may occur for numerous reasons, such as insertion by a retrotransposon or an unequal crossover event during meiosis, or a gene may decay within a species when its function is no longer required and selection pressure is removed. Unfortunately, there is no easy or reliable way to look for pseudogenes, except by using sequence comparisons to identify them based on similarity to functional reference genes.

2.6. RNA Genes

Genes encoding various RNA species can be more difficult to detect than protein-coding genes because of a lack of obvious signals. They do not have codons, so codon bias and third-base GC content are of no use. Instead, they must be

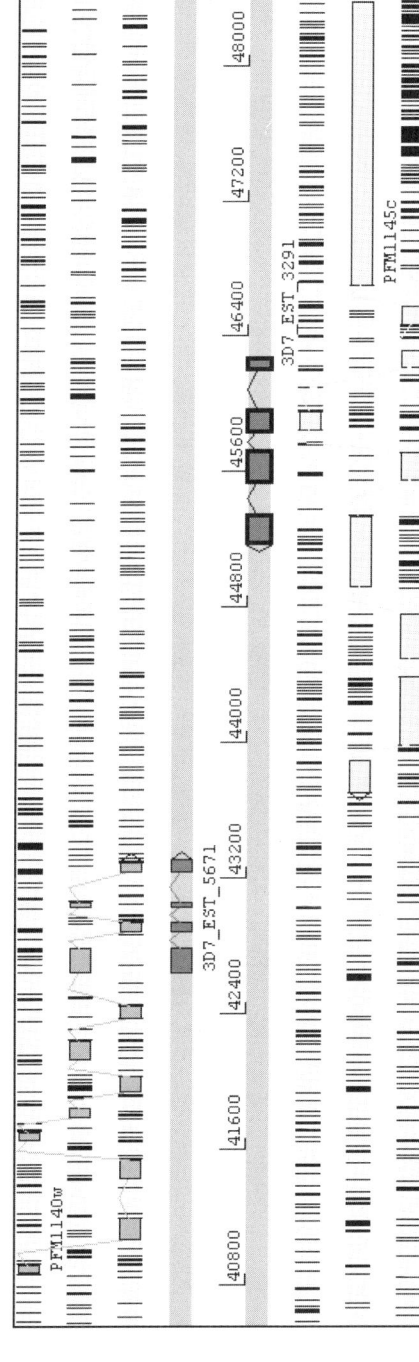

Fig. 4. Comparison of genome sequences with cDNA sequence data can help predict exons structures. A genomic sequence from *P. falciparum* and its six-frame translation is shown in Artemis. Identity to two ESTs is shown (in dark gray), which confirms the several intron/exon boundaries for two putative coding sequences (pale gray).

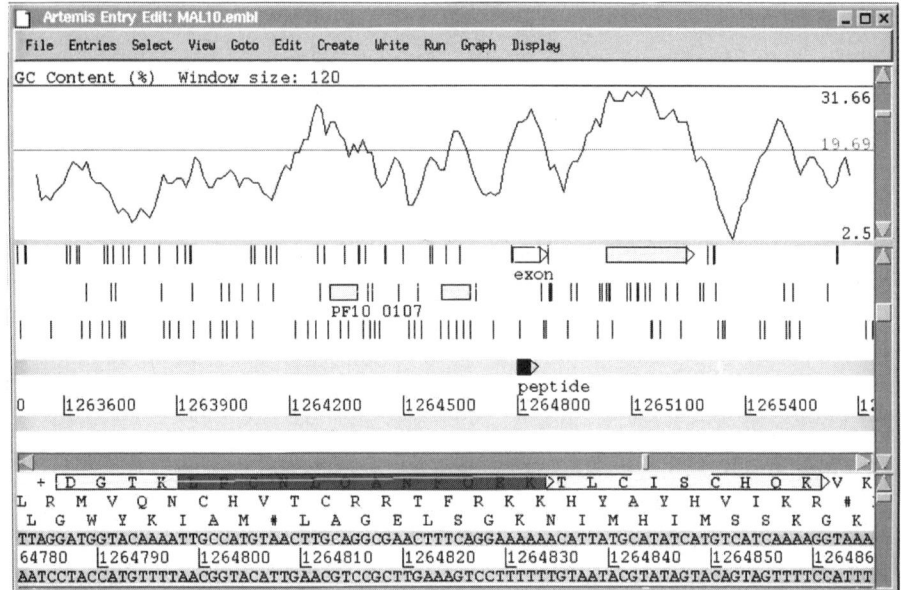

Fig. 5. Comparison of genome sequence with peptide sequence data can be used to predict exons *ab initio*. A putative three-exon coding sequence from *P. falciparum* is shown. Upon comparison with a sequenced peptide fragment, a missing exon is revealed (labeled "exon" in figure).

detected by the conservation between them (sequence comparisons are discussed later). Unfortunately, searches of this kind will work only when similar known sequences are available in a database, such as ribosomal RNA sequences. Often, the relationship between RNA species is structural rather than at the primary sequence level; thus, many RNA genes, such as those that encode small nucleolar RNA (snoRNA), can be very difficult to detect. Clear exceptions are transfer RNA (tRNA) genes. Because structure and function of tRNAs are well characterized, it is possible to find most—if not all—tRNAs in any organism using a computational approach. The standard program for this is tRNAscan-SE (http://www.genetics.wustl.edu/eddy/tRNAscan-SE/) *(10)*.

3. Predicting the Function of Genes

3.1. Sequence Comparisons

Comparing sequences is fundamental to bioinformatics. Although sequence conservation can be used to look for genes, it normally reveals only the approximate coordinates of coding sequences, even when closely related sequences are compared. However, sequence comparisons are at the hub of most predictions of gene function. Before discussing the assignment of functions, it is necessary first to consider some of the methods available for comparing sequences.

3.1.1. Sequence Alignment Algorithms

An optimal alignment of two sequences can be achieved using a computational method known as dynamic programming *(5)*. Here, every pair of characters between two aligned sequences is compared. The alignment includes matched as well as mismatched characters, and gaps are included to maximize the number of characters that are identical or related. The outcome of a particular alignment depends on the choice of numerous parameters, including the relative penalties for inserting or extending a gap, and on the scoring system for comparing related characters. For protein sequences, the scoring systems used in alignments include the Dayhoff percent accepted mutation matrices, based on an evolutionary model of proteins diverging, or the blocks amino acid substitution matrix (BLOSUM). The latter is not based on assumed evolutionary relationships; instead, a large set of sequences with conserved amino acid patterns were compared and the relative occurrence of each amino acid was used to determine the matrix. BLOSUM62—based on a comparison of sequences with an average of 62% identity—is most commonly used. Simple changes to the basic dynamic programming resulted in the Smith–Waterman *(11)* and Needleman–Wunsch *(12)* algorithms to perform local and global alignments, respectively. In a local alignment, the highest-scoring localized region between two sequences is matched, whereas a global alignment matches sequences along their entire lengths.

Though proven mathematically to perform optimal alignments, dynamic programming algorithms are very computationally intensive. Comparing a sequence against a whole database of other sequences is too slow and inefficient to be practically useful in most instances, so shortcuts must be taken. The two most common programs used for database searches are FASTA *(13)* and Basic Local Alignment Search Tool (BLAST) *(14)*, which respectively perform rapid global and local alignments against a database of sequences. FASTA produces an end-to-end alignment by first rapidly locating the best matching regions (or "words") between two sequences. Consecutive matching sequences are then joined into longer matching regions. Those that have scores above a certain threshold are joined further by the introduction of gaps, with penalties, until a final optimization step uses the Smith–Waterman algorithm to optimize alignments between the input query sequence and the best-scoring database sequences. BLAST also increases the speed of searches by initially looking for matching words but limits its search to finding only those that have the most statistically significant match. The algorithm then extends these matching words using sequences lying on either side. These locally aligned regions are called high-scoring pairs (HSPs) and are extended using gap penalties until the score no longer increases.

The statistical significance of alignments produced by most algorithms is expressed as *E*- or *p*-values, respectively. The *E*-value represents the expected number of matches with the same score that would be expected by chance using a random query sequence of the same length against the same database, and a *p*-value represents the probability of a match occurring by chance, again assuming a random query sequence searched against the same database. *E*-values and *p*-values are related and, for good alignments, are numerically similar, but the interpretation of *E*-values is somewhat more intuitive and, therefore, more commonly quoted. Unfortunately, assessing the statistical significance of an alignment is not straightforward. Parameters such as the choice of scoring matrix (e.g., BLOSUM 62) affect the score of an alignment, and, when the score is converted into a *p*-value, it is incorrectly assumed that the scores follow a normal distribution. Therefore, the statistics of sequence alignments should serve only as a guide to the true relative biological significance. In many traditional experimental situations, a *p*-value of less than 0.01 would be regarded as highly significant; with sequence alignments, this value lies in a "gray area," in which it is difficult to decide if sequences really are related. Often, cutoff values of $<10^{-5}$ or even $<10^{-10}$ are used, particularly in automated analyses. Again, even these apparently conservative cutoffs should be taken only as a rough guide. Extremely low (i.e., apparently significant) *p*-values can result from aligning sequences biased by their unusual length or sequence compositions. In particular, repetitive sequences can cause strong bias. If individual repeat units within two unrelated proteins show some short insignificant regions of similarity, the significance is artificially amplified by the recurrence of the repeats.

Although FASTA and BLAST often will produce similar results, the nature of global and local alignments gives the two programs different strengths. FASTA, for instance, gives a better view of the overall similarity of two sequences but may miss key patterns such as functional domains within a protein. The domain may be aligned incorrectly to achieve the best possible alignment over the remainder of the sequence. BLAST, on the other hand, is particularly good at detecting domains because it does not extend the alignment beyond regions of high similarity, instead reporting each similar region as a separate HSP. However, an investigator should be aware that the increased sensitivity comes at a price: the potential for false positives is greater with BLAST. Consider two proteins, each with two domains in common, which would be detected by BLAST as two HSPs. The BLAST results may lead to an incorrect assumption that the two proteins share a common function. However, conservation of a domain is not necessarily sufficient for conservation of function. The two domains could be regulatory in nature and not responsible for the overall function of the protein; that could be to the result of another, less conserved region in each protein sequence. FASTA would highlight these differences as a large mismatched region. Many BLAST interfaces that are now available via the Internet, such as the BLAST server at the National Center for Bioinformatics (http://www.ncbi.nlm.nih.gov/BLAST/), include a graphical view of HSPs positioned along the query sequence, which makes the interpretation of results far easier.

3.1.2. Database Searches

Searching across a sequence database is often the first step in trying to decide the likely function of a putative gene and is based on the assumption that similar sequences have been conserved in evolution to conserve their functions. The sequence databases (*see* Chapter 3) contain a rich source of biological information, but care must always be taken when using database search results to ascribe a function directly to an unknown gene. First, the quality of data within the database cannot be assured in many cases; it is neither policed nor peer-reviewed. When a sequence is originally submitted, it may be annotated with a function that amounts to little more than an educated guess. Furthermore, if subsequent evidence reveals the true role of a gene to be different, it is the responsibility of the original submitting author to amend the database entry; unfortunately, many database entries are not updated. Sensible precautions can be taken, however. Wherever possible, sequences ascribed with functions based on experimental characterization should be used for comparisons. In this way, transitive annotation errors—in which incorrect descriptions are transferred from sequence to sequence—can be avoided. Under **Subheading 4.2.**, a new classification scheme (Gene Ontology) is discussed that allows information such as direct experimental confirmation to be recorded within genome annotation. Searching curated databases (*see* Chapter 3) also helps. In addition to being nonredundant, they are also checked manually for quality.

The significance of database hits must also be considered before deciding the most likely function of a sequence. Although extremely high-quality hits allow for relatively easy interpretation, it is much harder to interpret weaker hits to numerous sequences. Often, in these cases, the depth of annotation will be adjusted according to the amount of information that can be gleaned from the search results. For instance, several hits to chitinases, aminopeptidases, and oxaloacetases may suggest that a query sequence is a hydrolase, while not specifying the function further.

For related sequences of a similar size, global alignments allow amino acids from each sequence to be compared directly along the entire length of the sequences. Thus, for every amino acid in one sequence, a corresponding gap or amino acid is always shown in the other sequence. This can be particularly

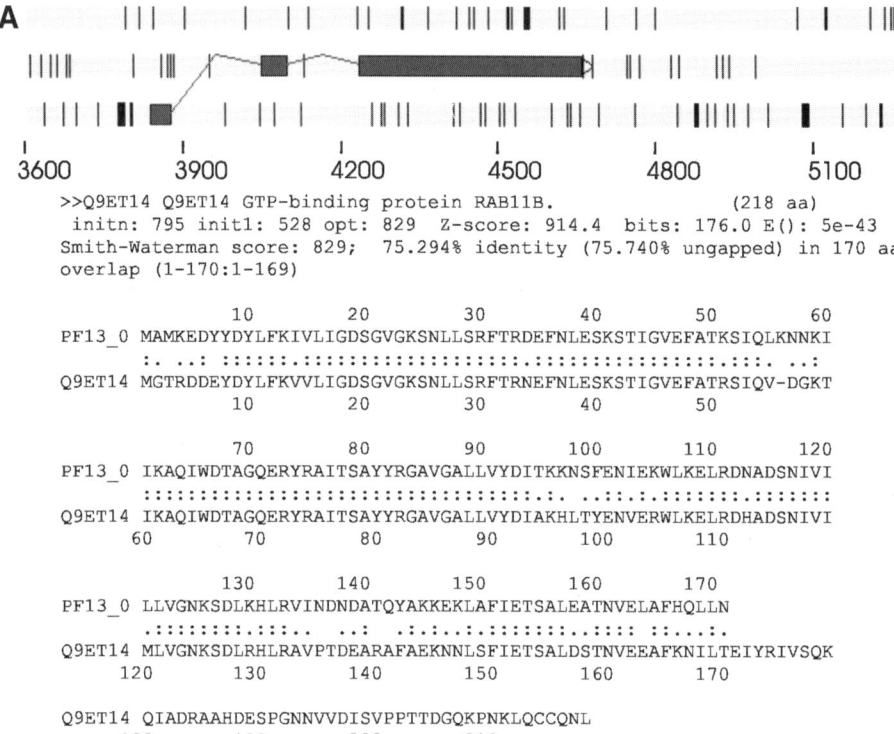

Fig. 6. Global alignments provide a good visual tool to refine gene predictions manually. A putative coding sequence is shown, translated to its amino acid sequence. (**A**) An initial global alignment reveals that the sequence is truncated.

useful when trying to decide if key amino acids are present, such as residues known to be important for the catalytic activity of an enzyme. Because its results are easy to interpret, FASTA is also useful for refining gene predictions (*see* **Subheading 2.**). A global alignment between highly related sequences can highlight potential errors. In **Fig. 6A**, a predicted coding sequence appears to be truncated based on a FASTA alignment with a similar sequence that extends beyond the C-terminus. After a visual inspection, further exons are found downstream to extend the predicted coding sequence. A subsequent FASTA search (**Fig. 6B**) indicates that the second model is more likely.

3.2. Domains

Sequence similarity searching often will reveal numerous sequences to which a query sequence could be related, but the results may be ambiguous. A more

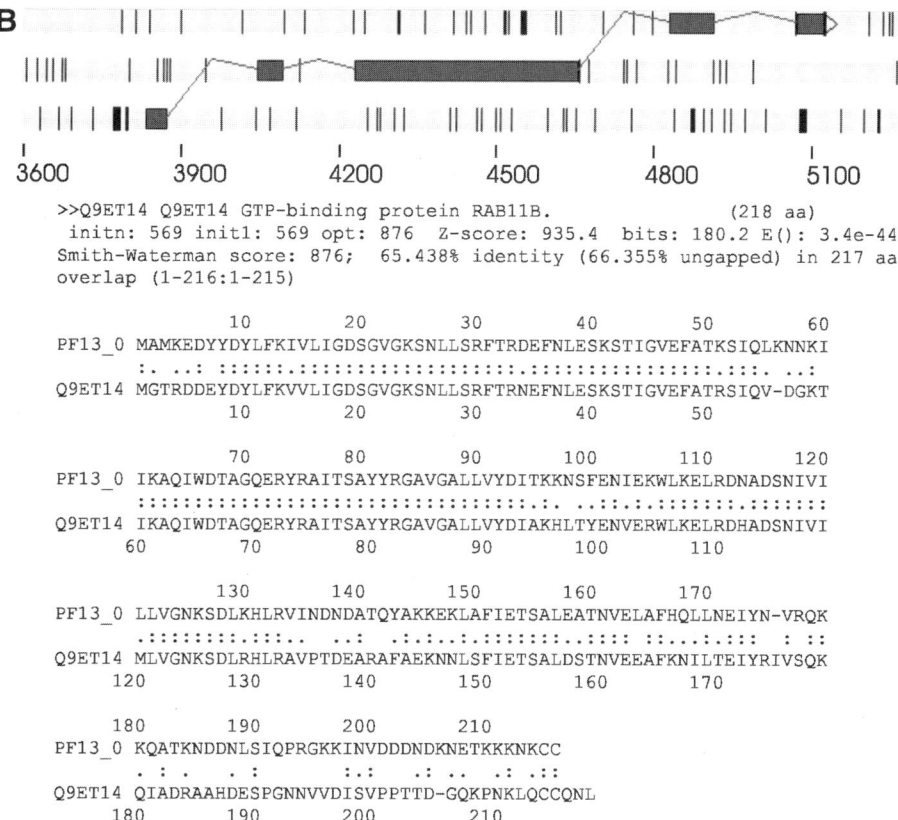

Fig. 6. **(B)** After inspecting the genome sequence, two possible additional exons are found. When added to the initial coding sequence, a more complete global alignment is obtained.

detailed analysis can reveal patterns of sequences that are common to a family or group of proteins. The simplest methods involve using a multiple alignment algorithm to reveal if a query sequence is indeed related or whether it represents an "outlier." A sequence could be similar to several members of a family but not similar enough for a clear inference of function to be made by a simple pairwise alignment. However, a multiple alignment often will show residues that are highly conserved throughout the family members and therefore assumed important. The Clustal W algorithm *(15)* is a commonly used multiple alignment algorithm and is available from several resources via the Internet.

Several tools take this process further and allow investigators to search using domains or motifs common to all members of protein families. The methods for collecting, grouping and then extracting signature motifs and domains differ among many of the available tools, but the principle remains the same: they allow searches that only consider evolutionarily conserved sequences that are important for the structure and function of proteins. Searching against domains allows the modular architecture of proteins to be considered in a way that surpasses simple sequence similarity searching. Individual regions within a query sequence may match different domain entries within databases to build up a composite picture of a protein's function. Furthermore, most tools available via the Internet, and described later, provide graphical viewers that illustrate the extent and organization of domains within sequences.

Prosite *(16)* is one of the simplest systems. It contains collections of motifs defined using a simple syntax based on regular expressions used in programming. For instance, many ATP- or GTP-binding proteins contain [AG]-x(4)-G-K-[ST] (Ala or Gly followed by any four amino acids, Gly, Lys, and either Ser or Thr). The method is highly sensitive but, depending on the individual motif, is not always very specific. Furthermore, it is too inflexible to define extended domains or diffuse features within sequences. Such motifs suffer from the inability to attach any weighting to the presence or absence of particular residues at specific positions. For instance, in the previous motif it is not possible to specify the relative occurrence of having an A rather than a G at the first position or an S or T at the last position. Profiles overcome this problem; they are probability matrices derived from aligned sequences that represent the relative likelihood of residues occurring at specific positions. When a query sequence is compared to a profile, its residues are scored using a weighting calculated from the frequency of matching residues occurring in the profile. The scores are combined for all the residues in the query sequence and an empirically determined score cutoff is used to determine whether the domain matches. Prosite has recently been extended to include profiles *(17)*.

An extension of profiles is the use of Profile HMMs. The algorithms again create models of domains based on the relative probabilities of residues at each position but treat inserted gaps differently because of the way they handle transitions from position to position in the aligned sequences *(4,18,19)*. The most popular method for generating HMMs from an alignment is HMMer (http://hmmer.wustl.edu/). HMMs have proved highly useful for complex sequence analysis, and now large collections exist. The HMMs are based on large multiple alignments and are tested to ensure that they detect only related proteins. Central to their usefulness is good curation; careful choice of sequences used in the construction of HMMs can influence their specificity and selectivity. One such collection is Pfam (http://www.sanger.ac.uk/Software/Pfam) *(20)*. It contains HMMs based on diverse sequences, designed to focus on major broad families of proteins, and is particularly useful for detecting distantly related family members. The Institute for Genome Research houses another collection of HMMs: TIGRFAMs (http://www.tigr.org/TIGRFAMs/). TIGRFAMs focus on more defined families, for instance, distinguishing between Stevor and Rif genes *(21)* in *P. falciparum*, which have closely related sequences. TIGRFAMS are thus excellent for defining some protein families to high resolution, whereas Pfams are more able to detect more distant relatives.

Several secondary databases group multiple domain collections together to define protein families. The SMART database (Simple Modular Architecture Research Tool; http://smart.embl-heidelberg.de/) *(22)* contains its own collection of HMM domains as well as several other domain collections including Pfam. Interpro (http://www.ebi.ac.uk/interpro/) is a curated database that groups domain definitions from several sources to provide a centralized overview of protein families.

3.3. Orthologs and Paralogs

When annotation is transferred from sequence to sequence, the type of evolutionary relationship between the sequences should be considered. Often, an annotator will take it for granted that two very similar sequences have very similar roles. This is often the case if homologous sequences (i.e., those that share a common ancestor) diverged after speciation. Such sequences are known as orthologs. In another type of homology relationship, known as paralogy, sequences diverge after gene duplication events within a species. Paralogous genes may perform the same role, or they may have diverged sufficiently to have different properties, such as binding to different surface molecules on host cells or having altered substrate specificities. Finding and distinguishing between orthologs and paralogs is an important part of the annotation process. Often, it will be assumed that a "very good hit" to a sequence in another organism is, in fact, to its ortholog and functional annotation will be transferred. Of course, deciding what qualifies as "very good" can be quite subjective, so more rigorous methods must be employed, especially for automated analyses in which no further checks will be applied.

One method uses reciprocal top hits from a sequence similarity search because orthologous genes should be more similar to each other than any other genes. In **Fig. 7**, all genes from two organisms are translated and used as queries in BLAST or FASTA searches against one another. In **Fig. 7A**, gene 1b is identified as a top hit for gene 1a, and, because the reciprocal BLAST search also identifies gene 1a as a top hit for gene 1b, the sequences are putative orthologs. In **Fig. 7B**, gene 1b is again identified as the top hit for gene 1a, but the reciprocal search reveals gene 2a as the top hit. Therefore, gene 1b is not the ortholog of 1a. Furthermore, 1a and 2a are likely to be paralogous genes. The possibility does exist, however, that true orthologs will not exist between two species.

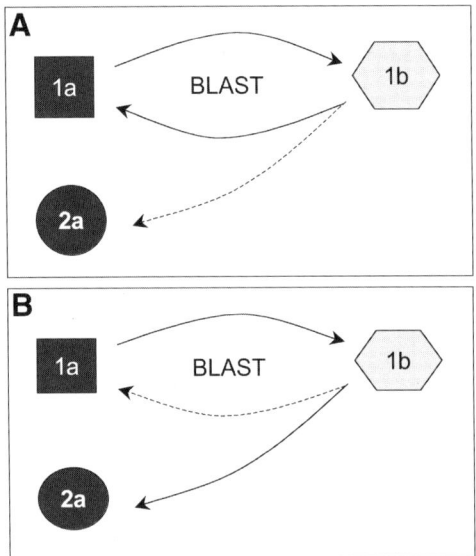

Fig. 7. Use of reciprocal BLAST searches to identify possible orthologous and paralogous genes. (**A**) A BLAST search using gene 1a identifies 1b in another organism as the top hit. The reciprocal BLAST, using 1b as a query, identifies gene 1a, thus the genes may be orthologs. (**B**) If the reciprocal BLAST identifies gene 2a as the top hit for gene 1b, gene 1a and 1b cannot be orthologs.

Consider the following example: in an ancestral cell line, gene X duplicates to form gene Y, which after subsequent divergence has a new function. If speciation occurs, but gene X is lost from one organism and gene Y is lost from the other, X and Y would incorrectly appear (by reciprocal top hits) to be orthologs, despite having very different functions.

Further evidence for orthology can come from analyzing large-scale synteny. This is where arrays of genes, in closely related organisms, share the same general organization as well as sequence similarity. Studying the evolutionarily conserved organization of genes as well as their sequence can provide further evidence of orthology *(23)*. Indeed, some orthologs may, in fact, have rather weak sequence similarity and, therefore, would be missed were it not for the preservation of genes around them.

3.4. Annotating Without Sequence Similarity

Where similarity data and good domain evidence exist, genes can be annotated with reasonable confidence. However, there are vast numbers of predicted genes—at least 50% in a typical parasite genome project—for which functional annotation cannot be assigned in this way. These genes fall into two groups that have either similarity to other sequences but only to those with no known function, or no similarity at all. Often, these groups are annotated with the descriptions of "conserved hypothetical protein" and "hypothetical protein," respectively. However, it should be borne in mind that though their annotation is minimal, these groups will include novel genes encoding novel functions. In particular, the hypothetical proteins will undoubtedly include those that have parasite-specific functions that await characterization. Some additional information can be added, using numerous software tools, to provide clues to their functions based on the predicted cellular localization of the gene products. Membrane proteins sometimes can be found simply by searching for transmembrane helices. Two popular tools for this are TMHMM (http://www.cbs.dtu.dk/services/TMHMM/) *(24)* and TopPred (http://bioweb.pasteur.fr/seqanal/interfaces/toppred.html) *(25)*. Signal peptides and signal anchors can be predicted using SignalP (http://www.cbs.dtu.dk/services/SignalP/) *(26)*, and a more extensive analysis of possible targeting can be attempted using PSORT (http://psort.nibb.ac.jp/) *(27)*. However, the accuracy of predictions does vary with all of these tools. For instance, TMHMM often will falsely predict transmembrane helices in stretches of hydrophobic amino acids and cannot be used to provide additional evidence for the likelihood that a predicted coding region is real. Likewise, the accuracy of PSORT in predicting the localization of parasite gene products has not been fully evaluated; the tool has been developed using model organisms that contain different subcellular structures and may have different targeting sequences. Organism-specific tools can also be used to predict the localization of some gene products. For instance, the PlasmoAP *(28)* algorithm has been developed to predict which of the approx 5300 genes in *P. falciparum* encode products that are targeted to the parasite-specific organelle, the apicoplast. With the benefit of more data, tools like these can be increasingly honed to greater accuracy.

4. Classification and Large-Scale Analysis

The utility of any annotation can be increased greatly when genes are classified according to their inferred roles. Various aspects of an organism's biology, such as metabolism, often can be reconstructed from these role assignments *in silico*. Furthermore, during this reconstruction, further areas for analysis can be targeted by exposing missing areas of annotation.

4.1. Reconstructing Metabolism

One of the simplest classification systems is the use of Enzyme Commission (EC) numbers to annotate enzymes according to their catalytic activity. Initial EC numbers are assigned either by sequence similarity to other curated sequences or by the presence of conserved catalytic domains (e.g., InterPro/Pfam). A specialized database such as MetaCyc (http://biocyc.org/metacyc/) can also be used for information gathering or for assigning EC numbers by homology. Each EC number corresponds to a specific chemical reaction and, when viewed together, whole biochemical pathways can be reconstructed. The Kyoto Encyclopedia of Genes and Genomes (KEGG, http://www.genome.ad.jp/kegg/) includes a database of many known biochemical pathways and is excellent for this purpose. When initial annotated EC numbers from a genome project are entered, KEGG displays numerous pathways in which genes could be involved. Those maps with multiple EC number hits are most likely to reflect real pathways in an organism and can be used to identify missing enzymes.

Figure 8 shows an example from the *P. falciparum* genome project. Pyrimidine metabolism was identified based on five annotated EC numbers, and an initial map is drawn with their positions highlighted (**Fig. 8A**). For each EC number that may be missing from the annotation, representative example sequences are obtained from other organisms. Similarity searches against all predicted genes in *P. falciparum* reveal good candidates for genes encoding these enzyme activities. Many EC numbers are not assigned in conservative initial annotation—evidence may fall into one of the many "gray areas" where it is either weak or ambiguous. However, a given gene often will fit into only one of the available holes when pathways are reconstructed, so that many of the earlier ambiguities can be resolved. In *P. falciparum*, the redrawn map (**Fig. 8B**), with extra EC numbers from the refined annotation included, now shows *de novo* pyrimidine biosynthesis as a contiguous pathway.

The EC number classification system is an example of a strict hierarchy of terms that can be used for unambiguous functional descriptions covering one aspect of biology, namely, enzymatic reactions. Of course, methods of this kind rely on sequence similarity matches and, therefore, cannot be used to identify completely novel biochemical activities.

4.2. Gene Ontology

As the amount of available data from genome sequencing, gene expression, proteomics, and other large-scale experiments increases, different aspects of biology need to be described unambiguously to enable biologically relevant database queries. One key problem is that any given biological phenomenon can be described in many different ways. For example, if one database describes "translation," whereas another uses the phrase "protein synthesis," it will be difficult for a biologist—and even harder for a computer—to find equivalent terms and retrieve data from both databases.

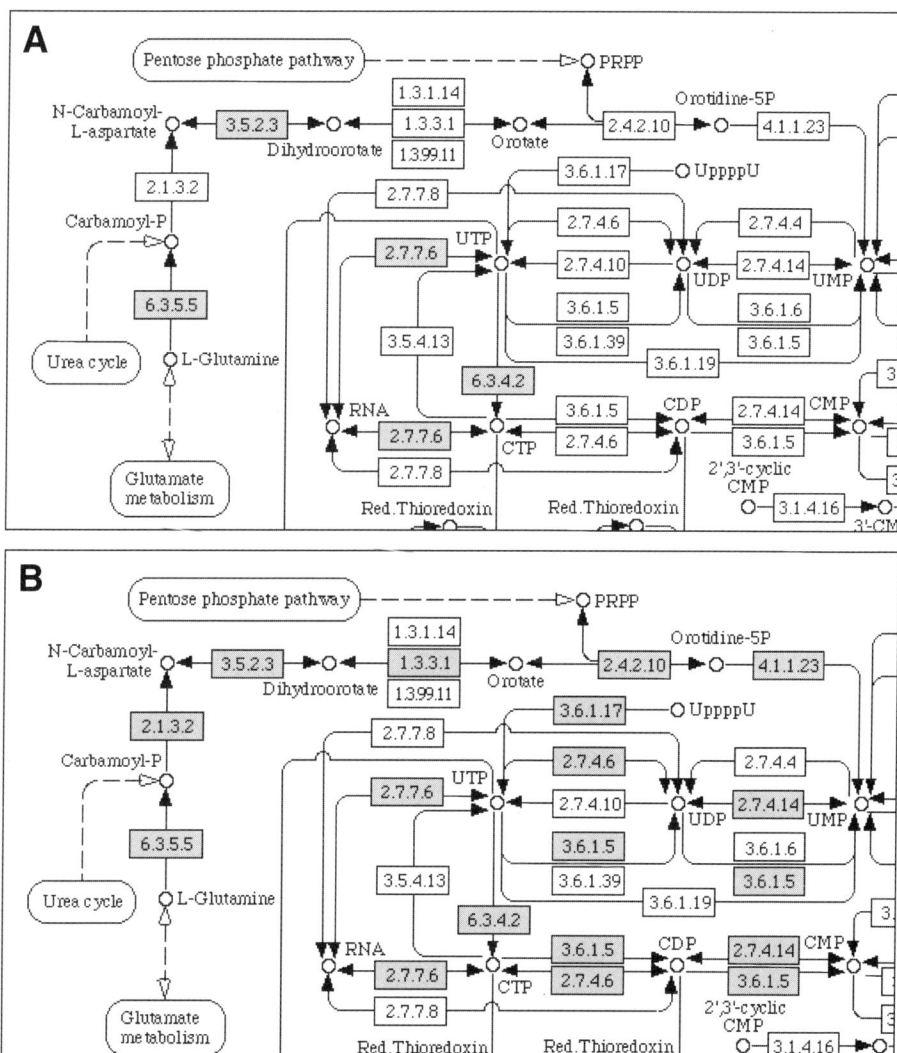

Fig. 8. Enzyme commission numbers can be used to reconstruct biochemical pathways and finding missing genes. (**A**) An initial query of KEGG (Kyoto Encyclopedia of Genes and Genomes) using annotated EC numbers from prelimary genome annotation reveals an incomplete *de novo* pyrimidine biosynthesis pathway in *P. falciparum*. By searching against the genome sequence using genes from other organism representing the gaps, more EC numbers can be added until candidate genes for every enzyme in the pathway have been identified (**B**).

The Gene Ontology (GO) project *(29)* was initiated to address the growing need for meaningful, consistent annotation of genes and their products in different organisms by providing a set of controlled vocabularies of terms. Originally, it was developed to describe a limited number of organisms. Since then, GO has been expanded, based on feedback from biologists, to cover diverse organisms such as parasites *(30)*. The vocabularies are three independent, structured networks of terms (known as ontologies) that describe attributes of gene products in terms of their "molecular function," "biological process," and "cellular component."

Molecular function (e.g., hexokinase activity) describes the activities or tasks performed by individual gene products at the molecular level; biological process (e.g., glycolysis) describes broad biological goals that are accomplished by ordered assemblies of molecular functions; and cellular component (e.g., glycosome) encompasses subcellular structures, locations, and macromolecular complexes.

Within each vocabulary, terms are organized hierarchically so a more general "parent" term can be related to a more specific "child." Thus, a child term is a subset of its parent(s). The vocabularies are structured as treelike directed acyclic graphs, wherein any term may have one or more parents as well as any number of children. An additional constraint is imposed on these parent–child relationships: every possible path from a specific term toward the root of the tree must be biologically accurate. Thus, in the following structure, annotating a gene product to "cytoadherence to microvasculature" would automatically imply its involvement in cell–cell adhesion:

GO:0016337 : cell–cell adhesion
 ∟ GO:0007157 : heterophilic cell adhesion
  ∟ GO:0020035 : cytoadherence to microvasculature

Numerous tools allow users to browse the GO vocabulary, including AmiGO (http://www.godatabase.org/), which allows branches of the tree to be expanded and contracted as different parent–child relationships are explored (*see* below).

4.2.1. GO Annotation of Gene Products

When gene products are annotated using GO, one of several evidence codes is included to provide a rough guide to how that gene product was annotated. For instance, annotations that are made based on sequence similarity bear the code inferred from sequence similarity, and automatic annotations made by computer programs bear the code inferred from electronic annotation. In this way, investigators can effectively weight the data by their own criteria, and some of the dangers of transitive annotation errors can be avoided. Furthermore, when terms from different depths in the ontology are combined with appropriate evidence codes, complex annotations can be built up. Consider the following example:

GO:0008236 (serine-type peptidase activity)   **I**nferred from **S**equence **S**imilarity
GO:0004175 (endopeptidase activity)   **I**nferred from **D**irect **A**ssay

Here, the annotation can convey that the gene product has experimentally confirmed endopeptidase activity but, based on its sequence, looks like it may have serine-type peptidase activity.

Alongside the evidence codes, every GO annotation is attributed to a source, which may be a literature reference, another database, or a computational analysis.

4.2.2. Querying GO

An obvious query is to retrieve all gene products for a given GO term. In **Fig. 9A**, Amigo allows an investigator to search or browse to find a GO term (e.g., enzyme activity) and then find gene products to which it has been annotated. Again, because of the ontology's structure, the query finds anything that has been annotated to the general description of enzyme activity or to more specific types of enzyme activity (e.g., hydrolase or peptidase activity). The list of genes that is displayed by the browser spans several species, which can be filtered (**Fig. 9B**) to show annotated genes from specific organism databases. It is then possible to use these GO annotations to identify similarly annotated gene products from other organisms.

4.2.3. Assigning GO Terms

Several methods exist for assigning GO terms, and it is beyond the scope of this chapter to review them all. By far, the most reliable method is to assign terms based on reading published literature. However, this extremely time-consuming approach can be used only for characterized genes. For uncharacterized genes, similarity to genes from other organisms that have been annotated with GO terms can be used; a BLAST interface included in the AmiGO browser can be used for this. In addition, the InterPro database (http://www.ebi.ac.uk/interpro/) contains many protein domains that have been mapped to GO terms.

4.2.4. Other Ontology Efforts

GO does not cover all aspects of molecular biology that would be of interest to someone with gene products to annotate. However, the GO consortium does support the development of other biological ontologies to complement GO through an effort called Global Open Biology Ontologies (http://www.geneontology.org/doc/gobo.html). The use of GO combined with other ontologies covering other aspects of biology should allow more comprehensive descrip-

A

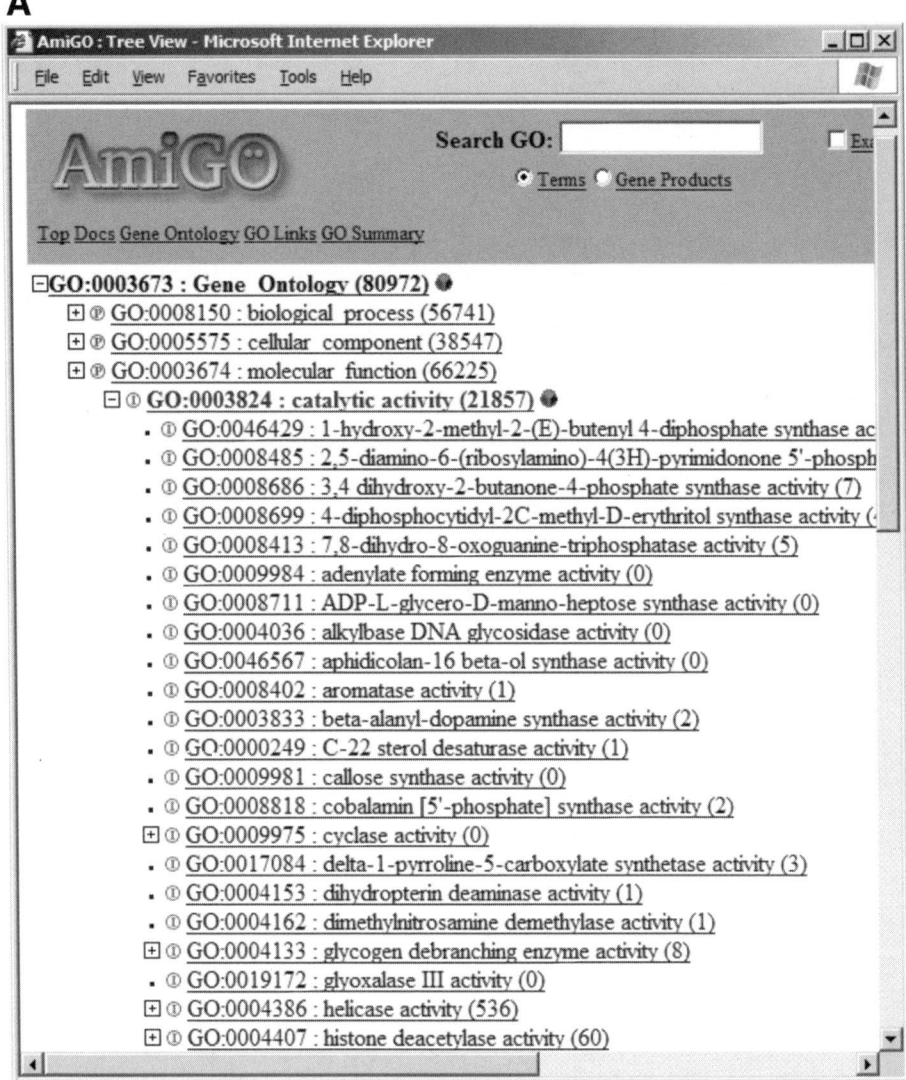

Fig. 9. Browsing gene ontology using the AmiGO browser. (**A**) Branches of the GO tree can be expanded or collapsed to find a term in the vocabulary. The number in parentheses shows the number of genes that have been annotated with that GO term.

tions to be built that can easily be accessed and queried by biologists. Phenotype ontologies would allow the effect of knockout and knockdown mutations to be described. The stage(s) in a life cycle in which a gene is active could be described using terms taken from a parasite life-cycle ontology.

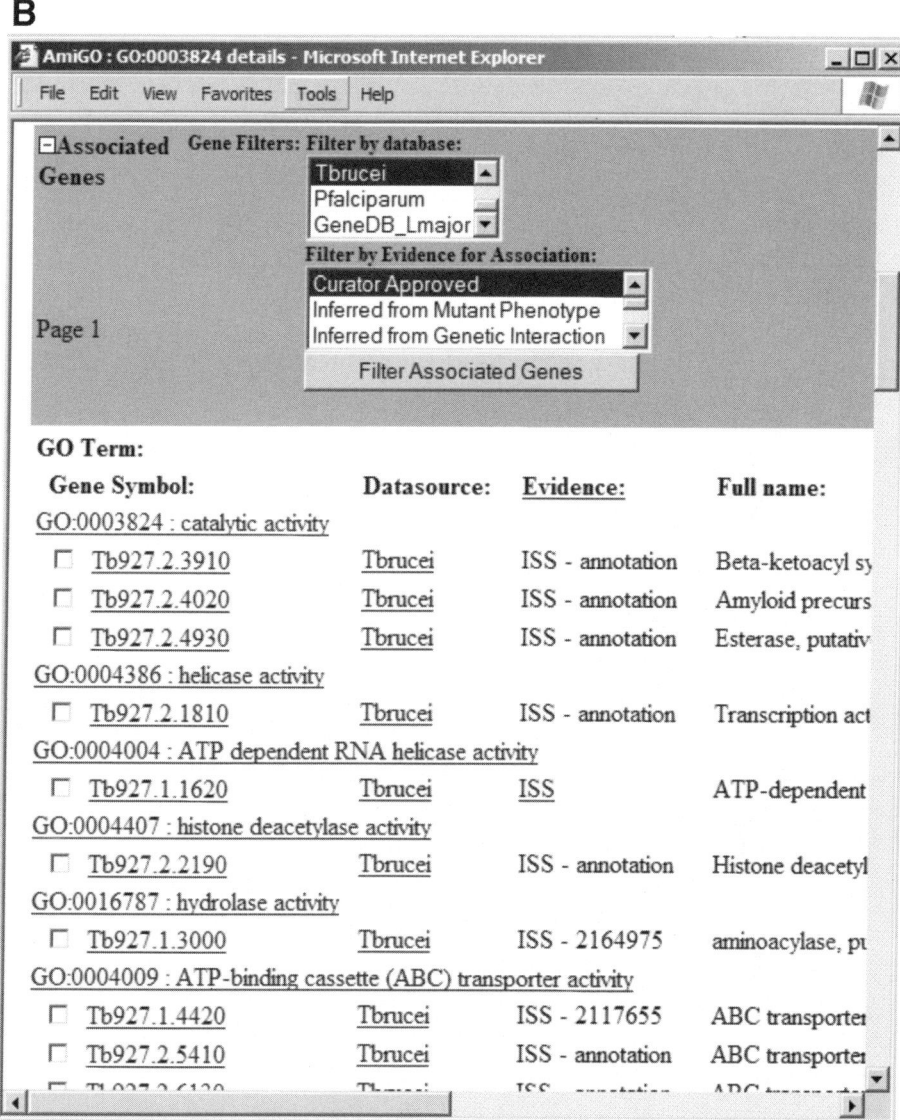

Fig. 9. (**B**) Clicking on a term provides a list of the genes to which it has been annotated (or to any terms below it in the ontology). The example shows those genes annotated to "catalytic activity" or below, filtered to show only those from *T. brucei*.

5. Summary

Annotation is an essential step in a parasite genome project if the maximum benefit is to be gained. The quality and usefulness of any genome annotation is related directly to the amount of effort invested in it. Ideally, all projects would be annotated to the highest level, but this is frequently not possible because of timing or financial considerations. Therefore, researchers should assess the reliability of annotation before they use it. Automated annotation provides a rough catalog of a genome, allowing genes to be short-listed quickly using "handles" based on their sometimes quite error-prone descriptions. Higher-quality manual annotation aims to go a step further and provides the user with more informative and accurate descriptions, more akin to an encyclopedia than a simple catalog. In the context of databases, gene descriptions ideally should be drawn from controlled vocabularies. These are allowing progressively more software tools to be written that can interrogate data and facilitate asking increasingly sophisticated questions before any laboratory experiments have even been performed.

References

1. Parkhill, J. (2002) Annotation of microbial genomes, in *Methods in Microbiology: Functional Microbial Genomics*, Vol. 33 (Brendan, W. and Dorrell, N., eds.), Academic Press, London, UK, pp. 3–26.
2. Berriman, M. and Rutherford, K. (2003) Annotation and visualisation of sequences using Artemis. *Brief Bioinform* **4(2)**, 124–132.
3. Rutherford, K., Parkhill, J., Crook, J., et al. (2000) Artemis: sequence visualization and annotation. *Bioinformatics* **16**, 944–945.
4. Krogh, A. (1998) An introduction to hidden Markov models for biological sequences, in *Computational Methods in Molecular Biology* (Salzberg, S. L., Searls, D. B., and Kasif, S., eds.), Elsevier, Amsterdam, pp. 45–63.
5. Mount, D. W. (2001) Bioinformatics: Sequence and Genome Analysis, Cold Spring Harbor Laboratory Press, Cold Spring Harbor, NY.
6. Salzberg, S. L., Pertea, M., Delcher, A. L., et al. (1999) Interpolated Markov models for eukaryotic gene finding. *Genomics* **59**, 24–31.
7. Cawley, S. E., Wirth, A. I., and Speed, T. P. (2001) Phat—a gene finding program for *Plasmodium falciparum. Mol. Biochem. Parasitol.* **118**, 167–174.
8. Gardner, M. J., Hall, N., Fung, E., et al. (2002) Genome sequence of the human malaria parasite *Plasmodium falciparum. Nature* **419**, 498–511.
9. Ishikawa, J. and Hotta, K. (1999) FramePlot: a new implementation of the frame analysis for predicting protein-coding regions in bacterial DNA with a high G + C content. *FEMS Microbiol. Lett.* **174**, 251–253.
10. Lowe, T. M. and Eddy, S. R. (1997) tRNAscan-SE: a program for improved detection of transfer RNA genes in genomic sequence. *Nucleic Acids Res.* **25**, 955–964.
11. Smith, T. F. and Waterman, M. S. (1981) Identification of common molecular subsequences. *J. Mol. Biol.* **147**, 195–197.
12. Needleman, S. B. and Wunsch, C. D. (1970) A general method applicable to the search for similarities in the amino acid sequence of two proteins. *J. Mol. Biol.* **48**, 443–453.

13. Pearson, W. R. and Lipman, D. J. (1988) Improved tools for biological sequence comparison. *Proc. Natl. Acad. Sci. USA* **85**, 2444–2448.
14. Altschul, S. F., Gish, W., Miller, W., et al. (1990) Basic local alignment search tool. *J. Mol. Biol.* **215**, 403–410.
15. Thompson, J. D., Higgins, D. G., and Gibson, T. J. (1994) CLUSTAL W: improving the sensitivity of progressive multiple sequence alignment through sequence weighting, position-specific gap penalties and weight matrix choice. *Nucleic Acids Res.* **22**, 4673–4680.
16. Bucher, P. and Bairoch, A. (1994) A generalized profile syntax for biomolecular sequence motifs and its function in automatic sequence interpretation. *Proc. Int. Conf. Intell. Syst. Mol. Biol.* **2**, 53–61.
17. Falquet, L., Pagni, M., Bucher, P., et al. (2002) The PROSITE database, its status in 2002. *Nucleic Acids Res.* **30**, 235–238.
18. Eddy, S. R. (1996) Hidden Markov models. *Curr. Opin. Struct. Biol.* **6**, 361–365.
19. Eddy, S. R. (1998) Profile hidden Markov models. *Bioinformatics* **14**, 755–763.
20. Sonnhammer, E. L., Eddy, S. R., Birney, E., et al. (1998) Pfam: multiple sequence alignments and HMM-profiles of protein domains. *Nucleic Acids Res.* **26**, 320–322.
21. Cheng, Q., Cloonan, N., Fischer, K., et al. (1998) stevor and rif are *Plasmodium falciparum* multicopy gene families which potentially encode variant antigens. *Mol. Biochem. Parasitol.* **97**, 161–176.
22. Schultz, J., Milpetz, F., Bork, P. et al. (1998) SMART, a simple modular architecture research tool: identification of signaling domains. *Proc. Natl. Acad. Sci. USA* **95**, 5857–5864.
23. Carlton, J. M., Angiuoli, S. V., Suh, B. B., et al. (2002) Genome sequence and comparative analysis of the model rodent malaria parasite *Plasmodium yoelii yoelii*. *Nature* **419**, 512–519.
24. Krogh, A., Larsson, B., von Heijne, G. et al. (2001) Predicting transmembrane protein topology with a hidden Markov model: application to complete genomes. *J. Mol. Biol.* **305**, 567–580.
25. Claros, M. G. and von Heijne, G. (1994) TopPred II: an improved software for membrane protein structure predictions. *Comput. Appl. Biosci.* **10**, 685–686.
26. Nielsen, H., Engelbrecht, J., Brunak, S., et al. (1997) Identification of prokaryotic and eukaryotic signal peptides and prediction of their cleavage sites. *Protein Eng.* **10**, 1–6.
27. Nakai, K. and Horton, P. (1999) PSORT: a program for detecting sorting signals in proteins and predicting their subcellular localization. *Trends Biochem. Sci.* **24**, 34–36.
28. Foth, B. J., Ralph, S. A., Tonkin, C. J., et al. (2003) Dissecting apicoplast targeting in the malaria parasite *Plasmodium falciparum*. *Science* **299**, 705–708.
29. Ashburner, M., Ball, C. A., Blake, J. A., et al. (2000) Gene ontology: tool for the unification of biology. The Gene Ontology Consortium. *Nat. Genet.* **25**, 25–29.
30. Berriman, M., Aslett, M., and Ivens, A. (2001) Parasites are GO. *Trends Parasitol.* **17**, 463–464.

3

Parasite Genome Databases and Web-Based Resources

Christiane Hertz-Fowler and Neil Hall

Summary

In the last decade, high-throughput genome sequencing and complementary techniques such as microarray and proteomics have generated, and will continue to generate, ever-increasing amounts of data. These technologies of gene discovery, expression, and functional analysis have been applied to a vast array of organisms, including parasites. In most instances, the data are freely available via the Internet, and researchers are becoming increasingly reliant on up-to-date, centralized data repositories to complement wet bench science.

This chapter presents an overview of resources relevant to researchers with an interest in parasite genomics and biology. After briefly touching on some of the publicly available nucleotide and protein sequence as well as domain databases, the focus turns to parasite genome projects and associated Web-based resources. A list of parasite sequencing projects current at the time of writing, including relevant Web site addresses, is provided. The available resources range from network sites and project pages at sequencing institutes to databases that integrate and curate sequence data and associated annotation with diverse biological datasets. Particular attention is given to three databases, GeneDB (http://www.genedb.org/), PlasmoDB (http://plasmodb.org/), and tigr db, detailing the scope of each database and the tools available for data querying and retrieval.

Key Words: Annotation; data curation; genome databases; parasite sequencing projects; Web resource.

1. Introduction

In the last decade, there has been a fundamental change in the landscape of molecular biological research with the advent of high-throughput genome sequencing and complementary techniques such as microarray and proteomics. These new technologies of gene discovery, expression, and functional analysis have been applied to a vast array of organisms, including parasites. Such projects have usually been funded as precompetitive resources for the scientific community, and data have been made freely available via the Internet, facilitating basic research. For researchers to use this new information effectively, an array of tools for data querying and retrieval has been developed. However, genomic information can be underused, as researchers may be unaware of the full extent of Web-based resources, the contents of datasets, or the available tools.

This overview aims to outline resources relevant to researchers with an interest in parasite genomics and biology. It briefly mentions some of the public nucleotide and protein sequence and domain databases, which act as repositories and provide analytical tools for multiple species. The latter part of the review focuses on parasite genome projects and associated resources. Particular attention is given to databases that integrate sequencing data with other biological data, thus contributing to an overall understanding of genome organization and function.

2. General Databases/Resources

The databases listed below (*see also* **Table 1**) are commonly used as repositories for sequence retrieval and submission as well as aiding in the characterization of new sequences. Such computational sequence analyses are particularly pertinent, considering the functions of only a small number of genes and proteins from parasitic organisms have been determined experimentally or can even be inferred by homology (e.g., 60% of predicted *Plasmodium falciparum* proteins were classified as hypothetical protein *[1]*).

2.1. Primary DNA and Protein Databases

European Molecular Biology Laboratory (EMBL)/GenBank/DNA Data Bank of Japan *(2–4)* is a collection of redundant, primary nucleotide sequence databases, updated and synchronized among the three collaborating databases on a daily basis. They contain sequences from direct submissions, genome sequencing projects, scientific literature, and patent applications. The format of entries is identical to that of the Translated EMBL (TrEMBL) and SWISS-PROT databases (*see* below), resulting in ease of use. Unfortunately, sequences and associated information may be out of date, as only the original submitter of a sequence can modify an existing entry.

TrEMBL *(5)* contains the translation as well as automated and, in some instances, manual annotation of all coding sequences (CDSs) in EMBL that are not yet incorporated into SWISS-PROT. It is subdivided into three sections: TrEMBL new contains the translations of recently submitted protein-coding genes; subsequently, TrEMBLnew data are moved to either SP-TrEMBL, storing entries that eventually will be incorporated into SWISS-PROT, or REM-TrEMBL (the remainder).

Table 1
Overview of Nucleotide and Protein Resources Useful for Sequence Analysis

Database	Web address	Description	Data access
EMBL (2)	http://www.ebi.ac.uk/embl/	Nucleotide database	Nucleotide sequence submission/search/retrieval
DDBJ (4)	http://www.ddbj.nig.ac.jp/	Nucleotide database	Nucleotide sequence submission/search/retrieval/alignment Protein sequence analytical tools
GenBank (3)	http://www.ncbi.nlm.nih.gov/Genbank	Nucleotide database	Nucleotide sequence submission/search/retrieval
TrEMBL (5)	http://www.ebi.ac.uk/trembl/	Protein sequence database of translated nucleotide sequences	Text and sequence search Sequence retrieval
PIR-PSD (38)	http://pir.georgetown.edu/pirwww/dbinfo/pirpsd.html	Annotated protein sequence database	Text and sequence search Sequence retrieval Pattern search/peptide match
SWISS-PROT (6)	http://www.ebi.ac.uk/swissprot/	Annotated protein sequence database	Text and sequence search Sequence retrieval Sequence submission
InterPro (7)	http://www.ebi.ac.uk/interpro/	Resource integrating protein signature databases	Text and sequence search
Pfam (39)	http://www.sanger.ac.uk/Software/Pfam/	Protein domain and family database, based on hidden Markov models, manually reviewed	DNA and protein sequence search Text search Browse Pfam entries
PRINTS (40)	http://www.bioinf.man.ac.uk/dbbrowser/PRINTS/	Protein motif database, based on groups of conserved structural motif used to characterize a protein family, manually reviewed	Text and sequence search Query by motifs

(continued)

Table 1 (Continued)

Database	Web address	Description	Data access
ProDOM (41)	http://prodes.toulouse.inra.fr/prodom/2002.1/html/home.php	Protein domain family database automatically generated from the SWISS-PROT and TrEMBL sequence databases	Text and sequence search Browse ProDom entries
PROSITE (42)	http://ca.expasy.org/prosite/	Protein family and domain database, based on pattern matching of single most conserved motif observed in a multiple sequence alignment of known homologs	Text search (including author, description) Sequence search Browse PROSITE entries Download of all entries Tools (e.g., ScanProsite, MotifScan)
SMART (43)	http://smart.embl-heidelberg.de/	Resource for domain structures of extracellular, signalling and chromatin-associated proteins, based on hidden Markov models	Text and sequence search Search domains in defined taxa Integrates Pfam and signal peptide predictions Mapped Gene Ontology associations to SMART domains Browse SMART entries
TIGRFAM (44)	http://www.tigr.org/TIGRFAMs/	Resource for protein families based on hidden Markov models	Text and sequence search Mapped to GO, EC #s, other protein domain databases Download of all entries

European Bioinformatics Institute (EBI) **(9)**	http://www.ebi.ac.uk/	Nucleic acid sequence database Protein sequence, domain, structure databases Whole genome databases Gene expression databases Development of controlled vocabulary (GO) Sequence analysis and data-mining tools
Institut Suisse de Bioinformatique (ISB)	http://www.isb-sib.ch/	Protein sequence, domain, structure, and proteomics databases Promoter database Analytical tools ExPASy Molecular Biology Server
National Center for Biotechnology Information (NCBI) **(8)**	http://www.ncbi.nlm.nih.gov/	Nucleic acid sequence and polymorphism databases Taxonomy browser Literature database Gene expression databases Ortholog resource Access to genetic and physical maps Genome-specific resources Sequence analysis and data-mining tools
South African National Bioinformatics Institute (SANBI)	http://www.sanbi.ac.za/	Human Genome Server Links to sequence retrieval and analysis tools

Text-based searching refers to searching the annotation via keywords or retrieval tools such as SRS. Sequence-based searching refers to either BLAST (local alignment, can report more than one high-scoring match) or FASTA (global alignment, reporting only one match) searches. BLAST, Basic Local Alignment Search Tool; SRS, Sequence Retrieval System.

2.2. Curated Protein and Protein Family Signature Databases

SWISS-PROT *(6)* is a manually annotated protein database that integrates the outcome of predictive software packages with additional information extracted from literature and provided by external experts. It is nonredundant, collating reports and submitted nucleotide sequences for a given protein into one entry, which is extensively cross-referenced to other databases. Recently, the SWISS-PROT, TrEMBL, and PIR protein database activities have been merged to form the Universal Protein Knowledgebase (UniProt) (http://www.uniprot.org).

As repositories, the above databases contain a wealth of information of interest to parasitologists while also supporting data manipulation. Taking *P. falciparum* as an example, SWISS-PROT release 41 contains 154 curated entries, whereas SP-TrEMBL houses 8280 entries, which will be incorporated into SWISS-PROT, removing redundancies. In contrast, EMBL stores 28,746 entries. However, similarity searches of nucleotide and protein databases can only infer a function/structure of an unknown query sequence based on "clear" similarity to a characterized gene/protein. Increasingly, with the growth of sequence databases, matches most often occur to diverse, often uncharacterized sequences. Therefore, rather than assigning a possible function based on sequence similarity, proteins can also be diagnosed as belonging to a particular family of proteins by searching various pattern databases. InterPro *(7)* is one such resource, which integrates numerous signature and sequence cluster-based databases of curated protein families, domains, and functional sites (*see* **Table 1** for a comprehensive list).

2.3. Bioinformatics Institutes

Numerous bioinformatics institutes provide tools for manipulation of biological data (**Table 1**). Particularly noteworthy are the National Center for Biotechnology Information (NCBI) *(8)* and the European Bioinformatics Institute (EBI) *(9)*. Both provide tools for data retrieval, analysis, and similarity searching as well as creating and maintaining nucleotide databases such as GenBank (NCBI) and EMBL (EBI), protein repositories, a literature database (PubMed at NCBI), and curated databases for expression profiling, gene orthologs (Homolo Gene), and reference sequences (RefSeq). NCBI also houses a resource for completed genomes, such as *P. falciparum*, and gives an overview of parasitic organisms with ongoing sequencing projects (*see* http://www.ncbi.nlm.nih.gov/PMGifs/Genomes/EG_T.html).

3. Parasite Genome Projects and Web-Based Resources
3.1. Overview

Since the early 1990s, a large number of protozoan and helminth parasite genome sequencing projects have been initiated, some as part of multinational

Table 2
List of Protozoan and Metazoan Parasite Sequencing Projects

Parasite (strain)	Genome size	Type project	Sequencing center/network	Database
Encephalitozoon cuniculi (**15**) (GB-M1)	2.9 Mb	◆	Gscope http://www.genoscope.cns.fr/externe/English/Projets/Projet_AD/AD.html	
Leishmania major (**45**) (Friedlin)	33.6 Mb		NW http://www.ebi.ac.uk/parasites/leish.html	GeneDB (**28**): http://www.genedb.org/
		◆	Sanger http://www.sanger.ac.uk/Projects/L_major/	TIGR Leishmania gene Index: http://www.tigr.org/tdb/tgi/lshgi
		◆	SBRI http://www.genome.sbri.org/lmjf/lmjfindex.htm	
		♣	GSC http://genome.wustl.edu/est/index.php?leishmania=1	
Trypanosoma brucei (**46**) (TREU927/4 GUTat 10.1)	25 Mb		NW http://parsun1.path.cam.ac.uk/	GeneDB
		◆	Sanger http://www.sanger.ac.uk/Projects/T_brucei/	TbGAD: http://www.tigr.org/tdb/e2k1/tba1/tba1.shtml
		◆	TIGR http://www.tigr.org/tdb/e2k1/tba1/	TIGR *T. brucei* Gene Index: http://www.tigr.org/tdb/tgi/tbgi/
				U-insertion/deletion edited sequence database: http://www.rna.ucla.edu/trypanosome/database.html
				Minicircle database: http://www.ebi.ac.uk/parasites/kDNA/P2atryp.html
				guide RNA database: http://biosun.bio.tudarmstadt.de/goringer/gRNA/gRNA.html
				TrypanoFAN functional genomics: http://www.trypanofan.org
T. congolense	25 Mb	◆	Sanger http://www.sanger.ac.uk/Projects/T_congolense/	

(*continued*)

Table 2 (Continued)

Parasite (strain)	Genome size	Type project	Sequencing center/network		Database
T. cruzi (CL-Brener)	87 Mb (total)	◆	NW	http://www.dbbm.fiocruz.br/TcruziDB/index.html	TIGR *T. cruzi* Gene Index: http://www.tigr.org/tdb/etl/tca1/tca1.shtml
		◆ ✤	KI	http://web.cgb.ki.se/	Minicircle database: http://www.ebi.ac.uk/parasites/kDNA/P2atryp.html
		◆ ✤	SBRI	http://www.genome.sbri.org/tcbr/tcbrindex.htm	guide RNA database: http://biosun.bio.tu-darmstadt.de/goringer/gRNA/gRNA.html
		◆ ✤	TIGR	http://www.tigr.org/tdb/e2k1/tca1/	GeneDB: *T. cruz* DB: http://tcruzidb.org
T. vivax	25 Mb	◆	Sanger	http://www.sanger.ac.uk/Projects/T_vivax/	
Neospora caninum	ND	●	GSC	http://www.genome.wustl.edu/est/index.php?neospora=1	TIGR *N. canium* Gene Index: http://www.tigr.org/tdb/tgi/ncgi
Sarcocystis neurona	ND	●	GSC	http://www.genome.wustl.edu/est/index.php?sarcocystis=1	TIGR *S. neurona* Gene Index: http://www.tigr.org/tdb/tgi/sngi/
Entamoeba histolytica	18–20 Mb	◆	Sanger	http://www.sanger.ac.uk/Projects/E_histolytica/ (see also for other Entamoeba species)	
		◆	TIGR	http://www.tigr.org/tdb/e2k1/eha1/	
		◆	MBL	http://jbpc.mbl.edu/Giardia-HTML/index2.html	
Giardia lamblia (**47**) (WB clone C6)	12 Mb	◆			GiardiaDB (**25**): http://tuscany.mbl.edu/GiardiaDB/giardiadb.html
Babesia bovis	9.4 Mb	●	Sanger	http://www.sanger.ac.uk/Projects/B_bovis/	
Cryptosporidium parvum (**48**)	10.4 Mb	● ✤	VCU	http://www.parvum.mic.vcu.edu/	TIGR *C. parvum* Gene Index: http://www.tigr.org/tdb/tgi/cpgi/
		●	UCSF	http://medsfgh.ucsf.edu/id/CpTags/est.html	CryptoDB: http://cryptodb.org
		◆	MCPG	http://www.cbc.umn.edu/ResearchProjects/AGAC/Cp/index.htm	
Eimeria tenella (**49**)	60 Mb	●	GSC	http://genome.wustl.edu/est/index.php?eimeria=1	TIGR *E. tenella* Gene Index: http://www.tigr.org/tdb/tgi/etgi/
		◆	Sanger	http://www.sanger.ac.uk/Projects/E_tenella/	

(Houghton)			UFL	http://parasite.vetmed.ufl.edu/berg.htm	GeneDB
Plasmodium berghei (**50**) (ANKA)	25–27 Mb	●♣ ♦	Sanger	http://www.sanger.ac.uk/Projects/P_berghei/	PlasmoDB (**32**): http://plasmodb.org/ TIGR P. berghei Gene Index: http://www.tigr.org/tdb/tgi/pbgi/
P. chabaudi (AS strain)	25–30 Mb	♣	Sanger GLA	http://www.sanger.ac.uk/Projects/P_chabaudi/ http://www.gla.ac.uk/ibls/II/mal/malaria.htm	GeneDB PlasmoDB
P. falciparum (**1**) (3D7)	23 Mb	● ♦ ♦ ♦	GSC Sanger Stanf. TIGR	http://genome.wustl.edu/est/index.php?plasmodium=1 http://www.sanger.ac.uk/Projects/P_falciparum/ http://sequence-www.stanford.edu/group/malaria/ http://www.tigr.org/tdb/e2k1/pfa1/	PlasmoDB GeneDB PfGAD: http://www.tigr.org/tdb/e2k1/pfa1/pfa1.shtml TIGR *P. falciparum* Gene Index: http://www.tigr.org/tdb/tgi/pfgi/
		●♣	UFL	http://parasite.vetmed.ufl.edu/falc.htm	NCBI Malaria Genetics and Genomics: http://www.ncbi.nlm.nih.gov/projects/Malaria/ WHO/TDR Malaria Database: http://www.wehi.edu.au/MalDB-www/who.html Metabolic Pathways: http://sites.huji.ac.il/malaria/
P. knowlesi (H strain)	25 Mb	♦	Sanger	http://www.sanger.ac.uk/Projects/P_knowlesi/	PlasmoDB
P. reichenowi	25–27 Mb	♦	Sanger	http://www.sanger.ac.uk/Projects/P_reichenowi/	
P. vivax (**50**) (Salvador I)	30 Mb	♦ ♣	Sanger TIGR UFL	http://www.sanger.ac.uk/Projects/P_vivax/ http://www.tigr.org/tdb/e2k1/pva1/ http://parasite.vetmed.ufl.edu/	PlasmoDB

(continued)

Table 2 (Continued)

Parasite (strain)	Genome size	Type project	Sequencing center/network		Database	
P. yoelii (**14**) (17XNL)	25 Mb	◆●	NMRC TIGR	http://www.nmrc.navy.mil/index.html http://www.tigr.org/tdb/e2k1/pya1/	PyGAD:	http://www.tigr.org/tdb/e2k1/pya1/pya1.shtml PlasmoDB TIGR *P. yoelii* Gene Index: http://www.tigr.org/tdb/tgi/pygi/
Theileria annulata (Ankara C9)	8 Mb	◆●	Sanger	http://www.sanger.ac.uk/Projects/T_annulata/		
Theileria parva (**51**)	9 Mb	♣	ILRI TIGR	http://www.cgiar.org/ilri/ http://www.tigr.org/tdb/e2k1/tpa1/		
Toxoplasma gondii (B7)	80 Mb	♣◆	GSC Sanger TIGR	http://genome.wustl.edu/est/index.php?toxoplasma=1 http://www.sanger.ac.uk/Projects/T_gondii/ http://www.tigr.org/tdb/e2k1/tga1/	ToxoDB (**52**):	http://toxodb.org/ToxoDB.shtml TIGR *T. gondii* Gene Index: http://www.tigr.org/tdb/tgi/tggi/
Ancylostoma canium	ND	●	GSC	http://nematode.net	NemaGene:	http://genome.wustl.edu/nn_dav/NemaGene/
Ancylostoma ceylanicum	ND	●	GSC	http://nematode.net		
Ancylostoma duodenale	ND	●	GSC	http://nematode.net		
Ascaris lumbricoides	320–680 Mb	●	EH/Sanger	http://www.nematodes.org/nematodeESTs/Ascaris/Ascaris1.html	Nembase (**27**):	http://nema.cap.ed.ac.uk/nematodeESTs/nembase.html
Ascaris suum	320–680 Mb	●	EH/Sanger GSC	http://nema.cap.ed.ac.uk/nematodeESTs/Ascaris/Ascaris.html http://nematode.net	Nembase	

Species	Size		Center	URL	Database
Brugia malayi (53,54)	110 Mb	◆ ❋	NW	http://nema.cap.ed.ac.uk/fgn/filgen1.html	Nembase
			EH	http://nema.cap.ed.ac.uk/nematodeESTs/Brugia/Brugia.html	
		●	GSC	http://nematode.net	
		●	Sanger	http://www.sanger.ac.uk/Projects/B_malayi/	
		●	SC	http://www.math.smith.edu/~sawlab/home.html	
		●	TIGR	http://www.tigr.org/tdb/e2k1/bma1/	
Haemonchus contortus	ND	●	EH/Sanger	http://nema.cap.ed.ac.uk/nematodeESTs/Haemonchus/Haemonchus.html	Nembase
Litomosoides sigmodontis	ND	●	EH	http://nema.cap.ed.ac.uk/nematodeESTs/clade3.html	
Loa loa	ND	●	EH	http://nema.cap.ed.ac.uk/Nest.html	Nembase
Necator americanus	ND	●	EH/Sanger	http://nema.cap.ed.ac.uk/nematodeESTs/Necator/Necator.html	Nembase
Nippostrongylus brasiliensis	ND	●	EH	http://nema.cap.ed.ac.uk/nematodeESTs/Nippostrongylus/Nippostrongylus.html	Nembase
Onchocerca ochengi	ND	●	NW	http://nema.cap.ed.ac.uk/fgn/OnchoNet/OnchoNet.html	
			SC	http://www.math.smith.edu/~sawlab/home.html	
Onchocerca volvulus (55)	ND	●	NW	http://nema.cap.ed.ac.uk/fgn/filgen1.html	Nembase
			NW	http://circuit.neb.com/fgn/OnchoNet/OnchoNet.html	
			SC	http://www.math.smith.edu/~sawlab/OnchoNet/OnchoNet.html	
Ostertagia ostertagi	ND	●	GSC	http://nematode.net	
Parastrongyloides trichosuri	ND	●	GSC	http://nematode.net	
Strongyloides ratti	ND	●	GSC	http://nematode.net	
Strongyloides stercoralis	ND	●	GSC	http://nematode.net	

(continued)

Table 2 (Continued)

Parasite (strain)	Genome size	Type project	Sequencing center/network	Database
Teladorsagia circumcincta	ND	●	EH/Sanger http://nema.cap.ed.ac.uk/nematodeESTs/clade5.html	Nembase
		●●●	GSC http://nematode.net	
Toxocara canis	ND	●●●	GSC http://nematode.net	
			EH http://nema.cap.ed.ac.uk/nematodeESTs/Toxocara/Toxocara.html	Nembase
Trichinella spiralis	270 Mb	●●●	GSC http://nematode.net	
		●●●	EH http://nema.cap.ed.ac.uk/nematodeESTs/clade1.html	
Trichuris muris	ND	●●●	EH/Sanger http://nema.cap.ed.ac.uk/nematodeESTs/Trichuris/Trichuris.html	Nembase
Wuchereria bancrofti	ND	●	NW http://circuit.neb.com/fgn/filgen1.html	
		●	GSC http://www.nematode.net	
Schistosoma japonicum	ND	●	NW http://www.nhm.ac.uk/hosted_sites/schisto/index.html	
		● ♣	CHGC http://www.chgc.sh.cn/	
Schistosoma mansoni (54)	270 Mb	♦	NW http://www.nhm.ac.uk/hosted_sites/schisto/index.html	
		♦	Sanger http://www.sanger.ac.uk/Projects/S_mansoni/	
		♣ ♦	TIGR http://www.tigr.org/tdb/e2k1/sma1/intro.shtml	TIGR *S. mansoni* Gene Index: http://www.tigr.org/tdb/tgi/smgi/

Parasites with ongoing sequencing projects are listed, referencing the relevant sequencing institutions, the sequencing strain (where known), and indicating the nature of the data generated (expressed sequence tags = ●, whole chromosome/genome = ♦, genome survey sequences/BAC end sequencing = ♣). Contents of the table were current at the time of writing and websites actively maintained at the time of writing are listed. All sequencing center pages support similarity searches (BLAST and/or FASTA). The table also lists parasite-specific resources that support data analysis beyond similarity searches, some of which are discussed further in the text and **Table 7**. Haploid genome sizes are listed (where available); unless otherwise stated, information derives from the relevant project pages; ND signifies that the genome size has not yet been determined to date.

Abbreviations: CHGC, Chinese National Human Genome Centre, Shanghai, China; GSC, Genome Sequencing Centre, Washington University Medical School, USA; EH, University of Edinburgh, Edinburgh, UK; GLA, University of Glasgow, Glasgow, UK; Gscope, Genoscope, Evry, France; ILRI, International Livestock Research Institute, Nairobi, Kenya; KI, Karolinska Institute, Stockholm, Sweden; MBL, Marine Biological Laboratory, Massachusetts, USA; MCPG, Minnesota C. parvum Genome Project, USA; NW, Network Web site; Sanger, The Wellcome Trust Sanger Institute, Hinxton, UK; SC, Smith College, Massachusetts, USA; Stanf., Stanford Genome Technology Center, California, USA; TIGR, The Institute for Genomic Research, Maryland, USA; UFL, University of Florida, Florida, USA; VCU, Virginia Commonwealth University, Virginia, USA.

consortia (**Table 2**; for a comprehensive list of completed and ongoing genome projects, *see* http://wit.integratedgenomics.com/GOLD/). This has resulted in a vast increase in genome data for parasites of the phyla Euglenozoa, Apicomplexa, and Nematoda.

The phrase "genome project" encompasses different types of sequencing approaches. Although relatively small in size, the varying degrees of ploidy (e.g., *Trypanosoma cruzi* and *Leishmania major*) and, in some instances, a particular nucleotide bias, make the genomes of parasitic organisms difficult to sequence. Briefly, there are two types of sequencing methodologies (*see also* Chapter 1).

1. Shotgun sequencing is the sequencing of fragmented or sheared DNA. In the case of genome survey sequences (GSS), these single-pass reads remain unassembled. Despite a potentially high error rate, GSSs are very effective for early gene discovery in and gene content assessment of small genomes, as they are representative of the genome and not subject to transcription bias *(10–13)*. For whole genome/chromosome or mapped bacterial artificial chromosome (BAC) or yeast artificial chromosome (YAC) clone sequencing projects, the overlapping sequence reads are assembled into contiguous sequences. Although this often is the fastest sequencing approach, assembly is computationally difficult, especially where repetitive sequences are concerned. Gaps are closed by customized primer design for the amplification of the unsequenced DNA fragments. Increasingly, related species are being sequenced to a lower coverage, and assembly is aided by comparison to a "fully" sequenced species (e.g., *Plasmodium* species *[14]*). To date, only few parasite genomes, or parts thereof, have been published *(1,15–19)*.

2. Expressed sequence tags (ESTs) are also short, single-pass reads generated from the 5'- or 3'-end of a complementary DNA (cDNA) library (*see also* Chapters 4 and 11). ESTs, like GSSs, have been widely used for parasite gene discovery *(20–23)*. Unlike GSSs, ESTs reflect the transcriptome and are useful to identify expressed genes. Furthermore, researchers working on organisms with intron/exon splicing might find that the lack of introns in the ESTs simplifies analysis. However, presence/absence of an EST is a consequence of representation in the cDNA library, which can vary depending on developmental stage and transcript levels *(24)*. GSSs and ESTs from the same species can be "clustered" on the basis of sequence similarity to derive consensus sequences, postulated to originate from the same gene. This not only can reduce redundancy, improve quality, and increase sequence length, it can also provide an idea of relative copy numbers without the expense/labor required to fully sequence and annotate a genome.

The type of sequencing approach clearly affects the quality, reliability, and utility of the dataset. Ongoing sequencing projects need careful evaluation, as they often represent incomplete datasets, can be prone to redundancy, and may be contaminated (e.g., from other chromosomes in the case of a chromosome shotgun sequencing project).

Most parasite sequencing projects disseminate information via dedicated websites. For reasons of brevity, only sequencing eukaryotic, nonfungal parasites infective to mammals are included. **Table 2** lists these, citing project pages at the respective sequencing centers and network sites. Network sites usually provide a good overview and background to the organism and the available biological resources, as well as the rationale behind sequencing projects, information on how libraries have been prepared, and community activities. Project pages, in contrast, outline sequencing strategies and inform users of the sequencing coverage and progress, though this is often restricted to the data generated at that site. They are predominantly oriented toward analysis. Most support access to the sequence data by sequence retrieval, keyword searches of sequence annotation, and homology searches, which is probably the most common way to query such resources. Increasingly, researchers now have access to both parasite and host sequence information (human/mouse, for example) and, in some instances, the parasite vector (for example, *Anopheles gambiae* as well as partial sequences of *Glossina morsitans* and *Biomphalaria glabrata* [*see* **Table 3**]).

In addition, an organism-independent site at the EBI (http://www.ebi.ac.uk/parasites/parasite-genome.html) provides links to organism-specific websites and analytical tools, including a parasite-specific BLAST server.

3.2. Examples of Parasite Resources

Web-based parasite genome resources are commonly used for gene discovery, as well as storage and integration of information about predicted and experimental knowledge of the organism's sequence features. Such resources (some of which are further discussed here):

1. Aim to facilitate access to the latest genome data (both "completed," annotated genomic sequence as well as unfinished data such as raw sequencing reads and draft contig assemblies) in a user-friendly and intuitive manner
2. Provide an overview of (often automated) bioinformatical analyses such as prediction of gene models and the results of similarity searches against sequence and protein pattern databases
3. Provide tools that support interactive analyses as well as offering graphical displays to aid interpretation of the data

In addition, some databases aim to consolidate the sequence data with the considerable information available on a particular organism's biology by:

1. Curating knowledge extracted from the literature and other publicly available resources (e.g., mutations, alleles, phenotypes) and feedback from researchers
2. Integrating this information with results of large-scale expression and phenotypic data

Table 3
List of Parasite Vector Genome Projects

Species	Type of resource	Websites
Mosquito spp.	Overview of mosquito genomics databases	http://klab.agsci.colostate.edu/
	Resource for genomic/biological information on anopheline mosquitoes, particularly *Anopheles gambiae*	AnoBase: http://www.anobase.org/
Anopheles gambiae		NCBI: http://www.ncbi.nlm.nih.gov/mapview/map_search.cgi?chr=agambiae.inf
Anopheles gambiae	Genome browser	Ensembl: http://www.ensembl.org/Anopheles_gambiae/
Biomphalaria glabrata	Overview of genome initiative	NW: http://biology.unm.edu/biomphalaria-genome/index.html
Glossina morsitans	Analysis of clustered ESTs	Sanger Institute: http://www.sanger.ac.uk/Projects/G_morsitans/ GeneDB: http://www.genedb.org

Table 4
Features of GiardiaDB (http://tuscany.mbl.edu/GiardiaDB/giardiadb.html)

	Description
Data content	Sequences and automated annotations generated by *G. lamblia* genome project
	Each putative gene prediction is displayed on a page with:
	Results of BLASTX searches, alignments and links to NCBI
	Link to ExPASy for protein sequence analysis
	Position of open reading frame within contig assemblies
Data access points	Sequence search
	BLAST against sequence reads, assembled contigs, and automatically predicted open reading frames
	Text search
	Sequence reads and clone names
	Description lines of precomputed BLASTX hits
	Browse
	Reads/contigs with BLASTX results and assembly statistics
Data downloads	Nucleotide sequence in FASTA format
Integration of genomic data	In progress for SAGE

Abbreviations: BLASTX, basic local alignment search tool X; NCBI, National Center for Biotechnology; SAGE, serial analysis of gene expression.

3. Linking extensively to other organism-specific resources and public databases to represent an integrated view of all available information
4. Housing datasets from related species (e.g., GeneDB and PlasmoDB)
5. Being closely linked to the respective research communities, relying on interaction with scientists to maintain correct and up-to-date information, and adding value and supporting evidence to gene predictions and annotation
6. Having underlying architectures that are amenable to incorporation of new datasets, so as to evolve with projects and reflect the level of continuing annotation/curation

3.2.1. GiardiaDB

GiardiaDB (25) (http://tuscany.mbl.edu/GiardiaDB/giardiadb.html) is an example of a single-organism database (*see* **Table 4**). It is designed to bring together information on the *Giardia lamblia* genome sequence and assembly with automated gene predictions, annotations, and information on gene expression. Manual annotation of putative open reading frames is in progress.

3.2.2. The Institute for Genomic Research (TIGR) Gene Index

TIGR Gene Index *(26)* (http://www.tigr.org/tdb/tgi/) is a compilation of all EST sequences retrieved from public databases for various parasitic organisms

Table 5
Features of TIGR Gene Index (hhttp://www.tigr.org/tdb/tgi/)

	Description
Data content	Automated analyses of clustered EST sequences retrieved from public databases
	Pages are generated for each consensus sequence with putative open reading frames marked up in all six frames with:
	Alignment of all ESTs
	GO annotations
	Information on cDNA library origin
	Results of similarity searches, links to NCBI
	Links to reconstructed metabolic pathways
	Ortholog analysis
Data access points	Sequence search
	BLAST
	Text search
	Gene product name
	Accession ids/clone names
	Consensus sequence ids
	Browse
	Automatically assigned gene ontology associations
Data downloads	Nucleotide sequence in FASTA format

The TIGR Gene Indices are maintained by The Institute for Genomic Research.
Abbreviations: TIGR, The Institute for Genomic Research; EST, expressed sequence tag; GO, gene ontology; cDNA, complementary DNA; NCBI, National Center for Biotechnology Information; BLAST, basic local alignment search tool.

(*see* **Table 5**). The ESTs are assembled into contiguous sequences to create nonredundant, unique consensus sequences, which only represent transcribed genes. Through similarity searches, these have been annotated in an automated manner to assign putative gene identities and products. Although each organism has its own gene index database, all the databases are included in the analysis of orthologous families.

3.2.3. NEMBASE

NEMBASE *(27)* (http://nema.cap.ed.ac.uk/nematodeESTs/nembase.html) is designed to display automated annotation of partial (EST, GSS) genome sequences and currently houses ESTs from collaborative sequencing projects between the nematode genomics group at the University of Edinburgh (UK) and the Pathogen Sequencing Unit at the Wellcome Trust Sanger Institute (UK) for nine

Table 6
Features of Nembase (http://nema.cap.ed.ac.uk/nematodeESTs/nemabse.html)

	Description
Data content	Clustered, contiguated sequences of nematode ESTs annotated with:
	Alignment of ESTs on contig, access to traces and retrieval of entry from public databases
	Information on cDNA library origin
	Protein sequence with predicted motifs, link to respective database
	Predicted physical properties of protein
	Results and alignments of similarity searches
Data access points	Sequence search
	BLAST
	Text search
	Clone and cluster name
	Description lines of precomputed BLASTX hits
	Clusters on the basis of library expression and similarity profiles
	Browse
	Clustered contigs with BLASTX results
Data download	Nucleotide sequence in FASTA format

Abbreviations: EST, expressed sequence tag; cDNA, complementary DNA; BLAST, basic local alignment search tool; BLASTX, basic local alignment search tool X.

nematode species (see **Table 6**). It is projected to host all nematode EST datasets including those generated at Genome Sequencing Center in St. Louis, Missouri (Washington University Medical School, USA).

3.2.4. tigr db

The database schema tigrdb contains the sequences and annotations of some of the parasitic organisms sequenced at TIGR and, in some instances, collaborating sequencing institutes (see **Table 7**). Each organism data set is housed in its own database. Annotated sequences are displayed based on their map position and are searchable by BAC clone name, gene name, assigned product, and sequence similarity.

3.2.5. GeneDB

GeneDB *(28)* (http://www.genedb.org/) houses sequences and annotations of fungal and parasitic genomes from both completed and ongoing projects (see **Table 7**). Because of the collaborative nature of most of the sequencing projects, GeneDB represents a central resource for genomic and postgenomic

Table 7
Data Contents of and Tools Available Through GeneDB
(http://www.genedb.org/), PlasmoDB (http://plasmodg.org/), and tigrdb

	GeneDB (28)	PlasmoDB (32)	tigr db
Protozoan species	L. major (whole genome) P. berghei (partial shotgun) P. chabaudi (partial shotgun) P. falciparuml (whole genome) T. brucei (whole genome) T. cruzi (whole genome) T. vivax T. annulata	P. falciparum (whole genome)	E. histolytica (whole genome) P. falciparum (whole genome) P. yoelii (partial shotgun) T. brucei (portion of the genome) T. cruzi (whole genome) In preparation for: S. mansoni
Content	Manual annotations of finished and automated annotations of unfinished sequences integrated with biological datasets Automated annotation of ESTs	Manual and automated annotations of finished and unfinished sequences integrated with biological datasets	Manual and automated annotations of finished and unfinished sequences
Feature page content	Gene and product name Chromosome information Physical properties of the protein Enzyme classification Predicted protein features Protein classification using GO Literature references Polymorphisms Similarity analysis Curated annotation	Gene name and description Gene predictions (various algorithms) Predicted protein features Enzyme classification Protein classification using GO Representation of BLAST results Polymorphisms	Gene and product name Chromosome information Physical properties of the protein Predicted protein features Protein classification using GO Similarity analysis
Data access			
Sequence search	BLAST/multispecies BLAST	BLAST/PSI-BLAST	BLAST
Text search	Keyword (gene name and product/description) Site-wide keyword search	Keyword (including output of pre-computed BLASTX description lines, available for all *Plasmodium* species)	Keyword (gene name and product/description) Locus search

(continued)

Table 7 (Continued)

	GeneDB (28)	PlasmoDB (32)	tigr db
Browsing	Product GO catalogs Chromosome/contig maps InterPro Pfam	Customizable contig maps	Chromosome/clone maps
Querying (Boolean)	Keyword Chromosome location Type of gene Signal peptide and transmembrane predictions Gene structure GO EC number Pfam protein family domains	Keyword (inc. of similarity searches) Chromosome location Type of gene Polymorphisms Signal peptide and transmembrane prediction Gene structure GO EC number Pfam protein family domains Expression-based queries Metabolic pathways	Keyword
Query history/ downloads	Available	Available	
Graphics	Gene structure Gene context and chromosome maps Protein feature map	Customizable gene context map Protein feature map Results of gene prediction algorithms (inc. EST/GSS/SAGE tag alignments and self-BLAST analysis) Microsatellite and optical map	Gene structure Gene context map Protein feature map
Links from feature pages	Neighboring genes Nucleotide and amino acid sequence Enzyme databases AmiGO (GO browser) Protein structure and motif databases	Neighboring genes AmiGO (GO browser) Protein motif databases Genbank/Entrez Protein Expression data Metabolic pathway analysis	Nucleotide and amino acid sequence Protein motif databases (inc. other proteins with a particular domain) Similarity search alignments

Category			
	Entrez (literature database)		GenPept protein database
	SWISS-PROT/EMBL		Local GO browser
	Phenotype database		
Tools	BLAST	BLAST	BLAST
	Peptide motif search	ePCR	Protein family search against TIGRFAMs and Pfam
	Peptide mass fingerprinting	Peptide motif searches	
	Artemis (sequence and annotation viewer) applet	Peptide mass fingerprinting	
		Clustering software	
Integrated data Curated information	Data extracted from literature, public database submissions, large-scale genomics experiments	Manual annotation provided by the sequencing centers	Manual annotation
	Community feedback		
Expression data		Microarray	
		SAGE	
		EST sequences	
		Proteomics	
Sequence retrieval	Nucleotide and amino acid sequence of CDS features	All available sequence (customizable)	Nucleotide and amino acid sequence of CDS features
	Contig assemblies from ftp directories	*Plasmodium berghei*	Contig assemblies from ftp directories
	User-specified range of sequence and annotated features	*P. chabaudi*, *P. falciparum*	
		P. knowlesi, *P. vivax*	
		P. yoelii	

Abbreviations: BLAST, basic local alignment search tool; GO, gene ontology; EMBL, European Molecular Biology Laboratory; GSS, genome survey sequences; Pfam, protein families; SAGE, serial analysis of gene expression; EST, expressed sequence tag; CDS, coding sequence; EC, enzyme commission; PSI-BLAST, position specific iterative BLAST.

information of organisms sequenced at different sites, allowing interpretation of data within the context of the whole genome. The current release houses datasets from 11 organisms, including *L. major* and *T. brucei*, 6 of which are curated and maintained by biologists. At the center of GeneDB is a feature page (**Fig. 1**) that displays the results of predictive software packages (gene finding, domains, similarity searches, signal peptide, transmembrane, and GPI anchor predictions) and the manual annotation and curation process in graphical and text-based formats. Additional sequence features and annotations can be viewed in an Artemis applet, a sequence browser and annotation tool available from the Sanger Institute (http://www.sanger.ac.uk/Software/Artemis/) *(29)*. Selected regions can also be downloaded either as sequence or as sequence with annotated features. Extensive cross-referencing to public resources such as sequence, expression, and phenotype databases allows retrieval of related information from external sources. For collaborative projects (such as *P. falciparum* and *T. brucei*), this includes links to the feature pages in the respective genome databases at TIGR. Thus, users can not only access the whole genome through GeneDB but also seamlessly transfer between databases.

In GeneDB, data can be accessed via searchable text indices and/or browsable chromosome/contig maps and Pfam and InterPro catalogs, providing levels of access suitable for novice and expert users. Genes can also be accessed using AmiGO, the official browser of the GO consortium integrated into GeneDB. An additional query interface supports a wide range of queries on sequence data and (curated) annotations. All queries are tracked via the history page, allowing further refinement of searches and downloading of search results. The data are frequently updated. This includes automated updates of unfinished chromosomes as well as information added during the curation process according to new/updated submissions to public database, publications and feedback from the research community. The use of structured syntax for such annotations facilitates data access and querying. For the kinetoplastids, GeneDB is also involved in implementing nomenclature guidelines *(30)*, aiding consistency in annotation and data retrieval.

Fig. 1. (*Opposite page*) The GeneDB feature page for the *T. brucei* PGKB gene and its product, phoshoglycerate kinase. The report provides location details, predicted peptide properties, and access to the sequence. A context map shows neighboring genes. The domain information is displayed both graphically and as text, linking to the relevant databases, and includes nonsynonymous single nucleotide polymorphism (SNP) information (where available). Literature cited for manually curated statements, gene ontology associations, and external database are cross-referenced. Numerous search tools, such as a BLAST server and complex querying (*see also* **Fig. 2**), are accessible from the navigation bar. For more information on this report, go to: http://www.genedb.org/genedb/Search?name=Tb927.1.710&organism=tryp.

3.2.6. PlasmoDB

PlasmoDB *(31,32)* (http://plasmodb.org/) is the Malaria Genome Sequencing Project Consortium Database. This resource integrates the *P. falciparum* sequence and manual annotation provided by the sequencing centers with both automated annotations and data generated via genomic-scale experiments (e.g., microarray, proteomics, polymorphisms) by the research community in a graphic and text format. Users are presented with the results of multiple lines of evidence (e.g., gene-finding algorithms, protein feature prediction algorithms), supporting the examination of potentially overlapping/competing evidence. PlasmoDB permits gene identification via text and sequence searches and also offers a wide range of queries, which cover, among other datasets, sequence features, annotations, and expression data and can be easily combined using Boolean operators. All search results, in addition to user-specified ranges of sequence, can be retrieved and downloaded. For a full listing of data types and data analysis tools available through PlasmoDB, *see* **Table 7**.

The resource also houses all available sequences (e.g., ESTs, GSSs) from other *Plasmodium* species, which are available for downloading, similarity searching, keyword searching of precomputed BLASTX description lines, and cross-species comparisons.

4. Database Mining

The above resources organize large quantities of data to maximize users' ability to query and understand the data and their underlying trends. This is clearly important if they are to facilitate research. The integration of genomics with biochemistry, for example, may map putative genes onto (metabolic) pathways or may provide clues as to alternate/novel pathways. A short list of candidate genes can then be tested experimentally. In addition, the representation of genomes from related species in a single resource allows cross-species analyses, which may point toward sequences present only in closely related species, representing possible phylogenetic or pathogenic adaptation.

A recent review has demonstrated the mining of nematode databases *(33)*. The *Schistosoma* genome project has been similarly mined *(34)*, and examples of how PlasmoDB can be queried to retrieve nuclear encoded apicoplast genes *(31)* and to reconstruct a metabolic pathway map of the apicoplast *(35)* have been published previously. **Figure 2** illustrates how users might query GeneDB. To retrieve a shortlist of putative cell surface transporters such as nucleoside transporters, users can frame a number of alternative queries by:

1. Combining searches (http://www.genedb.org/gusapp/servlet?page=boolq) for predicted protein domains (e.g., Pfam) with transmembrane predictions (**Fig. 2**).

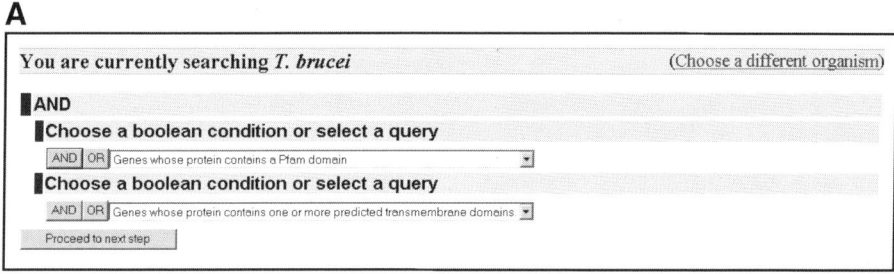
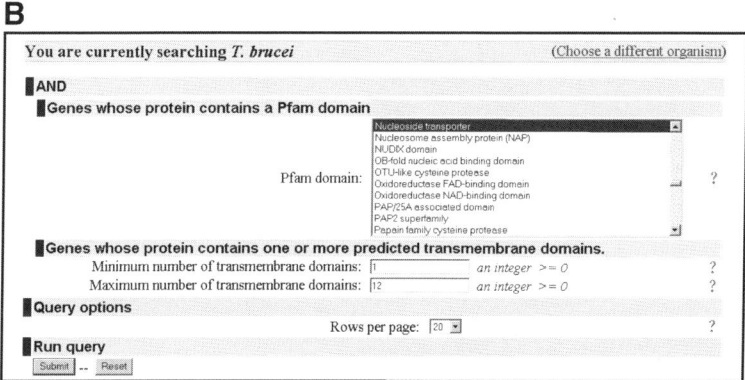
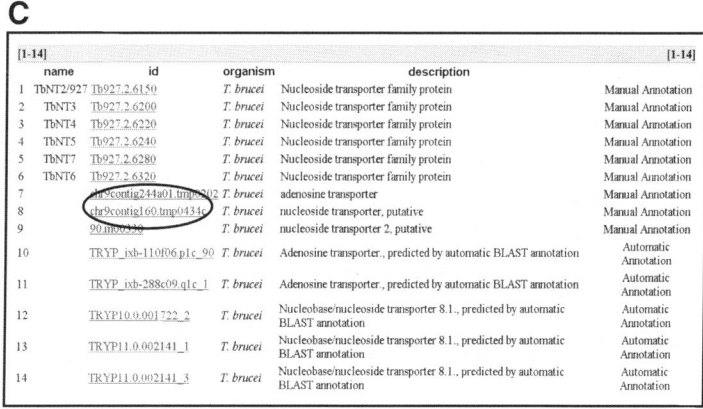

Fig. 2. Complex querying of GeneDB. GeneDB can be searched by combining search parameters with Boolean operators (http://www.genedb.org/gusapp/servlet?page=boolq &organism=tryp). To identify putative cell surface nucleoside transporters, a search for the nucleoside transporter domain (PF01733) can be combined with signal peptide and transmembrane predictions (**A,B**). The list returned (**C**) gives access to individual feature pages (**D**).

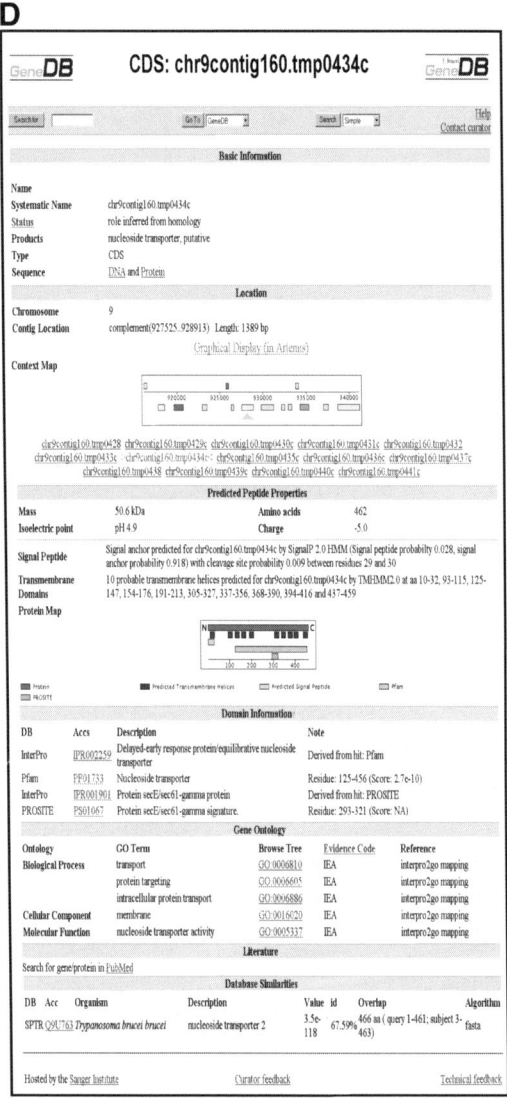

Fig. 2. (Continued)

Including the requirement for predicted signal peptides could further refine the search.
2. Browsing the gene ontology functional classification (http://www.genedb.org/genedb/GOfunction?organism=tryp&keywords=Browse) for GO:0005337 (nucleoside transporter).
3. String searching of curated and automated annotation, using a putative product such as "transporter" or even "nucleoside transporters" as a keyword.

4. String searching all of GeneDB, taking advantage of the site-wide indexing tool (http://www.genedb.org/genedb/search.jsp?organism=tryp).
5. Carrying out homology searches with transporter sequences from related species such as *L. major*.

5. Outlook

Researchers are becoming increasingly dependent on up-to-date, easily queryable, and centralized data repositories to complement wet-bench science. Numerous developments will aid users and database developers to achieve the common goal of a "truly useful" resource. One of these will be the drive toward integration of resources combined with transferability of tools and sharing of code among different databases. Efforts such as the generic model organism database (http://www.gmod.org), are underway to develop "building blocks." Such blocks would allow the design of a database from individual components (e.g., a common database schema, literature curation tool, or graphical interface), which, in turn, would reduce the cost and time required to set up a resource *(36)*. These tools will also be designed to integrate with existing components of a database. In addition, user interfaces will be standardized and become more familiar to both users and bioinformaticians *(37)*. Furthermore, the development of ontologies, such as gene ontology (*see* Chapter 2) and sequence ontology (*see* http://www.geneontology.org/doc/gobo.html for a comprehensive list of ontologies in development) will aid annotation and curation of genomes and facilitate the comparison between organisms based on shared biology.

This chapter illustrates the wealth of parasite-related Web resources. The sequencing of numerous parasite genomes is nearing completion, and there is increasing emphasis on comparative genome projects. As a consequence, researchers are now in a good position to assess gene content, function, and expression in a range of parasitic organisms. This should further our understanding of infectivity, immune-evasion, host–parasite interaction, and pathogenicity, still poorly understood in most parasites. It is increasingly important to integrate bioinformatical sequence analyses with all available biological data. As such, organism databases "have an interpretative and curatorial role that distinguishes them from...other strictly archival databases" *(36)*. Only by continuously updating and curating biological information is it possible to ensure these resources become valuable research tools and the stored data do not become obsolete.

References

1. Gardner, M. J., Hall, N., Fung, E., et al. (2002) Genome sequence of the human malaria parasite *Plasmodium falciparum*. *Nature* **419,** 498–511.
2. Stoesser, G., Baker, W., Van Den Broek, A., et al. (2003) The EMBL Nucleotide Sequence Database: major new developments. *Nucleic Acids Res.* **31,** 17–22.

3. Benson, D. A., Karsch-Mizrachi, I., Lipman, D. J., et al. (2003) GenBank. *Nucleic Acids Res.* **31**, 23–27.
4. Miyazaki, S., Sugawara, H., Gojobori, T., et al. (2003) DNA Data Bank of Japan (DDBJ) in XML. *Nucleic Acids Res.* **31**, 13–16.
5. O'Donovan, C., Martin, M. J., Gattiker, A., et al. (2002) High-quality protein knowledge resource: SWISS-PROT and TrEMBL. *Brief Bioinform* **3**, 275–284.
6. Boeckmann, B., Bairoch, A., Apweiler, R., et al. (2003) The SWISS-PROT protein knowledgebase and its supplement TrEMBL in 2003. *Nucleic Acids Res.* **31**, 365–370.
7. Mulder, N. J., Apweiler, R., Attwood, T. K., et al. (2003) The InterPro Database, 2003 brings increased coverage and new features. *Nucleic Acids Res.* **31**, 315–318.
8. Wheeler, D. L., Church, D. M., Federhen, S., et al. (2003) Database resources of the National Center for Biotechnology. *Nucleic Acids Res.* **31**, 28–33.
9. Brooksbank, C., Camon, E., Harris, M. A., et al. (2003) The European Bioinformatics Institute's data resources. *Nucleic Acids Res.* **31**, 43–50.
10. Janssen, C. S., Barrett, M. P., Lawson, D., et al. (2001) Gene discovery in *Plasmodium chabaudi* by genome survey sequencing. *Mol. Biochem. Parasitol.* **113**, 251–260.
11. El-Sayed, N. M. and Donelson, J. E. (1997) A survey of the *Trypanosoma brucei rhodesiense* genome using shotgun sequencing. *Mol. Biochem. Parasitol.* **84**, 167–178.
12. Liu, C., Vigdorovich, V., Kapur, V., et al. (1999) A random survey of the *Cryptosporidium parvum* genome. *Infect. Immun.* **67**, 3960–3969.
13. Smith, M. W., Aley, S. B., Sogin, M., et al. (1998) Sequence survey of the *Giardia lamblia* genome. *Mol. Biochem. Parasitol.* **95**, 267–280.
14. Carlton, J. M., Angiuoli, S. V., Suh, B. B., et al. (2002) Genome sequence and comparative analysis of the model rodent malaria parasite *Plasmodium yoelii yoelii*. *Nature* **419**, 512–519.
15. Katinka, M. D., Duprat, S., Cornillot, E., et al. (2001) Genome sequence and gene compaction of the eukaryote parasite *Encephalitozoon cuniculi*. *Nature* **414**, 450–453.
16. McDonagh, P. D., Myler, P. J., and Stuart, K. (2000) The unusual gene organization of *Leishmania major* chromosome 1 may reflect novel transcription processes. *Nucleic Acids Res.* **28**, 2800–2803.
17. Worthey, E. A., Martinez-Calvillo, S., Schnaufer, A., et al. (2003) *Leishmania major* chromosome 3 contains two long convergent polycistronic gene clusters separated by a tRNA gene. *Nucleic Acids Res.* **31**, 4201–4210.
18. Hall, N., Berriman, M., Lennard, N. J., et al. (2003) The DNA sequence of chromosome I of an African trypanosome: gene content, chromosome organisation, recombination and polymorphism. *Nucleic Acids Res.* **31**, 4864–4873.
19. El-Sayed, N. M., Ghedin, E., Song, J., et al. (2003) The sequence and analysis of *Trypanosoma brucei* chromosome II. *Nucleic Acids Res.* **31**, 4856–4863.
20. El-Sayed, N. M., Alarcon, C. M., Beck, J. C., et al. (1995) cDNA expressed sequence tags of *Trypanosoma brucei rhodesiense* provide new insights into the biology of the parasite. *Mol. Biochem. Parasitol.* **73**, 75–90.

21. Verdun, R. E., Di Paolo, N., Urmenyi, T. P., et al. (1998) Gene discovery through expressed sequence Tag sequencing in *Trypanosoma cruzi*. *Infect. Immun.* **66,** 5393–5398.
22. Blaxter, M., Daub, J., Guiliano, D., et al. (2002) The *Brugia malayi* genome project: expressed sequence tags and gene discovery. *Trans. R. Soc. Trop. Med. Hyg.* **96,** 7–17.
23. Parkinson, J., Whitton, C., Guiliano, D., et al. (2001) 200000 nematode expressed sequence tags on the Net. *Trends Parasitol.* **17,** 394–396.
24. Manger, I. D., Hehl, A., Parmley, S., et al. (1998) Expressed sequence tag analysis of the bradyzoite stage of *Toxoplasma gondii*: identification of developmentally regulated genes. *Infect. Immun.* **66,** 1632–1637.
25. McArthur, A. G., Morrison, H. G., Nixon, J. E., et al. (2000) The *Giardia* genome project database. *FEMS Microbiol. Lett.* **189,** 271–273.
26. Liang, F., Holt, I., Pertea, G., et al. (2000) An optimized protocol for analysis of EST sequences. *Nucleic Acids Res.* **28,** 3657–3665.
27. Parkinson, J., Mitreva, M., Hall, N., et al. (2003) 400000 nematode ESTs on the Net. *Trends Parasitol.* **19,** 283–286.
28. Hertz-Fowler, C. and Peacock, C. S. (2002) Introducing GeneDB: a generic database. *Trends Parasitol.* **18,** 465–467.
29. Rutherford, K., Parkhill, J., Crook, J., et al. (2000) Artemis: sequence visualization and annotation. *Bioinformatics* **16,** 944–945.
30. Clayton, C., Adams, M., Almeida, R., et al. (1998) Genetic nomenclature for *Trypanosoma* and *Leishmania*. *Mol. Biochem. Parasitol.* **97,** 221–224.
31. Kissinger, J. C., Brunk, B. P., Crabtree, J., et al. (2002) The *Plasmodium* genome database. *Nature* **419,** 490–492.
32. Bahl, A., Brunk, B., Crabtree, J., et al. (2003) PlasmoDB: the *Plasmodium* genome resource. A database integrating experimental and computational data. *Nucleic Acids Res.* **31,** 212–215.
33. Tarleton, R. L. and Kissinger, J. (2001) Parasite genomics: current status and future prospects. *Curr. Opin. Immunol.* **13,** 395–402.
34. Oliveira, G. and Johnston, D. A. (2001) Mining the schistosome DNA sequence database. *Trends Parasitol.* **17,** 501–503.
35. Roos, D. S., Crawford, M. J., Donald, R. G., et al. (2002) Mining the *Plasmodium* genome database to define organellar function: what does the apicoplast do? *Philos. Trans. R. Soc. Lond. B. Biol. Sci.* **357,** 35–46.
36. Stein, L. D., Mungall, C., Shu, S., et al. (2002) The generic genome browser: a building block for a model organism system database. *Genome Res.* **12,** 1599–1610.
37. Stein, L. D. (2002) Creating a bioinformatics nation. *Nature* **417,** 119–120.
38. Wu, C. H., Yeh, L. S., Huang, H., et al. (2003) The Protein Information Resource. *Nucleic Acids Res.* **31,** 345–347.
39. Bateman, A., Birney, E., Cerruti, L., et al. (2002) The Pfam protein families database. *Nucleic Acids Res.* **30,** 276–280.
40. Attwood, T. K., Bradley, P., Flower, D. R., et al. (2003) PRINTS and its automatic supplement, prePRINTS. *Nucleic Acids Res.* **31,** 400–402.

41. Servant, F., Bru, C., Carrere, S., et al. (2002) ProDom: automated clustering of homologous domains. *Brief Bioinform.* **3,** 246–251.
42. Sigrist, C. J., Cerutti, L., Hulo, N., et al. (2002) PROSITE: a documented database using patterns and profiles as motif descriptors. *Brief Bioinform.* **3,** 265–274.
43. Letunic, I., Goodstadt, L., Dickens, N. J., et al. (2002) Recent improvements to the SMART domain-based sequence annotation resource. *Nucleic Acids Res.* **30,** 242–244.
44. Haft, D. H., Selengut, J. D., and White, O. (2003) The TIGRFAMs database of protein families. *Nucleic Acids Res.* **31,** 371–373.
45. Myler, P. J., Beverley, S. M., Cruz, A. K., et al. (2001) The *Leishmania* genome project: new insights into gene organization and function. *Med. Microbiol. Immunol. (Berl.)* **190,** 9–12.
46. El-Sayed, N. M., Hegde, P., Quackenbush, J., et al. (2000) The African trypanosome genome. *Int. J. Parasitol.* **30,** 329–345.
47. Adam, R. D. (2000) The *Giardia lamblia* genome. *Int. J. Parasitol.* **30,** 475–484.
48. Spano, F. and Crisanti, A. (2000) *Cryptosporidium parvum*: the many secrets of a small genome. *Int. J. Parasitol.* **30,** 553–565.
49. Shirley, M. W. (2000) The genome of *Eimeria* spp., with special reference to *Eimeria tenella*–a coccidium from the chicken. *Int. J. Parasitol.* **30,** 485–493.
50. Carlton, J. M., Muller, R., Yowell, C. A., et al. (2001) Profiling the malaria genome: a gene survey of three species of malaria parasite with comparison to other apicomplexan species. *Mol. Biochem. Parasitol.* **118,** 201–210.
51. Nene, V., Bishop, R., Morzaria, S., et al. (2000) *Theileria parva* genomics reveals an atypical apicomplexan genome. *Int. J. Parasitol.* **30,** 465–474.
52. Kissinger, J. C., Gajria, B., Li, L., et al. (2003) ToxoDB: accessing the *Toxoplasma gondii* genome. *Nucleic Acids Res.* **31,** 234–236.
53. Blaxter, M., Aslett, M., Guiliano, D., et al. (1999) Parasitic helminth genomics. Filarial Genome Project. *Parasitology* **118(Suppl.),** S39–S51.
54. Williams, S. A. and Johnston, D. A. (1999) Helminth genome analysis: the current status of the filarial and schistosome genome projects. Filarial Genome Project. Schistosome Genome Project. *Parasitology* **118(Suppl.),** S19–S38.
55. Williams, S. A., Laney, S. J., Lizotte-Waniewski, M., et al. (2002) The River Blindness Genome Project. *Trends Parasitol.* **18,** 86–90.

4

Expressed Sequence Tags

Medium-Throughput Protocols

Claire Whitton, Jennifer Daub, Marian Thompson, and Mark Blaxter

Summary

Generating expressed sequence tags is a simple, cheap, and efficient way to sample the genome of a target organism. An expressed sequence tag (EST) is a single-pass sequence derived from a single complementary DNA (cDNA) clone, and the sequence serves to identify the gene from which it derives. We present a set of tested laboratory protocols for setting up and performing an EST analysis of any chosen species. These medium-throughput protocols do not require dedicated genomics equipment, such as robots, and focus on the use of microtiter plates and multichannels. Using these protocols, a single competent research worker should be able to generate 2000 ESTs in 1 mo. In a nonnormalized library, these 2000 ESTs should identify between 1000 and 1500 different genes, and thus possibly between 10 and 20% of the genes of any target parasite.

Key Words: Expressed sequence tags; medium-throughput genomics; parasite genomes.

1. Introduction

The genomes of eukaryotic parasites carry with them the signatures of their phylogenetic origin *(1)*. Thus, protozoan and fungal parasites usually have genomes from 15 to 60 Mb, organized as many small (<3 Mb) chromosomes, and carrying 5,000–10,000 protein-coding genes *(2–4)*. Metazoan parasites have larger genomes (80–1000 Mb), organized as larger chromosomes (>10 Mb) and carrying more genes (approx 15,000–20,000) *(5–7)*. The average size of the protein-coding part of a gene is approx 2000 bases, and thus, in all eukaryotic genomes, much of the genome is noncoding. In mammals, the noncoding portion reaches 98.5% *(8)*, whereas in nematodes it is approx 75% *(5)*, and in fungi it is approx 60% *(9–11)*. One goal of genome projects is to identify all the genes of the target organism, and thus, whole genome sequencing carries with it the "burden" of determining the sequence of the large proportion of the genome that is not genic. It would be much simpler, and possibly more efficient, if it was possible to sequence only the genes.

For some parasites, such as the trypanosomatids, gene density on the chromosomes is such that random sequencing of genomic DNA fragments is an efficient gene discovery tool *(12)*. In addition, as trypanosomatids have no introns, the identification of open reading frames (ORFs) coding for proteins is relatively simple *(13)*. This genome survey sequence (GSS) strategy is less applicable to organisms with larger noncoding regions of the genome and those that have introns. In some taxa, it is possible to enrich genomic DNA for genic fragments because of their different base composition. In malarial parasites, gene fragment libraries derived from mung bean nuclease-digested genomic DNA have been used for gene discovery *(14,15)*. One issue with GSS-based gene discovery is that it is often difficult to distinguish between active genes and pseudogenes rendered inactive by mutation.

The cell naturally selects coding regions from its genome for transcription, and this feature can be used to generate clone libraries that represent only coding DNA. Genes are transcribed into premessenger RNAs (mRNAs), which are then processed by the spliceosome to generate mature mRNA. mRNAs are marked by the cell through the addition of a 3'-poly(A) tail, from 30 to 300 bases long. This poly(A) tail is used by the cell to determine mRNA half-life and localization but is also then available to a molecular biologist as a (relatively) unique molecular tag to identify and isolate only the mRNAs from a cell lysate. The poly(A)-containing fraction of cellular RNA can be reverse-transcribed to DNA and the resulting double-stranded cDNA cloned to make a cDNA bank or library representing the expressed genes of the original cells (or organism or tissue). The cDNAs are cloned directionally, such that the 5'-end of the mRNA is always in the same orientation with respect to the vector. Sequencing of randomly selected clones from this cDNA clone bank will result in a sequence dataset sampled only from actively transcribed genes. In some parasitic taxa, all (trypanosomatids), many (nematodes), or some (platyhelminths) mRNAs are *trans*-spliced at their 5'-ends to a single or small family of short exons, called spliced leader (SL) exons *(16–20)*. SL exons offer another sequence-based unique marker of mRNAs in these taxa.

An EST is a single-pass sequence derived from a single cDNA clone *(21–23)*. The sequence is "single-pass" because no attempt is made to acquire the complete and correct sequence of the cDNA, just a "tag" that can be used to identify it uniquely. Different genes have different steady-state levels of transcription and thus will be differentially represented in the cDNA clone bank.

As ESTs usually are derived from randomly selected clones, the representation of a particular gene in an EST dataset will reflect its abundance in the starting mRNA. Thus, abundantly expressed genes will be readily identified, whereas genes expressed at low levels, or conditionally through developmental or environmental regulation, may never be sampled. One benefit of using a GSS strategy is that all genomic fragments are present at equimolar amounts (excluding those less favored by the peculiarities of the cloning system used), thus all genes are equally likely to be sampled. However, for EST sequencing, it is possible to perform normalization steps, either during library construction or upon the cDNA clone bank, to maximize the number of unique genes identified.

ESTs can be generated from either end of a cloned cDNA. The 3'-end of the cloned insert usually is marked by the poly(A) tail used to isolate and reverse transcribe the mRNA. This poly(A) (or poly[T] on the opposite strand) stretch often is problematic for thermostable polymerase sequencing, and sequencing through poly(T) can reduce the length and quality of the subsequent sequence. Thus, we prefer 5'-ESTs. Also, 5'-ESTs have the advantage of being more likely to include some of the ORF of the cDNA and, therefore, facilitate identification of the encoded product. The biochemistry of cDNA synthesis and cloning usually results in clones that are shortened at their 5'-ends compared to the starting mRNA. The population of clones deriving from a single gene's transcripts may differ in the position of truncation. With SL-containing taxa, the SL can be used as a (universal) 5'-tag to ensure the isolation of only full-length cDNAs *(24,25)*. For non-SL taxa (and non*trans*-spliced genes in SL-containing taxa), the 5'-trimethylguanosine cap structure of mRNAs can be exploited to direct synthesis of full-length cDNAs *(26)*. Also, 5'-cap libraries have the advantage of avoiding mitochondrial and ribosomal RNA transcripts.

The analysis of ESTs has to be carried out with an understanding of the preliminary nature of the data *(27,28)*. ESTs may contain sequencing errors and usually do not yield the full sequence of a mRNA. Therefore, several bioinformatics methods have to be used to compare ESTs to each other, to cluster them into sets putatively deriving from a single gene transcript, and to use these clusters to predict more robust consensus sequences. Multiple 5'-ESTs of a single gene may yield a consensus sequence that is significantly longer than any individual EST. Subsequent to clustering, the consensus sequences can be used to predict ORFs and encoded peptides, which can be functionally identified by database similarity search.

This chapter presents optimized protocols for generating ESTs using a medium-throughput system that does not rely on robotic technology *(27)* (*see* **Fig. 1** for an overview of the method). Given a good cDNA library and no distractions, a single competent research worker should be able to generate 2000 ESTs in 1 mo

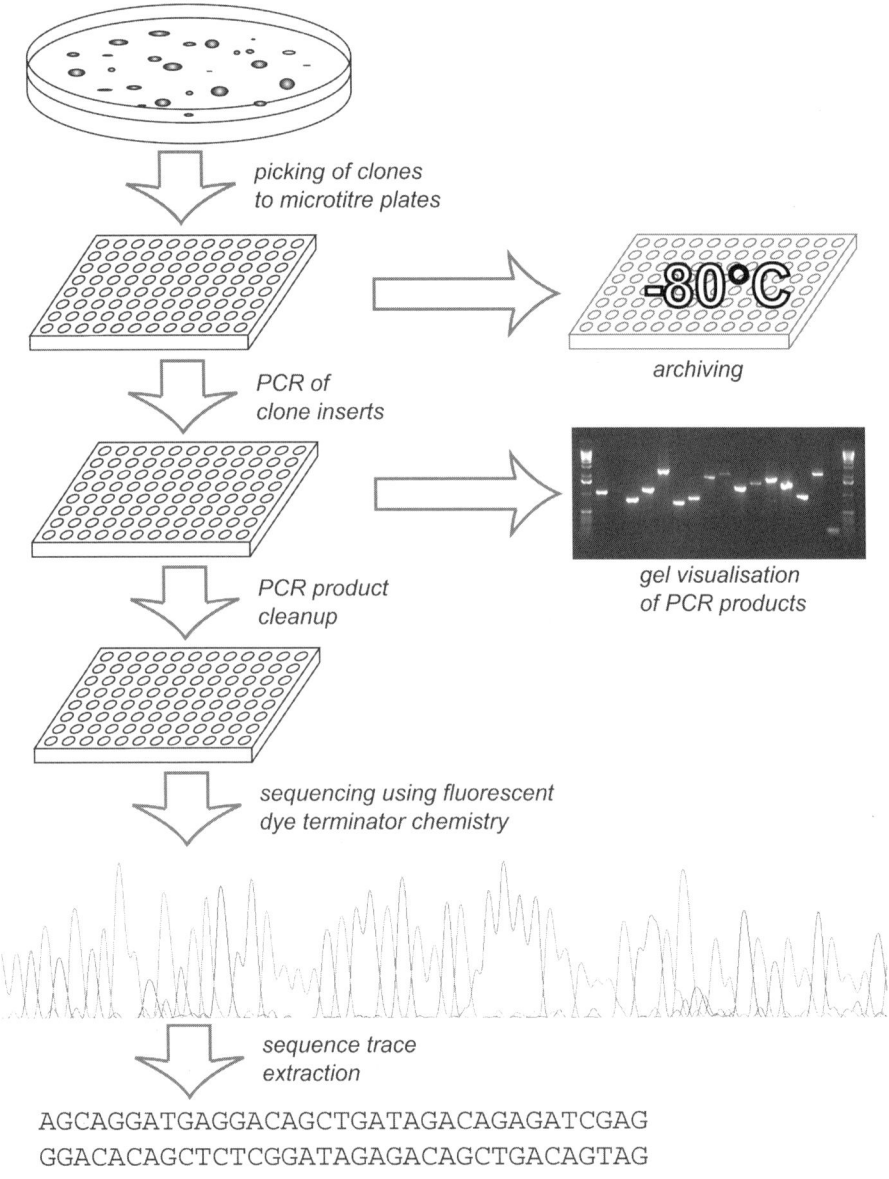

Fig. 1. Overview of the EST sequencing process. Clones are picked from Petri dishes into microtiter plates and archived for later use. All subsequent manipulations (PCR, clean-up, and sequencing) are carried out in microtiter plates to yield medium throughput.

(see **Note 1**). In a nonnormalized library, these 2000 ESTs should identify between 1000 and 1500 different genes and thus, possibly between 10 and 20% of the genes of any target parasite.

2. Materials

2.1. Clone Picking and Archiving

2.1.1. Lambda Phage Libraries

1. Your library plated out: Petri dishes with approx 150 plaque-forming units per 9-cm plate, grown overnight. One Petri dish for every microtiter plate you intend to pick.
2. Sterile V-bottom-well 96-well microtiter plates, two per plate you intend to pick.
3. Self-adhesive plate sealers: We recommend the "easy-peel" heat-sealable type, as these are robust to polymerase chain reaction (PCR) conditions. Their use requires a heat sealer.
4. Sterile phage buffer: 100 mM NaCl, 50 mM Tris-HCl, pH 7.5, 8 mM MgSO$_4$.
5. Dimethyl sulfoxide (DMSO) 20% in phage buffer.
6. Sterile 200-μL pipet tips ("yellow tips"), two boxes per plate you intend to pick.
7. Sterile inserts from 10-μL pipet tip boxes (the segment of the box that holds the tips).
8. Multichannel pipet (see **Note 2**).
9. 37°C incubator for bacterial growth.
10. −80°C freezer for archival storage of clones picked for sequencing.

2.1.2. Plasmid Clone Libraries

1. Your library plated out: Petri dishes with approx 150 colonies per 9-cm plate, grown overnight. One Petri dish for every microtiter plate you intend to pick.
2. Sterile V-bottom-well 96-well microtiter plates, two per plate you intend to pick.
3. Self-adhesive plate sealers (as above).
4. Sterile Luria Bertani (LB) broth with 8% glycerol.
5. Sterile 200-μL pipet tips ("yellow tips"), two boxes per plate you intend to pick.
6. Sterile inserts from 10-μL pipet tip boxes (the segment of the box that holds the tips).
7. Multichannel pipet (see **Note 2**).
8. 37°C incubator for bacterial growth.
9. −80°C freezer for archival storage of clones picked for sequencing.

2.2. Amplification of Clone Inserts by PCR, and Gel Analysis

1. PCR machine (see **Note 3**).
2. Sterile 0.2-mL 96-well propylene microtiter plates (e.g., ABgene AB-0700) (see **Note 3**).

3. Taq polymerase stock at 5 U/μL (e.g., Quiagen 201203, but *see* **Note 4**).
4. 10X enzyme buffer supplied with Taq polymerase.
5. Stock deoxynucleotide solutions at 100 mM (i.e., 100 mM dATP, 100 mM dGTP, 100 mM dCTP, and 100 mM dTTP; e.g., Promega U2140). Make a working stock of dNTP mix for PCR by adding each nucleotide to a final concentration of 2 mM in sterile distilled H_2O.
6. Oligonucleotide primers (1 and 2) complementary to the plasmid or phage vector either side of the cDNA inserts (approx 20 nucleotides). Prepare working solutions of 10 pmol/μL.
7. Sterile double distilled H_2O (ddH_2O).
8. Sterile 0.2-mL strip tubes and caps (strips of eight tubes with the same spacing as the wells of a microtiter plate) (e.g., ABgene AB0266).
9. Sterile 1000-, 200-, and 10-μL pipet tips.
10. "Easy-peel" plate sealers (e.g., ABgene AB0745) (as above).
11. Agarose powder for gels (e.g., Seakem LE Biowhittaker 50005).
12. 10X TBE buffer stock: 108 g Tris base, 55 g boric acid and 40 mL 0.5 M EDTA, pH 8.0, in distilled H_2O to 1 L.
13. Ethidium bromide (EtBr) stock solution at 10 mg/mL.
14. Gel electrophoresis setup (e.g., Hybaid Electro4 system with 18-well combs and 18-cm casting trays) and power pack.
15. Loading buffer: 20% glycerol, 8 mM EDTA, 0.05% bromophenol blue, and 0.05% xylene cyanol in H_2O.
16. DNA size markers (e.g., Invitrogen Readyload ladder 10381-010).
17. UV transilluminator and documentation system (e.g., Polaroid or digital camera).

2.3. Cleaning PCR Reactions for Sequencing

1. Shrimp alkaline phosphatase (SAP) (1 U/μL) (Amersham E70092Z).
2. Enzyme dilution buffer (supplied with the SAP).
3. Exonuclease I (10 U/μL) (Amersham E70073Z).
4. Sterile 10-μL pipet tips.
5. Sterile 1.5-mL Eppendorfs.
6. Sterile 0.2-μL strip tubes and caps (e.g., ABgene AB0266).
7. "Easy-peel" plate sealers (e.g., ABgene AB0745).

2.4. Sequencing Reactions

1. Sequencing kit (an enzyme, nucleotide, and buffer premix supplied by the manufacturer: either DyenamicET [Amersham U581050] or BigDye v3.1 [Applied Biosystems 4337455]).
2. Sterile 0.2-mL 96-well propylene microtiter plates (e.g., ABgene AB-0700).
3. Sterile 0.2-mL strip tubes and caps (e.g., ABgene AB0266).
4. Sterile distilled and deionized H_2O.
5. "Easy-peel" plate sealers (e.g., ABgene AB0745).
6. 96-Well spin column CLEANUP plate (e.g., Amersham Autoseq 96 27-5340-10).
7. Flat-bottomed 96-well (support) plate (e.g., BDH Nunc 402/0328/04).

8. Sequencing primer stock solution at 10 pmol/μL (for Amersham DyenamicET sequencing) or 1.6 pmol/μL (for Applied Biosystems BigDye 3.1 sequencing). This primer should be different from the PCR primers used to amplify the insert.
9. A benchtop centrifuge capable of spinning microtiter plates.
10. A vacuum concentrator with a microplate rotor.
11. Sequencing apparatus (*see* **Note 5**).

3. Methods

3.1. Clone Picking and Archiving

We prefer to use plasmid-based libraries, as the simplicity of the vector aids many downstream applications. Although lambda phage vector-based library construction can be much more efficient in terms of clones recovered per microgram of input cDNA, if input cDNA is not limiting, then we recommend a plasmid vector. Usually, it is feasible for a single researcher to pick and archive 12 microtiter plates (1152 clones) in a session.

3.1.1. Lambda Phage Library Picking and Archiving

1. Plate your phage library using standard procedures (*see* **Note 6**).
2. Plates should be incubated for 18 h maximum and kept at 4°C thereafter to minimize diffusion of phage particles between plaques.
3. For each plate you intend to pick, set up a conical-bottomed microtiter plate (the WORKING plate) with 25 μL of phage buffer in each well. Label the plate clearly (*see* **Note 7**). It is simplest to transfer 320 μL of phage buffer into each tube of a strip tube and then to use the eight-channel pipet to transfer 25 μL to each of columns 01–12. Fix a sterile pipet tip box insert over the top of the plate using tape. The insert will serve to support the tips you use to pick the plaques.
4. Using pipet tips (200-μL "yellow tips"), pick from the centers of well-separated plaques into each well of the microtiter plate. Stab into the center of the plaque with the tip, trying not to penetrate beyond the top agar. Place the tip in the well, and leave it there supported by the insert.
5. Once a full plate has been picked, set it aside for 20 min to allow the phage to diffuse into the phage buffer, and then carefully remove the insert and the tips. Discard the tips, and recycle the insert. The insert should be washed and autoclaved wrapped in foil before reuse.
6. Add 5 μL 20% DMSO to each well of a second ARCHIVE plate. It is easiest to aliquot 70 μL of 20% DMSO into each tube of a strip tube and then use the eight-channel pipet to transfer 5 μL of 20% DMSO to each column of wells of the ARCHIVE plate. Change the tips for each column.
7. Transfer 10 μL of the WORKING plate well contents to a second ARCHIVE plate using the multichannel (*see* **Note 7**). Mix the DMSO and phage buffer in the ARCHIVE wells by pipetting up and down once or twice. Change tips for each column of wells on the plate.

8. Take 2 μL of phage buffer from each well of the WORKING plate to a PCR plate (*see* **Subheading 3.2.**). The transfer should be done at this point, as repeated cycles of freeze–thaw can kill phage stocks.
9. Cover both plates with a self-adhesive plate sealer, check that they are uniquely labeled, and place at −80°C. The ARCHIVE plate should be kept as an archive backup; all manipulations usually should use samples from the WORKING plate.

3.1.2. Plasmid Library Picking and Archiving

1. Plate your plasmid library using standard procedures (*see* **Note 6**).
2. Plates should be incubated for 18 h maximum and kept at 4°C thereafter.
3. For each plate you intend to pick, set up a V-bottomed microtiter plate (the WORKING plate) with 200 μL of LB buffer with 8% glycerol and the appropriate antibiotic in each well. Label the plate clearly (*see* **Note 7**). It is simplest to use a sterile tray to hold the LB and transfer using a multichannel pipet. Fix a sterile pipet tip box insert over the top of the plate using tape. The insert will serve to support the tips you use to pick the plaques.
4. Using pipet tips (200 μL "yellow tips") pick from the centers of well-separated colonies into each well of the microtiter plate. Do not pick up any agar. Place the tip in the well, and leave it there supported by the insert.
5. Once a full plate has been picked, set it aside for 20 min. Gently agitate the tips in the wells using the insert, then carefully remove the insert and the tips. Discard the tips, and recycle the insert.
6. Incubate the plate at 37°C for 12–18 h. Usually, it is not necessary to agitate the plates, but it can help clone growth.
7. Mix the contents of each well by pipetting up and down with the eight-lane multichannel. Transfer 100 μL of the culture from each WORKING plate well to a second ARCHIVE plate using the multichannel pipet (*see* **Note 8**). Change tips for each column.
8. Take 2 μL of culture from each well of the WORKING plate to a PCR plate (for use in **Subheading 3.2.**).
9. Cover the working and archive plates with a self-adhesive plate sealer, check that they are uniquely labeled and place at −80°C. The ARCHIVE plate should be kept for emergencies; all manipulations should use samples from the WORKING plate.

3.2. Amplification of Clone Inserts by PCR, and Gel Analysis

The next stage in the process is to isolate DNA for sequencing from each clone picked. This can be achieved in many different ways. We suggest the use of PCR, as it is robust, cheap, scales easily to medium-throughput, and yields similar amounts of DNA for each clone. In addition, it permits insert size estimation for each clone. You could also use a plasmid miniprep procedure to generate template DNA for sequencing. For PCR, you should identify a robust PCR primer pair that will amplify from the vector used. Importantly, you should ensure that a nested 5'-primer site is also available for sequencing (*see* **Subheading 3.4.**).

1. Label a PCR microtiter plate for each plate you intend to amplify. We use 200 µL volume thin-wall microtiter plates.
2. Make up a PCR master mix in a 2-mL vial for each plate. On ice, mix: 200 µL 10X PCR buffer, 200 µL nucleotide mix (working stock), 1320 µL sterile ddH$_2$0, 40 µL primer 1, 40 µL primer 2, 8 µL thermostable polymerase (e.g., Taq polymerase) stock (*see* **Note 4**). This gives a total volume of 1808 µL.
3. Pipet 18 µL of the PCR master mix into each well of the PCR plate. To facilitate this, it is easiest to pipet 225 µL into each tube of a strip tube and then to use the eight-channel pipet to transfer 18 µL to each of columns 01–12. Try to avoid bubbles.
4. Add 2 µL of the target DNA sample from the WORKING plate to each well of the PCR plate (*see* **Note 8**) using the multichannel pipet. This can be either 2 µL of the phage buffer for phage libraries, or 2 µL of the LB broth plus glycerol for plasmid libraries. We have not noticed that the additional Mg^{2+} ions in phage buffer have an appreciable effect on the efficiency of PCR.
5. Cover the PCR plate with an "Easy-peel" heat sealer, and transfer to the 96-well PCR machine.
6. Run the optimized PCR cycle parameters for your primer pair (*see* **Note 9**).
7. Store the completed PCR amplified templates on ice or at –20°C until required.
8. Pour a 1.6% agarose gel in 0.5X TBE buffer with 0.2 µg/mL EtBr, using 18-well combs. The 18-well combs should have a start and finish marker lane and 16 wells with spacing of half a 96-well microtiter plate column (i.e., 4.25 mm centers). It is useful if six tiers of wells can be poured in a single gel, yielding 96 lanes (*see* **Note 10**). Leave to solidify, and carefully remove the combs. Fill each well on the gel with 0.5X TBE running buffer.
9. In a 0.2-mL well microtiter plate, the LOAD plate, place 4 µL of load buffer in each well. Add 48 µL of load buffer into each tube of a strip tube, and use the multichannel pipet to transfer 4 µL to columns 01–12.
10. Carefully remove the "Easy-peel" heat sealer from the PCR plate. Using the eight-lane multichannel pipet, transfer 4 µL of the PCR reaction from the first column (01) of the PCR plate to column 01 of the LOAD plate. Mix by pipetting up and down, and then carefully pipet all 8 µL of the mix into the first tier of wells in the agarose gel. Lanes 1 and 18 of the gel are reserved for DNA size markers. The spacing of the wells means that you will be loading lanes 2, 4, 6, 8, 10, 12, 14, and 16.
11. Discard the tips from the multichannel, recharge and load wells in column 02 of the PCR plate, this time filling lanes 3, 5, 7, 9, 11, 13, 15, and 17.
12. Repeat with the remaining columns of the PCR plate in tiers two to six (*see* **Figs. 1** and **2**).
13. Finally, load marker DNAs into lanes 1 and 18 of each tier (*see* **Note 11**). As the wells are filled with running buffer, it is possible to load the gel at the bench before transferring it to the electrophoresis rig. The LOAD plate may be rinsed, dried, and reused.
14. Electrophorese the gel at approx 8-V/cm length until the xylene cyanol marker (bright blue) has reached the wells of the tier below.

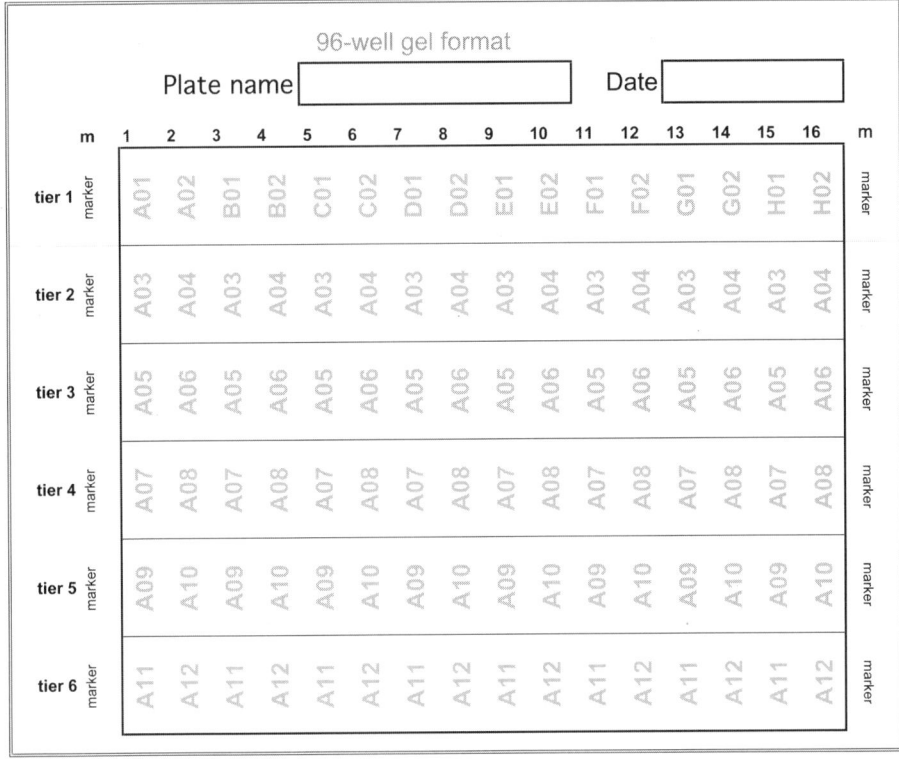

Fig. 2. Template for loading 96 wells of a microtiter plate on a 96-well agarose gel.

15. View and photograph the gel under UV illumination. The sizes of the PCR-amplified inserts may be estimated from the gel. As a segment of vector DNA will be present at each end of the insert fragment, its length will have to be subtracted from the estimated band size to give an insert size.
16. Select products that are suitable for sequencing (*see* **Note 12**). If you have products in a PCR microtiter plate, it is often useful to circle or annotate on the plate the wells that are to be excluded from sequencing. Alternatively, if large numbers of clones (>30) are to be rejected per plate, it is often easier for sequencing and downstream pipetting and processing if the successful PCR products are transferred to a fresh plate.
17. Prepare clear records of selected/rejected clones. This is very important, and we recommend the use of a 96-well paper "template" on which the selected/rejected clones can be recorded. (You can easily generate a 96-well template by photocopying a microtiter plate or by using the template in **Fig. 3**).

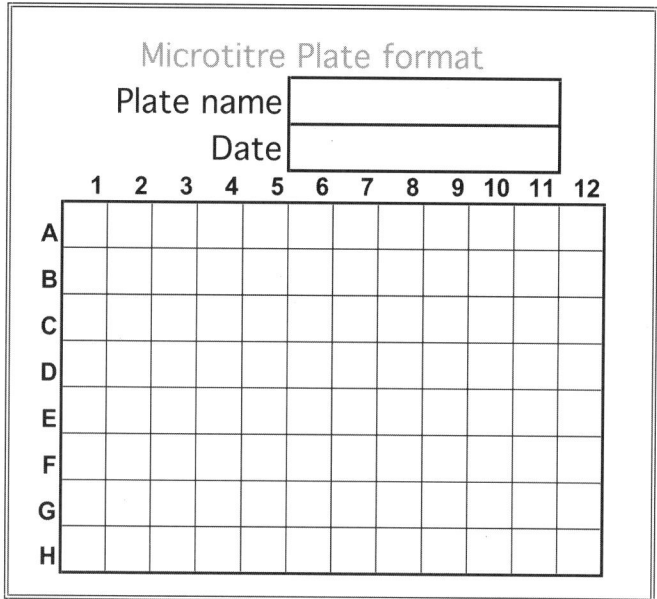

Fig. 3. Microtiter plate template.

3.3. Purifying PCR Reactions for Sequencing

To sequence the PCR products, excess nucleotides (dNTPs) and primers remaining from the PCR reaction must be removed. Our preferred method of cleaning is via the use of two enzymes: Exonuclease I (ExoI) and SAP. The ExoI digests the single-stranded primers into free nucleotides, and the SAP dephosphorylates free nucleotides, making them unavailable for polymerization. This method of PCR clean-up is only effective when there is only one amplified PCR product per cDNA clone. Double products, smears, or primer dimer in the reaction will not be removed by this clean-up method, and the sequencing will not be successful. Both enzymes are readily inactivated by heating at 80°C and so do not affect the downstream sequencing reaction. All of the manipulations should be carried out on ice.

1. Defrost the enzyme dilution buffer, mix by gentle vortexing or pipetting, and keep on ice. The SAP and ExoI enzymes should be kept on ice (never vortex the enzymes).
2. Prepare a SAP/ExoI master mix for 96 products: to a clean sterile Eppendorf tube on ice, add 155 µL enzyme dilution buffer (supplied with SAP), 100 µL SAP, 15 µL ExoI to make a total volume of 270 µL.

3. Mix the enzymes together well. Aliquot 33 μL of the SAP/ExoI master mix into each of the eight wells of a PCR strip tube (or suitable reservoir).
4. Use an eight-channel multipipet to transfer 2.5 μL of SAP/ExoI master mix to each well of the PCR plate containing the PCR products.
5. Ensure the enzymes are mixed well with the PCR product by gentle pipetting.
6. Seal the plate with an "Easy-peel" heat sealer, and incubate using the 96-well PCR machine, with hot lid on, at 37°C for 40 min. Inactivate the enzymes by incubating at 80°C for 15 min.
7. The cleaned PCR product can be cooled to room temperature and is now ready for use in a sequencing reaction.

3.4. Sequencing Reactions

To sequence the insert portion of the cDNAs, we recommend the use of a primer nested within the primers used for PCR. It is possible to use a PCR primer for sequencing, but, in our view, results are more reliable and significantly longer sequences are obtained with a new primer. The primer should be designed such that it ends approx 30–40 bases from the cloning site of the vector.

For automated sequencing, two major commercial products are available: Applied Biosystems' BigDye version 3.1 and Amersham's DyenamicET. Both use similar chemistries and, in our opinion, yield equivalent results. You should consult with the operators of your sequencing service to ascertain which chemistries their instruments are equipped to detect, as the Applied Biosystems and Amersham dye sets differ from each other. The protocol below is written for Amersham's DyenamicET, but we indicate the minor alterations required for use of Applied Biosystems' BigDye.

1. Label a 96-well 0.2-mL plate with the relevant details. This is the SEQUENCING plate.
2. Assemble the sequencing master mix. To 400 μL of DyenamicET enzyme/nucleotide mix as supplied by the manufacturer, add 50 μL of sequencing primer (from stock at 10 pmol/μL) (different for BigDye: *see* **Note 13**) and 50 μL of distilled and deionized H_2O. Mix well.
3. Aliquot 62 μL to each of the eight wells of a 0.2-mL strip tube. Use the eight-lane multichannel to transfer 5 μL of sequencing master mix to each well of the SEQUENCING plate (*see* **Note 14**).
4. Carefully remove the "Easy-peel" heat sealer from the PCR plate (containing SAP/ExoI cleaned products from **Subheading 3.3.**), and transfer 5 μL of each cleaned PCR product to the corresponding well of the SEQUENCING plate using the multichannel pipet (*see* **Note 8**).
5. Cover the SEQUENCING plate with an "Easy-peel" heat sealer and put it on the 96-well PCR machine. Run the following cycle regime (different for BigDye: *see* **Note 15**):25 cycles (95°C for 20 s, 50°C for 15 s, 60°C for 60 s); hold at 4°C.

Medium-Throughput ESTs

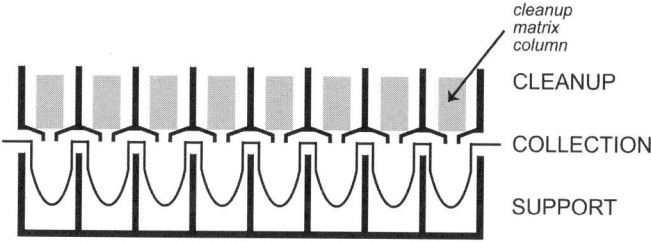

Fig. 4. Cartoon of the clean-up of sequencing reactions using spin columns.

6. Once the sequencing reaction is complete, remove the SEQUENCING plate from the PCR machine, and carefully remove the "Easy-peel" heat sealer. Add 10 µL distilled and deionized H_2O to each well.
7. Take a 96-well spin column CLEANUP plate and remove the upper and lower seals. Label the column CLEANUP plate. Place onto a flat-bottomed 96-well SUPPORT plate and centrifuge at approx $1000g$ for 5 min at 4°C. Discard the flowthrough that will collect in the SUPPORT plate.
8. Carefully pipet the 20 µL of diluted sequencing reaction from the SEQUENCING plate onto the top of the columns of matrix in the CLEANUP plate (*see* **Note 8**). Do not let the liquid spill down the sides of the columns.
9. Place a COLLECTION plate (a 0.2-mL PCR plate is ideal) onto the SUPPORT plate, and replace the CLEANUP plate (*see* **Note 16** and **Fig. 4**). Centrifuge at approx $1000g$ for 5 min at 4°C. Approximately 20 µL of solution (cleaned-up sequencing reactions) should be in each well.
10. Discard the CLEANUP plate. Place the COLLECTION plate on the SUPPORT plate in a vacuum concentrator, and evaporate the cleaned-up sequencing reaction components until dry.
11. The dried sequencing reactions in the COLLECTION plate can now be passed to your sequencing service for gel or capillary electrophoresis. Retain the SUPPORT plate for further cycles of use.

4. Notes

1. It is perfectly feasible for a single worker to carry out a significant EST project alone; we estimate at least 15,000 clones processed per year by the methods we outline here. If you have a more ambitious project in mind, or would like to get the data faster, we recommend approaching one of the larger sequencing centers or custom sequencing companies, as they will be able to use their robotic liquid handling and high-capacity sequencing to both produce good data fast and reduce per-sequence costs.
2. To speed manipulations, we strongly recommend the use of eight-channel multi-channel pipets. Although it is possible to perform all pipetting singly, the multi-

channel makes the process much more robust. Our current choice of pipet is the Finnpipette range, and we use 0.5–10-µL (ThermoLifesciences 4510000), 5–50-µL (4510020), and 50–300-µL (4510030) versions for our EST sequencing program.

3. All the manipulations are carried out in 96-well microtiter plates, and reactions are carried out using a 96-well PCR machine. We prefer PCR machines that avoid the use of oil overlays (so called "hot lid" machines), as many small-volume manipulations are rendered annoyingly difficult when oil is present. To minimize reagent use, 0.2-mL-reaction-volume tubes are recommended.
4. It does not matter which thermostable polymerase you use, as long as it is reasonably processive. It is not necessary to use an error-correcting polymerase, as the sequence will be determined from the bulk PCR product, and individual misincorporations will not interfere.
5. We presume that the researcher has access to an automated sequencer (fluorescent sequencer). These machines are rapid, efficient, and reliable, and many academic centers provide at-cost access to their core service machines. It is, of course, possible to perform EST sequencing using any sequencing technology.
6. The plates from which you pick clones should have clearly separated plaques or colonies and no condensation or other artifacts. If there is color selection in the vector you use, you should include color reagents (usually isoproylthiogalactoside [IPTG] and X-gal) when plating out the phage or bacteria.
7. One essential part of all medium-throughput projects is a robust and simple labeling system. The wells in a microtiter plate are labeled A–H in rows and 01–12 in columns, giving a unique identifier to each well and the clone it contains. Note that it is best to use a full two-digit identifier for columns 1–9 (i.e., 01, 02, . . . , 09), as this both ensures that each clone/well has a name of the same character length, and also guards against errors of transcription (in which case clone 11 could become clone 1). The naming of plates also needs to be carefully thought out. We suggest that a plate name should contain a species identifier, such as Bm for *Brugia malayi*; a library identifier, such as L3a for third-stage larval library version "a"; and a plate number, such as a simple 01–99 for plates 1–99 from a library. We also suggest you separate the different parts of a name with underscores, as this aids both human cognition and machine recognition later. Thus, a plate from the library used as the example above might be "Bm_L3a_78," and a clone from this plate might be "Bm_L3a_78A07." Sequences derived from the clone are most easily named by adding a symbol for the primer used at the end of the clone name, such as "Bm_L3a_78A07_T7."
8. When copying plates, or transferring material between plates, make sure that the plates are lined up with each other (A01 to A01 and H12 to H12). It is worryingly easy to reverse a plate.
9. The exact PCR conditions used will depend on your system. For primers of standard length and GC% on plasmid templates with inserts up to 3.5 kb, we use the following parameters. They may need to be optimized for your primers, vector, thermostable polymerase, and PCR machine.

Initial denaturation	95°C	5 min
35 cycles of amplification	94°C	0.5 min
	53°C	0.75 min
	72°C	2 min
One cycle to "polish" any unfinished product	72°C	10 min
Hold	8°C	Indefinitely

10. Several manufacturers can supply gel electrophoresis rigs specially designed for analysis of the products of 96-well microtiter plate reactions.
11. DNA size markers can be purchased from many suppliers or made in-house by restriction digestion of a plasmid template. You should aim to have a marker set that spans the expected range of PCR products (approx 200 bp to 5 kb). We recommend loading 0.5 µg of marker per lane.
12. For each insert, calculate the size of the insert DNA (i.e., fragment size minus the vector portions). We tend not to sequence inserts of less than 150 bp, though in the early stages of a project it is often informative to do so, as they may represent cDNAs of very short transcripts *(29)*.
13. For Applied Biosystems' BigDye, the primer stock should be at 1.6 pmol/µL.
14. These volumes are for a reaction that is half the manufacturer's recommended reaction volume. It is possible to reduce the amount of reagent mix added to one-quarter or one-eighth of that recommended, saving considerably. The lower reaction mix sizes do not yield as high-quality sequence as the half-volume ones but may be sufficient if optimized for your primer and insert sets.
15. Applied Biosystems' BigDye reaction cycling times are different: 25 cycles of (95°C for 30 s, 50°C for 20 s, 60°C for 240 s); hold at 4°C.
16. Make sure that the CLEANUP and COLLECTION plates are oriented identically (i.e., that well A01 is over A01).

References

1. Johnston, D. A., Blaxter, M. L., Degrave, W. M., et al. (1999) Genomics and the biology of parasites. *BioEssays* **21,** 131–147.
2. Blackwell, J. M. and Melville, S. E. (1999) Status of protozoan genome analysis: trypanosomatids. *Parasitology* **118(Suppl.),** S11–S14.
3. Myler, P. J., Sisk, E., McDonagh, P. D., et al. (2000) Genomic organization and gene function in *Leishmania. Biochem. Soc. Trans.* **28,** 527–531.
4. Gardner, M. J., Hall, N., Fung, E., et al. (2002) Genome sequence of the human malaria parasite *Plasmodium falciparum. Nature* **419,** 498–511.
5. The *C. elegans* Sequencing Consortium. (1998) Genome sequence of the nematode *C. elegans*: a platform for investigating biology. *Science* **282,** 2012–2018.
6. Blaxter, M. L. (2002) Parasite genomics, in *Molecular Medical Parasitology* (Koumeniecki, R., Marr, J., and Nilsen, T. W., eds.), Academic Press, New York, pp. 3–28.
7. Blaxter, M., Whitton, C., Thompson, M., et al. Comparative nematode genomics. *Nematology Monographs Perspect 2*, in press.

8. Lander, E. S., Linton, L. M., Birren, B., et al. (2001) Initial sequencing and analysis of the human genome. *Nature* **409**, 860–921.
9. Goffeau, A., Barrell, B. G., Bussey, H., et al. (1996) Life with 6000 genes. *Science* **274**, 546, 563–547.
10. Wood, V., Rutherford, K. M., Ivens, A., et al. (2001) A re-annotation of the *Saccharomyces cervisiae* genome. *Comp. Funct. Genomics* **2**, 143–154.
11. Wood, V., Gwilliam, R., Rajandream, M. A., et al. (2002) The genome sequence of *Schizosaccharomyces pombe*. *Nature* **415**, 871–880.
12. El-Sayed, N. M. and Donelson, J. E. (1997) A survey of the *Trypanosoma brucei rhodesiense* genome using shotgun sequencing. *Mol. Biochem. Parasitol.* **84**, 167–178.
13. Myler, P. J., Audleman, L., deVos, T., et al. (1999) *Leishmania major* Friedlin chromosome 1 has an unusual distribution of protein-coding genes. *Proc. Natl. Acad. Sci. USA* **96**, 2902–2906.
14. McCutchan, T. F., Hansen, J. L., Dame, J. B., et al. (1984) Mung bean nuclease cleaves *Plasmodium* genomic DNA at sites before and after genes. *Science* **225**, 625–628.
15. Reddy, G. R., Chakrabarti, D., Schuster, S. M., et al. (1993) Gene sequence tags from *Plasmodium falciparum* genomic DNA fragments prepared by the "genease" activity of mung bean nuclease. *Proc. Natl. Acad. Sci. USA* **90**, 9867–9871.
16. Donelson, J. E. and Zeng, W. (1990) A comparison of *trans*-RNA splicing in trypanosomes and nematodes. *Parasitol. Today* **6**, 327–334.
17. Davis, R. E. (1995) *Trans*-splicing in flatworms, in *Molecular Approaches to Parasitology*, Vol. 12 (Boothroyd, J. C. and Komuniecki, R., eds.), Wiley-Liss, New York, pp. 299–320.
18. Nilsen, T. W. (1995) *Trans*-splicing: an update. *Mol. Biochem. Parasitol.* **73**, 1–6.
19. Davis, R. L. (1996) Spliced leader RNA *trans*-splicing in metazoa. *Parasitol. Today* **12**, 33–40.
20. Davis, R. E. (1997) Surprising diversity and distribution of spliced leader RNAs in flatworms. *Mol. Biochem. Parasitol.* **87**, 29–48.
21. Adams, M. D., Kelley, J. M., Gocayne, J. D., et al. (1991) Complementary DNA sequencing: expressed sequence tags and the human genome project. *Science* **252**, 1651–1656.
22. McCombie, W. R., Adams, M. D., Kelley, J. M., et al. (1992) *Caenorhabditis elegans* expressed sequence tags identify gene families and potential disease gene homologues. *Nat. Genet.* **1**, 124–131.
23. Adams, M. D., Kerlavage, A. R., Fleischmann, R. D., et al. (1995) Initial assessment of human gene diversity and expression patterns based upon 83 million nucleotides of cDNA sequence. *Nature* **377(Suppl.)**, 3–174.
24. Blaxter, M. L., Raghavan, N., Ghosh, I., et al. (1996) Genes expressed in *Brugia malayi* infective third stage larvae. *Mol. Biochem. Parasitol.* **77**, 77–96.
25. Blackwell, J. M. (1997) Parasite genome analysis. Progress in the *Leishmania* genome project. *Trans. R. Soc. Trop. Med. Hyg.* **91**, 107–110.

26. Fernandez, C., Gregory, W. F., Loke, P., et al. (2002) Full-length-enriched cDNA libraries from *Echinococcus granulosus* contain separate populations of oligo-capped and *trans*-spliced transcripts and a high level of predicted signal peptide sequences. *Mol. Biochem. Parasitol.* **122,** 171–180.
27. Parkinson, J., Whitton, C., Guiliano, D., et al. (2001) 200,000 nematode ESTs on the net. *Trends Parasitol.* **17,** 394–396.
28. Parkinson, J., Guiliano, D., and Blaxter, M. (2002) Making sense of EST sequences by CLOBBing them. *BMC Bioinf.* **3,** 31.
29. Daub, J., Loukas, A., Pritchard, D. I., et al. (2000) A survey of genes expressed in adults of the human hookworm, *Necator americanus. Parasitology* **120,** 171–184.

5

Expressed Sequence Tags

Analysis and Annotation

John Parkinson and Mark Blaxter

Summary

Expressed sequence tags (ESTs) present a special set of problems for bioinformatic analysis. They are partial and error-prone, and large datasets can have significant internal redundancy. To facilitate analysis of small EST datasets from in-house projects, we present an integrated "pipeline" of tools that take EST data from sequence trace to database submission. These tools also can be used to provide clustering of ESTs into putative genes and to annotate these genes with preliminary sequence similarity searches. The systems are written to use the public-domain LINUX environment and other openly available analytical tools.

Key Words: BLAST; expressed sequence tags; LINUX annotation; sequence analysis.

1. Introduction

ESTs are single-pass DNA sequence reads derived from complementary DNA (cDNA) clones *(1,2)*. The EST strategy has been used by many parasitology programs for gene discovery, drug target, or vaccine candidate identification with significant success *(3–22)*. In addition, EST sequences, and the clones from which they derive, often are the input reagents for other methodologies such as immunoscreening or DNA microarray expression analysis. The sheer size of many EST datasets makes human curation of the data difficult or impossible, and thus computer-based bioinformatic methods have been developed to reduce the complexity of the datasets and analyze them efficiently for informative content. The reduction in complexity in the number of different sequence objects being analyzed is matched by an increase in usability of the data, in which each gene is represented by a single sequence object with much higher-quality sequence and annotation, and is essential for many downstream applications.

An EST dataset is a sample of the messenger RNAs (mRNAs) present in the original tissue used for construction of the cDNA bank. As different genes have very different patterns of steady-state mRNA levels, an EST dataset will have some genes represented by many ESTs, and others will not be represented at all. This differential representation is one of the benefits of EST analysis, because it permits inference of the expression levels of the genes, as well as one of its major analytical problems. Placing the ESTs into clusters that are inferred to derive from one gene can be achieved by grouping those with high levels of identity, but even this step can be difficult. As ESTs are single-pass reads, the actual sequence derived from the sequencing chromatograph may contain errors. In particular, the beginning and end of the sequence may be more prone to error (i.e., of lower quality) than the central portion. Therefore, different ESTs may disagree in their sequence because of sequencing error, and error rates may differ along each sequence.

Most cDNA banks are constructed using technology that does not guarantee that every clone will be full-length; therefore, different ESTs that derive from the same mRNA sequence may have different 5'- and 3'-ends. Also, the process of library construction sometimes can result in the construction of chimeric cDNAs that arise from the illegitimate ligation of two fragments deriving from different genes. Comparing the chimeric EST with a nonchimeric one, the sequences will appear to be identical for some of the sequence and then diverge significantly after the illegitimate join. Unfortunately, this phenotype is also a property of some real sequences. Many genes in eukaryotes give rise to alternatively spliced mRNAs in which particular exons can be included or not, depending on complex regulatory cues. ESTs derived from alternatively spliced versions of a mRNA will also diverge significantly after the "join."

When trying to attribute biological function to the genes that the ESTs represent, cross-species sequence comparisons are used. Prediction of open reading frames and encoded peptides from ESTs is compromised by the low quality of some of the sequence and, in particular, by the presence of insertions and deletions of bases that cause "virtual" frameshifts. Thus, functional annotation of ESTs must recognize the possibility of poor sequence and frameshifts. In addition, because many parasitic organisms are only distantly related to well-studied model organisms, the evolutionary distance separating the parasite gene from its nearest sequence neighbor may be so great as to obscure informative similarity. Careful attention needs to be paid to sequence similarity searches using ESTs so that interpretation of a match is tempered by understanding of the evolutionary history of the organisms being compared.

EST sequencing is over 10 yr of age *(1,2)*, and thus, these problems are not new. A large number of software solutions have been developed for each one. What we present here is an integrated suite of software solutions to EST analysis that can be run on local computers. The reliance on open-source or freely available software solutions, with easily customized parameters, allows us to propose a pipeline that can take ESTs from sequencer chromatograph to annotated database entry efficiently and cheaply (*see* **Note 1**).

2. Materials

2.1. Computing Hardware

The computing demands for analysis of even moderate-sized EST projects (approx 20,000 sequences) are not beyond the capabilities of today's modern desktop computers. The software we recommend is written for UNIX-based operating systems; thus, the bioinformatics machine should be capable of running the freely available LINUX version of UNIX. For the analyses outlined in this chapter, we recommend the following minimum setup: a PC with a Pentium III or IV processor (800 MHz to 2.7 GHz), with 80 GB of hard disk storage capacity and 512 MB of RAM; and an ethernet connection.

By the time this book is published, such specifications will be superceded by entry-level desktop PCs costing a few hundred dollars (*see* **Note 2**).

For an operating system, we recommend using Red Hat LINUX (current release 8.0) from http://www.redhat.com, although other flavors of LINUX should work equally well.

2.2. Software Sources *(see Note 3)*

LINUX, like other UNIX systems, works most efficiently if you take close note of where files are kept, particularly "executable" files or programs. We suggest that core resources such as the basic local alignment search tool (BLAST) and phred programs (*see* below) are stored in directories under "/usr" in the LINUX file hierarchy, such as "/usr/local/bin," or "/usr/ncbi" (for BLAST). The perl programs written by you or downloaded (for example, from our website, http://www.nematodes.org/scripts/) can be kept in a special file directory "/usr/local/bin/." The LINUX notation for the home directory of the current user is "~" (tilde). If you are organized about where programs are stored, it is possible to set your operating system login to know where to look for them (*see* **Note 4**).

1. perl. Many of the programs mentioned in this chapter rely on the use of the perl scripting language (current version 5.8, although the scripts also work on the previous release 5.6), which is usually bundled with the LINUX operating system. perl is freely downloadable from http://www.perl.org/.
2. perl scripts and programs written by us (rename_files.pl, trace2dbest, make_a_table.pl, fsa2clus, CLOBB.pl, pre_assemble.pl, multi_phrap, prepare4blast.pl, blast_5db.pl). These are available at our website, http://www.nematodes.org/scripts. We recommend that you store these in "/usr/local/bin/."

3. The trace2dbest software package (current version 2.01; *see* **Note 5**). This is available as an rpm for ease of installation. An rpm is a software bundle that is easily installed onto a host machine using the RPM package manager that comes with Red Hat LINUX distributions, but the scripts can be made available separately by e-mailing the authors. To install the package on a machine running Red Hat LINUX, simply download the file from the above site using the ftp capabilities of a Web browser. Once the file is downloaded, enter rpm-install trace2dbest.pl-1.0-2.i386.rpm.
   The executables are located in "/usr/local/bin" (*see* **Note 4**).
4. Our in-house EST clustering solution, cluster on basis of BLAST (CLOBB) *(23)*. This is a freely available perl program that relies only on the installation of the "blastall" executable (obtained as part of the BLAST package; *see* below). It is available from http://www.nematodes.org/scripts.
5. phred, phrap, and cross_match. These are available via a license (free to academics) from http://www.phrap.org *(24,25)*. Install these programs in "/usr/local/bin/" (*see* **Note 4**).
6. BLAST. This is freely available from the National Center for Biotechnology Information (NCBI) from their website, http://www.ncbi.nlm.nih.gov/BLAST/ *(26,27)*. We recommend storing the BLAST distribution in a directory, "/usr/ncbi/," and the executables in "/usr/ncbi/bin" (*see* **Note 4**).

3. Methods

In the following, we use as an example an EST project from the platyhelminth *Echinococcus granulosus*, and presume that all analyses are being performed in a directory, "~/ESTproject/," in your home directory.

3.1. Preparing ESTs for Database Submission and Downstream Analysis

3.1.1. Extracting DNA Sequence From Sequencer Chromatograms and Trimming Low-Quality Sequence

Different automated sequencers have their own proprietary associated software suites for processing the chromatographic information and predicting the sequence of bases for each sequencing read. Because these softwares are closely tied to each sequencing platform and can be costly, they are not ideal. However, software for the calling of bases from the chromatograms has been developed and refined in the public genome projects and is available for local installation. This can read the chromatograms produced by all the major sequencers (*see* **Note 6**) and allows users to process sequences according to their own criteria.

The program phred was developed by Phil Green and colleagues and is the "industry standard" (*see* **Note 7**). phred takes the raw chromatograms and uses a series of heuristics both to infer which base should be called at any particular position and to ascribe that base a quality score. The score is computed by assessing the relative height, separation, and shape of each fluorophore peak in relation to the signal from other fluorophores. This score is on a log scale, and a phred score of 20 (indicating an error probability of 1 in 100) usually is taken to indicate a good base call. The phred scores then can be used to remove low-quality bases from the predicted sequence.

Usually, as each EST is sequenced from a vector-directed primer, each sequencing trace will include a small piece of vector DNA sequence. At the vector–insert junction, there may also be sequence corresponding to linkers or adapters used in the construction of the cDNA bank. In the case of a short insert, the derived sequence may extend through the poly(A) tail into the vector on the other side. These sequences should also be trimmed from the EST before further analysis. It is usual to trim the poly(A) tail and note its presence in the EST submission data as a text comment. cDNA bank construction can also inadvertently result in the cloning of contaminant DNA, particularly DNA from *Escherichia coli* and its bacteriophages that gets into the reactions through contamination of recombinant enzymes. These sequences should also be removed. A free program, cross_match, is used to find and trim away any vector, linker/adapter, and contaminant sequences. A pattern-recognition script written in perl can be used to remove and log poly(A) stretches at the 3'-end of sequences.

3.1.2. Submitting ESTs to the Public Databases

The public EST database, dbEST, is a valuable resource for genomics and genetics, and currently contains over 10^7 sequences (mostly from humans and other mammals) (*see* http://www.ncbi.nlm.nih.gov/dbEST/index.html and http://www.ncbi.nlm.nih.gov/dbEST/dbEST_summary.html) *(28)*. Although not all parasite EST projects deposit their data in dbEST, we encourage sequencers to do so, as the benefits of collective access to such data are enormous. In particular, the presence of parasite EST data in public databases alerts researchers working on model organisms to the existence of homologs of their favorite genes in medically important taxa and has resulted in many fruitful collaborations. The public databases provide a mechanism for submitting sequences, thus being allocated database accession numbers but preventing immediate public release. This mechanism allows researchers to have their data ready for release on a given date or on publication of a descriptive paper.

dbEST has a simple but rigid submission format for EST data (*see* http://www.ncbi.nlm.nih.gov/dbEST/how_to_submit.html and **Table 1**). The format involves three files describing the cDNA bank from which the ESTs were derived, a contact person, and a publication. The EST sequence is submitted in a fourth file that quotes these three descriptors. As the format of all four file types is simply text (or flat file format), it is simple to use perl scripts to gener-

Table 1
Submission Files for ESTs[a]

Publication
    TYPE: Pub[b]
    MEDUID: Medline unique identifier
    TITLE: Title of article[b]
    AUTHORS: Author name[b]
    JOURNAL: Journal name
    VOLUME: Volume number
    SUPPL: Supplement number
    ISSUE: Issue number
    I_SUPPL: Issue supplement number
    PAGES: Pages
    YEAR: Year of publication[b]
    STATUS: 1=unpublished, 2=submitted, 3=in press, 4=published[b]
Library
    TYPE: Lib[b]
    NAME: Name of library[b]
    ORGANISM: Scientific name of organism[b]
    STRAIN: Organism strain
    CULTIVAR: Plant cultivar
    SEX: Sex of organism (female, male, hermaphrodite)
    ORGAN: Organ name
    TISSUE: Tissue type
    CELL_TYPE: Cell type
    CELL_LINE: Name of cell line
    STAGE: Developmental stage
    HOST: Laboratory host
    VECTOR: Name of vector
    V_TYPE: Type of vector (Cosmid, Phage, Plasmid, YAC, other)
    RE_1:   Restriction enzyme at site1 of vector
    RE_2:   Restriction enzyme at site2 of vector
    DESCR: Free text description of library preparation methods, vector, etc.
    Text starts on the line below the DESCR: tag.
Contact
    TYPE: Cont[b]
    NAME: Name of contact person submitting the EST[b]
    FAX: Fax number as string of digits
    TEL: Telephone number as string of digits
    EMAIL: E-mail address
    LAB: Laboratory providing EST
    INST: Institution name
    ADDR: Address string, comma delineation

Table 1 (Continued)

Sequence		
TYPE:	EST[b]	
STATUS:	"New" or "Update"[b]	
CONT_NAME:	Name of contact[b,c]	
CITATION:	Publication information[b,d]	
LIBRARY:	Library name[b,e]	
EST#:	EST identifier assigned by contact lab[b]	
CLONE:	Clone identifier	
SOURCE:	Institutional source of clone (e.g., ATCC)	
OTHER_EST:	Other ESTs from this clone	
PCR_F:	Forward PCR primer sequence	
PCR_B:	Backward PCR primer sequence	
INSERT:	Insert length (in bases)	
ERROR:	Estimated error in insert length (bases)	
PLATE:	Plate number or code	
ROW:	Row number or letter	
COLUMN:	Column number or letter	
SEQ_PRIMER:	Sequencing primer description or sequence	
P_END:	Which end sequenced (e.g., 5')	
HIQUAL_START:	First base of highest quality sequence	
HIQUAL_STOP:	Last base of highest quality sequence	
DNA_TYPE:	cDNA[b]	
PUBLIC:	Date of public release[b]	
PUT_ID:	Putative identification of sequence by submitter	
POLYA:	Y or N	
COMMENT:	Comments about EST. Starts on line below COMMENT: tag	
SEQUENCE:	Sequence starts on line below SEQUENCE: tag[b]	

[a]Based on the official dbEST submission instructions at http://www.ncbi.nlm.nih.gov/dbEST/how_to_submit.html. The full version of the submission formats includes additional fields unlikely to be relevant for "neglected" parasitic organisms.

[b]These tags are obligatory. All other tags are optional, but we recommend that as many are filled as is possible.

[c]The CONT_NAME must match the NAME in the Cont submission.

[d]The CITATION must match the TITLE in the Pub submission.

[e]The LIBRARY must match the NAME in the Lib submission.

ate EST submission files. Many researchers screen the databases using text-based searches (looking for annotation matches to "protease" or "kinase," for example); thus, adding some preliminary annotation information as a text comment in the EST submission is very useful. This can be achieved simply by performing a BLAST sequence similarity search (*see* below) against a protein database and using the top-scoring match definition line, if significant, to add a "similar to..." comment.

3.1.3. Automating Sequence Calling, Trimming, and Preparation of dbEST Submission Files

To simplify the process of preparing EST sequences for submission to dbEST, we have developed a freely available perl-based software package (trace2dbest) that uses phred, cross_match, and BLAST to batch process raw EST sequence trace files. The package includes two main programs: rename_files.pl is a simple script that can be used to quickly reformat the names of the sequence traces to follow the suggested naming conventions (*see* **Note 8**); and trace2dbest.pl is a text-based menu-driven software package that creates the dbEST submittable files from the raw sequence traces.

1. In the following, we presume that you have transferred the sequencer output files to your computer and have stored them in a directory "~/ESTproject/traces."

   cd ~/ESTproject

2. The chromatogram files need to have a uniform naming scheme for the subsequent processes to find and recognize them. The perl script rename_files.pl renames files into our favored format (*see* **Note 8**). rename_files.pl is run by specifying a series of arguments: "-dir" specifies which directory should be searched; "-add" specifies which text should be added to the front of each file name (e.g., the EST library identifier; *see* **Note 8**); "-txt" indicates text that should be removed from filenames where it is found; "-sub" indicates text that should be replaced for the text removed using "-txt" and "-format" tells the program to reformat the filenames assuming a 96-well plate layout. Thus, if the files moved to "~/ESTproject/traces" above were named "1.seq" to "96.seq," and we wished them named "Eg_ad1_01A01" to "Eg_ad1_01H12" (where Eg stands for *Echinococcus granulosus*, ad1 stands for adult library version 1, and the microtiter plate number is 01; *see* **Note 8**) we would enter /usr/local/bin/rename_files.pl -dir traces -add Eg_ad1_01 -txt .seq -format

   (This is entered as one line with no returns.)

3. To run trace2dbest enter the command /usr/local/bin/trace2dbest.pl

   If you have used the program before, you may be asked to clean up any directories that were created previously. Further instructions on the use of this package are provided as one of the menu options.

4. You will be offered a menu with a list of options from which to select. The first task is to select the library used to create the sequences. If you are sequencing from a library for the first time, you will need to enter the library details by selecting either option 2 or 3 from the menu (*see* **Fig. 1** for a description of the sort of information needed for each library).

5. After the library has been selected, you will be prompted to enter the name and path to the directory where all the traces are located (in the example here, this would

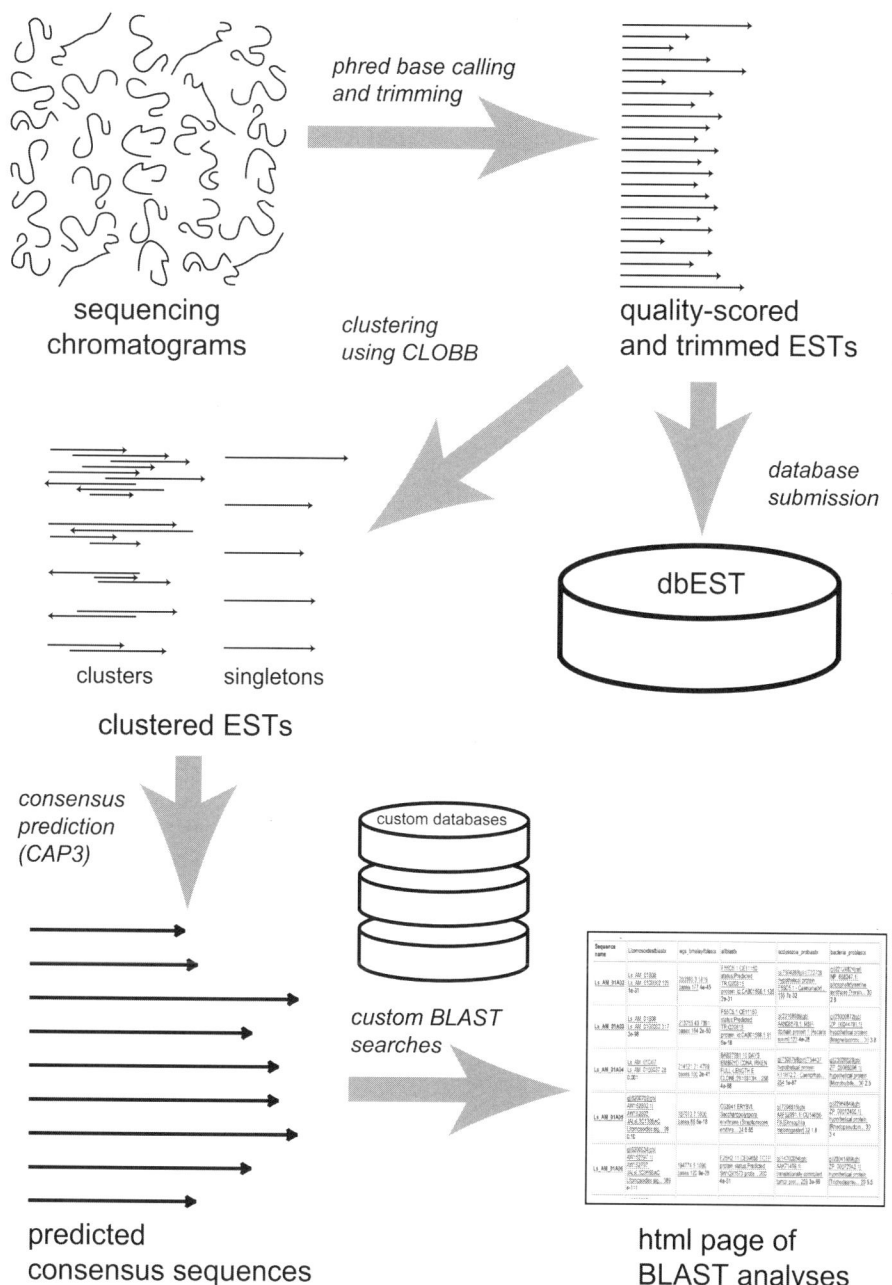

Fig. 1. Overview of the EST analysis process. Taking primary sequence reads through to an analyzed dataset requires only a few steps, which can be automated to handle thousands of sequences at a time. See the text for details of each step.

be "~/ESTproject/traces"). trace2dbest.pl will only process sequencer trace files with the name specified by the EST library identifier, which is formed from the species name tag and the library tag (*see* **Note 8** for information on naming conventions used).

6. You will then be asked to specify the sensitivity of vector trimming. This is very much related to the vector and primer combination used, and we recommend that you set the sensitivity to low unless you find that vector sequence is still present in the final submission files.
7. The next step asks if you would like to perform some preliminary annotation to include as a text comment in the EST submission file. Although there is an option to use the National Center for Biotechnology Information (NCBI) remote blast client (installed as part of the NCBI BLAST package), because of the time associated with this procedure, we recommend that you perform a BLASTX search against a locally installed copy of the nr protein database (*see* **Subheading 3.4.1.7.**).
8. After selecting these options, the program will then use phred to convert the raw trace files into usable sequence data and cross_match to perform vector trimming. Low-quality data and poly(A) tails are then discarded and the optional BLAST step may be undertaken. A series of comments relating to the progress of the process will appear on screen.
9. The dbEST submission files (*see* **Table 1**) are created in a directory, "subfiles," with the suffix ".sub." These files may be concatenated into one file using the simple UNIX command "cat." This may be e-mailed to the dbEST repository (batch_sub@ncbi.nlm.nih.gov). For example, cat~/ESTproject/subfiles/*.sub > ~/ESTproject/submission_file
10. After the trace2dbest process, the "~/ESTproject" directory, which used to have only a single directory, "traces," in it, will have numerous additional files and directories, visible by using the "ls" command as shown here:

The results of "ls" on "~/ESTproject/" after running format_trace2dbest	What this is:
fasta/	A directory of phred output sequence files in FASTA format (**Note 9**): for each input file in "/traces," there will be two new files, "xyz.seq" and "xyz.seqsc"
phd_dir/	A directory of phred output process files: for each input file in "/traces," there will be a new file, "xyz.phd"
qual/	A directory of phred output quality files: for each input file in "/traces," there will be a new file, "xyz.qual"
scf/	A directory of chromatograph files produced by phred: for each input file in "/traces," there will be a new file, "xyz.scf"

EST Analysis and Annotation

The results of "ls" on "~/ESTproject/" after running format_trace2dbest	What this is:
sequences	A directory of trace2dbest trimmed sequence files: for each input file in "/traces," there will be a new file with the same name
subfiles/	A directory of trace2dbest EST submission files: for each input file in "/traces," there will be a new file "xyz.sub"
submission_file	The result of the "cat" command, ready to send to dbEST
traces/	The directory of original trace files

3.2. Clustering of ESTs Into Putative Genes

As explained above, many ESTs may derive from the same gene. Therefore, it is advisable to group the sequences on the basis of sequence similarity into clusters, which can be used to derive consensus sequences. A cluster is defined as a unique set of sequences that share common sequence similarity. A cluster containing only one sequence is termed a singleton. Clustering both reduces the level of redundancy and increases the overall quality of the derived sequence. During production of ESTs, it is often useful to monitor the redundancy of the dataset to ensure that a particular cDNA library is not being oversampled. Initially, the level of redundancy from a newly sequenced library should be low but will increase as new ESTs are generated. From our experience, 10,000 sequences from a good-quality library should yield 4000–5000 clusters.

3.2.1. Cluster on Basis of BLAST

Although there are numerous EST clustering solutions available (*see* **Subheading 3.2.2.**), to monitor levels of sequence redundancy, we have developed a custom clustering solution, CLOBB *(23)*. An important feature of CLOBB is that cluster identifiers are retained between builds, allowing incremental analysis of an ongoing EST project (*see* **Fig. 2**).

1. Open a terminal and navigate to the project folder containing the sequences (in the example being followed, this would be "~/ESTproject/"). CLOBB expects to find the processed EST sequences in a directory, "sequences." CLOBB documentation is available using the command perldoc/usr/local/bin/CLOBB.pl
2. Run the program by specifying a three-letter cluster identifier to be used for creation of the clusters. We recommend that the taxon name be used, and we usually

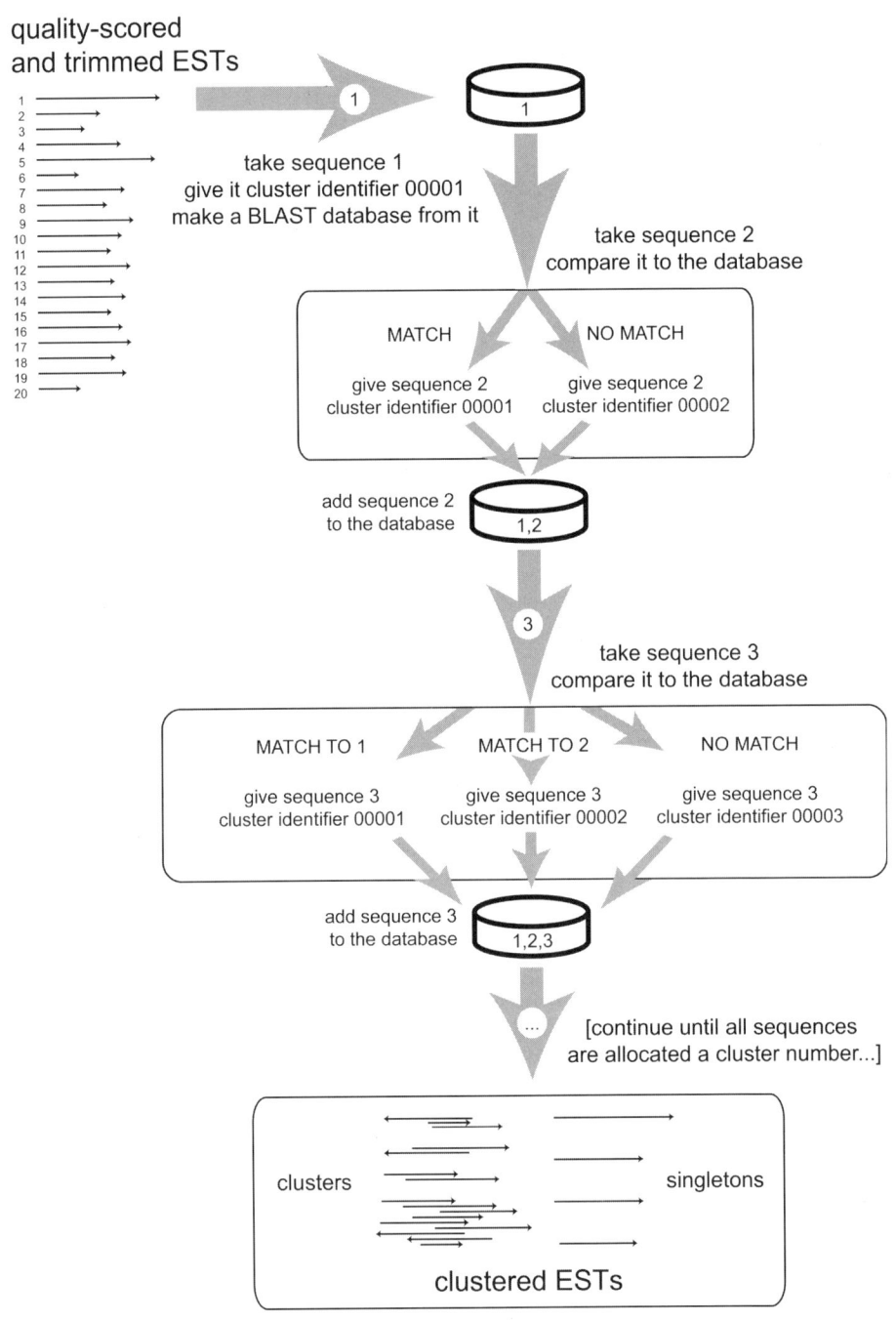

use "C" as the third letter to identify the resulting analysis as a cluster. For the species *Echinococcus granulosus*, the command could be cd ~/ESTproject/ usr/local/bin/CLOBB.pl EGC

3. The main output of the program is a single, multisequence FASTA (*see* **Note 9**) format file beginning with the cluster identifier followed by "EST.fsa" (in the example above, the file would be called "EGCEST.fsa"). In this file, each sequence has had added to its ">" header line a cluster identifier beginning with the three-letter code and followed by a five-digit number: thus, EGC00001, EGC00002, etc.

4. To convert this file into a list of separate cluster files, we provide the perl program fsa2clus. To run fsa2clus, you must specify the three-letter cluster identifier. Thus, for the above example the command would be /usr/local/bin/fsa2clus EGC

    This creates a directory, "Clus," which contains individual cluster files in FASTA format containing the sequences associated with each cluster (designated ABC00001, ABC00002, etc.), and a single FASTA file called "singletons.fasta," which contains all the sequences that did not group with any other sequence.

5. Numerous other files are created during the CLOBB process, including "merge," which contains a list of all the clusters that have been merged into a single cluster (as a result of a common sequence spanning between two clusters), and "super-clusters," which contains a list of clusters that are significantly related to each other, representing either alternative splice variants or chimeric clones. These files are available as aids to further annotation of the sequences and to record the history of changes between subsequent builds of the clusters. Using the "ls" command to view the project directory at this stage will reveal:

The results of "ls" on "~/ESTproject/" after running CLOBB and fsa2clus	What this is:
Clus/	The cluster files created by fsa2clus
EGCEST.fsa	The major CLOBB output file
EGCEST.fsa.nin, EGCFSA.fsa.nsq, EGCEST.fsa.nhr	BLAST data base files produced by the CLOBB process

Fig. 2. (*Opposite page*) Clustering using CLOBB. The process used by CLOBB to cluster ESTs relies on the use of custom-built BLAST databases. Each sequence is taken in turn and compared to all those previously examined. If a significant match is found, the cluster identifier of the matched sequence is attributed to the new one. If the new sequence is apparently novel, it is ascribed the next available cluster identifier. The CLOBB software is able to deal with more complex issues that can arise, such as a single new sequence matching more than one previous cluster (implying that clusters should be joined).

The results of "ls" on "~/ESTproject/" after running CLOBB and fsa2clus	What this is:
fasta/	A directory of phred output sequence files in FASTA format: for each input file in "/traces," there will be two new files, "xyz.seq" and "xyz.seqsc"
formatdb.log	A logfile associated with BLAST database construction
master	A CLOBB file used in cluster building
merge	A CLOBB record of clusters merged during the process
OUT/	A directory of CLOBB logfiles
phd_dir/	A directory of phred output process files: for each input file in "/traces," there will be a new file, "xyz.phd"
qual/	A directory of phred output quality files: for each input file, "xyz," in "/traces," there will be a new file, "xyz.qual"
scf/	A directory of chromatograph files produced by phred: for each input file, "xyz," in "/traces," there will be a new file, "xyz.scf"
sequences_done	A directory of sequence files that have been processed by CLOBB: for each input file, "xyz," in "/sequences," there will be two files, "xyz" and "xyz.old"
sequences	A directory of trace2dbest trimmed sequence files: for each input file, "xyz," in "/traces," there will be a new file with the same name
subfiles/	A directory of trace2dbest EST submission files: for each input file, "xyz," in "/traces," there will be a new file "xyz.sub"
submission_file	The result of the "cat" command, ready to send to dbEST
superclusters	A CLOBB record of clusters with significant similarity to each other
traces/	The directory of original trace files

3.2.2. Other Options

Other clustering solutions are available, including StackPACK *(29)*, ICA tools *(30,31)*, and the clustering suite of tools available from TIGR (*see* http://www.tigr.org/tdb/tgi/software/). However, of these, only StackPACK (available at http://www.sanbi.ac.za/Dbases.html), like CLOBB, offers the ability to maintain cluster identifiers between subsequent builds. This is an important criterion in being able to monitor library quality/redundancy.

3.3. Predicting the Consensus Sequence for Each EST Cluster

Once the sequences have been grouped into clusters, the next step is to obtain consensus sequences for each cluster that contains more than one sequence. Numerous programs are currently available that originally were developed to derive contiguous stretches of sequence (termed contigs) from genome sequencing initiatives. These programs may be applied readily to assemble contig sequences from clusters of ESTs. As the stringency for contig building differs between CLOBB and these programs, multiple contigs may result from each cluster. These may indicate the presence of alternative splices or alleles, which were not detected initially during the clustering step. This assembly process thus represents an additional clustering step. The use of two assembly programs, CAP3 and phrap, is outlined here.

CAP3, developed by Xiaoqiu Huang, makes use of base quality values (if available) in constructing an alignment of sequence reads and generating a consensus sequence for each contig. It clips poor 5'- and 3'-regions of reads and uses only good regions of reads in assembly. phrap is a program originally developed by Phil Green to assemble DNA shotgun sequence data *(24,32)*. phrap uses quality data generated during the base-calling step to create consensus sequences. It has the advantage over other programs such as CAP3 *(33)* in that built contigs include sequence data from regions of the alignment spanned by only a single sequence (i.e., end overhangs). For both processes, if a quality file is supplied, then it must have the same name as the cluster file with the additional suffix ".qual." Quality files are obtained from the "qual" directory created by phred in the base-calling procedure (as implemented in the trace2dbest program; *see* **step 3** in **Subheading 3.1.**).

If no quality file is supplied, the programs assume a relatively low score for the quality of each base.

1. To generate phrap-ready quality files from CLOBB output, we have developed a perl script called pre_assemble.pl. In this script, you specify the directory where the original ".qual" and sequence files are located. The program scans through the cluster files, identifies the appropriate quality and sequence files, and concatenates them to an appropriately named cluster quality file: all of these are stored in a new directory called "phrap." If the files are not found on the system, then the program creates a false entry in the cluster quality file in which each base is given a default quality value, which may be defined by the user. Because the original sequence files are being used, they must be trimmed for poly(A) tails and vector contamination.

```
cd ~/ESTproject/
/usr/local/bin/pre_assemble.pl
```

2. In most cases, it is preferable to run the assembly process as a batch process. To this end, we have created a perl script called multi_phrap:

```
cd ~/ESTproject/phrap/
/usr/local/bin/multi_phrap EGC
```

where EGC represents the three-letter cluster identifier. The program uses cross_match to remove any remaining vector contamination from the cluster sequence files and removes the corresponding quality value entries in the cluster quality file. It then uses phrap to assemble the clusters depending on the selection made by the user. If no consensus sequence can be derived, a consensus sequence is generated by selecting the longest sequence from the cluster sequence file.
3. Performing an "ls" command at this stage reveals the following within directory "phrap" for each input file (such as "EGC00001"):

EGC00001	The original CLOBB-derived file moved by the pre_assemble.pl script
EGC00001.ace	Part of the phrap output
EGC00001.qual	The original phred-derived files moved by the pre_assemble.pl script
EGC00001.contigs	A FASTA file of the contigs derived from EGC00001
EGC00001.contigs.qual	Quality files associated with each contig
EGC00001.singlets	Any singletons not used for building the contig
EGC00001.log	A logfile
EGC00001.out	Output associated with the running of the program
EGC00001.problems	Sequences that may have registered problems in assembly
EGC00001.problems.qual	The phred quality file associated with the ".problems" file

3.4. BLAST-Based Sequence Similarity Analysis

Functional annotation of sequences via bioinformatic tools is an important first step in the utilization of EST datasets for understanding the biology of parasites. Because genes evolve, it is possible to examine sequences from different species for conservation of sequence and then use this conservation to transfer functional annotation from one gene of known role to an otherwise uncharacterized EST cluster. The simplest way of doing this is to use sequence comparison tools that are attuned to the evolutionary changes known to occur in genes and their encoded proteins to search the EST consensus sequence against a database of annotated sequences. The best tool for this is BLAST *(26, 27)*. BLAST searches for exact matches to short subsequences of the query and then tries to extend these initial hits using a set of parameters that gives scores for matches and mismatches (*see also* Chapter 2). The significance of the final best hit (called a high-scoring sequence pair, or hsp, in BLAST) is calculated based on the probability of finding, by chance, a hit as good as the one found, given a query sequence of the same residue composition and length and a database of the same size and complexity. The output of a BLAST search is a table of matches, with both the score (given by summing the match rewards and mismatch penalties) and the probability of that score for each sequence with an hsp. Low-complexity sequence (such as peptide or nucleotide repeats, or homoresidue runs) can generate spuriously high-scoring matches, and thus, it is usual to mask these parts of a sequence before searching. Masking of the sequence allows the program to make the comparison based on the higher-complexity regions and return hsp of more likely biological significance. The masked regions are indicated by "X" in the output. Filtering for low complexity is on by default.

The BLAST family of programs can compare nucleotide with nucleotide (called BLASTN), nucleotide with protein (by translating the nucleotide sequence in all six frames; BLASTX), protein with protein (BLASTP), and protein back translated to nucleotide against nucleotide (TBLASTN). It is also possible to compare the six possible protein sequences from a nucleotide query with a six-frame translation of a nucleotide database (TBLASTX). BLASTN is optimized for finding very close matches and is less suited for comparisons across wide evolutionary distances. The protein–protein comparisons (BLASTP, BLASTX, and TBLASTX) use a matrix that gives a set of scores for each substitution, and these matrices can be tuned for looking at even very distant similarity. For EST analysis, the most informative searches are BLASTX, asking if there are known proteins with similarity to the potential translation of the EST, and BLASTN, asking if the gene or one very closely related has been sequenced before.

Web-based and client-server processes are available for performing all versions of BLAST search (for examples, see http://www.ncbi.nlm.nih.gov/blast/ and http://www.ebi.ac.uk/blast2/). Although these services are comprehensive, based on the complete, up-to-date public databases, they can be slow, particularly if many comparisons are required. For most purposes, it is faster and just as informative to use BLAST locally, using customized databases extracted from the public resources of GenBank, European Molecular Biology Laboratory (EMBL), and DNA Data Bank of Japan (DDBJ).

3.4.1. Local BLAST Databases

BLAST on a local machine can use either a remote database (client-server BLAST) or a local database. The use of local databases allows users to tailor searches to their needs and, in particular, to perform searches restricted by taxonomy. Thus, a search strategy for an EST project might include (a) a BLASTN search against nucleotide sequences from the same species (or the same genus), (b) a TBLASTX search against nucleotide sequences from the same major taxonomic group (class or phylum), excluding the species or genus of interest, and (c) a BLASTX search against a nonredundant protein database from all organisms. Building these local databases is relatively easy, using the ENTREZ system provided by the NCBI.

1. Launch a Web browser such as Netscape or Explorer, and go to the NCBI home page, http://www.ncbi.nlm.nih.gov/.
2. In the ENTREZ search bar near the top of the page, change the database (default is Nucleotide) to Taxonomy using the pull-down menu, and enter the name of the taxon of interest (for example, *Echinococcus granulosus*). Click on the "Go" button. The database server at NCBI will look for the term entered and return a page with matched entries: select the taxon you are interested in and navigate until you have the page listing its attributes as recorded by NCBI. This taxonomy database entry will start with a Taxonomy ID code, a number, which you should record (*see* **Note 10**).
3. From the small table on the right of the page, select the dataset you would like to download (nucleotide or protein) (*see* **Note 11**). The ENTREZ system will return you a page showing the first 20 sequences of the set selected. In the top ENTREZ search bar, you will see the taxonomy ID of your chosen taxon, with a qualifier "[Organism]." (There may be additional qualifiers after organism, but they are not relevant here.)
4. In the Display bar, change the format you wish to see from "Summary" to "FASTA" (*see* **Note 9**), and the "Send to" option from "Text" to "File." Click on the "Send to" button.
5. The browser will ask if you want to download all the sequences selected. Reply yes. It will then display the first sequence in FASTA format, and a dialog will appear asking you where the program should save the file and what name it should be given. We recommend that you use a name that is informative of the taxon source, the sequence type, and the format (such as "`E_granulosus_nuc.fsa`"). The sequences will be transferred to your computer. In addition, it is best to save all BLAST database files in a single location so that you do not have to remember where each is. We suggest making a "`localdb`" directory in your home directory (*see* below).
6. To get databases with a more complex inclusion of sequences, use the ENTREZ system at the NCBI. Go to the NCBI home page, and, in the ENTREZ search bar at the top, enter the query you wish to search. You can combine searches with "OR" or "NOT." Thus, to generate a database of *Echinococcus* (txid 6209) and *Schistosoma* (txid 6181) sequences, you could enter "`txid6209`[Organism] OR `txid6181` [Organism]." To build a database of platyhelminth (txid 6157) sequences excluding *Echinococcus*, the query would be "`txid6157`[Organism] NOT `txid6209` [Organism]." Select the database from which you want to download (Protein or Nucleotide), and then follow the same procedure as given in **step 5**.
7. The sequence file now has to be properly formatted for BLAST. A utility for this is provided with the BLAST programs. Launch a terminal window. Use the "`mkdir`" command to make a directory called "`localdb`" in your home directory, and move the FASTA file(s) of sequences you have downloaded there. Type `cd ~/`, then `mkdir localdb`, then `mv *.fsa localdb/`

(This presumes you downloaded the sequences into the top level of your home directory.)

```
cd ~/localdb
```

Now run the formatdb command from the NCBI BLAST distribution. You use the "-i" modifier to tell it which file to process and the "-p" modifier to tell it whether the database is protein ("-p T") or nucleotide ("-p F"). For example, using the *Echinococcus* dataset downloaded above, one would type `/usr/ncbi/bin/formatdb -i E_granulosus_nuc.fsa -p F`

If you now list the contents of the directory (using the "ls" command), you will find that in addition to the original ".fsa" file there are now ".fsa.nin," ".fsa.nsq," and ".fsa.nhr" for nucleotide databases, or ".fsa.pin," ".fsa.psq," and ".fsa.phr" for protein databases. These additional files are indices used by BLAST to find matches. Repeat the formatdb process for each database downloaded.

8. The nonredundant protein database provided by the NCBI is very useful. It is generated by including only one representative of each set of proteins that are identical in sequence and is much smaller than the whole protein database. This nr protein database is computed by NCBI frequently and is available from their ftp site. In a Web browser, go to ftp://ftp.ncbi.nih.gov/blast/db. The nonredundant protein database is stored as "nr.Z": select this and download (*see* **Note 12**). Open a terminal, move "nr.Z" to the "localdb" directory, and uncompress it using the gunzip command. Format the database as you would any other.

```
mv nr.Z ~/localdb/
cd localdb
gunzip nr.Z
rm nr.Z
mv nr nr.fsa
/usr/ncbi/bin/formatdb -i nr.fsa -p T
```

3.4.2. Basic BLAST

The BLAST family of programs is easily run interactively from the command line and can also be called by perl programs. The ability to call for BLAST to process a file identified by a perl program allows the user to perform thousands of custom BLAST searches automatically, saving a lot of time. It is useful to understand the possible options for command-line operation of the BLAST family of programs (*see* **Note 13**).

1. Open a terminal. The basic command line interface for BLAST is `/usr/ncbi/bin/blastall -p program -d database -i query -o outfile` (This is entered as one line with no returns.)

    The "-p" argument allows you to choose from one of the five flavors of BLAST (blastn, blastp, blastx, tblastn or tblastx; note that the program names are entered in lowercase). The "-d" argument asks you for the location of the database you want to search (such as "~/localdb/E_granulosus_nuc.fsa"). The "-i" argument identifies the sequence you wish to compare, known as the query sequence. The query should be a text file in FASTA format (*see* **Note 9**). The "-o" option tells the program to which filename to write the file of results. Additional arguments are available, including those that set the program running with parameters (such as substitution matrix) different from those set as default (*see* **Note 14**). One useful argument to know is "-T." Using "-T T" will yield hypertext-marked output (html), ready for viewing in a Web browser, whereas "-T F" will produce plain text output.

Thus, a command to search a nucleotide sequence, EGC00001, against the *Echinococcus* database, searching for matches because of encoded proteins might read:

```
/usr/ncbi/bin/blastall -p tblastx -d ~/localdb/
E_granulosus_nuc.fsa -i EGC00001 -o EGC00001.out -T F
```

(This is entered as one line with no returns.)

3.4.3. Running Multiple BLAST Searches

To run BLAST searches of a large number of sequences against a database, you can either catenate all the sequences together into one large FASTA file and use it as the query or use a perl script that will take each file and perform a search. In the first case, BLAST understands that a multiple-sequence FASTA file should be searched as a series of individual sequences but saves the results of all the searches to one (possibly very big) file. In the second, BLAST will save one search for each file processed. We have written a perl script (blast_5db.pl) that will take all the sequence files in a single directory and perform up to five BLAST searches (all the same BLAST type) against different databases, saving the results as either plain text or html (*see* http://www.nematodes.org/scripts).

To facilitate BLAST analysis of the output of the EST sequence processing and clustering process outlined earlier, we have written perl scripts that prepare the cluster contigs for BLAST analysis (prepare4blast.pl) and perform the BLAST searches (blast_5db.pl)

1. Open a terminal, and in the "~/ESTproject/" directory, run prepare4blast.pl. The script assumes that you have completed the trace2dbest, CLOBB, and consensus sequence prediction procedures outlined above and collects the required sequences from the project directory into a new directory, "~/ESTproject/blast."

   ```
   cd ~/ESTproject
   /usr/local/bin/prepare4blast.pl
   ```

2. Decide which databases you wish to search and which variety of BLAST you wish to use. The program "blast_5db.pl" only does one sort of BLAST at a time, on up to five databases. Locate the databases, and check that they have been formatted.

blast_5db.pl is instructed where to look for sequences and databases from the command line as follows: the first argument is which program to use, the second is where the sequences are to be found, the third is what format of output is required ("H" for html, "T" for text) and these are followed by a list of the databases to be searched. Thus, to search a nucleotide database in "~/localdb/" using BLASTN from CLOBB cluster consensus sequences in directory "~/ESTproject/blast/sequences" (as collected by prepare4blast.pl), with html output the commands would be cd ~/ESTproject/blast/ /usr/local/bin/blast_5db.pl blastn sequences H ~/localdb/E_granulosus_nuc.fsa

(The BLAST command is entered as one line with no returns. In this example, only one database is searched: it is not necessary to give the program five databases every time. To search additional databases, enter the path to each database at the end of the command.)

3. The program will go through the sequences in the specified directory and use BLAST to search each against the specified databases, saving the results in directories named "databasename_searchprogram," named after the sequence but with the extension ".html" for html files and ".txt" for text files.

3.4.4. Parsing BLAST Outputs

As indicated above, the output from a BLAST search is a table of hsp matches for the query, ranked by the quality of the match. As with the actual performance of the BLAST searches, reviewing the results can be tedious if carried out piecemeal. The standard format of the BLAST output allows us to use text-processing algorithms to extract the relevant information and present it in more palatable form. Below, we describe the use of a simple method for extracting significant matches from a set of BLAST results and tabulating them in a Web browser-readable (html) format. This sort of simple extraction of data is most useful for medium-size datasets (up to approx 1000 ESTs or clustered EST consensus sequences). Because this method looks only at the top hit for each search, informative or intriguing matches scoring less highly will not be noted. For a fuller analysis of the BLAST results, a relational database is most useful, as described in **Subheading 3.6.**

A BLAST search output has the following features (*see* **Table 2**): (a) a header indicating which program was used, (b) identifiers for the query and the database, (c) a table of hsp's, and (d) alignments of the query with the hsp's. These features can be changed by using options in the BLAST command, but in the following we assume that the default settings are used. The table of matches and the alignments contain most useful information. In the table, each match is listed with a segment of the FASTA ">" definition line as descriptor, the raw (bit) score, and the E (expect) value. The table is usually sorted by E-value, though it can be reported sorted by score. Unfortunately, there are no hard-and-fast rules for what constitutes a real or biologically significant hsp match. Gener-

Table 2
BLASTX Search Output[a]

BLASTX output information	Comment
BLASTX 2.0.11 [Jan-20-2000] Reference: Altschul, Stephen F., Thomas L. Madden, Alejandro A. Schäffer, Jinghui Zhang, Zheng Zhang, Webb Miller, and David J. Lipman (1997), "Gapped BLAST and PSI-BLAST: a new generation of protein database search programs," Nucleic Acids Res. 25:3389-3402.	A description of the program used and the literature reference for the program

Query = HCC00002.Contig1
        1446 letters)
Database: swall-1; swall-2
  1,018,951 sequences; 323,619,622 total letters
	A description of the query sequence and the database searched

Searching..done Score E Sequences producing significant alignments: (bits) value Q25049 Q25049 CAA56353.1 CAA56353.1 CAA56353.1 880 0.0 CAA56353.1 Q9TX76 Q9TX76 Desc: Beta-tubulin. 878 0.0 Q25024 Q25024 AAA29170.1 Desc: Beta-tubulin. 877 0.0 Q9GT34 Q9GT34 AAM95343.1 AAM95346.1 AAM95340.1 873 0.0 AAM95341.1 Q26901 Q26901 CAA93249.1 Desc: Beta-tubulin. 872 0.0 Q9GT35 Q9GT35 AAG13954.1 AAK72123.1 AAM95347.1 872 0.0 Desc: Beta- Q8MV65 Q8MV65 AAM95344.1 Desc: Beta-tubulin Cyca-1b. 872 0.0 Q8MV63 Q8MV63 AAM95349.1 Desc: Beta-tubulin Cci-1b. 871 0.0 Q9GT32 Q9GT32 AAG13961.1 Desc: Beta-tubulin 871 0.0 isoform 1-3.	A table of the high-scoring sequence pairs (hsp's), giving a brief identification (from the FASTA ">" header), the score of each match, and the computed probability (E-value)

```
Q9N611  Q9N611 AAF26294.1 AAF26293.1                              870  0.0
        Desc: Beta-tubulin.
Q8MV60  Q8MV60 AAM95353.1 Desc: Beta-tubulin Ccr-1b.              870  0.0
Q8MV62  Q8MV62 AAM95350.1 Desc: Beta-tubulin Cce-1a.              870  0.0
Q8MV64  Q8MV64 AAM95345.1 Desc: Beta-tubulin Cyca-2a.             869  0.0
Q8MV61  Q8MV61 AAM95351.1 Desc: Beta-tubulin Cce-1b.              869  0.0
Q8MV66  Q8MV66 AAM95338.1 Desc: Beta-tubulin Cyp-1a.              867  0.0
Q9GT33  Q9GT33 AAG13960.1 Desc: Beta-tubulin                      862  0.0
        isoform 1-2.
Q26900  Q26900 AAA30100.1 Desc: Beta-tubulin.                     838  0.0
Q25022  Q25022 AAA29168.1 Desc: Beta-tubulin.                     836  0.0
Q25023  Q25023 AAA29169.1 Desc: Beta-tubulin.                     835  0.0
Q18817  Q18817 CAB00853.1 Desc: C54C6.2 protein.                  830  0.0

Q25049 Q25049 CAA56353.1 CAA56353.1 CAA56353.1 Desc: TUB-1
       gene exon 1.
       Length = 448
Score = 880 bits (2248), Expect = 0.0
Identities = 427/448 (95%), Positives = 428/448 (95%)
Frame = +2
Query: 41  MREIVHVQAGQCGNQIGSKFWEVISDEHGIQPDGTYKGESDLQLERINVYYNEAHGGKYV 220
           MREIVHVQAGQCGNQIGSKFWEVISDEHGIQPDGTYKGESDLQLERINVYYNEAHGGKYV
Sbjct: 1   MREIVHVQAGQCGNQIGSKFWEVISDEHGIQPDGTYKGESDLQLERINVYYNEAHGGKYV 60
Query: 221 PRAVLVDLEPGTMDSVRSGPYGQLFRPDNYVFGQSGAGNNWAKGHYTEGAELVDNVLDVV 400
           PRAVLVDLEPGTMDSVRSGPYGQLFRPDNYVFGQSGAGNNWAKGHYTEGAELVDNVLDVV
Sbjct: 61  PRAVLVDLEPGTMDSVRSGPYGQLFRPDNYVFGQSGAGNNWAKGHYTEGAELVDNVLDVV 120
(etc etc...)
```

For each hsp an alignment is given, with scores and percent identities (and similarities for protein searches)

^aOther varieties of BLAST yield very similar output.

ally, in protein–protein comparisons (i.e., BLASTP, BLASTX, and TBLASTX), a score of greater than 80 bits and an *E*-value of less than e^{-6} is considered necessary before an hsp would be examined further. Matches scored at between e^{-6} and e^{-8} often are to short domains, which may occur by chance. In nucleotide–nucleotide comparisons, one is searching for bit scores of greater than 400 and *E*-values of less than e^{-50}.

The table of significant hsp's always starts with the text, "Sequences producing significant alignments," and this can be used as a tag to identify where a process should start looking for the top hit. We have written a perl program (make_a_table.pl) that builds html-marked-up tables of BLAST searches (*see* **Fig. 3** for an example). The perl program, on being told where to find the relevant sequence and BLAST result files, extracts from each the useful information and formats it for the Web. Each cell is hyperlinked to the original BLAST result file (or sequence), and it is simple to scroll down the table examining significant matches and ignoring the searches that returned no significant hits.

1. Open a terminal and navigate to the directory where your BLAST searches were carried out ("`blast`" from the example above). List the contents of the directory using the "`ls`" command to remind you of the names of the BLAST output directories.
2. Run the perl program make_a_table.pl. The program needs to know what the project name is (to place in the header and title of each html table), where the sequence files are, and where each blast search output is to be found. It is set by default to split the results into groups of 50 sequences, because some Web browsers have problems displaying html tables longer than about 50 rows.

Fig. 3. (*Opposite page*) An example of a table of EST BLAST results generated with make_a_table.pl. This table is the result of an analysis of primary EST sequence reads from the parasitic nematode *Litomosoides sigmodontis*. The first column provides links to the sequences analysed. The next five columns present the results of five different BLAST searches, with each cell linked to the relevant, saved BLAST search output. In the first column are searches using BLASTN against *L. sigmodontis* sequences to identify if the gene has been identified before. In the second column are searches against genomic sequence from *Brugia malayi*, a related nematode, using TBLASTN. In the third column are BLASTX searches against the nr protein database. In the last two columns are BLASTX searches against custom protein databases from "ecdysozoans" (nematodes, arthropods, and allies; to identify genes from this superphylum) and bacteria (to identify possible endosymbiont genes). In each BLAST result cell, one can quickly review the bit score and E-value to identify hits. Thus there are no hits to bacteria, but all five sequences have matches in the *B. malayi* sequence dataset.

Sequence name (in this case a *Litomosoides sigmodontis* nematode EST dataset)	Hit to database 1 (cognate ESTs using BLASTN)	Hit to database 2 (related nematode using TBLASTX)	Hit to database 3 (nr protein using BLASTX)	Hit to database 4 (proteins of arthropods and allies using BLASTX)	Hit to database 5 (bacterial proteins using BLASTX)
Sequence name	Litomosoidestblastx	wgs_bmalayitblastx	allblastx	ecdysozoa_problastx	bacteria_problastx
Ls_AM_01A02	Ls_AM_01B08 Ls_AM_0100002_125 1e-31	203595 3 1819 bases 177 4e-45	F55C5.1 CE11150 status:Predicted TR:Q20815 protein_id:CAB01566.1 135 2e-31	gi|7504268|pir|T22709 hypothetical protein F55C5.1 - Caenorhabd... 135 7e-32	gi|22124824|ref| NP_668247.1| phosphatidylserine synthase [Yersin... 30 2.6
Ls_AM_01A03	Ls_AM_01B08 Ls_AM_0100002_317 3e-98	213753 43 7391 bases 164 2e-50	F55C5.1 CE11150 status:Predicted TR:Q20815 protein_id:CAB01566.1 91 5e-18	gi|23168989|gb| AAN08879.1| MSP-domain protein 1 [Ascaris suum] 123 4e-28	gi|23000872|gb| ZP_00044791.1| hypothetical protein [Magnetococcu... 30 3.8
Ls_AM_01A04	Ls_AM_01G07 Ls_AM_0100037_28 0.001	214121 21 4799 bases 100 2e-41	BAB27981 10 DAYS EMBRYO CDNA, RIKEN FULL-LENGTH E CLONE:2610313H... 256 4e-68	gi|7505798|pir|T34437 hypothetical protein K11H12.2 - Caenorhab... 254 1e-67	gi|23026623|gb| ZP_00065096.1| hypothetical protein [Microbulbife... 30 2.5
Ls_AM_01A05	gi|6200702|gb| AW152802.1| AW152802 JALsL3C130SAC Litomosoides_sig... 26 0.10	187013 7 1600 bases 88 5e-18	O33941 ERYBVI. Saccharopolyspora erythraea (Streptomyces erythra... 34 0.65	gi|7296815|gb| AAF52091.1| CG14656-PA [Drosophila melanogaster] 32 1.8	gi|22964849|gb| ZP_00012455.1| hypothetical protein [Rhodopseudom... 30 3.4
Ls_AM_01A06	gi|6200634|gb| AW152797.1| AW152797 JALsL3C055SAC Litomosoides_sig... 389 e-111	194774 5 1090 bases 120 9e-28	F25H2.11 CE09656 TCTP protein status:Predicted SW:Q93573 prote... 200 4e-51	gi|14700054|gb| AAK71499.1| translationally controlled tumor prot... 259 3e-69	gi|23041489|gb| ZP_00072943.1| hypothetical protein [Trichodesmiu... 29 5.5

117

3. Thus, using the mock example above, "~/bin/blast_5db.pl blastn sequences H ~/localdb/E_granulosus_nuc.fsa" (the results of the "ls" command are shown in italics below),

    ```
    cd ~/ESTproject/blast/
    ls
            proteins_1_blastx
            proteins_2_blastx
            sequences
    ~/bin/make_a_table.pl Project_name sequences
                        E_granulosus_nuc_blastn
    ```

    (This command should be entered as one line with no returns. You can enter up to five directories containing BLAST output at the end of the command. The program will return some processing comments.)

    ```
    ls
            Project_name_top_page.html
            Project_name_table1.html
            Project_name_table2.html
            proteins_1_blastx
            proteins_2_blastx
            sequences
    ```

4. Open the document "Project_name_top_page.html" in a Web browser. It will have a title and links to the table pages, which will look like **Fig. 3**.

3.5. Advanced Analyses

3.5.1. Relational Databases for Storing EST Datasets

For large-scale EST projects, it may not be appropriate to scan through many tables of html output looking for specific genes of interest. Furthermore, you may wish to undertake large-scale analyses to identify groups of sequences sharing specific criteria (e.g., similar expression profiles). To facilitate the use of larger datasets, we recommend the use of a relational database. Relational databases organize data into sets of tables related to each other through common values. Queries can then be readily formulated to extract data of interest from these tables. The more popular database schemes tend to be based on the structured query language (SQL). Many different public domain and commercial SQL database solutions exist, including mySQL (http://www.mysql.com/) and Oracle (http://www.oracle.com/). Other non-SQL database solutions can also be used to store your data (e.g., Filemaker, http://www.filemaker.com/; Microsoft Access, http://www.microsoft.com/office/access/default.asp; and ACeDB, http://www.acedb.org/). However, because of its performance and low cost (it is in the public domain), we recommend the use of postgreSQL (latest version 7.3.2). The postgreSQL website (http://www.postgresql.org/) contains full instructions on how to download and implement postgreSQL on your workstation.

Once you have successfully installed postgreSQL on your system, you will need to create a database to store your data. It is beyond the scope of this chapter to describe the full implementation of an EST project database. However, as a minimum, we recommend the use of three tables: a "cluster" table to store information on the consensus sequences and number of ESTs associated with each cluster; an "EST" table containing information on each EST (origin, sequence, etc.); and a "BLAST" table containing information extracted from the BLAST searches associated with each cluster. perl scripts can then be written using the "Pg" module (installed as standard with perl) to automatically populate the database from the previously generated flat files.

If you wish to serve the data to the wider community, then you may wish to set up a Web server. Of the many types of Web servers available, Apache (current version 2.0) (http://www.apache.org/) is free and offers a well-documented and supported solution. Html pages are relatively easy to write and, by incorporating cgi-scripts or using the embedded Web scripting language php (http://ww.php.net/), remote users can connect and query your database (for an example, *see* NEMBASE: http://www.nematodes.org/nematodeESTs/nembase.html) (*see* **Fig. 4**).

Given the wealth of sequence data that EST projects generate, it is often useful to have tools that are able to identify sequences with unique properties. Relational databases aid such analyses, and we currently implement numerous strategies to identify genes of interest. The simplest of these involves extracting the text from the definition lines of the BLAST output obtained for the clusters and using the database to search for clusters that display specific keywords (e.g., "kinase" or "cysteine protease"). Another approach involves the use of sequence similarity to identify particular gene families. This involves generating a BLAST-formatted database from the clusters and searching against a sequence of interest. Expression profiles can also be used as a method for obtaining genes of interest. If your EST dataset is derived from several libraries, it is possible to derive those clusters that have a particular pattern of expression, such as "Which clusters contain less than X ESTs from library A and more than Y ESTs from library B?"

3.5.2. SimiTri—A Viewer for Analysis of Similarity Datasets

A novel approach to identifying interesting sequences from the dataset involves the use of similarity profiles. SimiTri (available at http://www.nema-

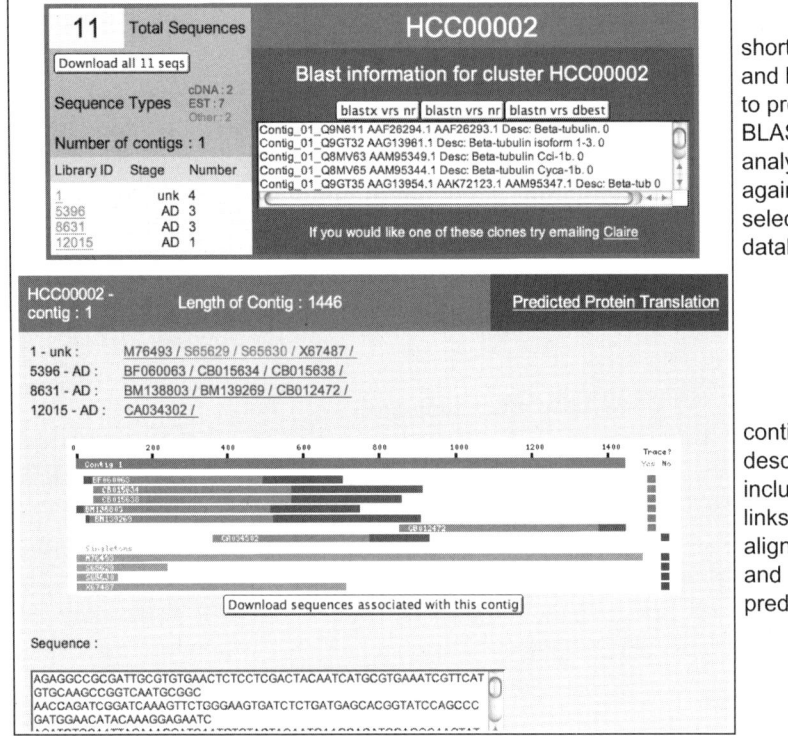

Fig. 4. NEMBASE: a relational database for EST datasets. This is a screen snapshot of one entry in a database of nematode ESTs, NEMBASE *(21)*. Each cluster of ESTs has been annotated with information such as the number and source of ESTs that it includes, the consensus sequences, and a set of precomputed blast searches against selected databases. The page also has links out to additional analyses, such as BLAST, an alignment of the ESTs, and putative protein predictions. The database is running in the background, using a relational database management software called postgreSQL, and the Web page is generated dynamically using the Web–database interface language php.

todes.org/SimiTri) is a java-based tool that was developed to display the phylogenetic profiles for a large number of clusters on one graphic *(34)*. For each cluster, the consensus sequence is BLAST-searched against three different databases. BLAST bit scores in excess of 50 (a relaxed "significance cutoff") are extracted and used to compute the relative position of the cluster within triangular phase space. SimiTri provides a graphical view of this data, enabling the identification of sequences that possess similarity profiles that deviate from the norm (suggesting an atypical mode of evolution). SimiTri is freely available from http://www.nematodes.org/SimiTri, and information on its use and implementation is found in the file SimiTri.txt.

3.5.3. Predicting Proteins From EST Consensus Sequences

Because ESTs represent the expressed portion of a genome, many of them will code for proteins. Therefore, you may wish to identify these proteins to enable further downstream analyses. The prediction of proteins from EST-derived sequence data is not a trivial exercise. The main problem arises from the fact that ESTs are unverified sequences and, therefore, are prone to base-calling and sequencing errors. This can lead to frameshift errors when attempting to predict the peptide sequence. At present, three software packages, ESTscan *(35)*, Decoder *(36)*, and DIANA-EST *(37)*, have been developed that attempt to tackle this issue. The following are our current rules for best practice.

1. If there is close homolog in the protein database, then use the prediction from the BLASTX hit as an initial starting point for the prediction.
2. If there is no homolog, then you will need to use one or more of the programs outlined above, all of which have their advantages and disadvantages. We recommend the use of DECODER, which involves the use of the phred-derived quality scores.
3. In certain cases when the predictions do not cover large portions of the sequence under consideration, you may wish to undertake a simple six-frame translation to "fill in" the missing areas.

4. Notes

1. We have described an informatic solution that relies on the open-source LINUX operating system, which is installable on most modern PC processors. LINUX includes several excellent graphical user interfaces (point-and-click), and there are many free programs available for standard computing needs such as word processing, databasing, and presentation. It is also possible to run much of the suggested software on non-LINUX platforms (Windows, MacOSX, other flavors of UNIX), but we have highlighted the LINUX platform because it has become the environment of choice for many bioinformatics researchers and thus is well supported and rapidly growing in diversity. We have recently produced an integrated version of the programs described in this chapter. Please see Parkinson, J., Anthony A., Wasmith, J., Hedley, A., and Blaxter, M. (2004) PartiGene—constructing partial genomes. *Bioinformatics,* in press.
2. As the LINUX machine will be networked, you may wish to discuss the versions of LINUX supported by your local computing resource people, and take their advice as to issues of machine purchase, adaptation, and security.
3. There are many commercial software solutions to the issues of analysis of ESTs.

4. In LINUX, the PATH environmental variable describes a list of directories that are automatically searched when you try to run a program (programs are also termed executables) from a command-line prompt. If the executable is located in a directory not listed in your PATH, then it can only be run by including the full directory path of the program. For example, if you have the executable "blastall" in a directory called "/usr/ncbi/bin" and the directory "/usr/ncbi/bin" is not in your PATH, then to run that executable you would have to type /usr/ncbi/bin/blastall at the command line prompt. With "/usr/ncbi/bin" in your PATH, this simplifies to blastall.

    The "PATH" variable can be updated by editing the ".cshrc" or ".bashrc" file in your home directory. The file that is present will depend on the login shell that you use (see the LINUX manual pages on "bash" or "tcsh" for more information). This chapter gives the full path to the executable in the examples, and we assume you have installed executables and scripts in the places recommended in **Subheading 2.2.**

5. The trace2dbest package is undergoing continual development. Hence, the operation of future releases may differ from that outlined here. Therefore, the user is asked to review the documentation that is bundled with each release (normally available as a menu option from the trace2dbest.pl executable).

6. The standard format for output of data from a sequencing instrument is the scf or standard chromatographic format. All the major instruments can output in this format.

7. Indeed, Phil Green's phred is now at the heart of many commercial solutions to the base-calling problem, and "phred scores" is a standard piece of jargon in genome sequencing worldwide.

8. Naming conventions. It is vital that you develop a robust naming protocol for your clones and sequences. We suggest the following:
    a. A unique, two-letter species code (thus, "Eg" for *Echinococcus granulosus*)
    b. A simple, unique, library identifier code (thus, "psc" for "protoscolex," or "ad1" for "adult, version 1")
    c. The unique microtiter plate address for the clone (thus, "01A03" for plate 01, column A, row 03)
    d. Letters indicating the sequencing primer used (thus, "T7" or "M13F")

    These can be combined most usefully with an underscore ("_") separating the different data fields. For example, a sequence derived using the T7 primer from clone A07 from plate 13 of an *Echinococcus granulosus* adult library (first version) could be named

    Eg_ad1_13A01_T7

    This format facilitates the use of perl scripts to extract from a sequence name relevant information, such as species, library, etc., without having to use a complex look-up table to attribute sequences to sources. The software we have used assumes that the above sequence naming scheme has been adopted. For example, trace2dbest assumes that the first two sections ("Eg_ad1" in the above example)

uniquely identify a cDNA library and the sequences derived from it. If you use a different scheme, it will be necessary to edit the perl scripts to match the patterns of names you use.
9. FASTA format is very simple: each sequence definition line starts with a ">" and is followed by free text. After the next line return, all the characters are expected to be part of the sequence, until the next line beginning with a ">" or the end of the file is reached. The sequence itself can be in lines of any length and can spread over many lines separated by line returns. For example here are some sequences from the platyhelminth *Echinococcus granulosus*. Note that the FASTA ">" header line has been used to store additional information as well as the sequence name:

>gi|28395298|gb|AY187811.1|Echinococcus granulosus Hox5 mRNA, partial cds
GAGCTGGAGAAGGAGTTCCATTTCAACCGGTACCTCACGCGTCGGCGGAGGATAGAGAT AGCCCACGCGC
TTTGCCTATCTGAGCGACAGATCAAAATCTGGTTCCAAAACAGCCG
>gi|28395296|gb|AY187810.1|Echinococcus granulosus Hox3 mRNA,partial cds
GAGTTTGAGAAGGAGTTCCACTTTAACAGGTACCTGTGCCGCCCGCGGCGCGTCGAGAT AGCCAACCTCC
TGAATCTCACCGAGCGCCAAATAAAGATCTGGTTCCAAAACCGCCG
>gi|28268125|gb|CB219933.1|CB219933 EgP 38D7 signal sequence trap (SST)
GCGATACAATTAATAAAGGGAATAGAGTGAACGTTCGGCCGGTTGATTTACAGCTGACC GCAGTCAGTAA
TCAGTACTTCTTGGGAAGTCTTGCCACTACTGCTACCGACGGCCTCCATATCCTTAACA ACATCTTCGCC
ACTTTCTACCTCACCAAAGACAACATGCTTCCCATCAAGCCAGCTGGTGACGGCGGTAG TGATGAAGAAT
TGCGAGCCATTGGTGTTCTTACCCGCATTCGCCATCGAGAGCATCATCGGCTTGCTGTG CTTGTGATTGA
AATTTTCATCCTCAAATTTGCTCCCGTATATGCTCTTGCCACCGGTACCATTCCCGGCA GTAAAATCACC
ACCTTGGCACATAAAACCGGA

10. The taxonomy ID remains constant for each taxon in the NCBI database, so you can use this number to recall the sequence set by adding "txid####[Organism]" to your query in the search bar on the NCBI home page.
11. We suggest that, for most parasites, it makes most sense to download the nucleotide dataset, as these datasets will encompass the highest diversity of genes identified. For model organisms and organisms, such as *Plasmodium falciparum*, that have had their whole genomes sequenced and annotated, it is more efficient in terms of search time to download the protein dataset. However, if you are working on a close relative of a fully sequenced organism, you should also search the

nucleotide sequence of the genome as your ESTs may correspond to genes missed in the annotation of the fully sequenced species. For the remainder of the sequenced biosphere, the nr protein dataset is the only one feasibly accessible to local search.
12. It is not necessary to download the full nr protein dataset every time you update it, as GenBank provides an update facility. *See* ftp://ftp.ncbi.nih.gov/blast/db/README for instructions.
13. A very useful tutorial on the use of BLAST, the sorts of parameter values that should and can be used, and the interpretation of BLAST outputs is available at the NCBI Web site: http://www.ncbi.nlm.nih.gov/Education/BLASTinfo/information3.html. A detailed BLAST course by S. F. Altschul, the creator of BLAST, is also available at http://www.ncbi.nlm.nih.gov/BLAST/tutorial/Altschul-1.html.
14. A complete list of possible arguments available for the blastall command is revealed by typing "`/usr/ncbi/bin/blastall -`." The current list is extensive (greater than 30 possible modifiers).

References

1. Adams, M. D., Kelley, J. M., Gocayne, J. D., et al. (1991) Complementary DNA sequencing: expressed sequence tags and the human genome project. *Science* **252,** 1651–1656.
2. McCombie, W. R., Adams, M. D., Kelley, J. M., et al. (1992) *Caenorhabditis elegans* expressed sequence tags identify gene families and potential disease gene homologues. *Nat. Genet.* **1,** 124–131.
3. El-Sayed, N. M., Alarcon, C. M., Beck, J. C., et al. (1995) cDNA expressed sequence tags of *Trypanosoma brucei rhodesiense* provide new insights into the biology of the parasite. *Mol. Biochem. Parasitol.* **73,** 75–90.
4. Wan, K.-L., Blackwell, J. M., and Ajioka, J. W. (1995) *Toxoplasma gondii* expressed sequence tags: insight into tachyzoite gene expression. *Mol. Biochem. Parasitol.* **75,** 179–186.
5. Blaxter, M. L., Raghavan, N., Ghosh, I., et al. (1996) Genes expressed in *Brugia malayi* infective third stage larvae. *Mol. Biochem. Parasitol.* **77,** 77–96.
6. Ivens, A. C. and Blackwell, J. M. (1996) Unravelling the *Leishmania* genome. *Curr. Opin. Genet. Dev.* **6,** 704–710.
7. Levick, M. P., Blackwell, J. M., Connor, V., et al. (1996) An expressed sequence tag analysis of a full length, spliced-leader cDNA library from *Leishmania major* promastigotes. *Mol. Biochem. Parasitol.* **76,** 345–348.
8. Ajioka, J. W., Boothroyd, J. C., Brunk, B. P., et al. (1998) Gene discovery by EST sequencing in *Toxoplasma gondii* reveals sequences restricted to the Apicomplexa. *Genome Res.* **8,** 18–28.
9. Djikeng, A., Agufa, C., Donelson, J. E., et al. (1998) Generation of expressed sequence tags as physical landmarks in the genome of *Trypanosoma brucei. Gene* **221,** 93–106.
10. Manger, I. D., Hehl, A., Parmley, S., et al. (1998) Expressed sequence tag analysis of the bradyzoite stage of *Toxoplasma gondii*: identification of developmentally regulated genes. *Infect. Immun.* **66,** 1632–1637.

11. Verdun, R. E., Di Paolo, N., Urmenyi, T. P., et al. (1998) Gene discovery through expressed sequence tag sequencing in *Trypanosoma cruzi. Infect. Immun.* **66,** 5393–5398.
12. Ivens, A. C. and Blackwell, J. M. (1999) The *Leishmania* genome comes of age. *Parasitol. Today* **15,** 225–231.
13. Johnston, D. A., Blaxter, M. L., Degrave, W. M., et al. (1999) Genomics and the biology of parasites. *BioEssays* **21,** 131–147.
14. Santos, T. M., Johnston, D. A., Azevedo, V., et al. (1999) Analysis of the gene expression profile of *Schistosoma mansoni cercariae* using the expressed sequence tag approach. *Mol. Biochem. Parasitol.* **103,** 79–97.
15. Urmenyi, T. P., Bonaldo, M. F., Soares, M. B., et al. (1999) Construction of a normalized cDNA library for the *Trypanosoma cruzi* genome project. *J. Eukaryot. Microbiol.* **46,** 542–544.
16. Williams, S. A. and Johnston, D. A. (1999) Helminth genome analysis: the current status of the filarial and schistosome genome projects. Filarial Genome Project. Schistosome Genome Project. *Parasitology* **118,** S19–S38.
17. Daub, J., Loukas, A., Pritchard, D. I., et al. (2000) A survey of genes expressed in adults of the human hookworm, *Necator americanus. Parasitology* **120,** 171–184.
18. McCarter, J. P., Abad, J., Jones, J. T., et al. (2000) Rapid gene discovery in plant parasitic nematodes via expressed sequence tags. *Nematology* **2,** 719–731.
19. Williams, S. A., Lizotte-Waniewski, M. R., Foster, J., et al. (2000) The filarial genome project: analysis of the nuclear, mitochondrial and endosymbiont genomes of *Brugia malayi. Int. J. Parasitol.* **30,** 411–419.
20. Degrave, W. M., Melville, S., Ivens, A., et al. (2001) Parasite genome initiatives. *Int. J. Parasitol.* **31,** 532–536.
21. Parkinson, J., Whitton, C., Guiliano, D., et al. (2001) 200,000 nematode ESTs on the net. *Trends Parasitol.* **17,** 394–396.
22. McCarter, J. P., Clifton, S. W., Bird, D. M., et al. (2002) Nematode gene sequences, Update for June 2002. *J. Nematol.* **34,** 71–74.
23. Parkinson, J., Guiliano, D., and Blaxter, M. (2002) Making sense of EST sequences by CLOBBing them. *BMC Bioinf.* **3,** 31.
24. Ewing, B. and Green, P. (1998) Base-calling of automated sequencer traces using phred. II. Error probabilities. *Genome Res.* **8,** 186–194.
25. Ewing, B., Hillier, L., Wendl, M. C., et al. (1998) Base-calling of automated sequencer traces using phred. I. Accuracy assessment. *Genome Res.* **8,** 175–185.
26. Altschul, S. F., Gish, W., Miller, W., et al. (1990) Basic local alignment search tool. *J. Mol. Biol.* **215,** 403–410.
27. Altschul, S. F., Madden, T. L., Schaffer, A. A., et al. (1997) Gapped BLAST and PSI-BLAST: a new generation of protein database search programs. *Nucleic Acids Res.* **25,** 3389–3402.
28. Boguski, M. S., Lowe, T. M., and Tolstoshev, C. M. (1993) dbEST—database for "expressed sequence tags." *Nat. Genet.* **4,** 332–333.
29. Christoffels, A., van Gelder, A., Greyling, G., et al. (2001) STACK: Sequence Tag Alignment and Consensus Knowledgebase. *Nucleic Acids Res.* **29,** 234–238.

30. Parsons, J. D., Brenner, S., and Bishop, M. J. (1992) Clustering cDNA sequences. *Comput. Appl. Biosci.* **8,** 461–466.
31. Parsons, J. D. (1995) Improved tools for DNA comparison and clustering. *Comput. Appl. Biosci.* **11,** 603–613.
32. Gordon, D., Abajian, C., and Green, P. (1998) Consed: a graphical tool for sequence finishing. *Genome Res.* **8,** 195–202.
33. Huang, X. and Madan, A. (1999) CAP3: A DNA sequence assembly program. *Genome Res.* **9,** 868–877.
34. Parkinson, J. and Blaxter, M. L. (2002) SimiTri—visualising similarity relationships for large groups of sequences. *Bioinformatics* **19,** 390–395.
35. Iseli, C., Jongeneel, C. V., and Bucher, P. (1999) in *Proc. Int. Conf. Intell. Syst. Mol. Biol.,* pp. 138-148.
36. Fukunishi, Y. and Hayashizaki, Y. (2001) Amino acid translation program for full-length cDNA sequences with frameshift errors. *Physiol. Genomics* **5,** 81–87.
37. Hatzigeorgiou, A. G., Fiziev, P., and Reczko, M. (2001) DIANA-EST: a statistical analysis. *Bioinformatics* **17,** 913–919.

6

Positive Selection Scanning of Parasite DNA Sequences

Winston A. Hide and Raphael D. Isokpehi

Summary

Parasites successfully exist within the host as a result of highly specific genetic adaptations. Therefore, detecting genes that contain relevant adaptive mutations can provide a guide to biological processes that are potentially essential to the parasite. Random genetic mutations that confer selective advantage can act to alter amino acids so as to confer gain of function that has positive impact on the survival of the parasite. Directional selection of advantageous mutations results from an elevated rate of nonsynonymous substitutions in rapidly evolving genes. Genes on which this positive selection operates are considered to have an evolutionary characteristic such that the normalized number of nonsynonymous (dn) substitutions is greater than that of synonymous (ds) substitutions. By searching in a statistically robust way for genes that contain this characteristic, it is possible to apply a stringent method to identify genes that may be under positive selection and thus to identify biological processes involving those genes that are essential to the survival of the parasite. Genes detected typically class into those under host immune surveillance and those intrinsic to pathways essential to survival of the parasite within the host. Depending on their function and location of protein expression, such genes have the potential to provide exceptional vaccine and drug candidates.

Key Words: Drug targets; parasites; pathogenicity; positive selection; vaccine candidates; virulence.

1. Introduction

Successful adaptation to the host is the key process that enables a parasite to reproduce successfully. To move from a free-living state to one within the host, or to adapt to a host niche, an organism must be able to exploit mechanisms that involve a change of function of existing genes or must gain genetic material to enable a change in function. Mechanisms that involve a change of function depend on the occurrence of random mutations of pre-existing genes, which enable a change to optimal fitness through a process of adaptation to the novel host environment. The search for genes that may have adapted or under-

From: *Methods in Molecular Biology, Vol. 270: Parasite Genomics Protocols*
Edited by: S. E. Melville © Humana Press Inc., Totowa, NJ

gone positive selection involves the detection of a signal for a higher rate of amino acid replacement or nonsynonymous substitutions (dn) compared to synonymous substitutions (ds) *(1,2)*. Thus, a dn/ds ratio > 1 is indicative of positive selection *(3)*. The difference between dn and ds can be tested for statistical significance using several statistical approaches. Here, we select the one-tail Z-test *(4,5)*. Large-scale systematic searches for genes under positive selection have previously revealed well-documented adaptive evolution cases *(6–8)*. However, sensitive methodologies that can deliver genes that appear to be under adaptive selection in the context of pathogens have yet to be detailed broadly. Nevertheless, the potential exists for such methods to provide for further study lists of gene candidates that appear to have unusually high polymorphism and that may in some cases represent adaptive selection. Interestingly, two classes of candidates tend to result from this approach: genes with products that are exposed to host immune surveillance and genes that appear to accrue gain of function mutations *(9)*. Large-scale *in silico* scanning for signals of positive selection in parasites has become more relevant with the growing availability of complete or partial genomes of medically and veterinary important parasites. Examples of available genomes include *Plasmodium falciparum*, *Plasmodium yoelii*, *Trypanosoma brucei*, and *Theileria parva* *(10–13)*. Furthermore, there is need to search for new drug targets and vaccine candidates using genetic signals encoded in the genomes.

The method for detection of genes that may be under selection requires that the genes compared be orthologous (share a recent common ancestor and are present in separate genomes) and exist in closely related species (*see* **Note 1**). The domain we have chosen for illustration of the method provides challenges to using this approach, as choosing true orthologs means that a complete genome must be available for each closely related species. Using *Plasmodium* as an example demonstrates that we can either find complete genomes for comparison, but only between pairs of genomes that are too distant for meaningful analysis, or we can use selected genes from more closely related strains or species.

Thus, the key methods presented in this chapter are identification of orthologous genes (**Fig. 1**) and positive selection scanning (**Fig. 2**). We illustrate the identification of orthologs from two genomes using the published genome sequences of a human malaria parasite, *P. falciparum* *(12)*, and a rodent malaria parasite, *P. yoelii yoelii* *(13)*. We illustrate positive selection scanning with selected sequences from published studies on natural selection in *P. falciparum* *(14)* and *P. chabaudi* *(15)*. The computational pipelines we have developed for scanning genomes are also presented.

The simple ortholog identification strategy presented is based on the reciprocal basic local alignment search tool (BLAST) best-match method, which

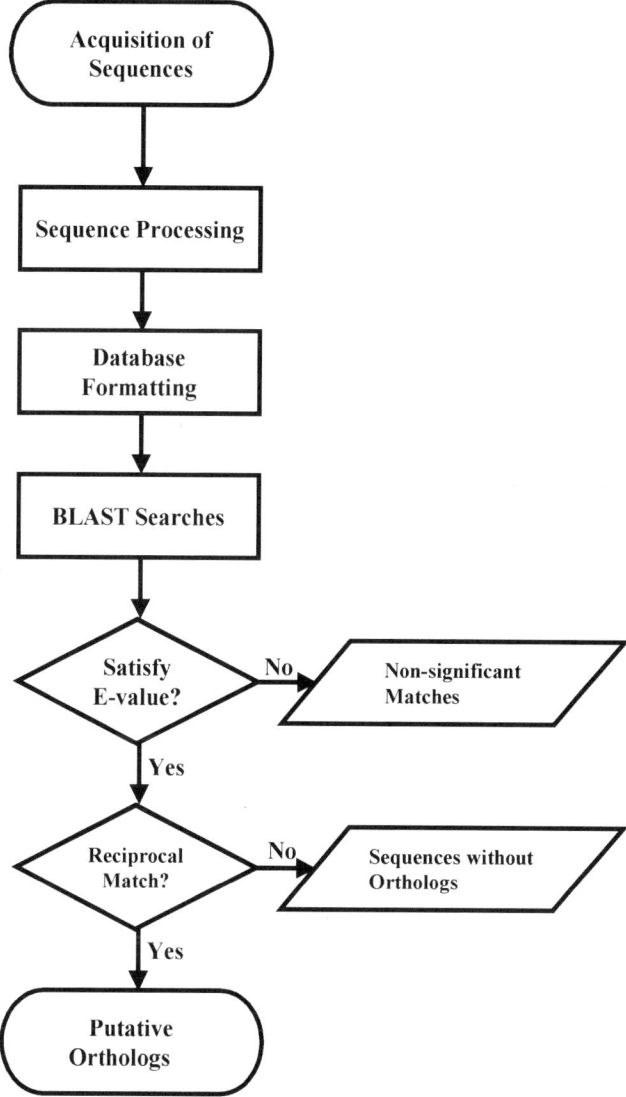

Fig. 1. Steps for identification of orthologs sequences using reciprocal BLAST best match method. See text for description of each step.

identifies two protein-coding sequences as orthologs if they mutually identify each other in BLAST searches *(16)*. Positive selection scanning starts with a codon-based alignment of the protein sequence to the coding sequence. This alignment is the input data for calculation of rates of nucleotide substitutions.

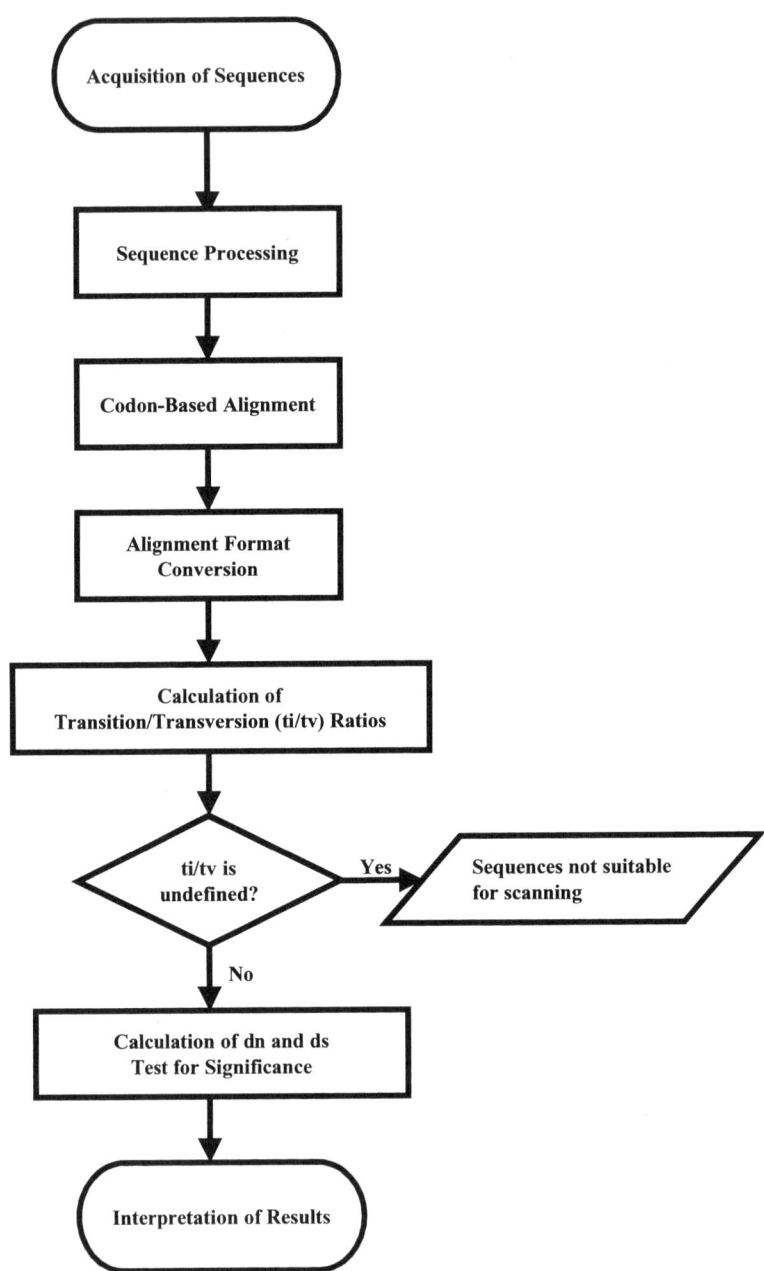

Fig. 2. Steps for positive selection scanning of orthologous protein sequences. See text for description of each step.

Table 1
Internet Locations of Software and Databases

Software/database	Location
EMBOSS	http://www.emboss.org/
CLUSTAL W	ftp://ftp.ebi.ac.uk/pub/software/unix/clustalw/
NG-NEW	http://mep.bio.psu.edu/readme.htm
READSEQ	http://iubio.bio.indiana.edu/soft/molbio/readseq/
PAUP* 4.0	http://paup.csit.fsu.edu/downl.html
PERL	http://www.perl.com/pub/a/language/info/software.html
Fisher	http://mep.bio.psu.edu/readme.htm
CodonW	http://www.molbiol.ox.ac.uk/cu/
Blastall	ftp://ftp.ncbi.nlm.nih.gov/blast/executables
Formatdb	ftp://ftp.ncbi.nlm.nih.gov/blast/executables
GenBank	http://www.ncbi.nlm.nih.gov
BLOSUM62	ftp://ftp.ncbi.nih.gov/blast/matrices
PlasmoDB	http://www.plasmodb.org

The method completes with the interpretation of results according to observed nucleotide substitution rates and calculated codon usage indices. The steps for identification of orthologs are (a) sequence retrieval and processing, (b) BLAST searches, and (c) selection of reciprocal best matches. Steps for positive selection scanning are (a) sequence retrieval and processing, (b) generation of codon-based alignment of coding sequence to aligned protein sequence, (c) calculation of nucleotide substitution rates, and (d) interpretation of results. All methods are performed using the UNIX operating system command line and are thus amenable to large-scale analysis of genomes. A key aspect of the approach is its dependence on the choice of gene regions for comparison between genomes. Are whole genes compared? Or is a window compared between sequences? How are the dimensions of such a window chosen? In this study, we will simply use the whole gene as a unit of comparison to demonstrate the most stringent application of the method.

2. Materials

The materials sections contain references to data and software. The software is entirely in the public domain, and pointers to where various of these applications can be obtained are presented in **Table 1**. Programs from the European Molecular Biology Open Software Suite (EMBOSS) are used extensively in our methodology. Processed datasets and perl scripts described here can be obtained from the SANBI FTP site: ftp://ftp.sanbi.ac.za/pub/trop_dis/selection/.

Each laboratory is likely to have its own preferences in terms of software applications. A primary problem in the area of selection scanning is that there is no dedicated package to provide consistent results, and, therefore, we have been forced to develop a series of scripts that join appropriate approaches together. By necessity, this results in numerous format changes of files between applications. As the application of the techniques become more mainstream, tools and packages will appear that provide simpler process.

2.1. Identification of Orthologs

2.1.1. Sequence Data and Processing

Protein-coding DNA and predicted protein sequences in FASTA format (concatenated sequences separated by a single header, which start with a greater than symbol [">"] followed by the sequence name and description *[17]*) from two complete genomes. We illustrate this analysis with available genomes of *P. falciparum* strain 3D7 *(12)* and *P. yoelii yoelii (13)*. Perl scripts for processing are

1. `proc_pfal.pl`
2. `proc_pyoe.pl`

2.1.2. BLAST Searches

1. NCBI formatdb
2. NCBI blastall

2.1.3. Retrieval of Reciprocal Best Matches

1. `pickbestmatch.pl`
2. `pickblastcutoff.pl`
3. `pickrecimatch.pl`

2.2. Positive Selection Scanning

2.2.1. Sequence Data Retrieval and Processing

We illustrate our pipeline for calculation of nucleotide substitution rates using pairs of complete protein-coding regions obtained from GenBank and used by published studies on natural selection in *P. falciparum* *(14)* and *P. chabaudi (15)*.

2.2.2. Codon-Based Alignment

1. CLUSTAL W
2. EMBOSS tranalign
3. `picksequences.pl`
4. `makecodonalign.pl`

2.2.3. Calculation of Nucleotide Substitution Rates

2.2.3.1. SEQUENCE FORMAT CONVERSION

1. READSEQ
2. EMBOSS seqret
3. makenexphyfiles.pl
4. makengnewfiles.pl
5. treecon2ngnew.pl
6. removegapsngnew.pl
7. makedegapngnewfiles.pl

2.2.3.2. CALCULATION OF TRANSITION TO TRANSVERSION RATIOS

1. PAUP*4.0 pairdiff
2. maketitivfiles.pl
3. picktitv1.pl
4. picktitv2.pl
5. picktitvdefined.pl

2.2.3.3. CALCULATION OF DN AND DS

1. NG-NEW
2. run-ngnew.pl
3. proc_outfile1.pl
4. proc_outfile2.pl
5. picknotsaturated.pl

2.2.3.4. CALCULATION OF DN/DS AND TEST OF SIGNIFICANCE

1. calcdndssign.pl
2. calcztest.pl

2.2.4. Interpretation of Results

1. CodonW
2. Fisher
3. makecodonwfile.pl
4. classdnds.pl
5. calcfisherinput.pl

3. Methods

The UNIX command lines and perl scripts for the steps to achieve the methods of ortholog identification and positive selection scanning are described in **Subheadings 3.1.** and **3.2.**, respectively. We recommend all analysis for each method be done in a separate file directory. However, it may be necessary to manage the contents of these directories by moving or deleting files that are no longer required in the computational pipeline. This process is simplified by the consistent filename extensions that enable UNIX special characters (wildcards) to be used.

The steps for identification of orthologs are (a) sequence retrieval and processing (b) BLAST searches, and (c) selection of reciprocal best matches. Steps for positive selection scanning are (a) sequence retrieval and processing (b) generation of codon-based alignment of coding sequence to aligned protein sequence, (c) calculation of nucleotide substitution rates, and (d) interpretation of results. We describe the steps to achieve each method using genome sequences and orthologous genes obtained from GenBank.

3.1. Identification of Orthologs

3.1.1. Sequence Retrieval and Processing

A database of protein coding and predicted protein sequences in FASTA format for *P. falciparum* and *P. yoelii yoelii* was retrieved from the download sections of http://www.plasmodb.org and http://www.tigr.org/tdb/e2k1/pya1/, respectively. The downloaded sequence data required various levels of processing for conversion to a FASTA format in which the sequence header contains the ">" symbol and the locus name of the gene (e.g., >PF08_0003 or >PY03625). perl scripts, proc_pfal.pl and proc_pyoe.pl (see command line below) were written to achieve this processing for *P. falciparum* and *P. yoelii yoelii*, respectively. These processed datasets can be obtained from ftp://ftp.sanbi.ac.za/pub/trop_dis/selection/.

The command lines for processing *P. falciparum* and *P. yoelii yoelii* sequences are

```
proc_pfal.pl pfdb.pep > dbpep1
proc_pyoe.pl pydb.pep > dbpep2
proc_pfal.pl pfdb.cds > dbcds1
proc_pyoe.pl pydb.cds > dbcds2
```

3.1.2. BLAST Searches

Identification of orthologous genes using reciprocal BLAST best match requires both databases to be formatted for use by National Center for Biotechnology Information (NCBI) blastall *(18)*. This is achieved using the NCBI formatdb program. Each database is then queried with the other database using the BLASTP implementation with BLOSUM62 matrix for scoring the protein alignments *(19)*. The file containing this matrix must be in the same directory for the BLAST search.

1. Format the sequence database as BLAST database.

Positive Selection Scanning

```
formatdb -i dbpep1 -p T -l dbpep1.log
formatdb -i dbpep2 -p T -l dbpep2.log
```

2. Perform BLAST searches.

```
blastall -p blastp -d dbpep1 -i dbpep2 -o
  dbpep2_v_dbpep1 -a 2 -v 3 -b 3
blastall -p·blastp -d dbpep2 -i dbpep1 -o
  dbpep1_v_dbpep2 -a 2 -v 3 -b 3
```

3.1.3. Retrieval of Reciprocal Best Matches

The best match for each query sequence is retrieved from the BLAST output file and checked to determine if it satisfies the expectation (E) value threshold. The E-value, which is dependent on the size of the database and length of the query sequence, corresponds to the number of sequences that is expected to occur in a database of random sequences. An estimate of the threshold can be obtained from a histogram that shows the frequency of negative log of the E-values *(20)*. In this example, we use a threshold of 10^{-15}, following the estimate of Carlton et al. *(13)*. The best matches for each database search are compared to determine if they meet the reciprocal best-match criterion. The command lines to achieve these steps described are outlined below. A comparative analysis of results obtained with that of **ref. 13** is presented in **Note 2**.

1. Select best match from BLAST output file.

```
pickbestmatch.pl dbpep1_v_dbpep2 >
  dbpep1_v_dbpep2.bestmatch
pickbestmatch.pl dbpep2_v_dbpep1 >
  dbpep2_v_dbpep1.bestmatch
```

2. Select matches that satisfy threshold. The script `pickblastcutoff.pl` can either retrieve matches below (significant) or above (not significant) the threshold by including 1 or 2, respectively, in the command line.

```
pickblastcutoff.pl dbpep1_v_dbpep2.bestmatch e-15 1 >
  dbpep1_v_dbpep2.cutoff1
pickblastcutoff.pl dbpep2_v_dbpep1.bestmatch e-15 1 >
  dbpep2_v_dbpep1.cutoff1
pickblastcutoff.pl dbpep1_v_dbpep2.bestmatch e-15 2 >
  dbpep1_v_dbpep2.cutoff2
pickblastcutoff.pl dbpep2_v_dbpep1.bestmatch e-15 2 >
  dbpep2_v_dbpep1.cutoff2
```

3. Select matches that satisfy threshold.

```
pickrecimatch.pl dbpep1_v_dbpep2.cutoff1
  dbpep2_v_dbpep1.cutoff1 > db1db2.ortholist
```

Table 2
**Pairs of Orthologous Genes
of *Plasmodium* Species Used for Positive Selection Scanning**

Organism and Gene	Accession number	
	seq1	seq2
Plasmodium falciparum		
Apical Membrane Antigen-1 (PfAMA-1)	M27133	U65407
	AF061332	U65407
	AF061332	M27133
Sexual-stage-specific surface antigen 48/45 (Pfs48/45)	Z22145	AF206630
Ookinete surface protein 25 (Pfs25)	AF179423	AF154117
Plasmodium chabaudi		
Merozoite surface protein 4/5 (PcMSP4/5)	AF080447	AF080446
Adenylosuccinate lyase (PcASL)	AF080447	AF080446

3.2. Positive Selection Scanning

3.2.1. Sequence Data Retrieval and Processing

Pairs of orthologous protein-coding DNA sequences and corresponding predicted protein sequences described previously *(15,16)* were retrieved using the gene accession numbers from GenBank and saved as GenBank format in different files (*see* **Note 3**). The organisms and accession numbers of gene pairs are shown in **Table 2**. The coding sequence and corresponding predicted protein in FASTA format are retrieved from the files using EMBOSS coderet (*see* **Note 4**). The sequence header of the coding sequence and its protein sequence are modified to provide unique and identical headers using EMBOSS descseq. The text of the headers should not contain spaces or nonalphanumeric characters. The products of this method are two files (DNA and protein) containing two sequences. There may be need to perform pairwise comparison for three or more strains or field isolates of a gene (e.g., PfAMA-1, **Table 2**). The number of combinations (NC) is the sum of integers from 0 to (NC–1). For example, the number of combinations for five strains is sum of integers 0 to 4: 0 + 1 + 2 + 3 + 4 = 10. A perl script, generate_comb.pl, which receives as input a file with a list of the sequence names, is used to generate the combinations for pairwise comparisons.

1. Retrieve coding and protein sequence from files containing orthologous genes (seq1.gb and seq2.gb).

    ```
    coderet -auto -seqall seq1.gb seq_ant -notranslation
     - nomrna
    ```

```
coderet -auto -seqall seq2.gb seq_bnt -notranslation
   -nomrna
coderet -auto -seqall seq1.gb seq_apt -nocds -nomrna
coderet -auto -seqall seq2.gb seq_bpt -nocds -nomrna
```

2. Change description of header line and create a new file.

   ```
   descseq -auto -seq seq_ant -out seq_ant.new -name "seq1"
   descseq -auto -seq seq_bnt -out seq_bnt.new -name "seq2"
   descseq -auto -seq seq_apt -out seq_apt.new -name "seq1"
   descseq -auto -seq seq_bpt -out seq_bpt.new -name "seq2"
   ```

3. Merge protein sequences into one file.

   ```
   cat seq_?pt.new > seq12.pep
   ```

4. Merge protein-coding sequences into one file.

   ```
   cat seq_?nt.new > seq12.cds
   ```

3.2.2. Codon-Based Alignment

Protein sequences are aligned using CLUSTAL W with alignment output specified as "GDE" format and sequence characters in uppercase. GDE stands for Genetic Data Environment and is a legacy format used by a still freely available X-window-based editor (http://www.bioafrica.net/GDElinux/index.html). The sequence headers in the GDE alignment file created by CLUSTAL W start with the percent symbol (%). This symbol is replaced by the "greater than" symbol (>) to convert the sequence format to FASTA. The replacement is achieved using the *sed* program of UNIX, and a new file is created. The codon-based alignment of the nucleotide sequences to this protein alignment is generated with EMBOSS tranalign and serves as input alignment for subsequent methods (**Fig. 3**). A second alignment generated with CLUSTAL W but using the default output in CLUSTAL format is used to determine the nature of amino acid replacements (*see* **Note 5** and **Fig. 4**).

Steps for an ortholog pair:

1. Perform sequence alignment of protein sequences using CLUSTAL W.

   ```
   clustalw seq12.pep /OUTPUT=GDE /CASE=UPPER
   clustalw seq12.pep /CASE=UPPER /SEQNOS=ON
   ```

2. Replace "%" of sequence header with ">."

   ```
   sed 's/%/>/g' seq12.gde > seq12.fasta
   ```

3. Perform codon-based alignment of coding sequences to aligned protein sequences.

   ```
   tranalign -auto seq12.cds seq12.fasta seq12.ptnt
   ```

   Steps for multiple ortholog pairs:

```
>seq1
atgaagatcgcaaattatttatcagcaattaatctctttgttgttttgttaatgtcattt
aaaaatgcacatgatacttctttaattaatacaaacaaatgtgcacataatatggtgggt
aataacagaattttaggaaatatccctctgataatacagaagatagttcgaatcaagtt
gaaggtaatgctgattcaaacgaagttactagcgaagaagctaaaaacacatcaaatcag
ggaaatggcgaaatgcaaatcaagctggaactccagctgaacccgctaaagctgacgct
actcaaaatgcagcagcacagccagataacacaaatggtactggaagtaaaactaccgac
gcaaatgttacccttctttctagtgaagaagacgacgaagaagaagatgatgataccgaa
tatgaagaaggcaattgtgaagtgaataatggaggatgtggaccaaatctaaatgtgaa
aaaatgccaaatggcgctataaaatgctcatgcccatctggatttaaattacagggtact
gattgtatcgcattgttcagtgcagattcagtgaattacagttttttgcggaattataatg
gttgctattgctataatagcattattatat
>seq2
atgaagatcgcaaattatttatcagcaattaatctctttgttgttttgttaatgtcattt
aaaaatgcacatgatacttctctaattaatacaaacaaatgtgcgcataatatggtgggt
aataacagaattttaggaaatatccctctgataatacagaagatagttcgaatcaagtt
gaaggtaatgctgattcaaacgaagttactagcgaagaagctaaaaacacatcaaatcag
gcaaatggcgaaactgcaaatcaagctggaactccagctgaacccgctaaagctgacgct
actcaaaattcaccagcacagccagataacacaaatggtactgaaagtaaaactacaggc
gaaaaagctacccttctttctagtgaagaagacgacgaagaagaagatgatgatgacgaa
cttgaagacggcgattgttcagtggataatggaggatgtggagaaaatctaatatgtgaa
aaaatggaaatggtagaataaaatgctcatgcccatctggatataaattacatggtact
gattgtatcgcattgctcagtgcagattcagtgaattacagttttttgcggaattataatg
gttgctattgctataatagcattattatat
```

Fig. 3. Codon-based alignment of protein-coding sequence to aligned protein sequence of MSP4/5 genes of *P. chabaudi chabaudi* 96V (seq1) and *P. chabaudi adami* DS (seq2) and generated by EMBOSS tranalign.

1. Retrieve coding sequence of the orthologues from database and save in separate files.

   `picksequences.pl nt db1db2.ortholist dbcds1 dbcds2`

2. Retrieve protein sequence of the orthologs from database and save in separate files.

   `picksequences.pl pt db1db2.ortholist dbpep1 dbpep2`

3. Obtain a list of all files storing protein sequences.

   `ls -1 *.pt > ptfilelist`

4. Generate files with codon-based alignments.

   `makecodonalign.pl ptfilelist`

3.2.3. Calculation of Nucleotide Substitution Rates

This step consists of (a) conversion of codon-based alignment to formats that can be used by the programs PAUP*4.0 and NG-NEW, (b) calculation of

```
seq1  MKIANYLSAINLFVVLLMSFKNAHDTSLINTNKCAHNMVGNNRILGNIPSDNTEDSSNQV  60
seq2  MKIANYLSAINLFVVLLMSFKNAHDTSLINTNKCAHNMVGNNRILGNIPSDNTEDSSNQV  60
      ************************************************************

seq1  EGNADSNEVTSEEAKNTSNQGNGENANQAGTPAEPAKADATQNAAAQPDNTNGTGSKTTD  120
seq2  EGNADSNEVTSEEAKNTSNQANGETANQAGTPAEPAKADATQNSPAQPDNTNGTESKTTG  120
      ********************.***.******************: .********* ****.

seq1  ANVTLLSSEEDDEEEDDDTEYEEGNCEVNNGGCGPNLKCEKMPNGAIKCSCPSGFKLQGT  180
seq2  EKATLLSSEEDDEEEDDDDELEDGDCSVDNGGCGENLICEKMENGRIKCSCPSGYKLHGT  180
      :.************** * *:*:*.*:***** ** **** ** ********:**:**

seq1  DCIALFSADSVNYSFCGIIMVAIAIIALLY  210
seq2  DCIALLSADSVNYSFCGIIMVAIAIIALLY  210
      *****:************************
```

Fig. 4. CLUSTAL W (version 1.82) alignment of MSP4/5 predicted protein sequences of *P. chabaudi chabaudi* 96V (seq1) and *P. c. adami* DS (seq2). The alignment of positions are classified by symbols under the sequences into no replacement (*), conservative replacement (:, .) and radical replacement denoted by a gap According to **ref. 26**, ":" indicates that one of the following "strong" groups is fully conserved: STA NEQK NHQK NDEQ QHRK MILV MILF HY FYW; "." indicates that one of the following "weaker" groups is fully conserved: CSA ATV SAG STNK STPA SGND SNDEQK (*see* **Note 5**).

rate of transitional nucleotide change (ti) (transitions) to transversional nucleotide change (tv) (transversions) ratio, (c) calculation of dn and ds, and (d) calculation of dn/ds ratio and test for significance.

3.2.3.1. CONVERSION OF CODON-BASED ALIGNMENT
TO NEXUS FORMAT AND NG-NEW FORMAT

The NEXUS file format is required by PAUP*4.0's pairdiff program. This program is used to calculate the transition to transversion (ti/tv) ratio. The alignment generated in **Subheading 3.2.2.** is converted to NEXUS format using READSEQ. The file format required by the program NG-NEW *(5)* is a modified TREECON format. We refer to this format as NG-NEW format. Conversion to NG-NEW format is achieved by first converting the codon-based alignment to PHYLIP format with READSEQ and then converting the PHYLIP format to TREECON format with EMBOSS seqret. The NG-NEW format requires the first line of the file to start with the number of sequences, followed by four spaces and the sequence length. TREECON format has only the alignment length. Thus, a perl script, `treecon2ngnew.pl`, has been developed to capture the alignment length and replace this first line with the number of sequences (2), four spaces, and the alignment length to yield the NG-NEW format.

Steps for an ortholog pair:

1. Convert codon-based alignment to NEXUS and PHYLIP formats.

   ```
   readseq seq12.ptnt -a -f17 -output=seq12.nexus
   readseq seq12.ptnt -a -f12 -output=seq12.phylip
   ```

2. Convert PHYLIP format to TREECON format.

   ```
   seqret -auto seq12.phylip seq12.treecon -osf treecon
   ```

3. Convert TREECON format to NG-NEW format.

   ```
   treecon2ngnew.pl seq12.treecon seq12.ngnew1
   ```

4. Remove gaps in NG-NEW format alignment.

   ```
   removegapsngnew.pl seq12.ngnew1 > seq12.ngnew2
   ```

Steps for multiple ortholog pairs:

1. Obtain a list of all files storing codon-based alignments of sequences.

   ```
   ls -1 *.ptnt > ptntfilelist
   ```

2. Generate NEXUS and PHYLIP formats of codon-based alignments.

   ```
   makenexphyfiles.pl ptntfilelist
   ```

3. Generate NG-NEW format for codon-based alignments.

   ```
   ls -1 *.phylip > phylipfilelist
   makengnewfiles.pl phylipfilelist
   ```

4. Remove gaps in NG-NEW format alignment.

   ```
   ls -1 *.ngnew1 > ngnew1filelist
   makedegapngnewfiles.pl ngnew1filelist
   ```

3.2.3.2. CALCULATION OF TRANSITION TO TRANSVERSION RATIOS

NG-NEW takes into account the rate of ti to tv to calculate dn and ds, because synonymous mutations are more often transitions than transversions *(21)*. The Nei-Gojobori method *(1)* gives biased estimates of dn and ds when the ti/tv is greater than 0.5, because it underestimates the number of synonymous sites (S) and overestimates the number of nonsynonymous sites (N) *(5)*. The pairdiff program of PAUP*4.0 is used to calculate the pairwise difference between aligned sequences. The *long* format output of pairdiff is used to facilitate extraction and parsing of the output using perl scripts, picktitv1.pl (multiple ortholog pairs) and picktitv2.pl (ortholog pair). Both scripts retrieve the calculated ti/tv ratio before adjusting for gaps. If there are no sites differing (i.e., 0 for transitions and transversions) or zero transversions, the ti/tv ratio is undefined. The ortholog pair should not be evaluated further, as the two protein-coding sequences are identical. Thus, a perl script, picktitvdefined.pl is applied to the output of picktitv1.pl to yield a list of orthologs suitable for calculation of nucleotide substitution rates.

Steps for an ortholog pair:

1. Copy file pairdiff.cmd to the current directory. This file has three lines; the first and third are blank, whereas the second line has the following command:

    pairdiff /shortfmt=no, longfmt=yes;

2. Run pairdiff command on the alignment in NEXUS format.

    paup -n -u seq12.nexus -1 seq12.titv1 < pairdiff.cmd

3. Retrieve ti/tv ratio from PAUP pairdiff output file (seq12.titv1) to seq12.titv2.

    picktitv2.pl seq12.titv1 > seq12.titv2

Steps for multiple ortholog pairs:

1. Copy file pairdiff.cmd to the current directory. This file has three lines; the first and third are blank, whereas the second line has the following:

    pairdiff /shortfmt=no, longfmt=yes;

2. Calculate ti/tv ratio.

    ls -1 *.nexus > nexusfilelist
    maketitvfiles.pl nexusfilelist

3. Retrieve ti/tv ratios from files.

    ls *.titv > titvfilelist
    picktitv1.pl titvfilelist > db1db2.titv1
    picktitvdefined.pl db1db2.titv1 > db1db2.titv2

3.2.3.3. CALCULATION OF DN AND DS

Codon-based alignment in NG-NEW format and the ti/tv ratio are used as input to the command line of NG-NEW. A modification to the NG-NEW source code is performed to allow for the ti/tv ratio to be included in the command line argument with the sequence alignment file name. This step enables automated processing. If the ti/tv ratio is not included, the program prompts the user to specify or accept the default ti/tv ratio of 0.5 (5). The output files created by NG-NEW are renamed.

We have observed in our implementation of NG-NEW that the program fails to generate an output for alignments greater than 3999 bp. Thus, in the pipeline for multiple ortholog pairs, we check with a script, pickngnewlist.pl, the length of the codon-based alignment (NG-NEW format without gaps) generated in **Subheading 3.2.3.1.** using the output file from **Subheading 3.2.3.2.** as input file. This yields a list for the calculation of dn and ds. The command line for this step is

```
pickngnewlist.pl db1db2.titv2 > db1db2.ngnewlist
```

The following estimates in the "outfile" generated by NG-NEW are useful for interpreting the evidence of positive selection:

a. Number of nonsynonymous sites (N)
b. Number of synonymous sites (S)
c. Number of nonsynonymous differences (n)
d. Number of synonymous differences (s)
e. Jukes-Cantor distances of nonsynonymous differences (dn)
f. Standard errors of dn (dnSE)
g. Jukes-Cantor distances of synonymous difference (ds)
h. Standard errors of ds (dsSE)

When dn or ds is reported as nan (not a number), it means the Jukes-Cantor corrected distances cannot be calculated because the value of proportion of nonsynonymous differences or proportion of synonymous differences is greater than or equal to 0.75. In this case, the dn/ds ratio is not calculated, as there is excess accumulation of point substitutions (mutational saturation) in the aligned sequences. Instances in which dn or ds is greater than 1 are also excluded from further analysis because of mutational saturation.

Steps for an ortholog pair:

1. Calculate dn and ds.

   ```
   ng-new seq12.ngnew2 < seq12.titv2
   mv outfile seq12.outfile
   ```

   Use unix remove file (rm) command to remove other files: n.dis, s.dis, and sn.rst.

2. Retrieve estimates from "outfile."

   ```
   proc_outfile2.pl seq12.outfile > seq12.dnds1
   ```

3. Retrieve orthologs that satisfy criteria for calculating dn/ds ratio.

   ```
   picknotsaturated.pl seq12.dnds1 > seq12.dnds2
   ```

Steps for multiple ortholog pairs:

1. Calculate dn and ds.

   ```
   run_ngnew.pl db1db2.ngnewlist
   ```

2. Retrieve estimates from "outfile."

   ```
   ls -1 *.outfile > outfilefilelist
   proc_outfile1.pl outfilefilelist > db1db2.dnds1
   ```

3. Retrieve orthologs that satisfy criteria for calculating dn/ds ratio.

   ```
   picknotsaturated.pl db1db2.dnds1 > db1db2.dnds2
   ```

3.2.3.4. CALCULATION OF DN/DS RATIO AND TEST OF SIGNIFICANCE

After estimating dn and ds, the next steps are calculation of the ratio of dn to ds and test for significance of difference between dn and ds. The output from **Subheading 3.2.3.3.** provide the input for scripts (*see* below) that extract the dn, dnSE, ds, dsSE, then calculates dn/ds and tests for significance of difference between dn and ds by testing if the value of the test statistic (Z-score) is greater than 1.645 and $p < 0.05$ in one-tail Z-test *(4,5)*. A perl script, calcztest, accepts dn, dnSE, ds, and dsSE on the command line and calculates the above estimates. This last script is useful as a Z-score significance calculator.

1. Command line for multiple ortholog pairs:

   ```
   calcdndssign.pl db1db2.dnds2 > db1db2.ztest
   ```

2. Command line for an ortholog pair:

   ```
   calcdndssign.pl seq12.dnds2 > seq12.ztest
   ```

3. Command line for Z-test calculator:

   ```
   calcztest <dn> <dnSE> <ds> <dsSE>
   ```

3.3. Interpretation of Results

Results are interpreted based on dn/ds as (*see* **Note 6**):

1. dn/ds < 1; purifying selection
2. dn/ds = 1; neutral evolution
3. dn/ds > 1 and dn–ds is not significant ($p < 0.05$); candidate gene under positive selection
4. dn/ds > 1 and dn–ds is significant ($p < 0.05$); candidate gene under positive selection

Command lines for retrieving these results from those generated in **Subheading 3.2.3.4.** are:

```
classdnds.pl db1db2.ztest 1 > db1db2.dnds_lt1
classdnds.pl db1db2.ztest 2 > db1db2.dnds_eq1
classdnds.pl db1db2.ztest 3 > db1db2.dnds_gt1notsign
classdnds.pl db1db2.ztest 4 > db1db2.dnds_gt1sign
```

In the case of dn/ds > 1, confounding factors that can affect nucleotide substitutions must be tested. Thus, CodonW, a program for calculating codon usage indices, is implemented with the command line below to calculate guanine + cytosine (GC) content, GC content at the third codon position (GC3), and a measure of codon usage bias (effective number of codons, ENc) using the codon-based alignment file (*see* **Note 7**).

1. Command line for multiple ortholog pairs:

   ```
   makecodonwfile.pl ptntfilelist
   ```

2. Command line for an ortholog pair:

   ```
   codonw seq12.ptnt seq12.codon -noblk -nomenu -nowarn
       -silent -gc -gc3s -sil_base -enc
   ```

To provide additional statistical support to guide further analysis as well as when numbers of nonsynonymous (n) and synonymous (s) differences is less than 10 (e.g., Pfs4845, **Table 3**), the Fisher exact test for positive selection *(5)* implemented with program Fisher is performed. The values of N, S, n, and s are extracted from NG-NEW outfile (`seq12.outfile`) and converted to integers. The input values (which have to be entered manually during the program run) and their coordinates (row, column) for the 2 × 2 table are n (1, 1) and s (1, 2), N − n (2, 1), and S − s (2, 2). A summary of the input values, dn/ds ratios, and statistical significance for the pairs of orthologous genes retrieved from GenBank is presented in **Table 3**.

1. Command line for multiple ortholog pairs:

   ```
   calcfisherinput.pl db1db2.dnds2 > db1db2.fisvals
   ```

2. Command line for orthologs retrieved from GenBank:

   ```
   calcfisherinput.pl seq12.dnds2 > seq12.fisvals
   ```

4. Notes

1. Orthologous genes share a recent common ancestor (speciation) and are present in separate genomes. The detection of signal of positive selection from two protein-coding sequences is dependent on comparing true orthologs as opposed to comparing paralogs, which are products of gene duplication and have a higher rate of evolution when compared with the constant rate observed in orthologs *(14)*. The signal for positive selection cannot be confidently resolved between distantly related orthologs because of possible multiple substitutions at nucleotide sites (mutational saturation) during their evolution

2. Carlton et al. *(13)* identified 3310 orthologs from 5878 *P. yoelii yoelii* and 5268 *P. falciparum* complete genes as well as a set of 713 pairs of genes that did not satisfy the reciprocal BLAST best match. Our implementation of this method using 5334 *P. falciparum* and 7861 *P. yoelii yoelli* protein sequences yielded 3746 orthologs. Of these, 3096 overlapped with the previously defined orthologs *(13)*. Furthermore, 414 orthologs from our set matched the dataset without reciprocal BLAST best match evidence (**Fig. 5**). This comparative analysis yielded 236 putative orthologs not present in the two datasets of **ref. *13***. The excess is the result of

Table 3
Output Values and Statistical Significance for Parasite Sequences[a]

Gene and sequence pair	N	S	n	s	ti/tv[b]	dn	dnSE	ds	dnSE	dn/ds[c]	Z-score[d]	Fisher test[e]
PfAMA-1: M27133 × U65407	1440	425	10	1	0.8333	0.0070	0.0024	0.0024	0.0024	2.92	1.36	0.24
PfAMA-1: AF061332 × U65407	1463	396	30	2	0.6000	0.0208	0.0040	0.0051	0.0031	4.08	3.10*	0.02
PfAMA-1: AF061332 × M27133	1478	381	26	1	0.5000	0.0178	0.0038	0.0026	0.0019	6.85	3.58*	0.02
Pfs48/45: Z22145 × AF206630	1063	280	3	0	0.5000	0.0028	0.0016	0.0000	0.0000	—	ND	0.49
Pfs25: AF179423 × AF154117	477	173	0	0	—	ND	ND	ND	ND	ND	ND	1.00
PcMSP4/5: AF080466 × AF080447	490	139	28	4	0.5714	0.0601	0.0136	0.0343	0.0155	1.75	1.25	0.12
PcASL: AF080466 × AF080447	984	389	0	16	3.0000	0.0000	0.0000	0.0421	0.0106	0.00	-3.97	1.00

[a]For description of abbreviations of column headings, see **Subheading 3.2.3.3.**
[b]—, ti/tv ratio is undefined, as ti and tv equals 0.
[c]—, dn/ds ratio is not calculated, as ds equals 0.
[d]Asterix (*) indicate dn–ds is significant at $p < 0.05$.
[e]Bold values indicate Fisher exact test is significant at $p < 0.05$ (one-tail).
ND, not determined.

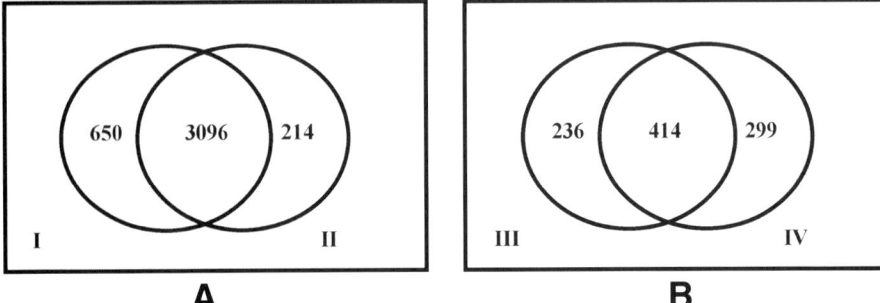

Fig. 5. Comparative analysis of detected *P. falciparum* and *P. yoelii yoelii* orthologs. **(A)** I (this study): 3746 orthologs using 7861 *P. yoelii yoelii* and 5334 *P. falciparum* protein sequences. II (Carlton et al. *[13]*): 3310 orthologs using 5878 *P. yoelii yoelii* and 5268 *P. falciparum* sequences. **(B)** III (this study): 650 orthologs not found in **ref.** *13* ortholog dataset. IV: 713 Pairs from **ref.** *13* that did not satisfy criterion of reciprocal best match.

inclusion of partial sequences in our *P. yoelii yoelii* dataset. Lists of putative orthologs from this study and a processed list from **ref.** *13* are available at ftp://ftp.sanbi.ac.za/trop_dis/selection. We recommend, when possible, to obtain additional evidence of orthology using methods such as protein domain organization, phylogenetics, and cluster analysis *(22–24)*. Thus, we extracted orthologs classified as enzymes using the Gene Ontology (GO) annotation of the *P. falciparum* gene available at the PlasmoDB Web site. We worked with enzymes in the dataset, as they are likely to be well-conserved. We obtained 10 genes, of which 8 showed additional evidence as orthologs (**Table 4**) based on identical annotations and/or protein domain, obtained from the Web pages of the ortholog of each species. Protein domains present in a sequence can be determined using InterProScan, which is available as a Web-based implementation (http://www.ebi.ac.uk/interpro) and a standalone perl-based package *(25)*. This software queries INTERPRO, an integrated database of protein families, domains, and functional sites.

3. If sequences are not retrieved from GenBank, the method starts from **step 2**. It is important that files containing the DNA and protein sequences are saved with names used in **step 2**; seq_ant, seq_bnt, seq_apt, and seq_bpt. Other methods for sequence retrieval include from (1) a file containing the sequence, (2) a file containing more than one sequence, and (3) the GenBank/European Molecular Biology Laboratory (EMBL) database stored on a local computer. EMBOSS seqret can be used for these alternatives. Generic commands that can be customized for these alternatives are listed below. A variation of alternative (1) is the use of EMBOSS nthseq (*see* **Note 4**).

```
seqret <file> [seq_ant|seq_bnt|seq_apt|seq_bpt]
seqret <file:entry> [seq_ant|seq_bnt|seq_apt|seq_bpt]
seqret <database:entry> [seq_ant|seq_bnt|seq_apt|seq_bpt]
```

Table 4
New Putative Orthologous Enzymes Identified
in Genomes of *P. falciparum* and *P. yoelii yoelii*

Locus name		Protein description (PlasmoDB)
P. falciparum	*P. yoelii yoelii*	
MAL13P1.335	PY06104	phosphatidylserine synthase i; putative
MAL6P1.182	PY06379	*N*-acetylglucosaminyl- phosphatidylinositol de-n-acetylase, putative
PF13_0109	PY07448	*N*2,*N*2-dimethylguanosine tRNA methyltransferase, putative
PFB0355c	PY00291	cysteine protease, putative
PFC0060c	PY03326	serine/threonine protein kinase, putative
PFD0965w	PY06820	hypothetical protein
PFE0170c	PY04701	protein kinase, putative
PFI0415c	PY05388	ribosomal RNA methyltransferase, putative

Abbreviations: RNA, ribonucleic acid; tRNA, transfer RNA.

4. The two genes of *P. chabaudi* analyzed here are stored under one GenBank accession number in both strains. Retrieval of coding sequence and protein sequence for each gene involves, first, retrieval of all the coding and protein sequences, followed by retrieval of the specific sequences with EMBOSS nthseq. The number of genes stored in AF080446 and AF080447 are two and three, respectively. In AF080446, the MSP4/5 is the first gene, whereas ASL is the second gene. In the case of AF080447, MSP4/5 is the second gene, whereas ASL is the third gene. The exact order of genes is how they are stored in the output of coderet and required by nthseq for retrieval of the sequences. Thus, a modified **step 1** of **Subheading 3.2.1.** for MSP4/5 is outlined below.

```
coderet -auto -seqall seq1.gb seq1.nt -notranslation
   -nomrna
coderet -auto -seqall seq2.gb seq2.nt -notranslation
   -nomrna
coderet -auto -seqall seq1.gb seq1.pt -nocds -nomrna
coderet -auto -seqall seq2.gb seq2.pt -nocds -nomrna
nthseq -auto seq1.nt -number 2 -outseq seq_ant
nthseq -auto seq2.nt -number 1 -outseq seq_bnt
nthseq -auto seq1.pt -number 2 -outseq seq_apt
nthseq -auto seq2.pt -number 1 -outseq seq_bpt
```

5. Positions with conservative and radical replacements *(26)* can be useful in interpreting the biological roles of specific regions under selection and can be correlated

Table 5
Codon Usage Bias and GC Content of Two *P. chabaudi* Genes

Organism and sequence name[a]	ENc	GC3	GC
PcMSP4/5			
seq1	41.72	0.180	0.356
seq2	36.18	0.176	0.357
PcASL			
seq1	37.60	0.129	0.261
seq2	36.30	0.116	0.256

[a]seq1, *P. chabaudi chabaudi* 96V; seq2, *P. c. adami* DS.
ENc, effective number of codons; GC3, GC content at third codon position; GC, GC content.

with studies on site-directed mutagenesis, protein domains, or structure *(9)*. In the case of the *P. chabaudi* MSP4/5 sequences, 8 of the 22 nonsynonymous substitutions are in the epidermal growth factor (EGF)-like domain (residues 145–183), consistent with observations in *(15)*. The cysteine residues responsible for disulfide bond formation are conserved in both strains; however, amino acid replacements within the domain, when correlated with other studies, may explain difference in virulence or ability to elicit immune response of the two strains.

6. Genes with dn/ds < 1 are candidate genes subject to functional constraints. These may participate in basic biological processes common to the two species compared. Genes with dn/ds > 1 but for which dn–ds is not significant ($p < 0.05$) can be studied with methods that scan for regions within a gene that may be under selection *(3)*.

7. GC3 and codon usage bias of the gene can reduce ds, resulting in an elevated dn/ds ratio *(27)*. These values should be compared with other genes (such as "housekeeping" genes) that are known to be under functional constraint and not likely to be subject to positive selection. This is illustrated in **Table 5** with PcMSP4/5 and the housekeeping gene PcASL, an enzyme involved in the purine salvage pathway and located adjacent to PcMSP4/5. Both genes in the two strains have an ENc that indicates absence of nucleotide bias, because values range from 20 (greatest bias) to 61 (least bias). Furthermore, consistent with observations in *(15)*, GC3 bias (AT-richness at the third codon position) cannot account for elevated dn/ds since GC3 of PcASL in both strains is lower than that of PcMSP4/5, the gene with evidence for selection.

In conclusion, positive selection scanning can serve as a rapid scan of genomes to prioritize genes for further studies and, in the case of tropical disease parasites, accelerate efforts toward disease control.

Acknowledgments

The authors are grateful to Cathal Seoighe and Junaid Gamieldien for useful suggestions and Paul Husler for help in modifying the source code of NG-NEW.

R. D. Isokpehi thanks the National Research Foundation (South Africa) and The Claude Harris Leon Foundation for postdoctoral fellowships.

References

1. Nei, M. and Gojobori, T. (1986) Simple methods for estimating the numbers of synonymous and nonsynonymous nucleotide substitutions. *Mol. Biol. Evol.* **3**, 418–426.
2. Hughes, A. L. (1992) Positive selection and intra-allelic recombination at the merozoite surface antigen-1 (MSA-1) locus of *Plasmodium falciparum*. *Mol. Biol. Evol.* **9**, 381–393.
3. Yang, Z. and Bielawski, J. P. (2000) Statistical methods for detecting molecular adaptation. *Trends Ecol. Evol.* **15**, 496–503.
4. Zhang, J., Kumar, S., and Nei, M. (1997) Small-sample tests of episodic adaptive evolution: a case study of primate lysozymes. *Mol. Biol. Evol.* **14**, 1335–1338.
5. Zhang, J., Rosenberg, H. F., and Nei, M. (1998) Positive Darwinian selection after gene duplication in primate ribonuclease genes. *Proc. Natl. Acad. Sci. USA* **95**, 3708–3713.
6. Endo, T., Ikeo, K., and Gojobori, T. (1996) Large-scale search for genes on which positive selection may operate. *Mol. Biol. Evol.* **13**, 685–690.
7. Benner, S. A., Chamberlin, S. G., Liberles, D. A., et al. (2000) Functional inferences from reconstructed evolutionary biology involving rectified databases—an evolutionarily grounded approach to functional genomics. *Res. Microbiol.* **151**, 97–106.
8. Liberles, D. A., Schreiber, D. R., Govindarajan, S., et al. (2001) The adaptive evolution database (TAED). *Genome Biol.* **2**, RESEARCH0028.
9. Davids, W., Gamieldien, J., Liberles, D. A., et al. (2002) Positive selection scanning reveals decoupling of enzymatic activities of carbamoyl phosphate synthetase in *Helicobacter pylori*. *J. Mol. Evol.* **54**, 458–464.
10. Coppel, R. L. (2001) Bioinformatics and the malaria genome: facilitating access and exploitation of sequence information. *Mol. Biochem. Parasitol.* **118**, 139–145.
11. Prichard, R. and Tait, A (2001) The role of molecular biology in veterinary parasitology. *Vet. Parasitol.* **98**, 169–194.
12. Gardner, M. J., Hall, N., Fung, E., et al. (2002) Genome sequence of the human malaria parasite *Plasmodium falciparum*. *Nature* **419**, 498–511.
13. Carlton, J. M., Angiuoli, S. V., Suh, B. B., et al. (2002) Genome sequence and comparative analysis of the model rodent malaria parasite *Plasmodium yoelii yoelii*. *Nature* **419**, 512–519.
14. Escalante, A. A., Lal, A. A., and Ayala, F. (1998) Genetic polymorphism and natural selection in the malaria parasite *Plasmodium falciparum*. *Genetics* **149**, 189–202.
15. Black, C. G. and Coppel, R. L. (2000) Synonymous and nonsynonymous mutations in a region of the *Plasmodium chabaudi* genome and evidence for selection acting on a malaria vaccine candidate. *Mol. Biochem. Parasitol.* **111**, 447–451.

16. Tatusov, R. L., Galperin, M. Y., Natale, D. A., et al. (2000) The COG database: a tool for genome-scale analysis of protein functions and evolution. *Nucleic Acids Res.* **28,** 33–36.
17. Pearson, W. R. (1990) Rapid and sensitive comparison with FASTA and FASTP. *Methods Enzymol.* **183,** 63–68.
18. Altschul, S. F., Madden. T. L., Schaffer, A. A., et al. (1997) Gapped BLAST and PSI-BLAST: a new generation of protein database search programs. *Nucleic Acids Res.* **25,** 338–3402.
19. Heinkoff, S. and Heinkoff, C. (1993) Performance evaluation of amino acid substitution matrices. *Proteins* **17,** 49–61.
20. Clarke, B., Lambrecht, M., and Rhee, S. Y. (2003) *Arabidopsis* genomic information for interpreting wheat EST sequences. *Funct. Integr. Genomics* **3,** 33–38.
21. Fay, J. C. and Wu, C.-I. (2003) Sequence divergence, functional constraint and selection in protein evolution. *Annu. Rev. Genomics Hum. Genet.* **4,** 213–235.
22. Remm, M., Storm, C. V., and Sonnhammer, E. L. L. (2001) Automatic clustering of orthologs and in-paralogs from pairwise species comparisons. *J. Mol. Biol.* **314,** 1041–1052.
23. Eisen, J. A. and Wu, M. (2002) Phylogenetic analysis and gene functional predictions: phylogenomics in action. *Theor. Popul. Biol.* **61,** 481–487.
24. Storm, C. E. and Sonnhammer, E. L. (2002) Automated ortholog inference from phylogenetic trees and calculation of orthology reliability. *Bioinformatics* **18,** 92–99.
25. Zdobnov, E. M. and Apweiler, R. (2001) InterProScan—an integration platform for the signature-recognition methods in InterPro. *Bioinformatics* **17,** 847–848.
26. Thompson, J. D., Gibson, T. J., Plewniak, F., et al. (1997) The CLUSTAL_X windows interface: flexible strategies for multiple sequence alignment aided by quality analysis tools. *Nucleic Acids Res.* **25,** 4876–4882.
27. Bielawski, J. P., Dunn, K. A., and Yang, Z. (2000) Rates of nucleotide substitution and mammalian nuclear gene evolution: approximate and maximum-likelihood methods lead to different conclusions. *Genetics* **156,** 1299–1308.

7

RACE and RAGE Cloning in Parasitic Microbial Eukaryotes

Bryony A. P. Williams and Robert P. Hirt

Summary

Many gene-cloning strategies and gene survey often provide partial sequence data. To exploit the information from these partial sequences numerous PCR-based approaches have been developed to clone full-length open reading frames. These approaches can be successful using small quantities of cDNA or genomic DNA as starting material and avoid the need to go through the complex and tedious process of constructing and screening gene libraries. Here we present two of these approaches, called RACE and RAGE, we used to successfully clone partial and full-length ORFs from amitochondriate parasitic microbial eukaryotes. The RACE approach uses cDNA as template for PCR cloning whereas RAGE uses genomic DNA. These two approaches were used to complement each other to provide full-length genes. The amitochondriate microbial eukaryotes we are investigating are of interest from both evolutionary and biomedical perspectives. We have investigated genes of mitochondrial origins in the obligate intracellular parasite called microsporidia. In these organisms spores are the only source of material that can be isolated from host cells and typically yield small amount of mRNA and genomic DNA for cloning. A full-length mitochondrial Hsp70 could be cloned and sequenced and specific antibody raised against a fusion protein. The highly specific antibody allowed us to demonstrate for the first time the presence of mitochondrial-like organelles in microsporidia.

Key Words: RAGE; RACE; *Trachipleistophora hominis*; *Trichomonas vaginalis*; mitochondria; amitochondriate eukaryotes; Hsp70; molecular evolution; phylogeny; gene family; DNA and RNA purification.

1. Introduction

Expressed sequence tags (ESTs), genome sequence surveys (GSSs) and many gene cloning strategies frequently result in partial sequence data. Fortunately, numerous polymerase chain reaction (PCR)-based approaches have been developed to clone full-length open reading frames (ORFs) based on the information from partial sequences. These include methods such as rapid amplification of cDNA ends (RACE), rapid amplification of genomic ends (RAGE), and inverse PCR *(1–3)*. Such approaches can be successful using small quantities of complementary DNA (cDNA) or genomic DNA as starting material and avoid the need to go through the complex and tedious process of constructing and screening gene libraries. A major advantage of these methods is that they use less starting material than the more demanding traditional cloning strategies, and this is particularly relevant for work with unicellular organisms in which RNA and DNA may be in short supply. We have used both the RACE *(2)* (**Fig. 1**) and the RAGE *(4)* (**Fig. 1**) approaches to clone partial and full-length ORFs from amitochondriate parasitic microbial eukaryotes. A RAGE cloning approach was used to complement a RACE cloning strategy when this did not allow us to obtain a complete ORF with flanking regions and, hence, infer the likely starting codon of a gene of interest.

The amitochondriate microbial eukaryotes we are investigating are of interest from both evolutionary and biomedical perspectives *(5–7)*. We describe below gene cloning results from work with two human parasites that were formally members of the Archezoa kingdom *(7)*. The kingdom Archezoa was created to encompass a heterogeneous grouping of amitochondriate microbial eukaryotes, which were hypothesized to represent early branching and primitively

---

Fig. 1 (*Opposite page*) Overview of RACE and RAGE cloning strategies. (**A**) Schematic organization of a protein-coding gene (genomic DNA: gDNA) with the ORF (boxed), the start codon (ATG), and the stop codon (TAA). The position of the transcription initiation site is also indicated (thick vertical bar). Both RACE and RAGE cloning strategies depend on pre-existing sequence data (in the region of the gray box), or conserved domains, which allow the design of both forward (F) and reverse (R) primers for use in PCR amplifications. Restriction enzyme sites (RS, thin vertical bar) for a given enzyme on both sides of the start codon are required for RAGE cloning. In many cases, obtaining the 5'-end of a cDNA can be fraught with difficulties and, hence, a RAGE cloning approach can represent an alternative solution to obtain the missing sequence data. RAGE and RACE PCR products are shown below the gene in dark and light gray boxes, respectively. The mRNA and examples of complete or partial cDNAs are shown below the gene: (1) a complete mRNA molecule with (2) the corresponding cDNA that would be produced in an ideal case, however, less ideal partial cDNA sequences are often produced (3) that lack the start codon. This may be the result of mRNA degradation, or secondary structures that inhibit cDNA synthesis. (**B**) The major steps are indicated for the RACE and RAGE approaches, from nucleic acid extraction to cloning. Both cloning approaches can focus on either the 5'- or the 3'-end of an ORF. The time frame for the different steps is also shown. The protocols presented in this chapter describe the steps undertaken on days 1 and 2, i.e., the production of RACE and RAGE fragments for subsequent cloning and sequencing.

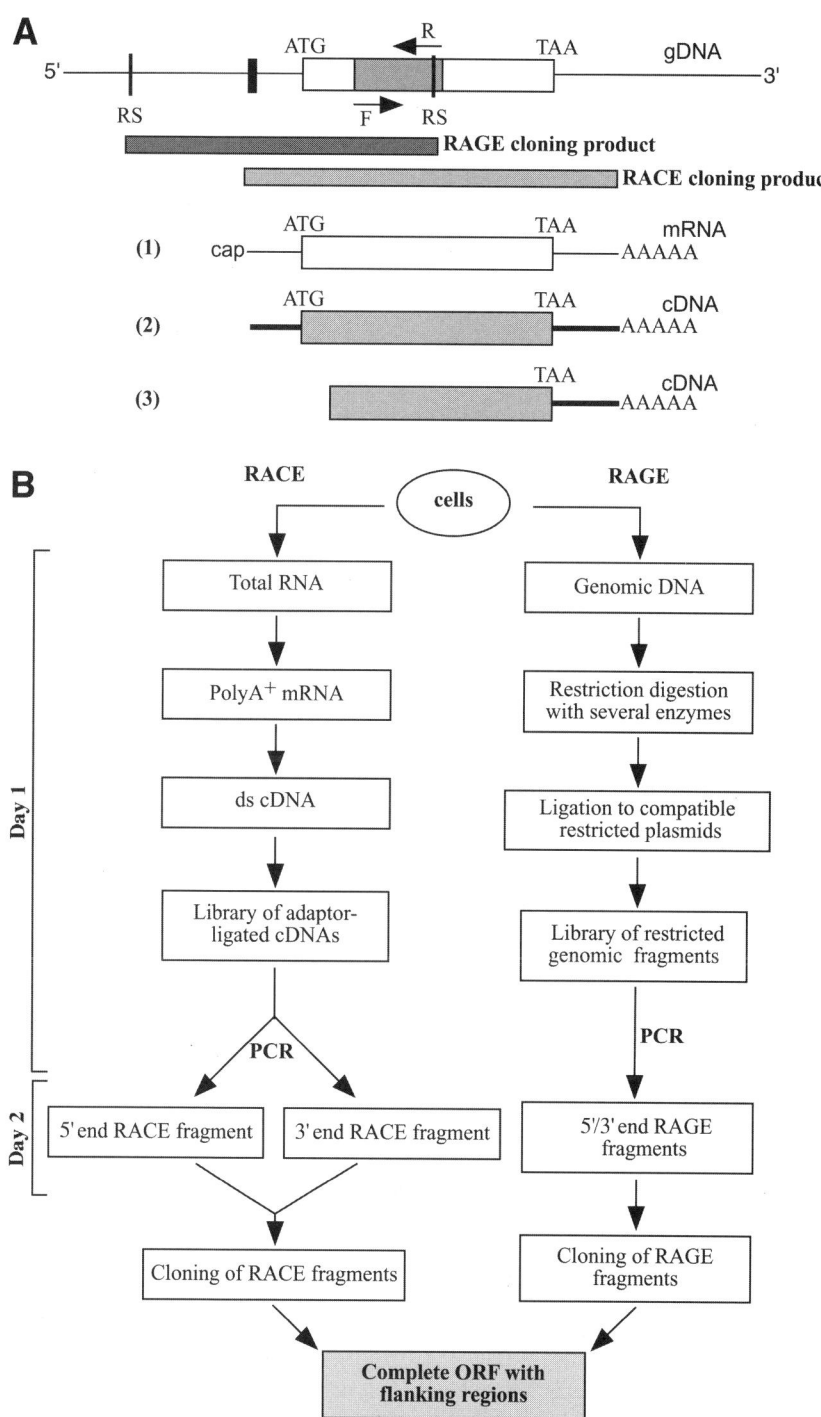

mitochondriate eukaryotes. They would have diverged from the main eukaryotic trunk prior to the acquisition of the bacterial endosymbiont that gave rise to mitochondria *(7)*. Data from several laboratories have shown clearly that none of the known eukaryotes are Archezoa as defined originally *(5,6)*. All extant investigated eukaryotes possess genes that suggest the past presence of mitochondria, and it is possible that all possess some form of mitochondria that may currently lie cryptic and undiscovered within their cells. In addition, some former Archezoa do not represent early taxa at all and are, in fact, sister taxa to organisms containing mitochondria *(5,6)*.

Trachipleistophora hominis is a member of the Microspora phylum *(8)*. Microspora, or microsporidia, are a large group of obligate intracellular parasites infecting all animal phyla and some protists. *T. hominis* was isolated from patients with acquired immunodeficiency syndrome (AIDS) and can be grown in a rabbit kidney cell line, making it a tractable experimental system *(9,10)*. *Trichomonas vaginalis* is a member of the Parabasalia taxon, a diverse group of flagellated protozoa that are mainly parasitic or symbionts of animals *(11)*. *T. vaginalis* is the most common nonviral sexually transmitted pathogen in humans, and it can be grown in vitro *(12)*. *T. vaginalis* and *T. hominis* are used in our laboratory as model systems to investigate the molecular and cellular evolution of amitochondriate microbial eukaryotes *(5)*. These organisms are of particular interest to study mitochondrial-derived organelles and gene function and evolution *(13–16)*. They do not possess organelles with the enzymatic machinery for the tricarboxylic acid (TCA) cycle and adenosine triphosphate (ATP) production based on oxidative phosphorylation, but they possess mitochondrial-derived organelles whose function and structure are being actively investigated *(14,16–19)*. In addition, little is known about the molecular and cellular organization of their secretory and endocytic pathways, despite the importance of these cellular processes in the pathogenicity of these parasites.

We have used RACE and RAGE cloning approaches to investigate particular genes of potential mitochondrial origin, or involved in membrane trafficking, to obtain molecular cues to study the evolution and cellular organization of these parasites. Cloning strategies in these organisms are simplified by the absence of introns (when using genomic DNA as starting material) in most of their protein-coding genes but is complicated by typically short 5'-untranslated regions (5'-UTR) *(20)*. This can make the identification of starting codons difficult from cDNA sequences.

We describe later the protocols we used to clone a gene encoding a mitochondrial Hsp70 (mitHsp70) from the microsporidian *T. hominis* using a combination of RACE and RAGE methods. The cloned gene was subsequently used for the expression in *Escherichia coli* of a recombinant protein that allowed the production of a highly specific polyclonal antibody. This specific antiserum allowed us to investigate mitHsp70 protein expression and its cellular localization, demonstrating at the cellular level the presence of mitochondrial-like organelles in microsporidia *(13)*. We also outline a RACE cloning approach that we have used to investigate the diversity of a small GTP-binding protein family in *T. vaginalis* and that could be applied to the cloning of most other protein families.

2. Materials

2.1. Preparation of DNA, Total RNA, and Messenger RNA (mRNA)

2.1.1. Extraction of T. hominis DNA (see **Note 1**)

1. Saturated phenol solution, pH 7.6 (BDH Laboratory Supplies, cat. no. 4367546).
2. Chloroform (BDH Laboratory Supplies, cat. no. 10077W).
3. Ribonuclease A (RNAase A) from bovine pancreas (Sigma, cat. no. 83831).
4. Microcon 100 filtration columns (Millipore, cat. no. UFC7PCR50).
5. Tris-EDTA (TE) (10:1), pH 7.5. 10 mM Tris-HCl, 1 mM EDTA, pH 8.0, pH adjusted to 7.5 with 1 M HCl.
6. 10% Sodium dodecyl sulfate (SDS) solution (Sigma, cat. no. 05030).
7. 0.10- to 0.11-mm diameter glass beads (Braun Biotech International, cat. no. 854-140/0).
8. Mikrodismembrator U. B. Bead beater (Braun Biotech International, type 853172/2).

2.1.2. Extraction of T. hominis Total RNA (see **Note 1**)

1. Lysis buffer: 62.5 g guanidine thiocyanate, 4.4 mL of 0.75 M sodium citrate, 30% (w/v), 2.2 mL Sarkosyl, 77.5 mL DEPC-dH$_2$O.
2. Saturated phenol solution, pH 6.6 (BDH Laboratory Supplies, cat. no. 4367546).
3. Chloroform (BDH Laboratory Supplies, cat. no. 10077WF).
4. Agarose gel electrophoresis equipment and 10-mg/mL ethidium bromide stock.
5. DEPC-dH$_2$O.
6. Phosphate-buffered saline (PBS), 10X concentrate (Sigma, cat. no. P7059).
7. Agarose (Sigma, cat. no. A4804).
8. 1X TBE: 10.8 g Trizma base, 5 g boric acid, 4 mL of 0.5 M EDTA, distilled H$_2$O to 1 liter.

2.1.3. mRNA Extraction

1. Dynabeads mRNA purification kit (Dynal, cat. no. 610.06).

2.2. RACE of the T. hominis Hsp70 Gene

1. Marathon cDNA amplification kit (Clontech, cat. no. K 1802-1).
2. PLATINUM *Taq* DNA polymerase (Invitrogen, cat. no. 10966).
3. Gene-specific oligonucleotide primers (*see* **Note 2**).
4. Microcon 100 filtration columns (Millipore, cat. no. UFC7PCR50).

2.3. 5'-RAGE of the T. hominis *mitHsp70* Gene

2.3.1. Preparation of Vector

1. Hind III restriction endonuclease and appropriate 10X buffer (New England Biolabs, cat. no. R0104T).
2. pPCR-Script Amp cloning kit (Stratagene, cat. no. 211188).
3. Shrimp alkaline phosphatase and appropriate 10X buffer (Promega, cat. no. 9PIM 820).
4. Microcon 100 filtration columns (Millipore, cat. no. UFC7PCR50).
5. T4 DNA ligase (Stratagene, cat. no. 600011).

2.3.2. Preparation of the T. hominis DNA Amplification Product

1. Hind III restriction endonuclease and appropriate 10X buffer (New England Biolabs, cat. no. R0104T).
2. *Taq* polymerase, 5 U/μL (Bioline, cat. no. M95801B), supplied with 10X NH_4 reaction buffer and 50 mM $MgCl_2$.
3. Gene- and vector-specific primers (*see* **Note 2**).
4. Microcon 100 filtration columns (Millipore, cat. no. UFC7PCR50).

2.4. General: PCR, Sequencing, and Cloning

1. *Taq* polymerase, 5 U/μL (Bioline, cat. no. M95801B), supplied with 10X NH_4 reaction buffer and 50 mM $MgCl_2$.
2. BSA, 10 mg/mL (Promega, cat. no. R3961).
3. dNTP 400 mM, (dATP, dGTP, dCTP, and dTTP) (Invitrogen, cat. no. 10297-018).
4. Centriflex gel filtration cartridges (Edge Biosystems, cat. no. 42453).
5. QIAquick gel extraction kit (Qiagen, cat. no. 28706).
6. pGEM-T-EASY vector system (Promega, cat. no. A1360).
7. T4 DNA ligase (Stratagene, cat. no. 600011).
8. Maximum-efficiency DH5α-competent *E. coli* cells (Life Technologies, cat. no. 18258012).
9. X-gal (5-bromo-4-chloro-3-indolyl-β-D-galactopyranoside), 50 mg/mL (Promega, cat. no. V3941).
10. IPTG (isopropyl β-D-thiogalactopyranoside), 100 mM (Sigma, cat. no. I5502).
11. Ampicilin sodium salt (Sigma, cat. no. A2804).
12. Material and solution to grow DH5α *E. coli*.
13. Material and solution for standard TBE agarose gel, including a gel imaging system, e.g., UVP imager (Ultra Violet Products, Cambridge, UK).
14. Hyperladder DNA ladder (Bioline, cat. no. HYPL1200).
15. BigDye Terminator v 3.0 sequencing standard (ABI PRISM, cat. no. 4390303).

3. Methods

Here, we describe how to perform extraction of genomic DNA and RNA from the eukaryotic organisms used in our research. We then give detailed protocols for the generation of cDNA from mRNA and the amplification of the ends of selected cDNAs. This approach is complemented by the description of a method for the amplification of the ends of genomic DNA fragments, using a *T. hominis* gene as an example. Amplification products may be cloned and sequenced as appropriate.

3.1. Preparation of Genomic DNA, Total RNA, and mRNA

3.1.1. Extraction of T. hominis DNA

T. hominis cells are maintained in cultured rabbit kidney cells and spores purified from these cultures, as previously described *(10)*. Purified spores are used as starting material for both mRNA and genomic DNA purification. Because spores are surrounded by a thick spore wall and, therefore, are difficult to break, we use bead beating (glass beads are added to the sample, which is shaken at high speed) to disrupt the cells prior to nucleic acid extraction.

1. Centrifuge purified spores at 3500g for 5 min (we used between 1.8×10^8 and 2.25×10^8 spores per extraction).
2. Discard supernatant and resuspend spores in 1 mL TE 10:1, pH 7.5.
3. Transfer spore suspension to a sterile 2-mL screw-cap tube and centrifuge at 3500g for 5 min. Discard the supernatant.
4. Add 400 µL TE 10:1, 10 µL 10% SDS, 2 µL of RNAase A, and 400 µL of 0.10- to 0.11-mm diameter glass beads to the tube. Bead-beat (as described above) for 45 s at 2000 shakes per minute.
5. Add 800 µL of phenol, pH 8.0, to the bead-beaten cells, mix by inversion, and centrifuge at 10,000g for 10 min.
6. Recover the supernatant and add to it 400 µL of chloroform. Centrifuge at 10,000g for 10 min.
7. Recover the supernatant and concentrate the samples using microcon 100 cartridges, according to manufacturer's instructions. Test a small aliquot for presence of DNA using an agarose gel. This material can be used for either PCR or the RAGE protocol described in **Subheading 3.3.**

3.1.2. Extraction of T. hominis Total RNA

To prepare cDNA we first need to purify total RNA, from which mRNA is purified. The latter is used for cDNA synthesis that will be the template DNA for the RACE PCR reaction.

1. Centrifuge culture medium of infected cells with abundant spores for 10 min at 200g so that spores and other cells form a pellet. Wash twice with double-distilled sterile H$_2$O and resuspend in PBS and use directly, or store at –80°C until RNA extraction.
2. Wash spores with RNAase-free sodium phosphate (20 mM sodium phosphate at pH 8) and then bead-beat in 400 µL of lysis buffer for 1 min at 2000 shakes per minute. Centrifuge sample at 10,000g for 10 min on a benchtop centrifuge.

3. Remove supernatant and rinse beads with 300 µL of RNase-free 20 mM sodium phosphate at pH 8.
4. Bead-beat spores again and centrifuge (as in **step 2**) and pool the supernatants.
5. Add 700 µL of phenol, pH 6.6, to the pooled supernatant and invert to mix the two solutions. Centrifuge for 10 min at 10,000g and retain the aqueous layer.
6. To this, add 700 µL of phenol–chloroform–isoamyl alcohol (25:24:1 v/v) and centrifuge the mixture at 10,000g for 5 min.
7. Keep the aqueous layer and add 700 µL of chloroform–isoamyl alcohol (24:1 v/v), mix, and then centrifuge at 10,000g for 5 min.
8. Precipitate the RNA by adding 2.5 vol of 100% ethanol to the extract and placing it at –20°C for 1 h. Then centrifuge this at 10,000g for 30 min at 4°C.
9. Wash the pellet with 70% (v/v) ethanol and resuspend in 50 µL TE 10:0.1 pH 7.5 (or DEPC-dH$_2$O).
10. Run a small aliquot on a 1.2% 1X TBE-agarose gel to check for the presence of RNA (**Fig. 2A**).
11. Preferably, one should proceed immediately to the next step to reduce chances of RNA degradation. Alternatively, store at –80°C until use.

Fig. 2. (*Opposite page*) Schematic representation of two RACE cloning strategies. cDNA are shown in black thick lines, mRNA as a thin line. Gray lines correspond to the adaptors (from the kit) ligated to the double-stranded cDNA. Horizontal arrows are the primers used for PCR (**A**) or cDNA synthesis and PCR (**B**). (**A**) The first approach corresponds to the methodology implemented by the Marathon cDNA amplification kit. This approach has the advantage of allowing both 5'- and 3'-RACE cloning from the same cDNA preparation. It can lead to the cloning of a complete ORF if the mRNA transcripts are not degraded during their purification, and if cDNA synthesis reaches the 5'-end. If the mRNA is characterized by a long 5'-UTR, then it makes the identification of the start codon more likely. As *T. hominis* and *T. vaginalis* both have short 5'-UTR, and *Trichomonas* has high levels of nuclease activity, many RACE clones do not contain the start codon. (**B**) A second approach, as implemented by the Invitrogen Life Technologies 5'-RACE kit, increases the chance of cloning the 5'-end by focusing the cDNA synthesis on the particular gene of interest by using a GSP (GSP-R) to initiate the cDNA synthesis. Performing shortened cDNA synthesis also reduces the likelihood of inhibition of cDNA synthesis because of potential severe secondary structures along the mRNA. A primary PCR is performed with an anchor primer and a GSP (GSP-R2). A secondary PCR typically is needed to obtain a PCR product (using GSP-R3). It is useful to sequence several clones to obtain as much sequence data as possible from the 5'-end. This approach was used successfully to clone several full-length ORFs in *T. vaginalis* (*29,30*).

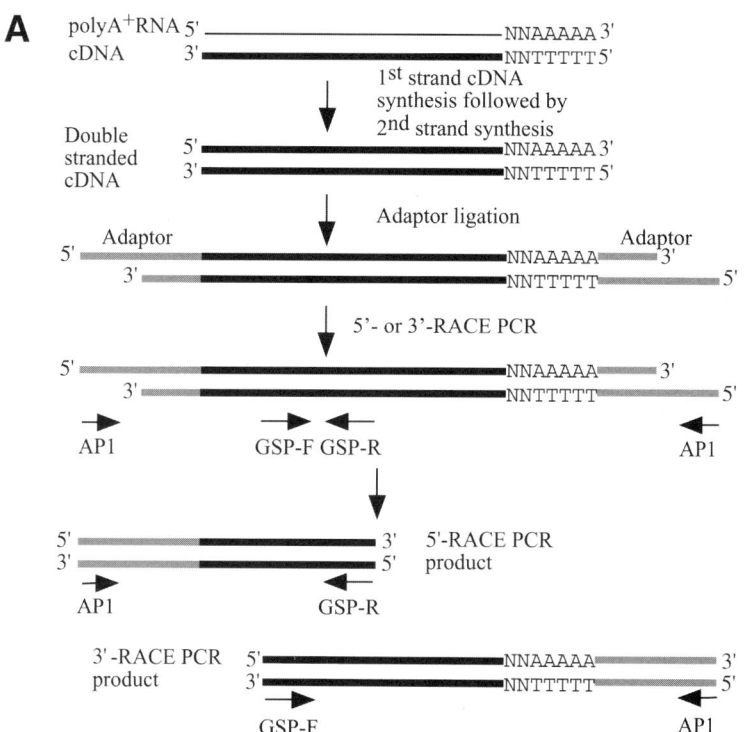
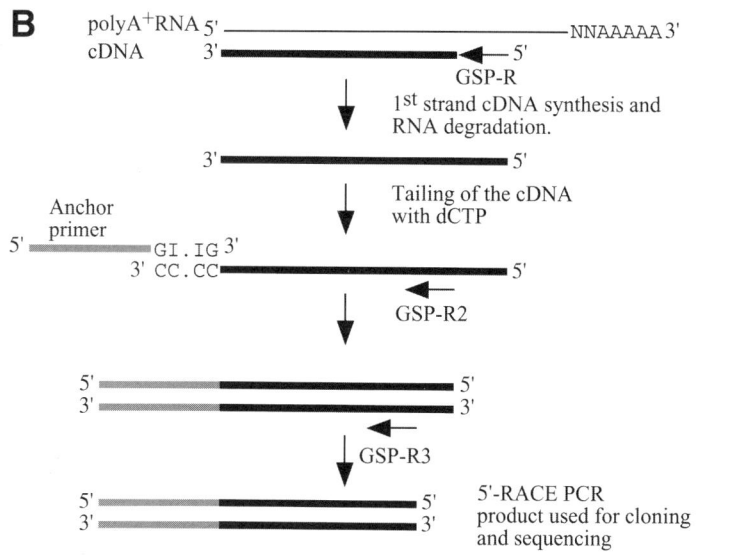

3.1.3. Extraction of mRNA

1. Extract poly-A$^+$ mRNA from total RNA using Dynal Dynabeads® following the manufacturer's protocol (*see* **Note 3**).
2. Proceed directly to the cDNA synthesis step described below to avoid mRNA degradation.

3.2. RACE of the T. hominis *Hsp70* Gene

To clone a mitHsp70 gene, we used the Clontech Marathon cDNA amplification kit for RACE cloning, following carefully the manufacturer's protocol, allowing both 5'- and 3'-RACE from the same cDNA preparation (**Fig. 2A**). In brief, this involves the following steps:

1. To generate the cDNA, reverse-transcribe the mRNA using an oligo d(T) primer (provided in the kit), which binds to the polyA tail of the 3'-end. The second strand is produced from this according to the Gubler and Hoffman method *(21)*.
2. Ligate the adapters from the kit by blunt-end ligation to each end of the cDNAs using the protocol provided by the manufacturer. The library of adaptor-ligated cDNA is now ready for RACE PCRs (**Fig. 2A**).
3. Design gene-specific primers (*see* **Note 2**) from previously sequenced areas of the gene that you are attempting to amplify. Design a nested primer to use in case no product is amplified using your first PCR primer. In our work, gene-specific primers were designed to a fragment of the *T. hominis* mitHsp70 gene, obtained by PCR from genomic DNA, using degenerate primers as described previously *(22)*. The gene-specific primers (GSPs) were ThHsp70-421F (GSPF1), ThHsp70-542F (GSPF2), ThHsp70-864F (GSPF3) for the 3'-RACE, and ThHsp70-652R (GSPR1) and ThHsp70-258R (GSPR2) for the 5'-RACE (**Figs. 3** and **5** for primer positions in relation to the ORF). Use these primers in combination with the primer AP1, complementary to the adaptor, provided in the kit.
4. If no PCR product can be detected in an agarose gel from the first amplification, then it may be necessary to carry out a nested PCR, which requires the design of two additional GSPs used with the AP2 primer (provided with the kit), an "internal" (in relation to AP1) adaptor-specific primer. For cloning of the *T. hominis* mitHsp70 a two-step amplification procedure was needed, because both primary 5'- and 3'-RACE PCRs (GSPF/Rs and AP1) did not yield detectable PCR products (**Fig. 3B**). A second round of successful nested 5'- and 3'-RACE PCR was carried out using the "internal" GSPs (GSPF/R2) with the "internal" AP2 adapter primer provided by the kit (**Fig. 3C,D**).
5. Carry out PCR amplifications using Invitrogen high-fidelity Platinum *Taq* and a Perkin Elmer GeneAmp Systems 9600 thermal cycler using the cycling conditions described below. Mix all ingredients as detailed in the Marathon kit and start the cycles. The Marathon kit suggests a touchdown PCR for primers with high *Tm* ($\geq 70°C$), to reduce back ground. This may help to increase the specificity of the PCR, but it was not needed for our work on *T. hominis* mitHsp70.
    RACE cloning PCR conditions:

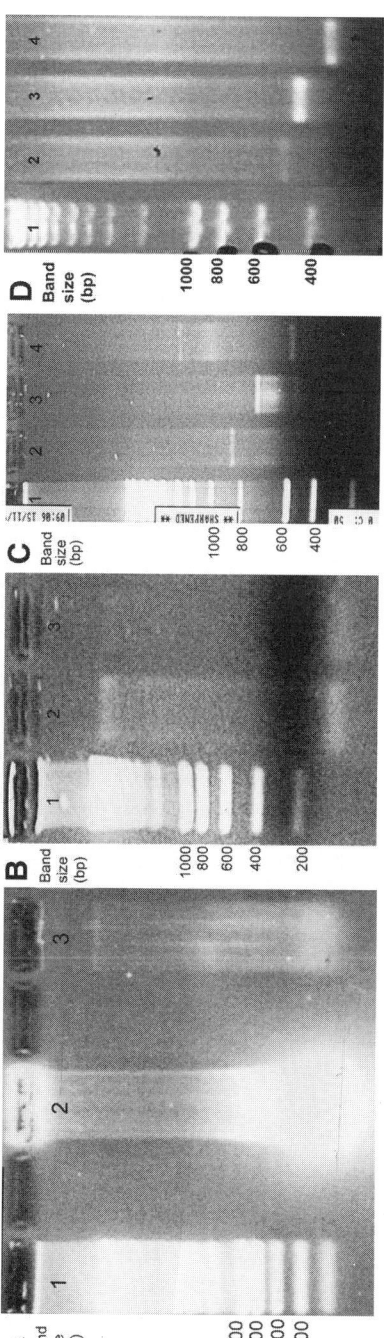

Fig. 3. RACE cloning of 5'- and 3'-ends of a mitochondrial Hsp70 ortholog from the microsporidium *T. hominis*. (**A**) Total RNA was purified from spores, the only cellular stage that can be readily isolated from the infected rabbit kidney cell line. Following spore isolation and lysis, total RNA was extracted. A small aliquot of the purified material was electrophoresed in an agarose gel (lane 3) in parallel with *E. coli* rRNA (lane 2) extract as a control for the presence of RNA. This total RNA preparation was used for mRNA purification, which was then used for cDNA synthesis (using the Marathon cDNA amplification kit). (**B**) A primary 3'-RACE PCR amplification was performed using either of the gene-specific primers ThHsp70-421F (lane 2) and ThHsp70-542F (lane 3) with adapter-specific primer AP1, producing no detectable product. Consequently, a secondary amplification was performed on an aliquot of the primary RACE-PCR after removal of the original primers (*see* **Subheading 3.2.**) using the internal adapter primer AP2 and internal gene-specific primers. (**C**) Secondary 3'-RACE PCR amplification with primers AP2 and ThHsp70-542F (lane 2) or ThHsp70-864F (lane 3 and 4) using either primary amplification shown in (**B**). Fragments were then gel-purified, cloned and sequenced, to obtain data on the 3'-end of the mtHsp70 ORF. (**D**) The primary 5'-RACE amplification did not produce a visible product (data not shown) and, therefore, a secondary RACE-PCR was carried out using the primer AP2 with the gene-specific primers ThHsp70-718R (lane 1), ThHsp70-652R (lane 2), and ThHsp70-258R (lane 3). One specific primer yielded an amplicon of the size expected if it were to contain the 5'-end of the ORF (lane 3). That fragment was gel-purified, cloned, and sequenced, revealing a potential start codon but no 5'-UTR sequence data.

Initial denaturation: 30 s at 94°C
Followed by 25–30 cycles: 5 s at 94°C
4 min 68°C

6. Electrophorese an aliquot of the PCR products on 1% TBE agarose gel to check for the presence of PCR product. If there is a detectable band, then electrophorese the remaining material on a gel and elute the band of interest using the QIAquick gel elution kit as described in the user manual. Quantify the eluted material on a gel by running a small aliquot of the eluate in parallel to a molecular size maker with a known amount of DNA for comparison.

7. Ligate the eluted DNA band to the pGEM-T-EASY vector using the following protocol, and mix:

T4 DNA Ligase 10X buffer	1 μL
pGEM-T-EASY vector, 50 ng/μL	1 μL
PCR product (28–85 ng)	x μL (molar ratio 3:1–1:1 to the vector)
T4 DNA ligase (1 Weiss unit/μL)	1 μL
Analar water to	10 μL

Perform a negative control ligation without PCR product; from this, no colonies—or very few—should be obtained here. As a positive control, one can use a control insert provided by the pGEM-T-EASY kit. Incubate the ligation overnight at 4°C.

8. Use 3 μL of the ligation mix to transform competent *E. coli* DH5α. The remainder of the ligation is stored at –20°C if this step has to be repeated. Incubate plates overnight at 37°C to allow *E. coli* colonies to grow. Typically, two dilutions of cells are plated 1:10 and 9:10 to provide well-spread colonies.

9. Pick up colonies for plasmid minipreps and perform restriction digestion or PCR directly from bacterial colonies to identify clones with appropriate inserts; 3–5 clones (per PCR product) inserts are typically sequenced.

The combination of the sequence data for the 5'- and 3'-RACE PCR products provide an ORF encoding a potentially complete mitHsp70 but no 5'-UTR. Sometimes it is not possible to obtain any 5'-UTR regions and hence infer the position of the potential start codon (*see* **Note 4**). In this case, a complementary RAGE approach may be used to provide the missing data.

3.3. 5'-RAGE of the T. hominis *mitHsp70 Gene*

To complement the 5'-end sequence data obtained by the RACE, we have used a RAGE cloning approach. The protocol given here broadly follows that of Mizobuchi and Frohman *(4)* (but *see* **Note 5**). The result of the RAGE cloning approach is shown in **Fig. 4**. A two-step amplification procedure is often needed to obtain a PCR product, and direct sequencing is required to confirm that the product contains the relevant sequence. The combined cDNA and genomic DNA sequence data allowed us to obtain the complete ORF for *T. hominis* mitHsp70 (**Fig. 5**) *(13)*.

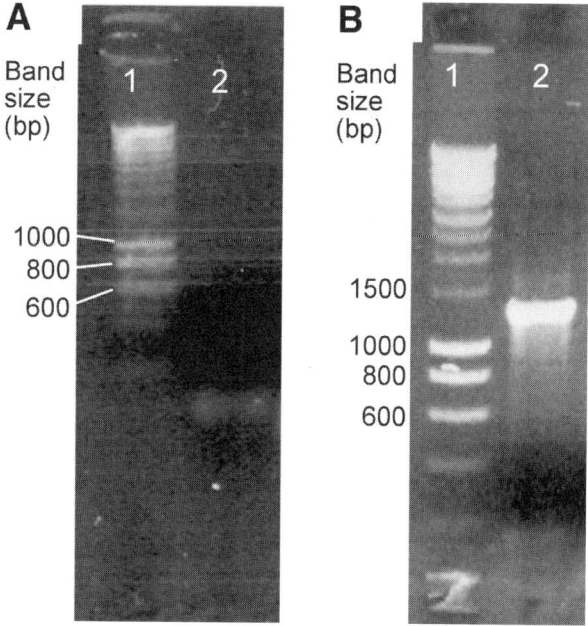

Fig. 4. RAGE cloning of the 5'-end of the *T. hominis* mitHsp70 ORF and the corresponding 5'-UTR. (**A**) Primary RAGE PCR. Lane 1, of an ethidium bromide-stained agarose gel, contains a Bioline Hyperladder Mw marker. Lane 2, the result of RAGE amplification using primers ThHsp70-258R and plasmid primer T3. No PCR amplification product was detected, and this material was used as template for a secondary PCR amplification. (**B**) Successful secondary, nested RAGE PCR. Lane 1, Bioline Hyperladder. Lane 2, secondary PCR of PCR products from (**A**) using primers ThHsp70-139R and vector primer T3. The amplicon was directly sequenced (partially) with mitHsp70 gene-specific primers to obtain the 5'-end of the mitHsp70 ORF (*see* **Fig. 5**).

3.3.1. Preparation of the Vector

1. Select an appropriate cloning vector and digest with a restriction enzyme. This enzyme should cut your known fragment of sequence at a single site. In our experiments, Stratagene pPCR Script was selected as vector and was digested with HindIII for 80 min at 37°C.
2. Digest approx 500 µg of plasmid with 100 U of restriction enzyme, adding appropriate restriction enzyme buffer to a final concentration of 1X buffer. Add an appropriate volume of double-distilled H_2O so that the restriction enzyme is not more than 10% of the total reaction volume. Check the level of digestion on an agarose gel. A single band (linear DNA) should be obtained for a linearized plasmid.
3. Remove the restriction enzymes from the mix using a Microcon microconcentrator, according to the manufacturer's instructions.

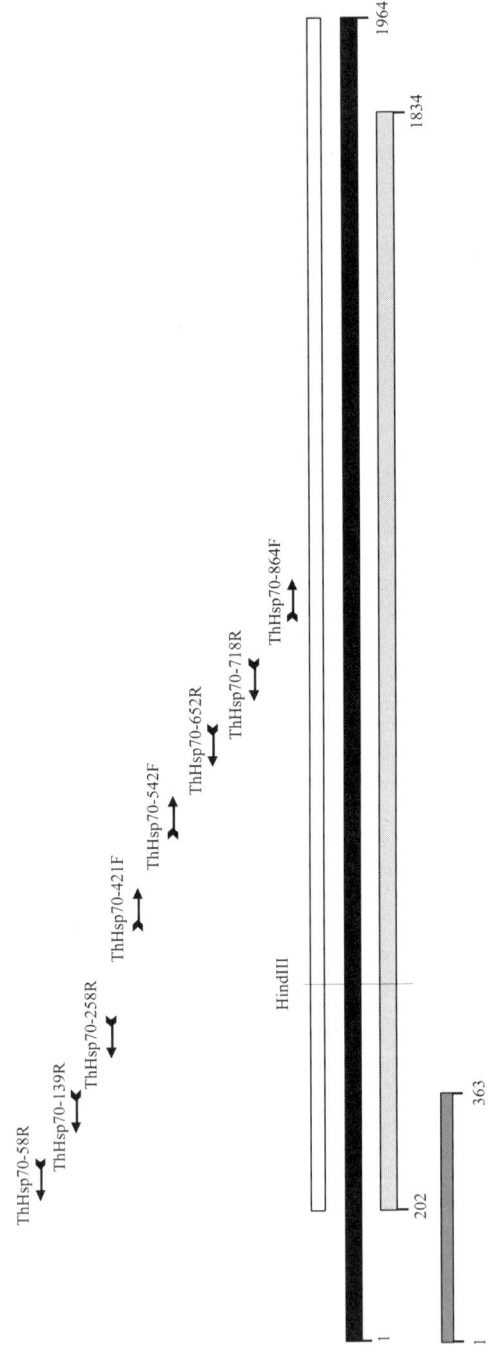

Fig. 5. The position of primers used for PCR relative to the sequenced *T. hominis* genomic DNA and cDNA encoding mitHsp70. In addition to the GSPs (some are not shown here for simplicity), cloning vector-specific primers were used for double-strand sequencing of the entire ORF and flanking regions. The black box shows the total DNA fragment amplified and sequenced from genomic DNA (*see* GenBank entry AF492453 for full sequence including the 5'- and 3'-UTRs). The light gray box shows the complete ORF with a total of 1632 base pairs. The white box shows the DNA fragment amplified and sequenced from cDNA using both 5'- and 3'-RACE. The darker gray box shows the segment sequenced from 5'-RAGE cloning. The position of the restriction site HindIII, used to digest the genomic DNA, is also shown. For Hsp70 primer sequences, *see* **Note 6**.

4. To prevent self-religation of the vector ends, treat the cut vector with alkaline phosphatase. Incubate approx 1.2 µg of the digested vector with 0.7 U of shrimp alkaline phosphatase (SAP) for 1 h at 37°C. In our experiment, 1.2 µg of vector corresponded to a volume of 12 µL, and a 20-µL reaction was set up by adding 0.7 µL of SAP, 2 µL of the provided 10X reaction buffer and 5.3 µL of H_2O.
5. Heat-inactivate the phosphatase by incubation for 30 min at 65°C and remove the restriction enzymes using a Microcon 100 filtration column.
6. Assess the success of this treatment by ligation of the vector to itself using untreated cut vector as a control. Transform the ligation products into DH5α. No colonies should be seen in the alkaline phosphatase-treated vector in the absence of insert.

3.3.2. Preparation of the T. hominis DNA Amplification Product

Restriction enzyme sites were identified in the mitHsp70 fragment that had been previously amplified and sequenced. This fragment has a HindIII restriction site 340 bps 3' of the most 5' bp in the known sequence. Because of the small amount of DNA that can be extracted from spores and of the smaller size of the typical microsporidian genome relative to that of mammals, only 0.5 µg of DNA was digested with HindIII for a RAGE cloning attempt.

1. Digest genomic DNA (0.5 µg) to completion with 3.5 µL (35 U) of HindIII overnight at 37°C.
2. Remove the restriction enzyme with a Microcon 100 filtration column.
3. Ligate the genomic DNA to the restricted and phosphatase-treated vector overnight at 4°C using T4 ligase using a 1:1 ratio of insert to vector (*see* **step 5** in **Subheading 3.2.**).
4. Carry out PCR amplifications on the ligation mix the following day. Two rounds of PCR may be needed. PCR conditions are given in **Subheading 3.3.3.** (**Fig. 4**). For the first round use an external GSP combined with the plasmid-specific primers (T3 or T7 for pPCR Script); we used either ThHsp70-258R (GSP1) or ThHsp70-139R (GSP1b) (*see* **Figs. 4** and **5**). The plasmid-specific primers sit either side of the HindIII cloning site. It is necessary to use primers from both sides of the cloning site, as the potential insert orientation is unknown, making up a total of four primer combinations. In our case, we had: GSP1-T7, GSP1-T3, GSP1b-T7, and GSP1b-T3.
5. If no PCR product can be detected by gel electrophoresis, purify potentially amplified material from this first round of amplification using a Microcon 100 column to remove former primers and dilute 1/100 in Analar water. For the second round of PCR, use an internal gene-specific primer and 1 µL of the purified primary PCR. Here, we used two GSP primers, either ThHsp70-139R (GSP1b) or ThHsp70-58R (GSP2) with the plasmid-specific T3 or T7 primers and the diluted primary PCR reaction as template DNA. In our experiment, one combination of primers, including ThHsp70-139R, produced a PCR product (**Fig. 4**) that was shown by direct sequencing to correspond to the 5'-end of the mitHsp70 ORF (**Fig. 5**).
6. Electrophorese PCR products on a 1.2% 1X TBE agarose gel and gel-extract any band obtained by PCR using the QIAquick gel extraction kit. Sequence PCR products

directly or clone using, for example, pGEM-T-EASY (*see* **Subheading 3.2.**) before sequencing.

3.3.3. PCR Conditions

Here, we present the general protocol we use to perform the hot-start PCR mentioned in **Subheading 3.3.2., step 4**. It is assumed that the user is familiar with the danger of DNA contaminations that can lead to spurious PCR products; hence, a negative control containing no template DNA is always run in parallel with the samples.

1. Mix all components as listed below:
   5 µL *Taq* 10X reaction buffer
   4 µL MgCl$_2$ (100 mM)
   1 µL BSA (10 mg/mL)
   1 µL dNTPs (400 mM)
   6 µL primers (3.2 pmol each)
   6 µL template DNA
   0.5 µL *Taq* (5 U/µL) (added later, *see* **step 3**)
   H$_2$O to 50 µL
2. Heat mix 5 min 96°C (hence hot start)
3. Add 0.5 µL *Taq* (5 U/µL) and start cycles:
   1 min at 94°C ⎫
   1 min at 56.5°C ⎬ 35 cycles
   1.5 min 72°C   ⎭
   30 min at 72°C   1×
4. Run a small aliquot, typically 10% of the PCR reaction volume, on an agarose gel to check for the presence of a PCR product.

3.4. Cloning of Gene Family Members Using a RACE Approach

Numerous genes form complex families with many paralogs *(23,24)*. Standard PCR can be used to investigate their diversity, in which case two conserved domains allowing the design of degenerate primers typically are needed *(23)*. We have used a RACE cloning approach, as used for mammals *(25)*, to investigate in *Trichomonas* the members of the Rab small GTP-binding protein family to complement an ongoing *Trichomonas* EST project. These two approaches have identified with more than 30 distinct cDNAs so far, an unexpectedly large Rab gene family for a single-celled organism (Hirt et al., unpublished data). **Figure 6** gives an outline of the approach. We used the strategy

---

Fig. 6. (*Opposite page*) A RACE cloning strategy used to clone members of the Rab gene family in *T. vaginalis*. A schematic representation of a Rab GTP-binding protein is shown with conserved domains in gray boxes. Primers can be designed to hybridize to one or more conserved domains and used in a RACE cloning approach. Shown here

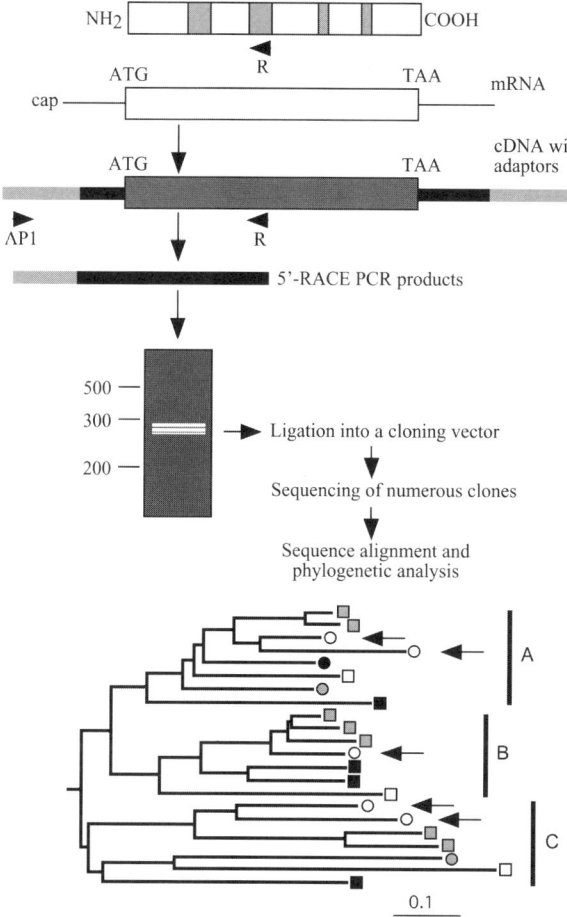

is the 5'-RACE cloning approach as used for Rab proteins *(25)*. Total mRNAs were purified and used for cDNA synthesis using the Marathon cDNA amplification kit (*see* **Fig. 1** and **Subheading 3.2.**). The primer AP1, specific for the ligated adaptor, was used with a degenerate reverse primer complementary to the conserved domain of the G-2 domain of Rab proteins *(25)*. The resulting RACE PCR amplicons of various sizes (ca. 250–300 bp), reflecting a combination of heterogeneity in paralog length and variable success in cDNA synthesis, were ligated into the pGEM-T-EASY plasmid. Sequencing of the clone fragments allows identification of Rab-encoding cDNAs by comparison to known sequences. A phylogenetic tree is used to differentiate between paralogs. The tree shown here suggests three major paralogs, A, B, and C, with the new sequences highlighted with arrows. Each symbol represents one species. Some organisms have additional specific paralogs because of duplications within taxa; *see* **ref. *31*** for a review discussing the interest of performing phylogenetic tree for comparing paralogs. Ultimately, all partial 5'-RACE clones will have to be confirmed by obtaining full-length cDNA and/or genomic DNA sequences.

described by Chavrier et al. *(25)* in combination with the Marathon cloning kit described above (**Fig. 2A, Subheading 3.2.**).

1. Choose an appropriately conserved protein region for your gene family of interest that allows the design of a degenerate primer for PCR (*see*, for example, **refs. 23** and **25**). The primer properties might be better for either a 5'- or 3'-RACE. If two conserved regions or primer directions are promising, then it may be useful to try alternative options to increase the chance of success in surveying the gene family diversity.
2. Perform either 5'- or 3'-RACE PCR (5'-RACE was performed for the Rab genes) and clone products into a plasmid vector as described in previous subheadings. (PCR products of expected size range were obtained for the chosen Rab-specific primer and amplicons were gel-purified and cloned into pGEM-T-EASY as described earlier.)
3. Prepare plasmid minipreps to analyze random clones by restriction digestion.
4. Sequence several clones that show distinct restriction patterns.
5. Compare the sequence by aligning them and perform phylogenetic analysis to differentiate them.

We could obtain, without difficulty, numerous pGEMT-RACE clones with distinct restriction fragments and identified among these by sequencing 10 distinct (partial sequence) Rab paralogs. The partial sequence data can then be used to clone full-length cDNA using, for example, a combination of 5'- and 3'-RACE cloning approach as presented earlier. Of these 10 5'-RACE cDNA clones, 8 have been confirmed by further cDNA cloning and sequencing, validating the approach.

4. Notes

1. For RNA purification, one can use a standard phenol extraction protocol. However, in some organisms, numerous nucleases are released, and these must be inhibited in some way. When working with the parabaslid *T. vaginalis*, one needs to take extra care to inhibit the numerous nucleases released during cell lysis and, hence, use methods that will efficiently destroy their activity *(27)*. We have used either a guanidium thiocyanate-based method (*see* Chapter 4 in **ref. 27**) or an RNA extraction kit. For the guanidium thiocyanate-based method, it is crucial to use fresh solutions as described *(28)*. Excellent results were also obtained for *Trichomonas* using the SV total RNA isolation system from Promega, cat. no. TM048.
2. Primers are key for any PCR-based cloning approach, and we recommend that researchers new to the technique carefully read more details in the literature. We have used OLIGO v. 5.0 for Macintosh (National Biosciences) and Primer3 on the Web http://www-genome.wi.mit.edu/cgi-bin/primer/primer3_www.cgi to design primers that are typically 20–30 nucleotides in length. The great majority of primers designed using these methods resulted in success. Specific information for primer design for RACE PCR is also given in the Marathon cDNA amplification

kit (*see* **Subheading 2.4.**). For degenerate primer design, the chance of success may be increased if you are able to take into account any available information about the codon usage of your organism. This may allow you to reduce primer degeneracy. For additional information on degenerate primer design see, for example, **ref. 23**.
3. Because of the small amount of available material for *T. hominis* preparations, we did not attempt to quantify either the mRNA or the cDNA. The extracted mRNA was processed directly for RACE PCR (**Fig. 3**). In the case of nonlimiting starting material, one should use the optimal amount of polyA$^+$ mRNA (1 µg) for the Marathon kit. This will provide an optimal amount of cDNA, and the more template DNA that can be used in the RACE PCR reactions the better. This might be a critical parameter for rare mRNAs. This was not a problem for *T. hominis* mitHsp70.
4. In the case of the *T. hominis* Hsp70 gene, this is probably to the result of short 5'-UTRs in microsporidia, which can have short intragenic sequences (*18*). Also, it may have been degraded during the purification, removing the most 5'-end of the mRNA including the start codon.
5. Mizobuchi and Frohman (*4*) suggest the use of at least three digestion enzymes and 5 µg of DNA for each digestion as detailed here. However, because of the small amount of DNA purified from spores, we made an attempt with a reduced amount of DNA (10 times less than recommended in **ref. 4**). This experiment was successful using a single digestion enzyme and only 0.5 µg of genomic DNA for the digestion. This success may be because of the fact that microsporidia have relatively small genomes sizes (*18,26*) and, therefore, a DNA extraction contains more copies of genome per microgram of DNA, compared to a DNA extraction from mammalian cells.
6. The sequences of the Hsp70 oligonucleotide primers (22–30-mers) used in PCR amplification (*see* **Fig. 5**) were as follows:

ThHsp70-58R	5'-GGCCACGTTGTCCTGCATTATAGAAAT-3'
ThHsp70-139R	5'-CGGTTTGCCAACTATCACGTTTTCACCCGA-3'
ThHsp70-258R	5'-AGTTCTCCGTTGTCGTCTATAA-3'
ThHsp70-421F	5'-TACACAGCGGGAGGAGACTAAAA-3'
ThHsp70-542F	5'-ACTTAGGCGGTGGTACGTTTGATA-3'
ThHsp70-652R	5'-GCCTAATTTTCGCCAGATCAATGTCAC-3'
ThHsp70-718R	5'-CAATCGTCACGGTTTCCTGTGTTGATAG-3'
ThHsp70-864F	5'-TTGCTCCACTGATAAAAAGAAC-3'

Acknowledgments

We wish to thank Dr G. M. Birdsey (University College London, UK) for advice to B.A.P. Williams on RAGE cloning. B.A.P. Williams was supported by a Wellcome Trust Biodiversity Studentship. R.P. Hirt was supported by a Wellcome Career Development Fellowship and is currently supported by a Wellcome Trust University Award.

References

1. Frohman, M. A. (1990) RACE: rapid amplification of cDNA ends, in *PCR Protocols: A Guide to Methods and Amplification* (Innis, M. A., Gelfand, D. H., Sninsky, J. J., and White, T. J., eds.), Academic Press, San Diego, pp. 28–38.
2. Zhang, Y. and Frohman, M. A. (1997) Using rapid amplification of cDNA ends (RACE) to obtain full-length cDNAs. *Meth. Mol. Biol.* **69,** 61–87.
3. Garces, J. A. and Gavin, R. H. (2001) Using an inverse PCR strategy to clone large, contiguous genomic DNA fragments. *Meth. Mol. Biol.* **161,** 3–8.
4. Mizobuchi, M. and Frohman, L. A. (1993) Rapid amplification of genomic DNA ends. *Biotechniques* **15,** 214–216.
5. Embley, T. M. and Hirt, R. P. (1998) Early branching eukaryotes? *Curr. Op. Genet. Dev.* **8,** 624–629.
6. Roger, A. J. (1999) Reconstructing early events in eukaryotic evolution. *Am. Nat.* **154(Suppl.),** S146–S163.
7. Cavalier-Smith, T. (1987) Eukaryotes with no mitochondria. *Nature* **326,** 332–333.
8. Wittner, M. and Weiss, L. M. (eds.). (1999) The *Microsporidia and Microsporidiosis*. American Society for Microbiology, Washington, DC.
9. Hollister, W. S., Canning, E. U., and Anderson, C. L. (1996) Identification of *Microsporidia* causing human disease. *J. Euk. Microbiol.* **43,** 104S–105S.
10. Hollister, W. S., Canning, E. U., Weidner, E., et al. (1996) Development and ultrastructure of *Trachipleistophora hominis* n.g., n.sp. after in vitro isolation from an AIDS patient and inoculation into athymic mice. *Parasitology* **112,** 143–154.
11. Viscogliosi, E., Edgcomb, V. P., Gerbod, D., et al. (1999) Molecular evolution inferred from small subunit rRNA sequences: what does it tell us about phylogenetic relationships and taxonomy of the parabasalids? *Parasite* **6,** 279–291.
12. Petrin, D., Delgaty, K., Bhatt, R., et al. (1998) Clinical and microbiological aspects of *Trichomonas vaginalis*. *Clin. Microbiol. Rev.* **11,** 300–317.
13. Williams, B. A., Hirt, R. P., Lucocq, J. M., et al. (2002) A mitochondrial remnant in the microsporidian *Trachipleistophora hominis*. *Nature* **418,** 865–869.
14. Roger, A. J. and Silberman, J. D. (2002) Cell evolution: mitochondria in hiding. *Nature* **418,** 827–829.
15. Martin, W. and Müller, M. (1998) The hydrogen hypothesis for the first eukaryote. *Nature* **392,** 37–41.
16. Embley, T. M., van Der Giezen, M., Horner, D. S., et al. (2003) Mitochondria and hydrogenosomes are two forms of the same fundamental organelle. *Phil. Trans. R. Soc. Lond. B* **358,** 191–203.
17. Tielens, A. G., Rotte, C., van Hellemond, J. J., et al. (2002) Mitochondria as we don't know them. *Trends Biochem. Sci.* **27,** 564–572.
18. Katinka, M. D., Duprat, S., Cornillot, E., et al. (2001) Genome sequence and gene compaction of the eukaryote parasite *Encephalitozoon cuniculi*. *Nature* **414,** 450–453.
19. Müller, M. (1998) Enzymes and compartmentation of core metabolism of anaerobic protists—a special case in eukaryotic evolution, in *Evolutionary Relationships Among Protozoa* (Coombs, G. H., Vickermann, K., Sleigh, M. A., and Warren, A., eds.), Chapmann & Hall—Kluwer Academy Publishers, London, pp. 109–132.

20. Liston, D. R. and Johnson, P. J. (1998) Gene transcription in *Trichomonas vaginalis*. *Parasitol. Today* **14,** 261–265.
21. Gubler, U. and Hoffman, B. J. (1983) A simple and very efficient method for generating cDNA libraries. *Gene* **25,** 263–269.
22. Hirt, R. P., Healy, B., Vossbrinck, C. R., et al. (1997) A mitochondrial Hsp70 orthologue in *Vairimorpha necatrix*: molecular evidence that microsporidia once contained mitochondria. *Curr. Biol.* **7,** 995–998.
23. Preston, G. M. (1997) Cloning gene family members using the polymerase chain reaction with degenerate oligonucleotide primers, in *cDNA Library Protocols, Vol. 69* (Cowell, I. G. and Austin C. A., eds.), Humana Press, Totowa, NJ, pp. 97–113.
24. Henikoff, S., Green, E. A., Pietrokovski, S., et al. (1997) Gene families: the taxonomy of protein paralogs and chimeras. *Science* **278,** 609–614.
25. Chavrier, P., Simons, K., and Zerial, M. (1992) The complexity of the Rab and Rho GTP-binding protein subfamilies revealed by a PCR cloning approach. *Gene* **112,** 261–264.
26. Biderre, C., Pages, M., Metenier, G., et al. (1994) On small genomes in eukaryotic organisms: molecular karyotypes of two microsporidian species (Protozoa) parasites of vertebrates. *C. R. Acad. Sci. Life Sci.* **317,** 399–404.
27. Chou, C.-F. and Tai, H. (1996) Simultaneous extraction of DNA and RNA from nuclease-rich pathogenic *Trichomonas vaginalis*. *Biotechniques* **20,** 790–791.
28. Ausubel, F. M., Brent, R., Kingston, R. E., et al. (eds.). (2001) *Current Protocols in Molecular Biology*, Wiley, Boston.
29. McKie, A. E., Edlind, T., Walker, J., et al. (1998) The primitive protozoon *Trichomonas vaginalis* contains two methionine gamma-lyase genes that encode members of the gamma-family of pyridoxal 5'-phosphate-dependent enzymes. *J. Biol. Chem.* **273,** 5549–5556.
30. Mallinson, D. J., Lockwood, B. C., Coombs, G. H., et al. (1994) Identification and molecular cloning of four cysteine proteinase genes from the pathogenic protozoan *Trichomonas vaginalis*. *Microbiology* **140,** 2725–2735.
31. Eisen, J. A. (1998) Phylogenomics: improving functional predictions for uncharacterized genes by evolutionary analysis. *Genome Res.* **8,** 163–167.

8

Amplified (Restriction) Fragment Length Polymorphism (AFLP) Analysis

Daniel K. Masiga and C. Michael R. Turner

Summary

The amplified (restriction) fragment length polymorphism (AFLP) technique is a method for DNA profiling that is now widely applied for assessing diversity among various organisms with varying genomic complexity, from small bacterial to large plant genomes. AFLP analysis combines the reliability of restriction enzyme digestion with the utility of the polymerase chain reaction. The technique can be applied to studies of DNA of any origin and complexity, without prior sequence knowledge. Therefore, it is very versatile and particularly valuable for organisms for which no substantive DNA sequence data are available. AFLP detects the presence of point mutations, insertions, deletions, and other genetic rearrangements. Typically, the fragments detected by AFLP are inherited in Mendelian fashion as co-dominant markers, making the technique amenable to tracking inheritance of genetic loci in progeny from crossed lines of organisms, and in studies of population genetics. This chapter describes the principles of AFLP and experimental procedures.

Key Words: AFLP; co-dominant markers; genotyping; Mendelian; PCR; restriction enzymes.

1. Introduction

Amplified (restriction) fragment length polymorphism (AFLP) analysis is a method for DNA fingerprinting that combines the reliability of restriction enzyme digestion with the utility of the polymerase chain reaction (PCR) *(1)*. Initially used widely in the world of plant science, AFLP is now recognized as a very powerful technique for assessing diversity among various organisms with varying genomic complexity, from small bacterial to large plant genomes *(2–17)*. AFLP can be used for DNA of any origin and complexity without prior sequence knowledge, making it very versatile and easy to use. It is particularly valuable in the study of organisms for which there is no substantive genome sequence database. However, it is essential that a single organism or clone is analyzed for each reaction. Hence, when parasite DNAs are analyzed, a pure clonal population must be used as starting material, and cell-free cultures are especially valuable.

Typically, AFLP is composed of four steps: (a) double digestion of genomic DNA to completion with two restriction enzymes, one recognizing a 6-bp recognition site (a hexacutter, e.g., *Eco*RI) and one tetracutter (e.g., *Mse*I); (b) ligation of double-stranded adapters to the termini of the restriction fragments; (c) PCR amplification using primers that anneal to the adapters; and (d) electrophoretic separation of the amplified fragments using a suitable matrix, often a denaturing 4–6% polyacrylamide gel. An outline of these procedures is shown diagrammatically in **Fig. 1**. The primers for PCR in step (c) are designed with nucleotide extensions that detect polymorphisms in the restriction fragments, enabling the amplification of only a subset of the fragments (**Fig. 2**). For genomes of the size range 10^6–10^7 bp, primers should not contain more than two selective nucleotides *(2–11)*. When related samples are compared, bands that are common to all the samples, as well as those that are present in only a subset of the samples, are detected (*see* **Fig. 3**). The latter subset are referred to as polymorphic bands or markers. AFLP detects DNA polymorphisms within or adjacent to specific restriction enzyme sites and enables these polymorphisms to be detected at multiple independent restriction sites simultaneously. The magnitude of changes observed when selective nucleotides are altered suggests that the most frequently detected AFLP polymorphisms are caused by sequence differences at the nucleotide level. Single-nucleotide polymorphisms will be detected by AFLP if a restriction enzyme site is affected, or if a nucleotide adjacent to the restriction site is affected, causing a mismatch of the primer and template at the 3'-end, hence preventing amplification (**Fig. 1**). The presence

---

Fig. 1. (*Opposite page*) A schematic presentation of the steps in the AFLP process. (**A**) Digestion of genomic DNA to completion with *Eco*RI (recognition sequence 5'-GAATTC-3') and *Mse*I (recognition sequence 5'-TTAA-3'). This process generates three types of fragments defined by the restriction site at the termini: fragments cut with *Eco*RI at each end, fragments cut at one end with *Eco*RI, and at the other end by *Mse*I, and fragments cut by *Mse*I at both termini. (**B**) Ligation of double-stranded adapters that have complementary ends to the restricted fragments. (**C**) Preamplification and selective amplification, the process that amplifies a subset of the restriction fragments. (**D**) Resolving the amplified fragments through 4–6% denaturing polyacrylamide, followed by autoradiography. Only [^{33}P]-labeled markers are detectable. Hence, the two types of markers that have an *Eco*RI terminus will be detected because only the *Eco*RI adapter is labeled. However, *Eco*RI-*Eco*RI fragments are likely to be large, because they occur every 4^6 (4096) bases (assuming random distribution of the four

AFLP Analysis

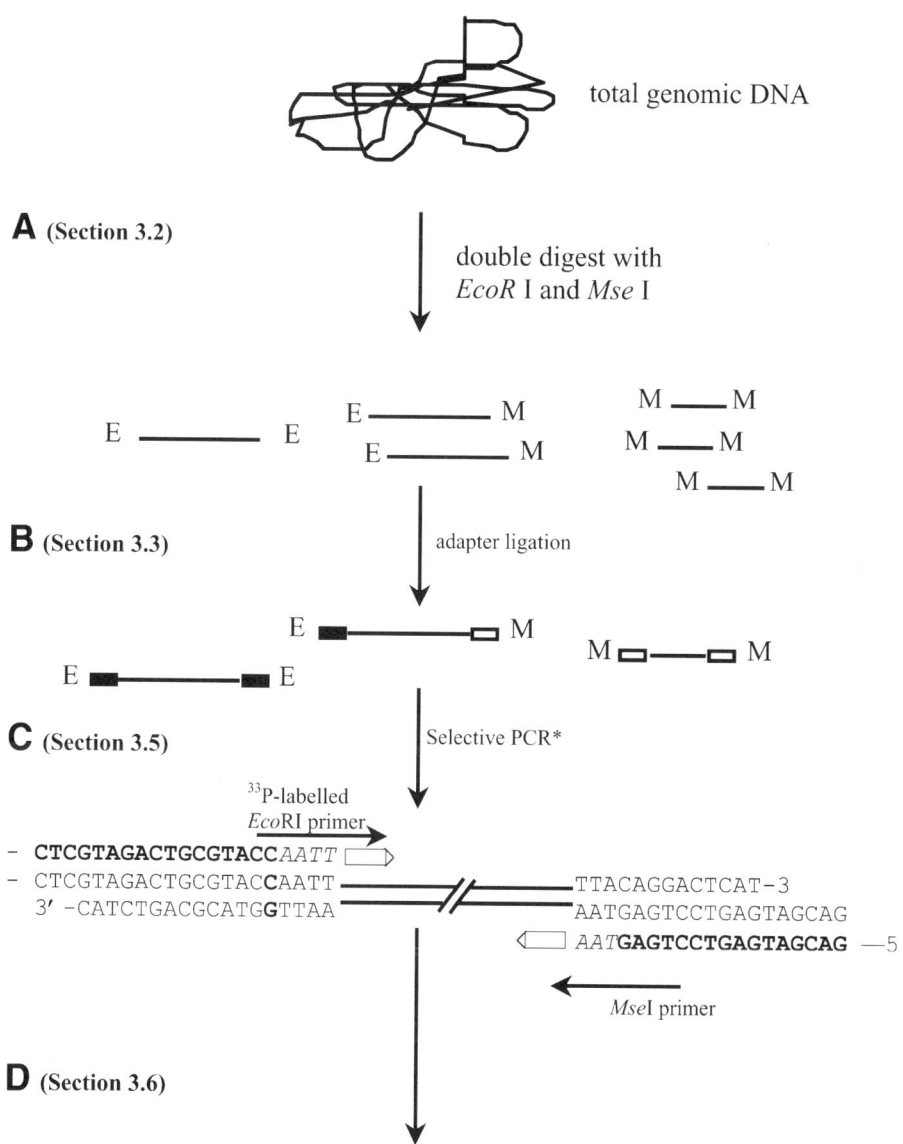

oligonucleotides) and these are too large to be resolved in the gels used. All labeled fragments of the correct size (<1000 bp) will be detected (mostly EcoRI-MseI). Fragments flanked only by MseI sites will occur more frequently—every 256 (4^4) bp, on average—but are unlabeled and will not be detected. *Can be preceded by a preamplification step to build up the reagents depending on the size of the project (see **Subheading 3.4.**).

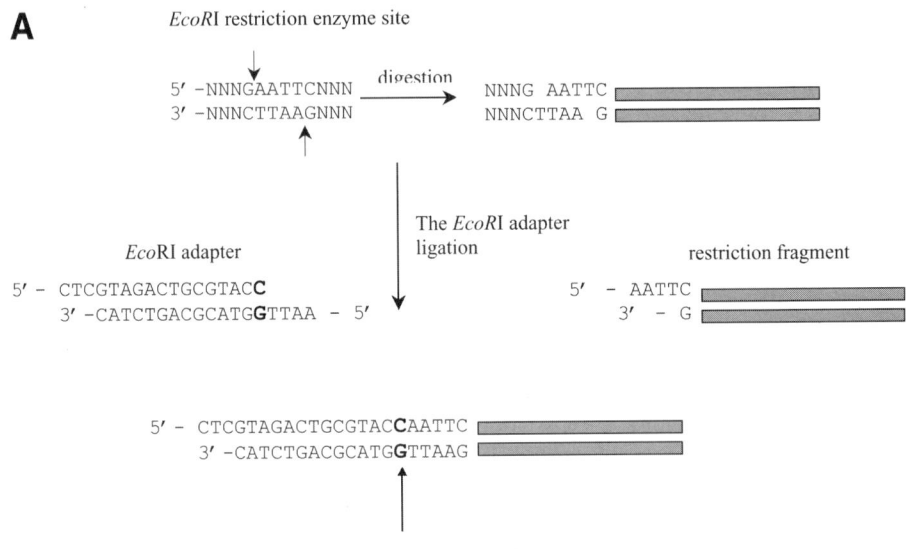

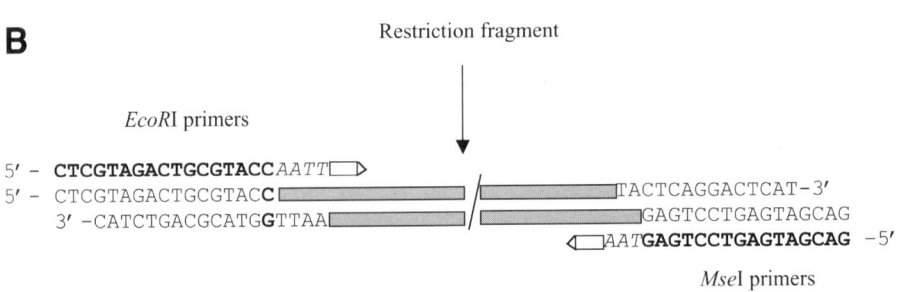

Fig. 2. **(A)** The ligation of adapters generates ends that are complementary to the termini of fragments generated by the restriction enzyme. This example shows the ligation of the *Eco*RI adapter to a restriction fragment. After ligation of adapter and restriction fragment, the site is no longer a substrate for *Eco*RI digestion because of the single nucleotide change shown. Hence, restriction enzyme digestion and ligation can proceed in the same simultaneously. **(B)** An illustration of three important features of the PCR primers used for selective amplification. The primer section in bold represents the region that is complementary to the adapter sequence; the bases in italics are complementary to the restriction enzyme site; whereas those represented by the clear box are selective nucleotides that determine the subset of fragments that are amplified. The arrowhead shows the direction of primer extension. The two central lines represent a genomic DNA fragment.

AFLP Analysis 177

of insertions, deletions, and other rearrangements may also affect the size, or the presence of a marker. The markers detected by AFLP are typically inherited in Mendelian fashion. They have been used for tracking inheritance of genetic loci in progeny from crossed lines of parasites *(4–7)*, and in studies of population genetics *(8–9,12)*.

The methods described here have been used for genotyping samples of trypanosomes (*Trypanosoma brucei* spp.), single-cell protozoans with a genome size of approx 30 Mb *(18)*. Therefore, these methods are appropriate for small genomes (10^6–10^7 bp). The most important consideration when analyzing genomes that are smaller or larger lies in the selection of the number of nucleotide extensions on the primers (selective nucleotides). The number of selective nucleotides should be increased with genome size: but the maximum recommended is three, because spurious amplifications can occur if more are used *(16)*. It is also possible to bias the selection of enzymes, if the base composition of the genome is known. For example, *Eco*RI (5'-GAATTC-3') and *Mse*I (5'-TTAA-3') would be a particularly suitable enzyme pair for *Plasmodium falciparum*, which has an AT content of about 80% *(19)*. This combination of enzymes has been used in *P. chabaudi* for marker identification for the analysis of inheritance *(8)*.

Here, we describe the protocols used for the AFLP analysis of *T. brucei* from the initial preparation of the DNA to amplification and analysis of polymorphic fragments. We also show how markers identified by this method may be isolated for further use.

2. Materials

1. *Eco*RI adapters: (oligonucleotide sequences: 5'-CTCGTAGACTGCGTACC-3' and 5'-AATTGGTACGCAGTCTAC-3').
2. *Mse*I adapters (oligonucleotide sequences: 5'-GACGATGAGTCCTGAG-3' and 5'-TACTCAGGACTCAT-3').
3. The following PCR primers: *Eco*RI preamplification primer (E-CORE primer): 5'-GACTGCGTACCAATTC-3'); *Eco*RI + TA (E-TA): 5'-GACTGCGTACCAAT TCTA-3'; *Mse*I + C (M+C): 5'-GATGAGTCCTGAGTAAC-3'; *Mse*I + CA (M-CA): 5'-GATGAGTCCTGAGTAACA-3'. Using the core sequence, additional primers can be made with different selective nucleotides.
4. Cell lysis solution (10 mM Tris, pH 8.0, 50 mM EDTA, 0.01% sodium dodecyl sulfate, 100ng/mL proteinase K).
5. TE-saturated phenol–chloroform with isoamyl alcohol (ratio 25:24:1) *(8)*.
6. Phosphate-buffered saline (PBS) (Sigma).
7. Tris-EDTA (TE) buffer (10 mM Tris-Cl, 1 mM EDTA, pH 8.0).
8. Restriction enzymes (*Eco*RI and *Mse*I) and appropriate 10X buffers.
9. *Taq* DNA polymerase + appropriate 10X buffer.
10. T4 DNA ligase + appropriate 10X buffer, with ATP.

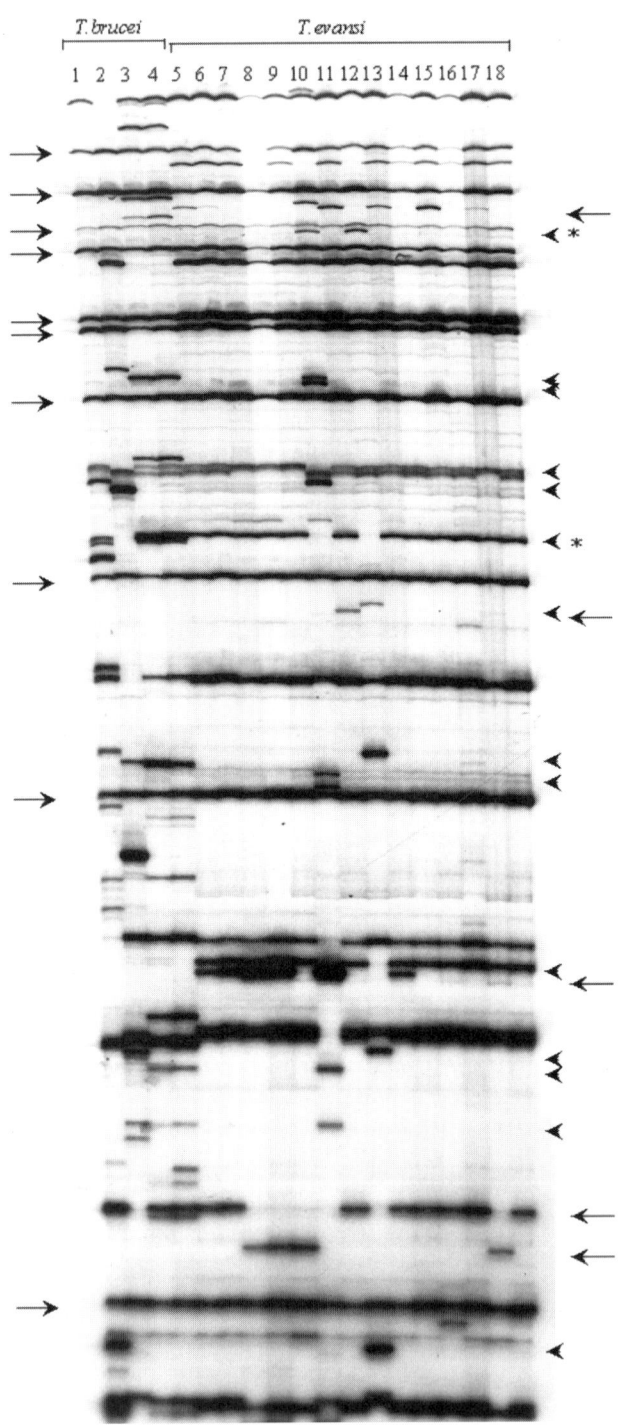

11. T4 polynucleotide kinase (PNK) + appropriate 10X buffer.
12. Tubes (1.5-mL microcentrifuge tubes, thin-walled 500–600-µL tubes for PCR).
13. Reagents for PCR (*Taq* DNA polymerase, 10X buffer, MgCl$_2$, deoxynucleoside triphosphates [dATP, dCTP, dGTP, dTTP], distilled H$_2$O).
14. Denaturing polyacrylamide gel reagents and equipment (premade solutions, or acrylamide, bis-acrylamide, 5X TBE buffer, urea, ammonium persulfate, TEMED, and distilled H$_2$O).
15. X-ray film, X-ray exposure cassette, and film development reagents or machine (Kodak).
16. Formamide loading dye: 98% formamide, 10 m*M* EDTA, pH 8.0, with bromophenol blue and xylene cyanol as tracking dyes.
17. 1X TBE buffer: 100 m*M* Tris, 100 m*M* Boric acid, 2 m*M* EDTA; prepare 5X stock solution.
18. [γ-^{33}P] dATP for radiolabeling *Eco*RI oligonucleotide primer.
19. Additional reagents for cloning: T/A cloning kit (e.g., pGEM-T from Promega, PCR-Script from Stratagene, or TOPO TA cloning kit from Invitrogen).
20. Transformation competent *Escherichia coli* cells, e.g., DH5α.
21. Luria Bertani broth: Bacto-tryptone, 10 g/L; yeast extract, 5 g/L; NaCl, 10 g/L in distilled H$_2$O.
22. Glycogen (molecular biology grade).

3. Methods

The methods described below outline the process of AFLP analysis in detail: (a) the preparation of DNA, (b) restriction enzyme digestion, (c) preparation and ligation of adapters, (d) preamplification, (e) end labeling of primer and selective amplification, (f) gel analysis, and (g) recovery of AFLP markers for downstream applications such as sequence determination.

3.1. DNA Preparation

Good-quality AFLP genotyping depends on complete digestion of DNA with restriction enzymes. Partial digestion will result in patterns that are not reproducible. Therefore, DNA should be prepared using methods that result in DNA of very high purity, such as standard ethanol precipitation procedures *(20)*. At least 10^7 trypanosomes, preferably an order of magnitude more, should be used to prepare DNA. The method used for parasite purification depends on

---

Fig. 3. (*Opposite page*) AFLP of a panel of independent isolates of *Trypanosoma evansi* and *T. brucei* using two primers, E-TA (*Eco*RI + TA-selective nucleotides) and M-CA (*Mse*I + CA). These primers allow the amplification of only a subset of restriction fragments in the reaction mixture, and result in clearly distinguishable fingerprints. In this example, arrows on the left side of the gel picture show some of the monomorphic markers (common to all), whereas the arrows on the right side show some of those that are polymorphic.

how the population is expanded (*see* **Note 1**). DNA from trypanosomes replicating in vitro should be prepared as follows.

1. Centrifuge the parasites at 2000*g* for 10 min. Resuspend fully in PBS, then centrifuge again as before.
2. Resuspend in 200 µL lysis buffer, transfer to a 1.5-mL microcentrifuge tube, and incubate at 37°C overnight.
3. Add an equal volume of phenol and roll gently for 5 min. Centrifuge in a microcentrifuge (at least 1000*g*) for 3 min and gently remove the aqueous layer. Add TE buffer (10 m*M* Tris-HCl, 1 m*M* EDTA, pH 8.0) to the phenol phase to extract any residual DNA from the interphase. Mix gently as before, repeat the centrifugation step, and pool the aqueous phases.
4. Add 1/10 vol of 3 *M* sodium acetate (pH 5.2), and 2 vol of absolute ethanol. Place at −20°C for at least 30 min.
5. Pellet at high speed in a (refrigerated) microcentrifuge at 11,000*g* to precipitate the DNA.
6. Wash with 1 mL of 70% ethanol, air-dry, and resuspend in 200 µL of TE buffer or sterile distilled H$_2$O.
7. Estimate the yield by spectrophotometry.

3.2. Restriction Endonuclease Digestion of Extracted Genomic DNA

Set up a digestion of 200–500 ng of genomic DNA with 10 u of *Eco*RI and 5 u of *Tru*9I (*Mse*I) in a reaction volume of 20 µL for 3 h at 37°C using a buffer that works well for both enzymes (e.g., Buffer A, Roche, Mannheim, Germany). Digestion is most convenient when using enzymes that work well in one buffer and in an identical optimum reaction temperature (*see* **Note 2**).

3.3. Preparation and Ligation of Adapters

The adapters are designed with a modification that changes the restriction site, allowing digestion and ligation to proceed noncompetitively (*see* **Fig. 2** and **Note 2**).

1. Add 50 µL containing 5 pmol of each *Eco*RI adapter oligonucleotide to a microcentrifuge tube and mix by light vortexing.
2. To a separate tube, add 50 µL of 50 pmol of each of the *Mse*I adapter oligonucleotides and mix by light vortexing. (The volumes depend on the scale of the project and can be scaled up or down.) *See* **Note 3** for estimation of adapter concentrations.
3. Place tubes in a floater rack and put the rack in a beaker containing water at 65°C. Transfer the beaker to a water bath at 65°C and incubate for 20 min.
4. Remove the beaker and allow to cool slowly to room temperature. The double-stranded adapter is now ready to be ligated to the restriction-digested DNA.
5. To the tube containing the restriction enzyme reaction mixture (from **Subheading 3.2.**), add 5 pmol *Eco*RI adapters, 50 pmol *Mse*I adapters, 3 µL 10X ligation buffer, 4 µL sterile distilled H$_2$O, and 1 µL T4 DNA ligase (final volume, 30 µL).

6. Incubate at 37°C for 2 h.
7. Dilute the sample 1:10 in TE buffer by adding 45 µL of buffer to 5 µL of sample. Store frozen at −20°C (see **Note 4**).

3.4. Preamplification

Preamplification is carried out for the purpose of expanding the amount of template DNA (restriction fragments ligated to adapters) available for subsequent reactions *(1)*. This is carried out in 50-µL reactions containing, as final concentrations: 1X PCR buffer, 0.1 µg of each primer (E-CORE primer and M+C), 1.5 mM MgCl$_2$, 200 µM dNTPs, 1.25 u of *Taq* polymerase, and 5 µL of the diluted template from **Subheading 3.3.5.** The following PCR cycling conditions should be used. Twenty cycles of denaturation at 94°C for 30 s, annealing at 65°C for 30 s, and extension at 72°C for 1 min. Following preamplification, the samples should be diluted 1:10 in TE buffer and stored at −20°C.

3.5. End Labeling of Primer and Selective Amplification

The three main features of the oligonucleotide primers for AFLP are shown in **Fig. 2**: one part is complementary to the adapter sequence; one part corresponds to a restriction enzyme site; and there is a 1- to 3-nucleotide extension at the 3'-end, which only amplifies fragments that have a sequence complementary to these nucleotides. The protocol described here works well with one primer radiolabeled (with [γ-^{33}P] dATP), which is safer to use than [^{32}P]-labeled nucleotides and gives very well resolved bands. Alterations to the protocol can be made to accommodate different strategies for resolving and detecting the AFLP amplicons (see **Note 5**).

End labeling is carried out using T4 PNK, which adds a phosphate group to the γ-position on nucleoside triphosphates *(21)*. Set the reaction up as follows.

1. To a tube containing 150 ng of the *Eco*RI-selective primer, add kinase buffer, 10 µL [γ-^{33}P] dATP (2000 Ci/mmol), and 20 u of PNK in a reaction volume of 50 µL. Incubate for 1 h at 37°C.
2. Heat-inactivate PNK by incubating at 70°C for 10 min.
3. Perform selective amplification in 20-µL reactions containing, as final concentrations, 1X PCR buffer, 30 ng of M-CA primer, 10 ng of E-TA primer (end-labeled with ^{33}P), 1.5 mM MgCl$_2$, 250 µM dNTPs, 1.25 u of *Taq* polymerase, and 1 µL of diluted template.
4. In a standard PCR machine, perform two cycles of denaturation at 94°C for 30 s, annealing at 65°C for 30 s, and extension at 72°C for 1 min. The same conditions for denaturation and extension are then repeated for 12 cycles, and the annealing temperature is stepped down by 0.7°C for each cycle, to 56°C. This is followed by 23 cycles, denaturing at 94°C for 30 s, annealing at 56°C for 30 s, and the extension step is at 72°C for 1 min.

5. Gel electrophoresis (*see* **Subheading 3.6.**) can be done immediately, or samples may be stored at −20°C for later analysis (*see* **Note 6**).

3.6. Preparation of Polyacrylamide Gel

1. Ensure that the glass plates are clean by washing with soap and plenty of running water, preferably warm. Air-dry by laying the plates in an angled position. Rinse with 95% ethanol.
2. Place spacers on the long plate and place the short plate on top, ensuring that the spacers extend to the bottom of the short plate.
3. Place binder clips on each side as close as possible to prevent leaks. Place the binder clips centrally over the spacer. A third spacer can be put at the bottom and more binder clips used to hold it in place. Alternatively, the bottom can be taped to prevent leaks (remember to remove tape before running gel).
4. To prepare a 60-mL gel, weigh 27 g urea into a beaker or cornical flask. Add 12 mL 5X TBE, 13 mL distilled H_2O, and stir with a magnet bar while warming slightly. If a flask is used, the urea can be dissolved by swirling lightly under warm running water. Add 9 mL of 40% acrylamide stock.
5. Filter the mix through a 0.45-mm, 25-mm-diameter membrane filter using a 60-mL syringe, or by pouring through a funnel containing filter paper (e.g., Whatman No. 1).
6. Add 320 µL of 10% APS (usually made fresh) and 30 µL TEMED. Swirl to mix, and immediately pour gel using a 60-mL syringe. Pour continuously to avoid forming bubbles.
7. Insert sharkstooth combs with teeth facing away from the gel to a depth of about 5 mm. Cover with extra acrylamide solution. Leave some in the container to monitor polymerization—usually complete in about 20 min, but it is advisable to wait about 1 h before proceeding to run the gel. Use one binder clip over each comb. Gels can be stored overnight at 4°C.

3.7. Gel Analysis

1. Pre-electrophorese the gel at constant power (55 W on a model S2 electrophoresis system, Invitrogen) for approx 20 min.
2. Add 20 µL formamide loading dye to 20 µL of the sample and mix. Heat-inactivate the mixture for 3 min at 95°C and immediately chill on ice.
3. Load 3 µL and run samples at 55 W until the xylene cyanol dye is about two-thirds down the gel (bromophenol blue will have run into the bottom buffer, 1X TBE).
4. Transfer the gels onto Whatman 3MM paper and dry using a heated vacuum slab dryer for 2 h at 80°C under vacuum.
5. Expose the dry gel to X-ray film (e.g., Kodak XAR) overnight at −70°C and then develop the film (*see* **Note 7**).

Autoradiograms can be read manually, or the bands can be digitized *(9)*. Further analysis depends on the use for which markers are generated. For example, polymorphic bands can be recovered from the gel and sequenced for development of specific markers for various purposes. The method for recovering markers is described in **Subheading 3.7.**

3.8. Recovery of DNA From Gels

You may wish to clone and sequence polymorphic fragments to develop a PCR-based marker or to determine genomic location. This is done by direct isolation of the band from the polyacrylamide gel that has been dried onto Whatman paper.

1. Determine the exact location of the band of interest on the dried gel by comparison to the autoradiograph.
2. Use a scalpel blade to cut the required band (with Whatman paper) into a 1.5-mL microcentrifuge tube.
3. Add 100 µL of distilled H_2O and soak at room temperature for 10 min.
4. Heat at 95°C for 20 min and allow to cool to room temperature.
5. Centrifuge for 1 min in a microcentrifuge at 11,000g.
6. Remove the solution to a clean tube, leaving paper and gel debris behind. Sufficient DNA will have dissolved in the water.
7. Add 0.1 vol of 3 M ammonium acetate, 10 ng glycogen, and 2.5 vol of absolute ethanol. Place at −70°C for 30 min.
8. Centrifuge in a microcentrifuge at 11,000g for 15 min.
9. Wash pellet with 80% ethanol, air-dry, and resuspend in 20 µL H_2O.
10. Use 5 µL of this solution for PCR using the two primers used in the preamplification process primers to recover the marker.

The PCR products may be cloned directly using PCR cloning vectors such as pGEM-T (Promega) and Topo-TA (Invitrogen) or PCR-Script (Stratagene).

4. Notes

1. Preparation of parasites. For trypanosomes, in vitro culture presents the best source of starting material. However, parasites grown in laboratory animals can be used, but should be purified through an anion-exchange column twice *(20)*. We recommend using at least 10^7 parasites, because losses will occur during purification. For *Plasmodium*, white blood cells should be removed by extensive washing *(6)*.
2. The success of AFLP depends on complete digestion of DNA. Addition of 5 mM spermidine hydrochloride enhances digestion and is recommended *(10,11)*. The primers for the adapters are designed to modify the restriction site, which allows digestion of genomic DNA and ligation of adapters to restriction fragments to proceed simultaneously. Therefore, it is not necessary to remove the restriction enzymes. The restriction enzyme *Tru*9I has the same recognition site as *Mse*I, but is much more cost friendly!
3. Prepare primer stocks at 1 µg/µL. For *Eco*RI adapters, mix 1.7 µL of adapter oligonucleotide 1 with 1.5 µg of adapter oligonucleotide 2. Add 60 µL of distilled H_2O to give a final concentration of 5 pmol/µL. For *Mse*I adapters, mix 16 µg

*Mse*I-oligonucleotide 1, 14 μg *Mse*I-oligonucleotide 2, and add 60 μL distilled H$_2$O to give a final concentration of 50 pmol/μL. These adapter preparations are sufficient for 60 ligations *(22)*.
4. Adapter ligation is carried out at 37°C, rather than at room temperature, to facilitate continuation of digestion. Complete digestion is central to the successful use of AFLP *(1,16,17)*. Different combinations of enzymes have been used, for example, *Hind*III and *Taq*I *(11)*.
5. Some authors recommend a much higher dilution (1:50). With good-quality DNA that is completely digested, a 1:10 dilution has given sharper bands in our hands. However, this can increase the background, as **Fig. 3** shows. We advise that you begin with a 1:20 dilution and make your adjustments appropriately.
6. There are numerous alterations to the use of radiolabeled oligonucleotides that may be useful in some circumstances. Fluorescent labeling generates patterns that are detectable by phosphorimaging, allowing data to be collected using an automated genotyper *(13)*. However, this is not suitable when recovery of the markers is desirable, and alternative detection methods that retain the bands on the gel should be used to facilitate this. This change in detection method can be limited to the specific primer combination containing the marker to be isolated. AFLP markers can also be detected on agarose gels, but this works well only when relatively large markers, for example, those generated using a single hexacutter. This has been described for analysis of bacterial genomes *(14)*, and we have some unpublished data from insects. Markers may also be detected by silver staining.
7. Development of X-ray films is most conveniently done in an X-ray film developer (e.g., X-OMAT from Kodak Instruments). However, development can also be done manually in a suitable darkroom using commercially available chemical from vendors of X-ray films.

References

1. Vos, P., Hogers, R., Bleeker, M., et al. (1995) AFLP: a new technique for DNA fingerprinting. *Nucleic Acids Res.* **11,** 4407–4414.
2. Lin, J. J., Kuo, J., and Ma, J. (1996) A PCR-based DNA fingerprinting technique: AFLP for molecular typing of bacteria. *Nucleic Acids Res.* **24, No. 18,** 3649–3650.
3. Janssen, P., Coopman, G., Huys, J., et al. (1996) Evaluation of the DNA fingerprinting method AFLP as a new tool for bacterial taxonomy. *Microbiology* **142,** 191–194.
4. Masiga, D. K., Tait, A., and Turner C. M. R. (2000) Amplified fragment length polymorphism in parasite genetics. *Parasitol. Today* **16,** 350–353.
5. Tait, A., Masiga, D., Ouma, J., et al. (2002) Genetic analysis of phenotype in *Trypanosoma brucei*: a classical approach to potentially complex traits. *Phil. Trans. R. Soc. Lond. B: Biol. Sci.* **357,** 89–99.
6. Grech, K., Martinelli, A., Pathirana, S., et al. (2002) Numerous, robust markers for *Plasmodium chabaudi* by the method of amplified fragment length polymorphism. *Mol. Biochem. Parasitol.* **123,** 95–104.

7. Shirley, M. W. and Harvey, D. A. (2000) A genetic linkage map of the apicomplexan parasite *Eimeria tenella*. *Genome Res.* **10**, 1587–1593.
8. Rubio, J. M., Berzosa, P. J., and Benito, A. (2001) Amplified fragment length polymorphism (AFLP) protocol for genotyping the malarial parasite *Plasmodium falciparum*. *Parasitology* **123**, 331–336.
9. Agbo, E. E. C., Majiwa, P. A. O., Claassen, H. J. H. M., et al. (2002) Molecular variation of *Trypanosoma brucei* subspecies as revealed by AFLP fingerprinting. *Parasitology* **124**, 349–358.
10. Gibson, J. R., Slater, E., Xerry, J., et al. (1998) Use of an Amplified Fragment Length Polymorphism technique to fingerprint and differentiate isolates of *Heliobacter pylori*. *J. Clin. Microbiol.* **362**, 2580–2585.
11. Ajmone-Marsan, P., Valentini, A., Cassandro, M., et al. (1997) AFLP markers for DNA fingerprinting in cattle. *Anim. Genet.* **28**, 418–426.
12. Koeleman, J. G., Stoof, J., Biesmans, D. J., et al. (1998) Comparison of amplified ribosomal DNA restriction analysis, random amplified polymorphic DNA analysis, and amplified fragment length polymorphism fingerprinting for identification of *Acinetobacter* genomic species and typing of *Acinetobacter baumannii*. *J. Clin. Microbiol.* **36**, 2522–2529
13. Folkertsma, R. T., Rouppe van der Voort, J. N., de Groot, K. E., et al. (1996) Gene pool similarities of potato cyst nematode populations assessed by AFLP analysis. *Mol. Plant–Microbe Interact.* **9**, 47–54.
14. Mueller, U. G. and Wolfenbarger, L. L. (1999) AFLP genotyping and fingerprinting. *TREE* **10**, 389–394.
15. Blears, M. J., De Grandis, S. A., Lee, H., et al. (1998) Amplified fragment length polymorphism (AFLP): a review of the procedure and its applications. *J. Ind. Microbiol. Biotechnol.* **21**, 99–114.
16. Savelkoul, P. H. M., Aarts, H. J. M., Haas, De J., et al. (1999). Amplified Fragment Length Polymorphism Analysis: the state of the art. *J. Clin. Microbiol.* **37**, 3083–3091.
17. Liu, Z., Nichols, A., Li, P., et al. (1998) Inheritance and usefulness of AFLP markers in channel catfish (*Ictalurus punctatus*), blue fish (*I. furcatus*), and their F1, F2 and backcross hybrids. *Mol. Gen. Genet.* **258**, 260–268.
18. http://www.genedb.org (database of trypanosomatid genomes, including *Leishmania major*, *Trypanosoma brucei*, and *T. cruzi*). Accessed 1-27-2004.
19. http://www.plasmodb.org (Web-accessible resource for the *Plasmodium falciparum* genome sequence). Accessed 1-27-2004.
20. Lanham, S. and Godfrey, D. G. (1970) Isolation of salivarian trypanosomes from man and other mammals using DEAE-cellulose. *Exp. Parasitol.* **28**, 521–534.
21. Sambrook, J., Fritsch, E. F., and Maniatis, T. (1989) Molecular Cloning: A Laboratory Manual, 2nd ed. Cold Spring Harbor Laboratory, Cold Spring Harbor, NY.
22. Liscum, M. and Oeller, P. AFLP: not only for fingerprinting, but for positional cloning. http://carnegiedpb.stanford.edu-publications-methods-aflp.html. Accessed 1-27-2004.

9

Minisatellites and MVR-PCR for the Individual Identification of Parasite Isolates

Annette MacLeod

Summary

In recent years, a wide variety of biochemical and molecular typing systems have been employed in the study of parasite diversity aimed at investigating the level of genetic diversity and delineating the relationships among different species and subspecies. Parasite sequence-specific polymerase chain reaction (PCR)-based genotyping systems are among the most useful tools employed to date, because they can be applied to very small quantities of host-contaminated parasite material and, using repeated loci such as mini- and microsatellites, allow the identification and tracking of individual strains as well as the determination of allele and genotype frequencies in populations. Although minisatellites have been used very successfully to study parasite populations, in particular *Trypanosoma brucei* populations *(13–15)*, there are some technical problems involved in the use of these markers. For example, minisatellite alleles tend to vary in a quasi-continuous fashion, making unambiguous allele identification difficult. The development of minisatellite variant repeat (MVR) mapping by the polymerase chain reaction (MVR-PCR) as a digital approach to DNA typing *(18)* has overcome many of the drawbacks of minisatellite length analysis. The system assays the dispersion patterns of MVRs within minisatellite alleles, producing an easily interpretable code for each allele. This technique not only allows unequivocal allele identification but also reveals cladistic information that can be used to determine the possible genetic relationships among the different strains and subspecies.

The MVR mapping technique has been applied successfully to minisatellites in the parasite *Plasmodium falciparum* *(20)* to uniquely identify strains, and more extensively in *Trypansoma brucei*, where it was used to determine population structure *(14)* and to examine the relationships among *T. brucei* subspecies, providing evidence for multiple origins of human infectively *(15)*.

In this chapter, the methods for genotyping of *T. brucei* parasites using both minisatellite allele length and MVR mapping are described in full and can be easily adapted to apply to minisatellites in other parasites.

Key Words: Genotyping; human infectivity; microsatellites; minisatellites; MVR-PCR; PCR; *Trypanosoma brucei*; VNTR.

From: *Methods in Molecular Biology, Vol. 270: Parasite Genomics Protocols*
Edited by: S. E. Melville © Humana Press Inc., Totowa, NJ

1. Introduction

Several different typing systems have been used to analyze variation in parasite populations, each method accessing different degrees of variation within and between populations. In the case of the African trypanosome, *Trypanosoma brucei*, the most extensively used techniques include isoenzyme analysis *(1–7)*, restriction fragment length polymorphism (RFLP) *(8)*, and RFLPs in repetitive ribosomal genes generating molecular fingerprints *(9)*. In recent years, the polymerase chain reaction (PCR)-based DNA typing system, randomly amplified polymorphic DNAs (RAPDs), has been increasingly applied to the study of parasite genetics and has been used to complement isoenzyme analysis *(10, 11)*. However, isoenzyme and RAPD analysis require the parasites to be purified from contaminating host material. In the case of African trypanosomes, this involves growth of the parasites in mice followed by purification of the parasites in a DEAE cellulose column to generate enough purified material for analysis, a procedure that may apply selection to mixed samples. The use of species-specific PCR can circumvent many of the drawbacks of the other techniques. Indeed, species-specific PCR amplification of multicopy repeats has proved useful at the species level for identifying different *Trypanosoma* spp. *(12)*. At the level of individual isolates, amplification of single-copy hypervariable loci such as mini- and microsatellites can uniquely identify strains *(13,14)*.

Hypervariable mini- and microsatellites, or variable number tandem repeat (VNTR) loci, are tandemly repeated regions of eukaryotic genomes, many of which show extreme levels of variation because of differences in the number of repeats within the tandem array. Hypervariable minisatellites have been identified in numerous parasite species, and these vary in a strain-specific manner and so provide a means of identifying and tracking individual strains as well as allowing the allele and genotype frequencies of populations to be determined. The use of locus-specific primers to amplify these regions by PCR enables the genotyping of parasites even when they are contaminated with large quantities of host DNA, in addition to allowing the analysis of small quantities of DNA (for example, **Fig. 1**). Because of their high level of polymorphism, minisatellite markers are particularly useful in determining variation among populations. Most work using minisatellites in *T. brucei* has focused on analyzing populations *(14,15)* and detecting heterogeneity within a sample (mixed stocks) *(13)*, although they have also proved useful in analyzing progeny from genetic crosses, providing evidence for Mendelian inheritance *(16)*. The high level of variability at minisatellite loci allows strains to be tracked through populations and can be used to identify trypanosomes that are responsible for epidemics. An example of the PCR amplification of one minisatellite from a series of field isolates is given in **Fig. 1C**.

A

```
consensus repeat unit         actgtgaaaccagtggcagtccctgaaccagccaaggcagar
allele 1   variant positions  gt            tt                  g      g   r
allele 2   variant positions  gt            t                          t   r
allele 3   variant positions                t                          t   r
```

B

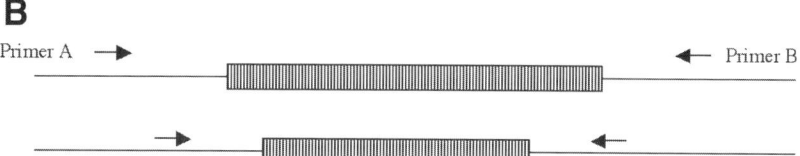

C

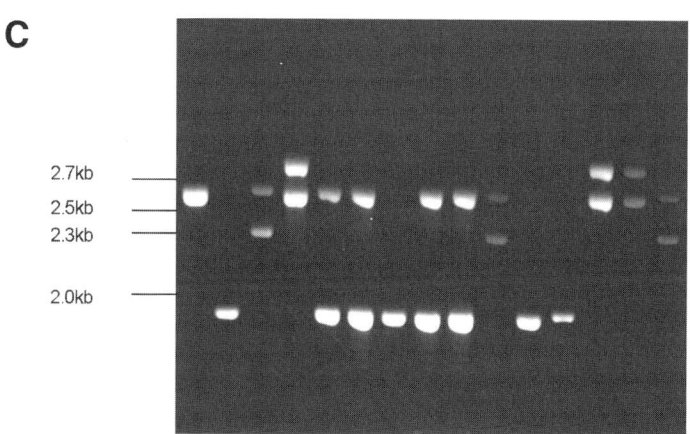

Fig. 1. (**A**) *MS42* repeat unit consensus sequence is shown. Positions that vary from this in three different alleles are indicated in bold. **r** = either **a** or **g** at that position. (**B**) A schematic diagram of two minisatellite alleles that vary in the number of tandem repeats. Positions of primers A and B are indicated by arrows flanking both alleles. (**C**) The PCR amplification of products of the *CRAM* minisatellite locus for a collection of *T. brucei* isolates from tsetse flies from Busoga, Uganda.

The identification of mini- and microsatellite loci has been facilitated by the ever-increasing number of genome sequences available and the development of computer programs that scan sequences and identify all tandem repeats (for example, Tandem Repeat Finder *[17]*).

However, despite the advantage of minisatellite markers accessing a higher degree of variation between isolates than any other loci, there are some technical difficulties involved in the use of these markers. For example, minisatellite alleles tend to vary in a quasi-continuous fashion, making unambiguous allele identification difficult. Thus, when size differences between alleles are small, allele length estimates lack accuracy and variation between gel runs can lead to a failure in identifying matching samples or the false matching of different samples, thus weakening the statistical power of population databases based on allele length. Also, alleles of the same size may differ in repeat structure (i.e., position of nucleotide polymorphisms), and so size may not indicate relatedness. The development of minisatellite variant repeat (MVR) mapping by the polymerase chain reaction (MVR-PCR) as a digital approach to DNA typing *(18)* has overcome many of the drawbacks of VNTR length analysis. The system assays the dispersion patterns of variant repeats within minisatellite arrays, producing an easily interpretable code for each allele. The simple and rapid MVR-PCR technique increases the level of information about each allele so that the ability to define differences is increased. Furthermore, the information is generated in a digital format ideal for computer-based analysis and for the production of population databases.

MVR mapping has been applied successfully to numerous minisatellites in humans *(18)*, mice *(19)*, and the protozoan parasites *Plasmodium falciparum (20)* and *Trypanosoma brucei (15)*. The study of variant repeats within minisatellite arrays has proved useful for individual/stock identification for the analysis of populations and has revealed cladistic information about evolutionary origins. For example, MVR-PCR analysis of a human minisatellite from samples from around the world has revealed that the non-African alleles are a subset of the much greater diversity found in the African population, supporting the "out of Africa" theory of human evolution *(21)*. In the African trypanosome *T. brucei*, MVR-PCR data support the theory of multiple origins of human infectivity *(15)*.

MVR-PCR has also been applied to the study of the mutation processes involved in the generation of new length alleles for numerous human minisatellites, providing evidence that interallelic as well as intra-allelic recombination may play an important role *(22,23)*, with the frequency of the mutation process apparently being modulated by *cis*-acting elements *(24)*. The discovery of minisatellite mutations that are associated with human disease has renewed interest in minisatellite mutation processes *(25)*.

The technique of MVR-PCR can be applied to any minisatellite that has one common variant repeat and has a constant repeat unit size. Also, to map variant repeats along the entire length of the repeat array, the minisatellite must be shorter that 5 kb, as only partial maps can be obtained from larger alleles. To illustrate the general principles of MVR-PCR, the method of analyzing one minisatellite in *T. brucei*, *MS42*, which satisfies these strict criteria, is presented here. *MS42* is a single-copy gene, located on chromosome I, containing a 42-bp repeat that varies in number from 16 repeats to 47 repeats. Sequence analysis of the *T. brucei* minisatellite *MS42* revealed the presence of variant repeats (*see* **Fig. 1A**), the most common being an A/G transition at the beginning of the repeat unit that does not affect the predicted amino acid sequence of the gene product. The repeat unit size is constant and all alleles are small and so can be easily amplified in their entirety using primers complementary to the regions flanking the tandem repeats (*see* **Fig. 1B,C**). Single alleles then can be purified and their internal structure determined by MVR-PCR. The MVR-PCR method accesses the dispersion pattern of the variant repeats within the repeat array of each allele of the minisatellite, *MS42*. Thus, each repeat can have either an A or G base at the first position of the repeat unit. The two classes of repeat unit were designated **a**-type and **g**-type repeats (**Fig. 2A**). To access this variation, PCR primers were generated that are specific for each repeat type. Using these MVR-specific primers and a specific primer located in the flanking DNA, a ladder of PCR products corresponding to the position of each **a**-type and each **g**-type repeat could be generated (**Fig. 2B,C**). Applied to single isolated alleles, a binary code of the distribution of **a**- and **g**-type repeats within the repeat array could be constructed; but if applied to total genomic DNA (including both alleles), a ternary code could be derived from the superimposed maps of the two alleles, which is individual-specific. From the ternary code, incomplete maps of each allele can be deduced (*see* **Note 9**). Unlike the human minisatellites *MS32* and *MS31* *(18)*, in which only partial allele maps can be obtained because of many alleles being greater that 10 kb long, all the alleles of the *MS42* variable region are small (less than 2.4 kb), and so every allele can be completely and readily amplified by PCR. This allows each allele to be analyzed in full, in a similar fashion to the human minisatellite *MS205* *(21,26)*.

MVR analysis can reveal a larger number of alleles at each minisatellite locus than can be detected by band size estimates, even assuming 100% accuracy of the band size measurements, as alleles that are identical in size can have very different internal structures. Therefore, using this system, unrelated alleles of the same size are no longer scored as matching alleles. This method is ideal for strain identification, providing a unique fingerprint for each isolate. Also, the distribution pattern of variant repeats within alleles contains cladistic information, as alleles with very similar patterns presumably share a recent common ancestor. This provides a rational method for grouping (or binning) alleles for population genetic analysis and evolutionary studies. In the case of *T. brucei*, alleles from human infective isolates from Uganda showed greater similarity to the local nonhuman infective isolates than to human infective isolates from a different focus in Zambia, indicating that there are at least two separate lineages of human infective trypanosomes *(15)* (for examples of MVR allele maps and groupings, *see* **Fig. 3**).

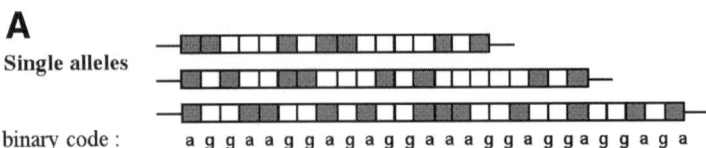

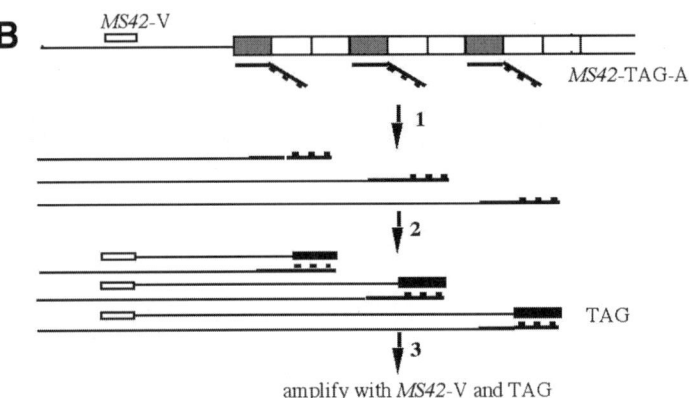

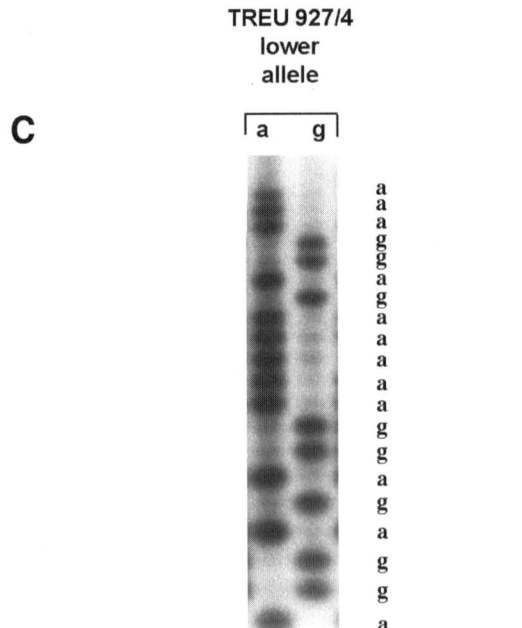

This protocol describes the methods used to (a) make crude lysates from unpurified parasite material, which is suitable for use as a template for PCR amplification, (b) amplify one hypervariable minisatellite locus, *MS42*, from *T. brucei* field isolates and determine the variant genotypes, and (c) further analyze the *MS42* alleles by MVR-PCR to reveal isolate-specific fingerprints and cladistic relationships between alleles.

2. Materials

1. Parasites or parasite DNA (*see* **Subheading 3.1.** for three alternative methods for extraction of *T. brucei* DNA) (*see* **Note 1**).
2. Phosphate-buffered saline (PBS).
3. Lysis buffer: 50 mM Tris-HCl, pH 8.0, 100 mM EDTA, pH 8, and 0.5% sodium dodecyl sulfate (SDS). Add 3.3 µL of 20-mg/mL proteinase K to every 100 µL lysis buffer.
4. Gene-specific oligonucleotide primers: (a) for amplification of entire minisatellite repeat region and (b) for amplification of variant repeats within the minisatellite. (*See* **Note 2** for examples relating to *T. brucei* minisatellites.)
5. 10X stock of PCR mix: 45 mM Tris-HCl (pH 8.8), 11 mM (NH$_4$)$_2$SO$_4$, 4.5 mM MgCl$_2$, 0.113 mg/mL bovine serum albumin (BSA), 4.4 mM EDTA, 1 mM each of dATP, dCTP, dGTP, and dTTP.
    Store at −20°C in 100-µL aliquots. Commercially available from ABgene as Custom PCR master mix CM116 (*see* **Note 3**).

Fig. 2. (*Opposite page*) The principles of minisatellite repeat coding. (**A**) Principle of digital coding. Minisatellite alleles consisting of interspersed arrays of two variant repeat units termed **a**-type (shaded boxes) and **g**-type (open boxes). Individual alleles can be encoded as a binary code extending from the first repeat unit. (**B**) The principle of MVR-PCR. Illustrated for a single allele amplified using *MS42*-TAG-A. 1. At low concentrations of *MS42*-TAG-A primer, the primer will anneal to approx one **a**-type repeat unit per target minisatellite molecule and extend into the flanking DNA. 2. Primer *MS42*-V then primes from the flanking DNA, creating a sequence complementary to the TAG sequence. 3. The newly synthesized fragments terminate in V at one end and the complement of TAG at the other. They can now be amplified using high concentrations of *MS42*-V and *MS42*-TAG primers to create a set of PCR products extending from the flanking *MS42*-V site to each **a**-type repeat unit. By using primer *MS42*-TAG-G in place of *MS42*-TAG-A, a complementary set of products terminating at each **g**-type repeat unit can be generated. This figure was adapted from Jeffreys et al., 1991 *(18)*. (**C**) MVR-PCR of *MS42* from the smaller allele of stock TREU 927/4. MVR mapping was achieved by amplification of a single allele (a PCR product isolated from a gel) with primers *MS42*-V, *MS42*-TAG, and *MS42*-TAG-A (left lane) or *MS42*-TAG-G (right lane), followed by agarose gel electrophoresis, Southern blotting, and hybridization to a *MS42* repeat probe. The derived internal map of **a**- and **g**-type repeats in this allele is shown to the right of the autoradiograph.

isolate	allele name	sequence
group 1		
U2274u	35d	aggagaggaagagagaaggagaaggagaaaaagagagagagg
K96u	35a	aggagaggaagagagaaggagaaggagaaaagaggagagagg
K2340-4	30e	aggagaggaagagagaaggagaaggagaaaagaggaagagg
K71	28	aggagaggaagagagaaggagaagggga---gaagagg
U8341	27d	aggagaggaagagagaaga----aaagaagagg
Upap601	26a	aggagaggaagagagaaga----aagaagagg
K927c11b1	20	aggagaggaagaaaagaggaaa
group 3		
U9341	22b	agaggaaagaa-ga-aaagaagaa
Z2181	42	agaggaaggaagagaaagagaagaaggagagagagaaggaagagg
Z218u	47	agaggaaggaagagaaagagaagaaggagagagagagagaaagagagg
Z194u	45	agaggaaggaagagaaagagaagaaggagagagagagaagaaagaagg
group 4		
K925c11	27b	-agaaaggagaaaaggaaaaa------aaagaga
K981un1	33a	-agaaaggggaaagaaaaaaaaaagaagaga
K997c11	25b	-agaaagagagaa-------aaaaaaaagaga
Z1941	32	aagaaagagagaaggaaagaaggaaaaggaga
Z222u	30f	aagaaagagagaaggaaggaaagagaagga
K925c1u	29	-agaaagagagaagaaggaaggagagaga
K981un3	30b	-agaaagagagagaaggaagagagagaa
group 6		
UB23u	38c	aaaaaaaaaaaaaaaaaanaannaaa-aaaaaaaaannnag
Upapo160u	38b	aaaaaaaaaaaaaaaaaaaannaaaannnaaa-aaaaaaaaannnag
UB25 u	38d	aaaaaaaaaaaaaaaaaaannaannaaaaaa-aaaaaannnag
UI155u	36b	aaaaaaaaaaaaaaaaannaaannaaanaannnnag
UI147u	35b	aaaaaaaaaaaaaaaaannaaanaa-aaaaaannnag
K2340-3	34	aaaaaaaaaaaaannaaanaa-aaaaaannnag
K2340u	38e	aaaaaaganaaannaaannaaaaaaaaannnaggagg

6. 0.2-mL PCR tubes (Stratagene).
7. *Taq* polymerase, 2.5 U/µL (ABgene).
8. Mineral oil (Sigma).
9. 0.1% SDS in PBS.
10. 5X TBE stock: 45 mM Tris-borate, 1 mM EDTA.
11. Agarose gel: 1% agarose in 0.5X TBE.
12. DNA markers: 1-kb and 100-bp ladders as supplied by ABgene.
13. Spin-X columns (Costar).
14. 20X SSC: 3 M NaCl, 0.3 M trisodium citrate, pH 7.0.
15. 3X SSC: 0.45 M NaCl, 0.045 M trisodium citrate, pH 7.0.
16. Denaturing solution: 0.5 M NaOH, 1.5 M NaCl.
17. Neutralizing solution: 3 M NaCl, 0.5 M Tris-HCl, pH 7.5.
18. Prime-it kit (Stratagene).
19. ^{32}P-dCTP.
20. Loading dye: 0.1% bromophenol blue, 40% ficoll, 5 mM EDTA.
21. Ethidium bromide stock solution, 10 mg/mL.
22. Magna membrane as supplied by MSI.
23. 3MM paper supplied by Whatman.
24. Hybridization solution: 0.5 M NaHPO$_4$, pH 7.2, 7% SDS, 1 mM EDTA.
25. High stringency washing solution: 0.1X SSC, 0.01% SDS.
26. X-ray film as supplied by Kodak.

3. Methods

3.1. Preparation of Parasite Genomic DNA

Here, we present three possible methods for the extraction of DNA from *Trypanosoma brucei*.

3.1.1. Preparation of Crude Lysate From Procyclic Culture

1. Pellet 1–5 mL of culture, by centrifuging at 2500 rpm for 5 min.
2. Remove supernatant and resuspend in 500 µL of PBS.
3. Centrifuge at 720g for 5 min in a standard microfuge.
4. Remove supernatant.
5. Resuspend pellet in 100 µL lysis buffer containing proteinase K.
6. Incubate for 3 h (or overnight) at 56°C.
7. Centrifuge briefly to remove condensation from lid. Store at 4°C or –20°C.

Fig. 3. (*Opposite page*) Examples of groups of aligned *MS42* alleles. Four allele groups are shown. For each allele, its place of isolation is given as a prefix to its stock number, Z, Zambia, K, Kenya, and U, Uganda. The derived MVR code (allele name) is given: **a** and **g** denote the positions of **a**-type and **g**-type repeats, respectively, **n** = **n**-type repeats (*see* **Subheading 3.4.**). Gaps (—) have been introduced to aid alignment.

8. Prior to PCR amplification, dilute 1/100 in PCR clean H₂O and inactivate the proteinase K by heating to 95°C for 5 min.
9. Use 1 µL as a template for subsequent PCR amplification.

3.1.2. Preparation of Crude Lysate From Bloodstream-Form Trypanosomes for Small Quantities

1. Starting material can be as little as three or four stabilate straws or as much as 1 mL of infected blood. The following method is for a few microliters, approx 20 µL, of infected blood. If using larger quantities, adjust volume accordingly or use the alternative method given in **Subheading 3.1.3.**
2. Thaw at room temperature.
3. If starting with stabilate straws, place straws in an Eppendorf tube and spin at low speed in a microfuge to flush them out.
4. Add 100 µL PBS.
5. Centrifuge once more at 720*g* in a microfuge for 5 min.
6. Discard the supernatant.
7. Resuspend pellet in 100 µL PBS.
8. Centrifuge at 720*g* in a microfuge for 5 min.
9. Discard supernatant.
10. Resuspend pellet in: 50 µL lysis buffer with 1.6 µL proteinase K solution (20 mg/mL).
11. Incubate overnight at 56°C.
12. Inactivate the proteinase K by incubating sample at 95°C for 5 min.
13. Make a 1:50 dilution of sample and use 1 µL as a template for PCR amplification.

3.1.3. Preparation of Crude Lysate From Bloodstream-Form Trypanosomes: Alternative Method for Larger Volumes

1. Add 9 vol of 0.1% SDS in PBS to 1 vol of infected blood.
2. Spin down at 1300*g* for 5 min in a benchtop centrifuge.
3. Discard the supernatant.
4. Resuspend pellet in: 200 µL lysis buffer with 6.4 µL proteinase K solution (20 mg/mL).
5. Continue from **step 10** to **step 13** in **Subheading 3.1.2.**

3.2. PCR Amplification of Minisatellite Locus MS42 From T. brucei DNA or Crude Lysate

The method below refers to one minisatellite, *MS42*. However, exactly the same protocol can be applied to minisatellites *CRAM* and *292 (13)*, including the same PCR conditions. Oligonucleotide primers for these loci are given in **Note 2**.

1. For each amplification reaction, prepare the following mix (but *see* **Note 4**):
   2 µL 10X PCR mix
   2 µL 10 µM primer (e.g., *MS42-V*)

Minisatellite Variant Repeat PCR

2 µL 10 µM primer (e.g., *MS42-F*)
0.2 µL *Taq* polymerase
12.8 µL Milli-Q H$_2$O (*see* **Note 5**)
1 µL template DNA (5 ng/µL)
Total volume = 20 µL.
(*See* **Note 6**.)

2. Overlay each reaction with 30 µL of mineral oil (*see* **Note 7**). Cycle under the appropriate conditions. For PCR amplification using a Stratagene robocycler, the following conditions were found to be optimal for all primer pairs given here (*see* **Note 2**): 95°C for 50 s, 64°C for 50 s, 66°C for 3 min for 30 cycles.
3. Run 10 µL of each PCR product on a 1% agarose gel in 0.5X TBE, 0.5 µg/mL ethidium bromide, and visualize under UV. Allele band sizes can be determined using software such as Kodak 1D image analysis software, on the basis of mobilities relative to a reference DNA standard (*see* **Note 8**).
4. The remaining PCR product is reserved, should the sample need to be electrophoresed again against different size standards or diluted if the concentration of the PCR product is so high as to distort its mobility during electrophoresis. Allele sizes are recorded as estimates of the number of repeat units in the tandem array.
5. Estimate allele frequencies by calculating the number of times a particular allele is observed divided by the total number of alleles in the population. Multilocus genotypes are the combined results of a number of minisatellite loci (for example, *MS42*, *CRAM*, and *292*).
6. Allele and multilocus frequencies can then be plotted on histograms and entered into any number of population genetics programs (for example, Genetic Data Analysis software).

3.3. MVR-PCR of Minisatellite MS42

3.3.1. Amplification by PCR of the Entire Minisatellite Repeat Array

1. Initially amplify the variable region of the *MS42* locus in each isolate to visible levels on an ethidium-stained gel, using the universal flanking primers *MS42*-V and *MS42*-F (as in **Subheading 3.2.**) (*see* **Fig. 1C** and **Note 9**).
2. Excise each PCR-generated band, corresponding to the different alleles in the sample, from the agarose gel under UV illumination.
3. Load the gel slice onto a Spin-X column (Costar) and centrifuge at 15,800*g* rpm for 5 min in a standard microfuge.
4. Store the filtrate (the purified allele) at −20°C as necessary.
5. Use a 1:50 dilution of each allele as a template for subsequent PCR reactions using MVR-specific primers.

3.3.2. Amplification of Repeat Types Within the Minisatellite

As each allele is composed of a mixture of two repeat unit types (**a**-types and **g**-types), which differ by one base, two separate PCR reactions are performed for each allele. The two different MVR-specific primers prime off either **a**- or g-type repeats in conjunction with the universal flanking primer (*MS42*-V) to generate two sets of products, one recording the products of repeats starting with **a** and the other repeat starting with **g** (**Fig. 2B**) (but *see* **Note 10**).

Reaction **a** for each allele: 1 μL 10X PCR mix; 1 μL 10 μ*M* primer *MS42*-V; 1 μL 10 μ*M* primer TAG (*see* **Note 10**); 1 μL 0.1 μ*M MS42*-TAG-A; 0.1 μL *Taq* polymerase; 4.9 μL Milli-Q H$_2$O; 1 μL template DNA (1/50 dilution of purified allele); total volume = 10 μL.

Reaction **g** for each allele: 1 μL 10X PCR mix; 1 μL 10 μ*M* primer *MS42*-V; 1 μL 10 μ*M* primer TAG (*see* **Note 10**); 1 μL 0.1 μ*M MS42*-TAG-G; 0.1 μL *Taq* polymerase; 4.9 μL Milli-Q H$_2$O; 1 μL template DNA (1/50 dilution of purified allele); total volume = 10 μL.

1. Perform PCR using the cycling conditions: 50 s at 95°C, 50 s at 65°C, and 3 min at 68°C for 18 cycles.
2. Separate PCR products by electrophoresis through a 40-cm-long 1% Seakem agarose gel in 0.5X TBE, 0.5 μg/mL ethidium bromide at 125 V for 17 h (*see* **Note 11**).
3. Transfer DNA by Southern blotting onto Magna membrane (MSI) using the following method.
4. Wash gel in denaturing solution, 0.5 *M* NaOH, 1.5 *M* NaCl, for 15 min, and for a further 15 min in fresh denaturing solution.
5. Submerge the gel in neutralization solution for 15 min, and for a further 15 min in fresh neutralization solution.
6. Transfer to standard blotting apparatus consisting of a reservoir of 20X SSC with a wick of 3MM paper. Cut Magna membrane to the size of the gel, rinse in H$_2$O, and place on top of the gel. A glass pipet can be used as a rolling pin to ensure no air bubbles are trapped between the gel and the membrane.
7. Lay 3MM paper on top of membrane, followed by 5 cm (depth) of paper towels, and a glass plate with a weight on top.
8. Replace wet towels with dry every few minutes for the first 15 min and then leave overnight.
9. Dismantle the blot and mark the top right-hand corner of the membrane.
10. Rinse the membrane in 3X SSC and bake in an 80°C oven for 2 h.
11. UV cross-link DNA to membrane using a UV Stratalinker, or place membrane on a UV transilluminator (DNA side down) for 40 s.
12. Label 25 ng of an *MS42* repeat probe (supplemented with 100 pg of marker DNA, typically 100-bp ladder [ABgene]) with ^{32}P-dCTP using the Prime-it kit (Stratagene) according to the manufacturer's instructions.
13. Hybridize probe and DNA on blot in hybridization solution at 65°C overnight.
14. Wash the membrane in high-stringency washing solution for 4×15 min at 65°C.
15. Expose to autoradiograph for 24 h, in the first instance.

3.4. Analysis of MVR-PCR Data

Southern blotting and hybridization of an *MS42* repeat probe will reveal a continuous ladder of complementary products. The number of rungs on the ladder corresponds to the number of repeat units within the array. This can be verified by comparing the number of repeats indicated by MVR-PCR analysis to the number of repeats in each allele estimated by allele size measurements, to confirm that the entire allele has been mapped. The largest allele detected (47 repeats in *MS42*) generates clearly separated products ranging in size from 319 to 2377 bp.

The MVR mapping autoradiograph for one allele from *T. brucei* stock TREU 927/4, illustrating the complementary ladder of PCR products generated by amplification using **a**- and **g**-type repeat-specific primers, is presented in **Fig. 2C**. Adjacent is the deduced MVR code generated from it. Occasionally, during MVR-PCR of *MS42*, some repeat units failed to amplify with either **a**- or **g**-type specific primers, indicating the presence of MVRs containing further sequence variation that prevents priming with the *MS42*TAG-A or *MS42*TAG-G primers. These "null" or **n**-type repeats appear as gaps in the MVR ladder, and can be scored as a "third state" for single-allele mapping. The MVR code for each allele can then be compared to a database of codes, and identical or similar alleles can be identified. Comparisons of different *MS42* alleles can be performed using the computer programs described in **ref. *21***. In addition, pairwise dot matrix analysis may be performed on some alleles, for example, by selecting for a match of 10 perfect repeats (*see* **Note 12**). The authenticities of selected matches and the final alignment of allele groups can be made by eye.

4. Notes

1. DNA may also be extracted using a Qiagen blood kit, cat. no. 51104.
2. The following oligonucleotide primers were designed to amplify the tandem repeats within the coding sequences of the *T. brucei* genes listed: *MS42* (Tb927.1.750 on chromosome I, cysteine-rich acidic membrane protein [*CRAM*], and gene *292*).

Primer	5' - 3' sequence
MS42-F	ttgtgcggtcgttaacgcgcgttcaa
MS42-V	cattattccacggacgcgaagcagc
MS42-TAG-A	tcatgcgtccatggtccggacagtccctgaaccagccaaggcagaa
MS42-TAG-G	tcatgcgtccatggtccggacagtccctgaaccagccaaggcagag
TAG	tcatgcgtccatggtccgga
CRAM-G	ctgctgatgccgtacatgatgatttc
CRAM-H	aactccctcccgatcgatcacaac
292-G	acaccccctctccacttcagatac
292-H	gctgaacctgtgggcccctcaattg

   TAG = nucleotide extension (*see* **Note 10**).

3. When amplifying minisatellite loci, there are two main types of unsatisfactory results, (a) a lack of product and (b) the amplicon is smeared. The first usually can be rectified by using the PCR buffer described in **Subheading 2.**, and not the buffer

supplied with the *Taq* polymerase. The components of the PCR buffer are critical. A high magnesium concentration ensures the maximum yield is obtained; however, this may result in mispriming, which is avoided by using a high annealing temperature in the cycling reaction. The second type of unsatisfactory result is the result of template concentration. The quantity of DNA is the major factor in obtaining reasonable amplification. Although both lysates and purified DNA can be used as templates for the PCR amplification, better yields usually are obtained from purified DNA than from lysates. The optimum amount of DNA per 10-µL reaction is 5 ng. More than this amount will result in "collapse" of the PCR product because of the repetitive nature of the amplicon. This occurs when incomplete PCR products, generated during the first few rounds of cycling, act as primers for subsequent PCR cycles and result in ever-shortening products. An example of the ladder of incomplete products can be seen below the lower allele in several lanes of **Fig. 1**. "Collapse" of PCR products can also occur if the cycle number is increased, resulting in smearing. If too much template is added or the number of cycles is increased, the lower allele of a heterozygote will be preferentially amplified, possibly resulting in a failure of the upper allele to amplify to visible levels.

4. For simplicity, the method given here applies to a single PCR reaction. For multiple reactions, make a master mix containing all the following components except the template DNA, multiply the volumes given by the number of samples to be typed (or by sample number +1, to allow for pipetting errors), and allowing for one negative control and one positive control sample. Dispense the mastermix into 9-µL aliquots then add 1 µL template DNA to each aliquot. This leads to better reproducibility.
5. Milli-Q H$_2$O should be stored in disposable plastic ware and not in laboratory glassware, to avoid possible contamination from the laboratory environment.
6. Thin-walled 0.2-mL PCR tubes are best used directly from the manufacturer, without any treatment. Autoclaving of tubes prior to use is unnecessary and increases the chances of contamination from the laboratory environment. Disposable plastic ware should be used at all times. Setting up of the PCR reactions should be performed in an area free of trypanosome DNA.
7. Even when using thermocyclers with heated lids, some evaporation occurs. Because such small reaction volumes are used, we recommend overlaying with mineral oil for all thermocyclers.
8. It is also useful to create a custom-made reference standard prepared from a well-characterized locus to give a collection of differently sized alleles.
9. MVR-PCR can be applied not only to single alleles (described here) but also to genomic DNA. However, from genomic DNA of a diploid organism, the resulting MVR map is a composite of both alleles, which provides a genetic fingerprint of the individual. It is possible to deduce partial maps for the two individual alleles from a diploid map, for example, where a band appears in the **a**-type lane only at a given position, both alleles have an **a**-type repeat unit at that position in the tandem array. However, where a band appears in both lanes, it is not possible to identify which allele has an **a**- or **g**-type repeat unit at that position.

10. Because the MVR-specific primers can prime internally from the initial PCR products, this can lead to progressive shortening of the PCR products with each cycle of amplification. To prevent this from occurring, each MVR-specific primer contains a 20-nucleotide extension, TAG. Thus, the MVR-tagged primers provide the specificity of the PCR reactions and are used at low concentrations, whereas the TAG primer is used at higher concentrations to amplify these products (**Fig. 2B**).
11. The 40-cm gel trays and tanks that we use are twice as long as standard apparatus and had to be custom-made. It is possible to use standard equipment, but the resolution for longer alleles is limited and may result in partial maps being obtained.
12. Matches of 10 repeats are selected, as it is expected that such matches will only appear approx once by chance in a dataset of the size of 1500 repeat units, based on the random distribution of the two types of repeat units (probability of a match is 1 in 2^{10} = 1 in 1024).

References

1. Gibson, W. C., Marshall, T. F. d. C., and Godfrey, D. G. (1980) Numerical analysis of enzyme polymorphism: a new approach to the epidemiology and taxonomy of trypanosomes of the subgenus Trypanozoon. *Adv. Parasitol.* **18,** 175–246.
2. Gibson, W. C. and Gashumba, J. K. (1983) Isoenzyme characterization of some Trypanozoon stocks from a recent trypanosomiasis epidemic in Uganda. *Trans. R. Soc. Trop. Med. Hyg.* **77,** 114–118.
3. Tait, A., Babiker, E. A., and LeRay, D. (1984) Enzyme variation in *T. brucei* spp. I. Evidence for the subspeciation of *T. b. gambiense*. *Parasitology* **89,** 311–326.
4. Tait, A., Barry, J. D., Wink, R., et al. (1985) Enzyme variation in *Trypanosome brucei* spp. II. Evidence for *T. b. rhodesiense* being a set of variants of *T. b. brucei*. *Parasitology* **90,** 89–100.
5. Godfrey, D. G., Baker, R. D., Rickman, L. R., et al. (1990) The distribution, relationships and identification of enzymatic variants within the subgenus Trypanozoon. *Adv. Parasitol.* **29,** 1–74.
6. Tait, A., Buchanan, N., Hide, G., et al. (1993) Genetic recombination and karyotype inheritance in *T. brucei* species, in *Genome Analysis of Protozoan Parasites* (Morzaria, S. P., ed.), ILRAD, Nairobi, Kenya, pp. 93–107.
7. Tait, A. (1990) Genetic exchange and evolutionary relationships in protozoan and helminth parasites. *Parasitology* **100,** 75–87.
8. Sternberg, J., Turner, C. M. R., Wells, J. M., et al. (1989) Gene exchange in African trypanosomes: frequency and allelic segregation. *Mol. Biochem. Parasitol.* **34,** 269–280.
9. Hide, G., Welburn, S. C., Tait, A., et al. (1994) Epidemiological relationships of *Trypanosoma brucei* stocks from South East Uganda: evidence for different population structures in human infective and non-human infective isolates. *Parasitology* **109,** 95–111.
10. Stevens, J. R. and Tibayrenc, M. (1995) Detection of linkage disequilibrium in *Trypanosoma brucei* isolated from tsetse flies and characterized by RAPD analysis and isoenzymes. *Parasitology* **110,** 181–186.

11. Gibson, W. C., Kanmogne, G., and Bailey, M. (1995) A successful backcross in *Trypanosoma brucei*. *Mol. Biochem. Parasitol.* **69,** 101–110.
12. Desquesnes, M. and Davila, A. M. R. (2002) Applications of PCR-based tools for detection and identification of animal trypanosomes: a review and perspectives. *Vet. Parasitol.* **109,** 213–231.
13. MacLeod, A., Turner, C. M. R., and Tait, A. (1999) A high level of mixed *Trypanosoma brucei* infections in tsetse flies detected by three hypervariable minisatellites. *Mol. Biochem. Parasitol.* **102,** 237–248.
14. MacLeod, A., Tweedie, A., Welburn, S. C., et al. (2000) Minisatellite marker analysis of *Trypanosoma brucei*: reconciliation of clonal, panmictic, and epidemic population genetic structures. *Proc. Natl. Acad. Sci. USA* **97(24),** 13,442–13,447.
15. MacLeod, A., Welburn, S. C., Maudlin, I., et al. (2001) Evidence for multiple origins of human infectivity in *Trypanosoma brucei* revealed by minisatellite variant repeat mapping. *J. Mol. Evol.* **52,** 290–301.
16. Tait, A., Masiga, D., Ouma, J., et al. (2002) Genetic analysis of phenotype in *Trypanosoma brucei*: a classical approach to potentially complex traits. *Phil. Trans. R. Soc. Lond. B* **357,** 89–99.
17. Benson, G. (1999) Tandem repeats finder: a program to analyze DNA sequences. *Nucleic Acids Res.* **27,** 573–580.
18. Jeffreys, A. J., MacLeod, A., Tamaki, K., et al. (1991) Minisatellite repeat coding as a digital approach to DNA typing. *Nature* **354,** 204–209.
19. Bois, P., Stead, J. D. H., Bakshi, S., et al. (1998) Isolation and characterization of mouse minisatellites. *Genomics* **50,** 317–330.
20. Arnot, D. E., Roper, C., and Bayoumi, R. A. L. (1993) Digital codes from hypervariable tandemly repeated DNA sequences in the *Plasmodium falciparum* circumsporozoite gene can genetically barcode isolates. *Mol. Biochem. Parasitol.* **61,** 15–24.
21. Armour, J. A. L., Anttinen, T., May, C. A., et al. (1996). Minisatellite diversity supports a recent African origin for modern humans. *Nat. Genet.* **13,** 154–160.
22. Armour, J. A. L., Harris, P. C., and Jeffreys, A. J. (1993) Allelic diversity at minisatellite MS205 (D16S309): evidence for polarized variability. *Hum. Mol. Genet.* **2(8),** 1137–1145.
23. Buard, J. and Vergnaud, G. (1994) Complex recombination events at hypervariable minisatellite CEB1 (D2S90). *EMBO J.* **13,** 3203–3210.
24. Monckton, D. G., Neumann, R., Guram, T., et al. (1994) Minisatellite mutation rate variation associated with a flanking DNA sequence polymorphism. *Nat. Genet.* **8,** 162–170.
25. Buard, J. and Jeffreys, A. J. (1997) Big, bad minisatellites. *Nat. Genet.* **15,** 327–328.
26. Armour, J. A. L., Monckton, D. G., Neil, D. L., et al. (1993) Mechanisms of mutations at human minisatellite loci, in *Genome Rearrangement and Stability* (Davies, K. E. and Warren, S. T., eds.) Cold Spring Harbor Laboratory Press, Cold Spring Harbor, NY, pp. 43–57.

10

Analysis of Differentially Expressed Parasite Genes and Proteins Using Transcriptomics and Proteomics

Daniel C. Gare

Summary

At any particular point in time, the full complement of transcribed RNAs and relevant proteins of a cell are known as the transcriptome and proteome, respectively. The composition of these two populations changes throughout the life cycle of a parasite or in response to environmental factors, such as drug treatments. Comparing the changes in the composition of the transcriptome and proteome between different life-cycle stages or in the same stage but under different conditions can be of particular interest, as it can allow the identification of potentially important differentially expressed genes and proteins. Combining the analysis of both the transcriptome and proteome in tandem allows changes in RNA transcripts to be followed right through to changes in the level of protein expression.

The protocols in this chapter describe methods for analyzing the transcriptome, by using suppression subtraction hybridization to construct subtracted complementary DNA libraries, and the proteome, by using two-dimensional sodium dodecyl sulfate-polyacrylamide gel electrophoresis. These two methods are then integrated to allow the global changes in RNA and protein expression to be examined. The protocols have been adapted for working on parasites and contain extensive notes.

Key Words: Differential expression; proteome; subtracted cDNA library; transcriptome; two-dimensional SDS-PAGE.

1. Introduction

The full complement of messenger RNAs (mRNAs) transcribed from a cell's genome at any particular point in time is termed the transcriptome, and the relevant protein complement is the proteome. The study of both of these populations, transcriptomics and proteomics, has developed at a very rapid rate over the last few years *(1–3)*. Studying the changes in gene and protein expression in parasite systems can be of particular interest, as it may allow the identification of genes that are involved in crucial stages of parasite development, such as penetration into host cells, for example *(4–11)*. Therefore, it is important to be able to identify the subset of genes and associated proteins that are differentially expressed at these times.

Transcriptomics includes various experimental approaches such as DNA arrays, which allow the global study of thousands of genes in one experiment, or perhaps a user-defined subset from any chosen developmental stage *(12)*. Although DNA arrays are tremendously powerful and provide a vast amount of data, they can have certain drawbacks, such as the need for prior knowledge of the sequences of the genes analyzed and the costs involved in preparing and screening the arrays. Producing and screening subtracted libraries is a powerful method to identify transcripts that are uniquely or selectively expressed, either in certain tissues or during development or in disease states (for both the host and parasite).

Proteomics is still largely centered on two-dimensional (2D) gel electrophoresis for the initial separation of complex mixtures of cellular proteins and normally is coupled to mass spectrometry to characterize separated proteins. This is now a very sensitive and reliable method for characterizing the protein profile of a particular sample.

This chapter's aim is to provide methods for the combined study of the parasite transcriptome and proteome by constructing subtracted complementary DNA (cDNA) libraries and carrying out 2D gel separation. The results of both of these analyses will then be integrated to relate changes seen in the transcriptome to those visualized in the proteome (**Fig. 1**).

2. Materials

Unless stated, use MilliQ H_2O throughout.

2.1. Construction of Subtracted cDNA Library

1. Trizol reagent (Invitrogen Life Technologies).
2. Message Maker mRNA Isolation System (Invitrogen Life Technologies).
3. Microfuge homogenisers (Anachem).
4. Polymerase chain reaction (PCR)-select cDNA Subtraction kit (Clontech).
5. PCR-select Differential Screening kit (Clontech).
6. Standard PCR machine.
7. AdvanTAge PCR cloning kit (Clontech).
8. DH5α electrocompetent cells (Promega).
9. Electroporation system and cuvets.
10. Agar plates with ampicillin, X-Gal, and IPTG (as required for protocol provided in AdvanTAge kit).

2.2. Separation of Proteins by Two-Dimensional Gel Electrophoresis

1. Methanol.
2. Chloroform.

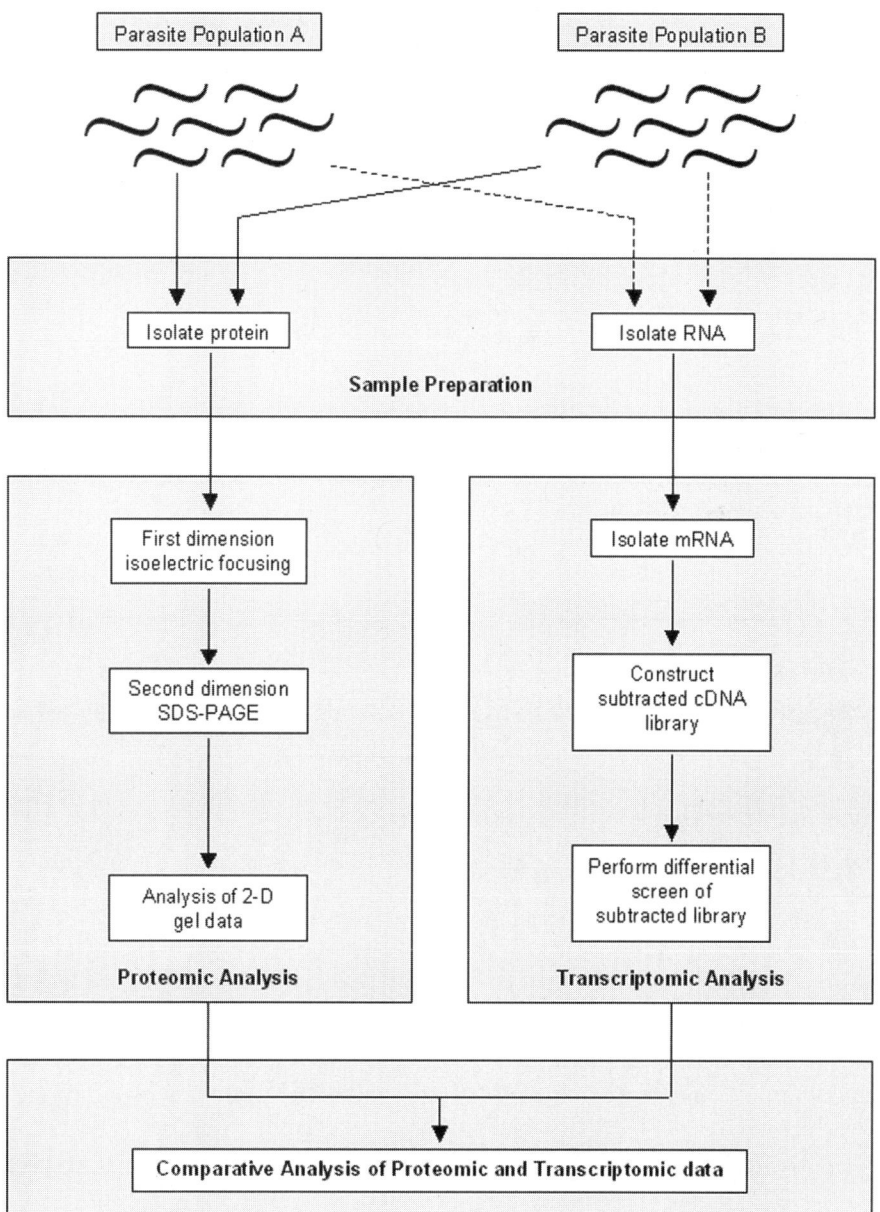

Fig. 1. Overview of the experimental approach described here. The starting point is two dissimilar parasite populations, for example, two different developmental stages from the same species or the same stage from related species. Proteins and RNA are isolated from each sample and then the proteome and transcriptome of each set of samples analyzed individually. Both datasets are then analyzed comparatively, allowing the study of expression of RNA and protein of a single gene.

3. 2D sample buffer: 9 M urea, 2% (w/v) CHAPS, 25 mM Tris-HCl, pH 7.5, 3 mM EDTA, 50 mM KCl, 50 mM dithiothreitol (DTT), 40 µM leupeptin, 2% resolyte ampholytes, pH 3.5–10. Prepare in 100 µL aliquots and store at –80°C; do not refreeze.
4. Microfuge homogenizers (Anachem).
5. 11-cm Immobiline DryStrips and relevant IPG buffer (APBiotech).
6. IPG reswelling stock buffer: 7 M urea, 2 M thiourea, 0.2% (w/v) CHAPS, 0.0002% bromophenol blue. Prepare in 2.5-mL aliquots, store at –20°C.
7. DryStrip Cover fluid (APBiotech).
8. Immobiline DryStrip Reswelling tray (APBiotech).
9. Immobiline DryStrip Kit (APBiotech).
10. Multiphor II Electrophoresis unit (APBiotech).
11. 12–14% ExcelGel precast gradient gels (APBiotech).
12. IPG equilibration buffer 1: 0.5 M Tris-HCl, pH 6.8, 6 M urea, 30% glycerol, 2% sodium dodecyl sulfate (SDS), 65 mM DTT. Omit DTT to prepare stock solution. Add DTT just prior to use.
13. IPG equilibration buffer 2: 0.5 M Tris-HCl, pH 6.8, 6 M urea, 30% glycerol, 2% SDS, 135 mM iodoacetamide. Omit iodoacetamide to prepare stock solution. Add iodoacetamide just prior to use.

3. Methods

The protocols below describe the individual protocols for preparing parasite samples (**Subheading 3.1.**), for examining the transcriptome (**Subheading 3.2.**), and proteome (**Subheading 3.3.**), and then combining the results of both to create an integrated approach (**Subheading 3.4.**).

3.1. Preparation of Parasite Material

Correct preparation of parasite starting material, for both the construction of the subtracted libraries and analysis by 2D gels, is crucial in ensuring efficient and meaningful results. All steps must be taken to ensure that parasite samples are as free from any contaminating material, such as host tissue, host cells, or bacteria, as is possible. Failure to do so will see the inclusion of the transcriptome and proteome of these contaminants in all downstream operations and analyses (*see* **Note 1**). For parasites for which an in vitro model is available, this obviously is a great advantage over having to isolate parasites from animal tissues (*see* **Note 2**).

3.2. Construction of cDNA-Subtracted Libraries Using Suppression Subtractive Hybridization

Suppression subtractive hybridization (SSH) is a PCR-based cDNA subtraction method and is a very powerful technique to compare two mRNA populations *(13)*. cDNAs of genes that are either overexpressed or exclusively expressed in one population and not the other are selectively amplified, whereas amplification of nondifferentially expressed targets is suppressed. Using only a single round of subtractive hybridization, SSH can achieve more than 1000-fold enrichment for differentially expressed cDNAs *(14)*. The population from which specific targets are to be found is called the tester cDNA, and the reference population is called the driver cDNA. To achieve the maximum of efficiency and sensitivity, contamination of target mRNA, and hence cDNA, by driver mRNA has to be completely avoided (*see* **Note 3**).

3.2.1. mRNA Isolation

Although methods for direct isolation of poly(A)$^+$ RNA are available, we have found isolating poly(A)$^+$ RNA from total RNA leads to greater yields and purity. Starting with high-quality mRNA is crucial for constructing a good subtracted library. Time invested in the early stages ensuring the quality of the RNA will pay dividends further downstream. The method outlined below uses TRIZOL reagent to extract total RNA from parasite tissues and the Message Maker kit to isolate the poly(A)$^+$ RNA fraction. As with all RNA work, prevention of degradation by RNases is vital, and great care should be taken to avoid the introduction of RNases. This can be helped by regularly changing disposable gloves, using sterile, RNase-free disposable plastics, and treating all non-Tris solutions with 0.1% (v/v) diethylpyrocarbonate (DEPC).

1. Once the tester and driver parasite samples have been isolated and purified away from all contaminating material, they should be thoroughly disrupted using a sterile homogenizer in the appropriate volume of TRIZOL reagent (1 mL/50–100 mg tissue) (*see* **Note 4**).
2. Incubate the samples at RT for 5 min, then add 200 μL of chloroform per 1 mL TRIZOL reagent and shake samples vigorously for 1 min. Incubate samples for 3 min at room temperature.
3. Centrifuge samples at 12,000g for 15 min at 4°C, then carefully transfer the aqueous phase to a fresh tube. Add 500 μL of isopropyl alcohol per 1 mL TRIZOL reagent and incubate samples for 15 min (*see* **Note 5**), then centrifuge at 12,000g for 10 min at 4°C.
4. The RNA should be visible as a gel-like pellet on the side and bottom of the tube. Remove the supernatant and wash the RNA pellet with at least 1 mL 75% ethanol per 1 mL TRIZOL reagent. Mix pellet and ethanol by vortexing and centrifuge at 7500g for 5 min at 4°C.
5. Air-dry RNA pellets for 5–10 min, then redissolve in an appropriate amount of DEPC-treated H$_2$O to a concentration of 0.55 mg/mL.
6. The poly(A)$^+$ RNA can then be isolated by passing the total RNA through the Message Maker mRNA isolation system. Heat up to 2 mg of total RNA (*see* **Note 6**) for 5 min at 65°C, then chill on ice. Adjust the salt concentration to 0.5 M NaCl by adding 0.1 vol 5 M NaCl.

7. Add the appropriate volume of oligo(dT) cellulose (100 μL per 100 mg total RNA) to the RNA sample and thoroughly mix by inversion. Incubate at 37°C for 10 min, then transfer the RNA/cellulose suspension to the filter syringe barrel and expel the liquid and unbound RNA by pushing the plunger to the bottom of the barrel.
8. Draw the appropriate amount of wash buffer 1 (in the kit: 3 vol per 1 vol of oligo [dT] cellulose added) into the syringe barrel and mix the contents well until all the cellulose is resuspended, then expel the liquid. Repeat with wash buffer 2 (in the kit).
9. To elute the poly(A)$^+$ RNA draw the appropriate amount of DEPC-treated H$_2$O, preheated to 65°C (equal to the volume of oligo[dT] cellulose added) into the syringe barrel and thoroughly mix the contents. Expel the single-selected poly(A)$^+$ RNA into a fresh tube.
10. To perform the second selection, incubate the eluted poly(A)$^+$ RNA at 65°C for 5 min, then chill on ice. Adjust the salt concentration to 0.5 M by adding 0.1 vol 5 M NaCl. Use the same oligo(dT) cellulose for the second selection but wash it first by drawing in the same amount of wash buffer 1 as in **step 8** and thoroughly resuspend the cellulose. Expel the wash buffer and draw in the single-selected poly(A)$^+$ RNA, resuspend completely, and incubate at room temperature for 10 min.
11. Expel the liquid and unbound RNA and repeat **steps 8** and **9** to wash and elute the double-selected poly(A)$^+$ RNA.
12. Determine the concentration of the poly(A)$^+$ RNA by spectrophotometry, and concentrate, if necessary, by ethanol precipitation. Check the quality by analyzing a small amount by agarose gel electrophoresis looking for a strong poly(A)$^+$ smear and absence of any ribosomal RNA.

3.2.2. Suppression Subtraction Hybridization

SSH is performed using the PCR-Select cDNA subtraction kit according to the manufacturer's protocol. Briefly:

1. AMV reverse transcriptase is used to synthesize first-strand cDNA from 2 μg of poly(A)$^+$ RNA from the tester population and 2 μg of poly(A)$^+$ RNA from the driver population (*see* **Note 7**).
2. DNA polymerase I is used to synthesize the second-strand cDNA, after which the double stranded tester and driver cDNA is restriction-digested with *Rsa*I. This generates shorter, blunt-ended fragments optimal for subtractive hybridization. Subtractions should be performed in both directions for each tester/driver pair. The forward subtraction will enrich for differentially expressed transcripts present in the tester but not in the driver, and the reverse subtraction will enrich for those sequences present in the driver and not in the tester.
3. Split the digested tester cDNAs into two separate tubes and ligate separately to the two different adapters that will allow the selective amplification of differentially expressed cDNAs.
4. Hybridize the two adapter-ligated tester cDNAs separately with excess driver cDNA, then combine the tester cDNAs and hybridize again with fresh driver cDNA.

5. Subject the entire population to two rounds of PCR to selectively amplify the differentially expressed sequences. The number of cycles in both the first and second rounds of PCR must be optimized to maximize the representation of target sequences.
6. Verify the efficiency of SSH by PCR by comparing the abundance of known cDNAs before and after subtraction. Two genes should be examined: one that is known not to be differentially expressed between the two RNA sources to ensure it has been subtracted and one that is present in the tester RNA but not in the driver RNA, to check for enrichment of differentially expressed genes (*see* **Note 8**).
7. After confirmation of the efficiency of SSH, clone 3 µL of the secondary PCR product from the forward and reverse subtraction into the AdvanTAge vector and transform into DH5α cells by electroporation.
8. After plating out on agar plates with ampicillin, X-Gal, and IPTG, recombinant clones (white colonies) are picked and used to inoculate LB medium in 96-well plates to provide a permanent collection of differentially expressed clones.

3.2.3. Differential Screening of Subtracted cDNA Library

Although the above method greatly enriches for differentially expressed sequences, the library still will contain some cDNAs that are common to both the tester and driver samples. This background depends partly on the quality of the starting RNA and on the efficiency of SSH but mainly arises when very few mRNA transcripts are differentially expressed between tester and driver populations. By using the PCR-select Differential Screening kit the background can be minimized and false positives removed before further detailed screening takes place. Briefly:

1. Randomly select recombinant, potentially differentially expressed clones from the 96-well plates of both the forward and reverse subtracted libraries and amplify the inserts by PCR using nested primers within the adapter sequences.
2. Array PCR products from each subtraction onto four identical nylon filters and prepare the filters for hybridization using standard dot-blot protocols.
3. Hybridize the following probes to each set of four filters: the tester-specific subtracted probe, the driver-specific subtracted probe, and the cDNA synthesized directly from tester and driver mRNA. When these probes are hybridized to the filters, truly differentially expressed targets should be present in the tester-specific subtracted probe and the tester cDNA but absent in the driver-specific subtracted probe and the driver cDNA (*see* **Note 9**).
4. Finally, confirm the differential expression of candidate clones identified by screening by hybridization to Northern blots, using standard protocols.

Of all the clones randomly selected from the forward and reverse subtracted libraries, normally only 10% or less will be found to be truly differentially expressed. It is not unusual at all to lose so many potential targets, which demonstrates the need for a screening step to minimize the number of false positives. To increase the number of true positives identified, it is necessary to screen more clones.

3.3. Separation of the Proteome on 2D Gels

Two-dimensional electrophoresis (2D electrophoresis) is a powerful and widely used method for the analysis of complex protein mixtures extracted from cells, tissues, or other biological samples. This technique sorts proteins according to two independent properties in two discrete steps: the first-dimension step, isoelectric focusing (IEF), separates proteins according to their isoelectric points (pI); the second-dimension step, SDS-polyacrylamide gel electrophoresis (SDS-PAGE), separates proteins according to their molecular weights (M_r, relative molecular weight). Each spot on the resulting two-dimensional array corresponds to a single protein species in the sample. Thousands of different proteins can be separated, and information such as the protein pI, the apparent molecular weight, and the amount of each protein is obtained. For comparison of the proteome with the transcriptome it will obviously be necessary to use the same samples on the 2D gels as for generating the tester and driver cDNA for the subtractive libraries.

3.3.1. Sample Preparation

Appropriate sample preparation is absolutely essential for good 2D gel analysis, and the optimal procedure must be determined for each sample type. Ideally, the process will result in the complete solubilization, disaggregation, denaturation, and reduction of the proteins in the sample. Great care must be taken to avoid contamination of the samples from either external sources (gloves must be worn at all times and changed frequently) or from the other samples. The method outlined below has been optimized for the analysis of soluble parasite proteins secreted into culture medium (*see* **Note 10**).

1. To precipitate proteins from medium, for 400 μL of sample add 400 μL of methanol and 100 μL chloroform. Vortex thoroughly for 30 s and then incubate on ice for 30 min.
2. Add 300 μL of H_2O and centrifuge for 2 min at 10,000g at 4°C. Proteins should precipitate at the interphase, so carefully remove the top aqueous layer and discard.
3. Add 500 μL methanol and vortex thoroughly, then centrifuge for 5 min at 10,000g at 4°C.
4. Remove the supernatant and air dry the pellet for 10 min, then resuspend in 25 μL of 2-D sample buffer.
5. Centrifuge the sample for 10 min at 15,000g to ensure that no particulate material is loaded onto the first dimension separation.

3.3.2. First-Dimension Isoelectric Focusing

There are a wide range of pH intervals available for the first-dimension separation, from a broad range for a wide overview of total protein distribution (pH 3.0–10.0), medium intervals for a more detailed overview (pH 3.0–7.0, 4.0–

7.0, 6.0–11.0, and 6.0–9.0), to narrow ranges for the highest resolution (pH 3.5–4.5, 4.0–5.0, 4.5–5.5, 5.0–6.0, and 5.5–6.7). The protocol below uses one of the medium intervals, pH 4.0–7.0, for a more detailed overview.

1. Prepare IPG reswelling stock buffer. Just prior to use, for 2.5 mL of stock buffer add 12.5 µL of the appropriate IPG buffer and 7 mg DTT.
2. Into a fresh tube, add the resuspended sample (50–300 µg) and the reswelling buffer to a final volume of 200 µL.
3. Apply the reswelling buffer including sample to the center of a slot on the reswelling tray, taking care not to introduce any bubbles. Remove the protective cover from the IPG strip starting at the acid (pointed) end, then position the strip, gel side down, with the pointed end of the strip against the sloped end of the slot. Gently lift and lower the strip and slide it back and forth along the surface of the buffer to help coat the entire strip. Overlay each strip with 3 mL of DryStrip cover fluid and leave to reswell at room temperature overnight.
4. Before the IPG strips are removed from the tray, clean all the components of the Immobiline DryStrip kit with detergent and rinse thoroughly with distilled H_2O and allow to dry. Set the cooling unit to 20°C. Add 4 mL of DryStrip cover fluid to the cooling plate on the Multiphor II unit and position the Immobiline DryStrip tray on the cooling plate so the red (anodic) electrode connection of the tray is positioned at the top of the plate near the cooling tubes. Remove any large bubbles between the tray and the cooling plate; small bubbles can be ignored. Connect the red and black electrode leads on the tray to the Multiphor II unit. Pour about 10 mL of DryStrip Cover Fluid into the Immobiline DryStrip tray and place the Immobiline DryStrip aligner, 12-groove-side-up, into the tray on top of the DryStrip Cover Fluid.
5. Cut two electrode strips to 11 cm and soak with 0.5 mL of distilled H_2O, then blot with tissue paper to remove excess H_2O.
6. Remove the IPG strips from the reswelling tray and immediately transfer to the adjacent grooves of the aligner in the Immobiline DryStrip tray. Place the strips with the pointed (acidic) end at the top of the tray near the anode.
7. Place the moistened electrode strips across the cathodic and anodic ends of the aligned IPG strips, then align the relevant electrode over the electrode strips and press down.
8. Apply DryStrip Cover Fluid to cover the IPG strips and run at the following conditions: Phase 1 = 300 V, 2 mA, 5 W for 1 min; Phase 2 = 3500 V, 2 mA, 5 W for 1.5 h; Phase 3 = 3500 V, 2 mA, 5 W for 3.5–5 h.
9. After completion of run, remove IPG strips to separate tubes and store at –70°C.

3.3.3. Second-Dimension SDS-PAGE

As with the first-dimension separation, there is a wide range of choice for the composition of the polyacrylamide gel for separation in the second dimension. When a gradient gel is used, the overall separation interval is wider and the linear separation interval is larger. In addition, sharper bands result because the decreasing pore size functions to minimize diffusion. Single-percentage gels offer better resolution for a particular M_r window. As with the choice of pH range in the first dimension, the protocol below uses a percentage range (12–14%) that will allow a more detailed overview of proteins (*see* **Note 11**).

1. Place IPG strips in individual tubes with the gel side facing inward and add 10 mL of IPG equilibration buffer 1 to each tube. Incubate the tubes on their sides on a rocker at room temperature for 15 min.
2. Remove the strips to a fresh tube containing 10 mL of IPG equilibration buffer 2 and incubate on the rocker again for 15 min.
3. Remove the strips from the tubes and eliminate excess fluid by blotting the sides of the strips onto tissue paper for at least 3 min, but do not leave the strips for longer than 10 min.
4. Set the cooling unit to 15°C and add 3 mL of DryStrip Cover Fluid to the Multiphor II cooling plate. Remove the gel from its package and position appropriately on the plate. A notch at the lower left-hand corner of the film identifies the 12.5 or 14% (i.e., anodic) end. Ensure the cover fluid spreads completely under the gel and gently smooth out any air bubbles. Remove the cover on the gel and allow to air-dry for 5 min.
5. Remove the protective foil from the colorless cathodic buffer strip and place smooth, narrow side downward, making sure the buffer strip is in complete contact with the gel. Repeat the procedure with the yellow anodic buffer strip.
6. Place the equilibrated IPG strips gel side down on the SDS gel so the cathodic strip and the IPG strips are parallel to each other and 2–3 mm apart.
7. Place one IEF sample application piece at the end of each IPG strip underneath the plastic tab formed by the overhanging gel support film at each end of the IPG strip, making sure the application pieces touch the ends of the IPG strip. Application pieces absorb H_2O that flows out of the IPG strips during electrophoresis. Make sure that the IPG strip is in full, direct contact with the SDS gel; remove any bubbles by stroking the plastic backing of the IPG strip gently with a spatula or forceps.
8. If loading marker proteins, place an extra application piece on the surface of the gel just beyond the end of the IPG strip and pipet 15–20 μL of the markers onto the extra sample application piece.
9. Position the electrodes on the IEF electrode holder so that they are aligned with the buffer strips, then plug in the electrode connectors and slowly lower the electrode holder onto the buffer strips.
10. Run the second-dimension gel at the following conditions: Phase 1 = 200 V, 20 mA, 30 W, 40 min; Phase 2 = 800 V, 40 mA, 40 W, 2 h 40 min. After the bromophenol blue dye front has moved approx 0.5 cm from the IPG strips, pause the power and remove the IPG strips and application pieces. Then move the cathodic buffer strip forward to cover the area of the removed IPG strip, also moving the cathodic electrode as well. Halt electrophoresis 5 min after the bromophenol blue dye front has reached the anodic buffer strip.

11. After completion of the run, remove the gels and stain with the appropriate method, for example, silver, colloidal Coomassie, or autoradiography. For picking spots for tryptic digestion and mass spectrometry, Coomassie staining is most appropriate.

3.3.4. Analysis of 2D Gel Data

In theory, the analysis of up to 15,000 proteins should be possible in one gel. In practice, however, 5000 detected protein spots is considered a very good separation. Evaluating high-resolution 2D gels by a manual comparison of two gels is not always possible. In large studies with patterns containing several thousand spots, it may be almost impossible to detect the appearance of a few new spots or the disappearance of single spots. Image collection hardware and image evaluation software (such as Phoretix 2D range, http://www.phoretix.com/index.htm) are necessary to detect these differences as well as to obtain maximum information from the gel patterns. Comparison between the protein profiles will lead to the identification of differentially expressed proteins between the two parasite samples (**Fig. 2**), be it from different developmental stages from the same species or between the same stages from very closely related species (*see* **Note 12**). For identification, these candidate spots can be excised from the gel and digested enzymatically. The resulting peptides can then be analyzed by mass spectrometry and the results used to screen protein databases for possible matches and thereby identify the protein spot. The success of finding a match in the database is obviously entirely dependent on the number of sequences deposited in there for the chosen species of parasite. Those parasite species with genome projects completed or in progress (such as *Plasmodium falciparum*, *Trypanosoma brucei*, and *Leishmania major*, http://www.sanger.ac.uk/Projects/Protozoa) will clearly have a distinct advantage over those that are not being investigated. One potential way round a lack of complete sequences in the database is to use any expressed sequence tags (ESTs) available (for example, *Trichuris muris* and *Trichinella spiralis* EST projects, http://nema.cap.ed.ac.uk/Nest.html). A custom database can be created from the EST data and the mass spectrometry data used to screen this. Obviously, this form of database has its limitations, as most ESTs are only 400–500 bp long, which will only yield a translated product of a few hundred residues and hence only provide a small target over which to match with the candidate spot removed from the gel. However, with these limitations taken into account, it can provide an excellent starting point for identification of spots from which further analysis can then take place.

3.4. Combining Transcriptomic and Proteomic Approaches

Each of the methods described above is a very powerful way of analyzing either the transcriptome or the proteome of a parasite at a specifically chosen

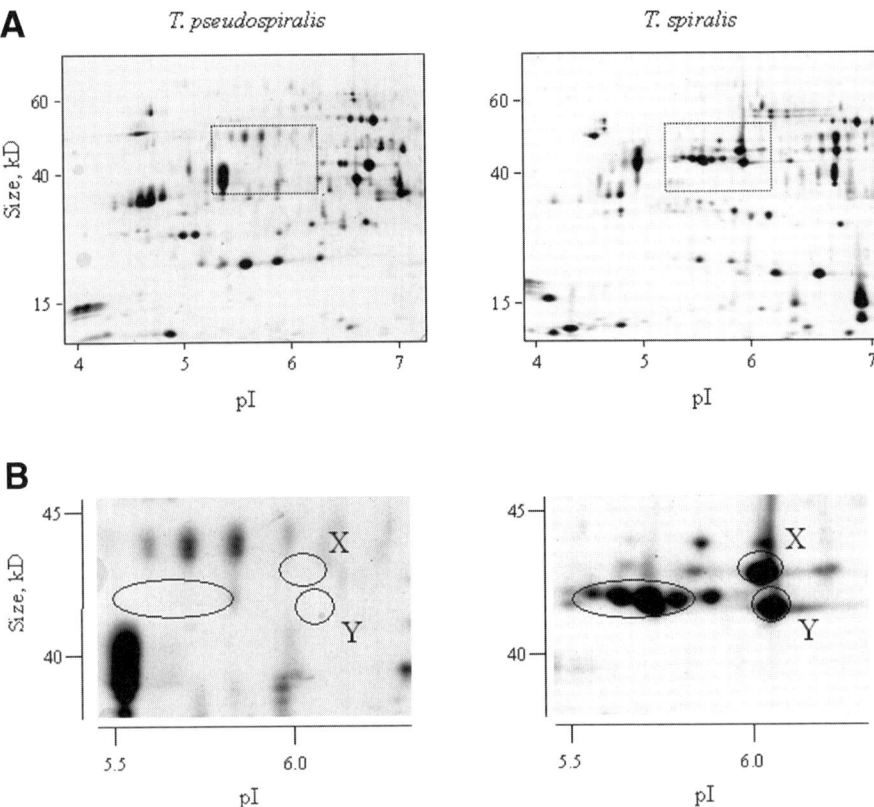

Fig. 2. 2D gel separation of excreted/secreted (ES) products from muscle larvae of *Trichinella pseudospiralis* and *T. spiralis*. ES products were collected as previously described *(15)* and separated by the protocol in this chapter (**A**). A region of interest (boxed) is shown in greater detail with examples of differential expression of peptides circled (**B**). Tryptic digest followed by mass spectrometry of spots X and Y identified them as a serine protease (AY028974) and the 43 kD secreted glycoprotein (M95499), respectively (Gare and Connolly, unpublished). Images reproduced by permission of B. Connolly.

position in its life cycle. Combining the two techniques allows us to relate changes seen in gene expression right through to changes in the level of protein expression. Once target clones have been selected and validated as being truly differentially expressed from the subtractive cDNA library, the inserts can be sequenced and the predicted amino acid sequence determined. From this the theoretical molecular weight and pI of the protein can easily be determined (e.g., Protein Parameter tool at ExPASy, http://ca.expasy.org/cgi-bin/protparam). Taking into account any glycosylations or other posttranslational modifications, the presence, or absence, of this protein can then be visualized on the relevant protein profile. The protein expression of the gene can then be examined and followed. Confirmation that the identified spot is the same as the theoretical protein from the library screen can be achieved by performing the enzymatic digest and mass spectrometry analysis on the candidate spot. It is then possible simply to compare the size of the fragments with a theoretical *in silico* digest (e.g., Peptide Mass at ExPASy, http://ca.expasy.org/tools/peptide-mass.html) of the amino acid sequence from the clone. So, combining the two approaches allows a global study of the changes in the transcriptome and proteome of a parasite at specified stages.

4. Notes

1. Various methods may be needed to purify the parasite material away from host tissues, which will need to be determined for each individual parasite. These may include sedimentation (either through a gradient such as sucrose or simply PBS) or filtering by size though cell filters. Multiple washes should always be carried out on the parasite sample to ensure elimination of all host material if at all possible.
2. When using samples generated from in vitro culture systems for proteomic analysis, it obviously is vital to know the exact composition of any medium used, as some components may remain in the sample and lead to false spots on gels.
3. SSH is usually performed on cDNA populations generated from tissues from the same species, as sequence differences, and hence reduced hybridization, in shared genes will result in falsely representing such common genes as differentially expressed. However, many parasites are very closely related or form species complexes, and sequence divergence in coding regions is extremely low. Performing SSH on two species is especially useful for identifying genes involved in parasitism when the other, closely related species does not exhibit this lifestyle. Obviously, there will be false positives, and the occurrence of these has to be considered in experimental design, for example, by increasing the number of clones screened and thoroughly validating differential expression by hybridization to Northern blot.
4. If there is a large starting quantity of parasite tissue, samples should be split into smaller aliquots and treated as separate extractions, as this will greatly increase the overall yield. It is always far more productive to keep the quantity of sample to 50 mg/1 mL TRIZOL. More efficient disruption is achieved by initially homogenizing in only 10% of the final volume of TRIZOL reagent, then adding the remainder after this.
5. If low yields are experienced, this incubation step can be increased to 30–40 min.
6. A minimum of 2 µg of poly(A)$^+$ RNA is required for first-strand cDNA synthesis. Depending on individual parasites and specific tissues, the quantity of total RNA to start with for poly(A)$^+$ selection to achieve this final amount will need to be determined. As a very rough guideline, the poly(A)$^+$ fraction normally represents about 5% of the total RNA population.

7. If it is not possible to isolate 2 μg of poly(A)$^+$ RNA from parasite samples, the SMART amplification technology (Clontech) can be used to preamplify cDNA from total RNA and then start the subtraction at the *Rsa*I digestion.
8. The primers supplied with the kit for checking subtraction efficiency are for glyceraldehyde-3-phosphate dehydrogenase (GAPDH), which works on human, rat, and mouse cDNA samples. Therefore, for parasite samples, primers will need to be designed against two genes, one differentially expressed and one not, for the verification of the efficiency of SSH. This can present some problems, as a certain amount of knowledge is required about genes before they can be used for this test, such as the level of expression in each RNA and the location of any *Rsa*I restriction sites. The latter is important, as there will be no amplification if there is an *Rsa*I site present between the primers because of the earlier digestion.
9. This is the optimum outcome, and results of the differential screen are not normally as clear cut as this. There is normally a spread of results from the hybridizations, and decisions on which clones are truly differentially expressed will need to be determined for individual screens.
10. For other tissue or cell samples, other steps will be needed in preparation. Gentle lysis methods such as repeated freeze/thaw cycles or detergent lysis can be used or, for more vigorous lysis, sonication or homogenization. Proteins can then be precipitated and resuspended in the same sample buffer to compare the protocol as described. However, before loading, samples should be centrifuged at 15,000g for 10 min to remove any particulate material.
11. The polyacrylamide gels used in this protocol are precast, offering numerous advantages. The main one of these, and something crucially important in 2D gel analysis, is reproducibility. Casting gradient gels is quite skillful, and making sure that the exact composition and acrylamide percentage is constant between gels is even harder. Precast gels overcome this problem and offer much more consistent conditions between runs.
12. As mentioned above for the subtractive cDNA library, comparing protein profiles between two separate species can be unreliable, and great care must be taken in selecting candidate differentially expressed proteins. The same protein may be present in both species but subtle sequence changes will result in a slightly different size or pH. However, as long as these problems are taken into account and factored into the analysis, it is possible to compare protein profiles from closely related species.

References

1. Patterson, S. D. and Aebersold, R. H. (2003) Proteomics: the first decade and beyond. *Nat. Genet.* **33,** 311–323
2. Aebersold, R. and Mann, M. (2003) Mass spectrometry-based proteomics. *Nature* **13,** 198–207.
3. Scheel, J., Von Brevern, M. C., Horlein, A., et al. (2002) Yellow pages to the transcriptome. *Pharmacogenomics* **3,** 791–807.

4. Gongora, R., Acestor, N., Quadroni, M., et al. (2003) Mapping the proteome of *Leishmania viannia* parasites using two-dimensional polyacrylamide gel electrophoresis and associated technologies. *Biomedica* **23**, 153–160.
5. Drummelsmith, J., Brochu, V., Girard, I., et al. (2003) Proteome mapping of the protozoan parasite *leishmania* and application to the study of drug targets and resistance mechanisms. *Mol. Cell. Proteomics* **2**, 146–155.
6. Carucci, D. J., Yates, J. R., and Florens, L. (2002) Exploring the proteome of *Plasmodium*. *Int. J. Parasitol.* **32**, 1539–1542.
7. Lasonder, E., Ishihama, Y., Andersen, J. S., et al. (2002) Analysis of the *Plasmodium falciparum* proteome by high-accuracy mass spectrometry. *Nature* **419**, 537–542.
8. Florens, L., Washburn, M. P., Raine, J. D., et al. (2002) A proteomic view of the *Plasmodium falciparum* life cycle. *Nature* **419**, 520–526.
9. El Fakhry, Y., Ouellette, M., and Papadopoulou, B. (2002) A proteomic approach to identify developmentally regulated proteins in *Leishmania infantum*. *Proteomics* **2**, 1007–1017.
10. Jefferies, J. R., Campbell, A. M., van Rossum, A. J., et al. (2001) Proteomic analysis of *Fasciola hepatica* excretory-secretory products. *Proteomics* **1**, 1128–1132.
11. Cohen, A. M., Rumpel, K., Coombs, G. H., et al. (2002) Characterisation of global protein expression by two-dimensional electrophoresis and mass spectrometry: proteomics of *Toxoplasma gondii*. *Int. J. Parasitol.* **32**, 39–51.
12. Bozdech, Z., Zhu, J., Joachimiak, M. P., et al. (2003) Expression profiling of the schizont and trophozoite stages of *Plasmodium falciparum* with a long-oligonucleotide microarray. *Genome Biol.* **4**, R9.
13. Diatchenko, L., Lau, Y. F., Campbell, A. P., et al. (1996) Suppression subtractive hybridization: a method for generating differentially regulated or tissue-specific cDNA probes and libraries. *Proc. Natl. Acad. Sci. USA* **93**, 6025–6030.
14. Diatchenko, L., Lukyanov, S., Lau, Y. F., et al. (1999) Suppression subtractive hybridisation: a versatile method for identifying differentially expressed genes. *Meth. Enzymol.* **303**, 349–380.
15. Gruden-Movsesijan, A., Ilic, N., and Sofronic-Milosavljevic, L. (2002) Lectin-blot analyses of *Trichinella spiralis* muscle larvae excretory-secretory components. *Parasitol. Res.* **88**, 1004–1007.

11

Gene Expression Studies Using Self-Fabricated Parasite cDNA Microarrays

Karl F. Hoffmann and Jennifer M. Fitzpatrick

Summary

DNA microarray platforms represent a functional genomics technology that uses structured information obtained from genomic sequencing efforts as a means to study transcriptional processes in a systematic and high-throughput manner. Specifically in this chapter, we outline the ordered processes involved in large-scale parasite gene expression studies including complementary (cDNA) microarray fabrication, total RNA isolation, cDNA labeling using fluorochromes, and DNA:DNA hybridization. Methods described herein were adapted for the study of schistosome sexual maturation and developmental biology but could be easily modified for the study of any additional parasitological system.

Key Words: cDNA; DNA microarray; functional genomics; gene expression; hybridization; parasite; RNA.

1. Introduction

Continued funding of parasite genome sequencing projects has led to the successful cataloging of several hundred thousand expressed sequence tags (ESTs) *(1)* from many different organisms. ESTs (short stretches of DNA sequence derived from cloned complementary DNAs [cDNAs]) have proved to be an invaluable starting resource for gene discovery projects *(2–7)* in many laboratories across the world and have contributed to our understanding of diverse biological processes. However, despite the obvious advantages of scanning large databases containing individual EST sequences for specific gene identification purposes, there is limited information to be obtained with this "piecemeal" approach for subsequent interpretation of gene expression studies in the modern genomic age. For example, it would require years of gene sequence analysis and several rather complicated experimental gene expression techniques (yeast 2 hybrid protocols, glutathione-*S*-transferase (GST)-capture assays, etc.) to dissect the related components of a signaling cascade in *Brugia malayi* from the starting point of a putative microfilarial chemokine receptor identified via the filarial EST databases (http://circuit.neb.com/fgn/filgen1.html). Additionally, if one were interested in the gene transcription pathways associated with sexual dimorphism in schistosome biology, one would find it difficult to ascertain this information via traditional experimental techniques (subtractive cDNA libraries, differential display, etc.) or querying the existing schistosome EST databases (http://verjol8.iq.usp.br/schisto/; http://www.nhm.ac.uk/hosted_sites/schisto/; http://schistosoma.chgc.sh.cn). Clearly, a more global approach to the study of gene expression is needed to relate ESTs to other gene products in complicated biological systems. With the advent and refinement of cDNA microarray technology, expression analysis and functional relationship studies can now be easily initiated, which, as a result, has led to a revolution in the way gene discovery projects are performed in the 21st century.

cDNA microarray platforms useful for gene expression studies are a vital component of functional genomics, as protein microarrays will be a component of functional proteomics. Functional genomics is a catch-all term used to describe a research area in which EST or genomic sequence information is used to decipher the *functional* meaning of *genomes*. A cDNA microarray containing different EST clones allows for a parallel, simultaneous, high-throughput screening strategy from very small amounts of starting material. Essentially, cDNAs identified through parasite EST sequencing efforts can be arrayed in spatially defined grid patterns. High densities of cDNA "spots" can be achieved on chemically modified glass microscope slides (although membranes can be used, the advantages of glass over membrane are discussed elsewhere *[8]*) through the use of a programmable robot. These cDNA microarrays are then used as a probe (by convention—probes are the spotted DNA on the array and target is the cDNA generated from each RNA pool) to assess the relative abundance of gene transcripts between two separate RNA target pools using Southern hybridization procedures *(9)*. In addition to single bimodal types of experimental assays, cDNA microarrays can also be used to study the longitudinal changes in parasite gene transcription in response to a physiological stress, drug, or immunological pressure.

This chapter describes the procedures (*see* **Fig. 1** for a flow diagram representation) essential for the successful design of cDNA microarray experiments

---

Fig. 1. (*Opposite page*) Flow diagram representation of processes and procedures involved in cDNA microarray gene-expression studies. Each **bold** number (3.1–3.5) indicates a subheading in **Section 3** in the Methods portion that contains detailed instructions for successful completion of the pictured or described procedure. Vector map of pBluescript II SK+ was obtained from Stratagene and individual parasite images from the World Wide Web.

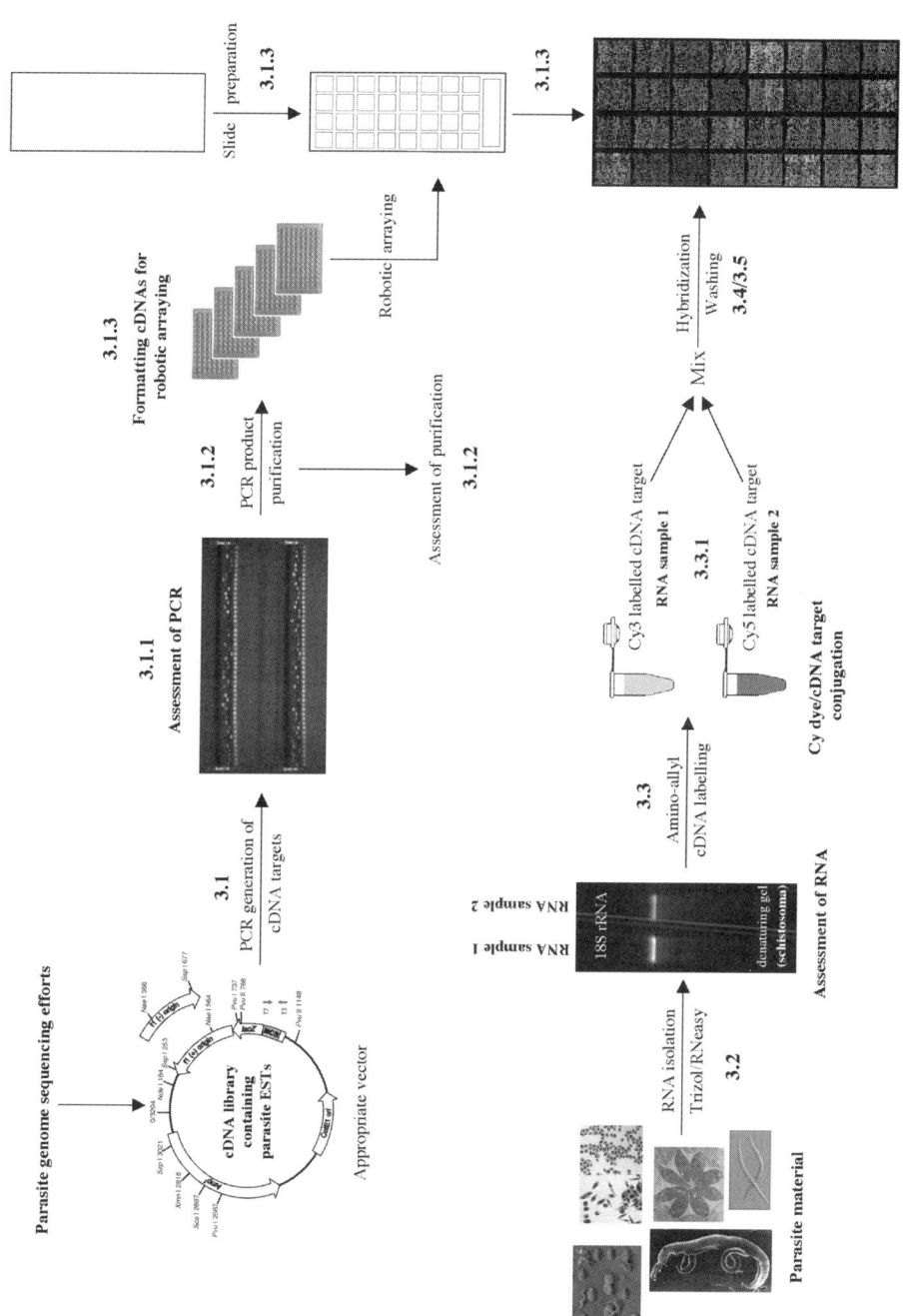

for use in parasitological-related gene expression studies. Techniques and protocols that have been used successfully in studies on schistosome-host immunobiology are included, although modifications to these protocols used by other investigators have been mentioned where applicable. The following techniques have been modified and adapted from protocols published within the cDNA microarray community and are cited as such. Intentionally excluded are the fine details of data mining and bioinformatics, as this in itself would require a separate discussion and is an ever-evolving mathematical and statistical discipline.

2. Materials

2.1. cDNA Probe Amplification, Purification, and Printing

1. Oligonucleotide polymerase chain reaction (PCR) primers to amplify cDNA probes (from any commercial source).
2. Standard PCR reagents and thermal cycler (capable of handling 96-well PCR plates).
3. Standard 96-well PCR plates.
4. Self-adhering PE foil (Macherey-Nagel 740676).
5. NucleoFast 96-well filter plates (Macherey-Nagel 7435000.4).
6. Vacuum manifold (e.g., Millipore/MultiScreen, Qiagen/QIAvac 96, Promega/Vac-Man 96, BioRad/Aurum).
7. Centrifuge alignment frame (Millipore MACF09604).
8. 96-well V-bottom plates (e.g., Nunc 442587).
9. 384-well V-bottom plates (e.g., Genetix X7022).
10. Bar-coded γ-amino propyl silane-coated slides (Corning 40003).
11. 4X Spotting buffer: 600 mM sodium phosphate, 0.04% sodium dodecyl sulfate (SDS). 48.9 mL 600 mM Na$_2$HPO$_4$, 1.1 mL 600 mM NaH$_2$PO4, 200 µL 10% SDS (*see* **Note 1** for alternative spotting buffers).
12. Printing tool (and pins).
13. Robotic printer: MicroGrid II, BioRobotics (biorobotics.co.uk); GeneTac G3, Genomics Solutions (also incorporating GeneMachines and Cartesian Technologies); Genetix "Q" Array (genetix.co.uk); LabMan HDMS (labman.co.uk).
14. UV stratalinker (Stratagene).
15. 80°C oven.

2.2. RNA Isolation (Everything Must Be RNAse-Free)

1. Trizol (Invitrogen 15596-018).
2. Rotor stator homogenizer (Tekmar TISSUMIZER or equivalent).
3. Chloroform.
4. Ethanol (EtOH).
5. RNeasy column chromatography kits (Qiagen: midi kit 75142, maxi kit 75162).
6. 3 M sodium acetate, pH 5.2.
7. 1 M Tris-HCl, pH 7.4.

8. RNase-free H$_2$O.
9. Filter barrier pipet tips.

2.3. cDNA Labeling, Target Cleanup, Hybridization, and Slide Washing

1. Superscript II reverse transcriptase, SSRT II (10,000 U/mL) (Invitrogen 18064-014).
2. Oligo-dTV primer (1 µM scale synthesis) 5'-TTT TTT TTT TTT TTT TTT TTV-3' (V designates any other deoxynucleotide except dT).
3. Microcon-YM30 concentrators (30,000 MW cutoff) (Amicon 42410).
4. Deoxynucleotides (100 mM each) (Pharmacia: dATP 272-050, dCTP 272-060, dGTP 272-070, and dTTP 272-080).
5. 5-(3-Aminoallyl)-2'-dUTP (1 mg) (Sigma A0410).
6. Cy3 mono-reactive dye pack (5 mg) (Amersham PA23001). Resuspend individual Cy dye tubes (5 tubes/pack) in 10 µL dimethyl sulfoxide (DMSO). Aliquot 1.25 µL to eight separate tubes. Evaporate dyes in Speed-Vac in dark. Store samples under vacuum in a dessicator at 4°C in dark.
7. Cy5 monoreactive dye pack (5 mg) (Amersham PA25001). Prepare as above prior to use.
8. 0.5 M EDTA, pH 8.0.
9. 1 M NaOH.
10. 20X saline sodium citrate (SSC).
11. DMSO.
12. 10% SDS.
13. 50X amino-allyl (aa)- dUTP/dNTP mix (2:3 ratio of aa-dUTP/dTTP): 10 µL dATP (100 mM stock), 10 µL dCTP (100 mM stock),10 µL dGTP (100 mM stock), 6 µL dTTP (100 mM stock), 4 µL aa-dUTP (100 mM stock).
14. Master mix for cDNA labeling (for 50 rxns): 300 µL SSRT II 1st strand cDNA buffer (supplied with SSRT II enzyme), 150 µL 0.1 M dithiothreitol (DTT) (supplied with SSRT II enzyme), 150 µL RNAse-free H$_2$O, 30 µL (50X) amino-allyl (aa)-dUTP/dNTP mix (2:3 ratio of aa-dUTP/dTTP).
15. 100 mM sodium bicarbonate (NaHCO$_3$).
16. 100 mM sodium carbonate anhydrous (Na$_2$CO$_3$). Mix 100 mM NaHCO$_3$ and 100 mM Na$_2$CO$_3$ together until final pH 8.5–9.0 is achieved.
17. 4 M hydroxylamine (Sigma H9876).
18. Qiaquick PCR purification kit (Qiagen 28104).
19. pd(A)$_{40-60}$ (Amersham 27-7988-01) (reconstituted at 8 mg/mL).
20. Yeast transfer RNA (tRNA) (Sigma R8759) (reconstituted at 4 mg/mL).
21. Bovine serum albumin (BSA) (Fraction V: Sigma).
22. 50X Denhardt's reagent.
23. Deionized formamide.
24. Sodium pyrophosphate.
25. 1 M Tris-HCl (pH 7.4).
26. Isopropanol.
27. Hybridization chambers (ArrayIt™ AHC-1 or equivalent).

28. Lifter-slips (Erie Scientific: size depends on cDNA array dimensions)/coverslips (size depends on cDNA array dimensions).
29. Coplin jars.
30. Slide swing handle rack (RA Lamb E102).
31. Cot-1 DNA; mouse and human (Invitrogen).

2.4. Scanning/Software

1. Scanner: e.g., GenePix 4000B, Axon (http://www.axon.com); ScanArray 3000, Packard BioChip Technologies/GSI Lumonics. (http://www.dssimage.com/microarraySystems.html); Avalanche, Molecular Dynamics (Amersham Biosciences) (http://www.mdyn.com); DNAscope, GeneFocus (http://www.genefocus.com).
2. Spot-finding software: e.g., GenePix Pro, Axon (http://www.axon.com); Microarray Suite, Scanalytics (http://www.scanalytics.com).
3. Data-mining software: e.g., Spotfire DecisionSite for Functional Genomics, Spotfire, (http://www.spotfire.com); Genespring, Silicon Genetics (http://www.silicongenetics.com); Vector Xpression 3, Informax Inc. (http://www.informaxinc.com/).

3. Methods

3.1. cDNA Probe Amplification, Purification, and Printing

3.1.1. Amplification

cDNA probes to be robotically printed onto chemically modified glass slides are generated according to standard PCR protocols (*see* **Note 2**). Briefly, a 100-μL reaction amplified for 35 cycles generally is sufficient to generate adequate probe material to be spotted onto several hundred glass slides. For ease in processing multiple samples, 96-well PCR plates are recommended for the amplification. The source of template used for each unique parasite cDNA typically is a phage suspension or a lysed bacterial colony (obtained from cDNA libraries or cloned individually). Usually, these clones have been sequence verified. There is no need to isolate purified DNA (containing parasite-specific cDNA) before proceeding with the PCR reactions unless problems arise with the amplification (*see* **Note 3**). PCR primers used in the amplification are either complementary to cDNA vector sequences, flanking the parasite-specific inserts in the starting library (e.g., T7/T3, M13R/M13F), or are gene-specific.

Analyze a 3-μL sample from each PCR reaction on a 1% agarose gel to verify size and number of products (*see* **Fig. 1**). The remaining PCR reactions may be stored at –80°C until purification can be performed.

3.1.2. Purification

Before the parasite-specific cDNA probes are suitable for printing, they must be purified away from all contaminating salts, dNTPs, and primers. An easy and efficient way to achieve high-throughput purification of multiple samples is via the use of NucleoFast® 96-well filter plates (Macherey-Nagel).

1. Transfer each 100-μL PCR reaction directly onto the membrane of the NucleoFast 96-well filter-plate placed on top of an appropriate vacuum manifold (with waste collection plate) (*see* **Note 4**).
2. To remove contaminants by ultrafiltration, apply a vacuum of –400 to –600 mbar^2 (reduction of atmospheric pressure), typically for 10–15 min (NO modifications to this protocol are necessary if mineral oil was used during the PCR reactions, although the actual filtration time may be longer).
3. Dispense 100 μL nuclease-free double-distilled H_2O (ddH_2O) to each well of the 96-well filter plate while applying the vacuum until the water has passed through the membrane. Repeat this wash two additional times and completely ventilate the manifold for 60–90 s.
4. To recover the DNA, add 100 μL nuclease-free ddH_2O (ensure no vacuum is applied) and transfer filter plate to a suitable microplate shaker with moderate shaking for 10 min (*see* **Note 5**). Gently remove the water from the filter plate wells with a multichannel pipet and dispense into the supplied elution 96-well plates. This will deliver the PCR product at the original concentration. A precipitation step may additionally be included at this point if the final concentration of DNA is required to be higher (*see* **Note 5**). It is advisable to check purification procedure by analyzing PCR products (approx 1 μL) on a 1% agarose gel. Purified products can be stored at –80°C in 96-well plates (source plate) that have been sealed with self-adhering PE foil and the lid safely taped to prevent evaporation.

3.1.3. Printing

1. A small aliquot (15 μL) of the total purified PCR product (from source plate) should be transferred to a new V-bottomed 96- or 384-well plate (print plate).
2. Add 5 μL (4X) spotting buffer to each well and mix thoroughly.
3. Print PCR products onto γ-amino propyl silane-coated slides according to robotic arrayer instructions (*see* **Note 6**). Be sure to include appropriate controls (which can be added to empty wells of print plate) including, but not all-inclusive: genomic DNA (positive control), yeast tRNA (negative control), buffer only (negative control), plasmid and phage DNA from which PCR products were amplified (negative controls).
4. After printing, slides should be allowed to dry at room temperature, baked for 2 h at 80°C and UV cross-linked (600-mJ dose) to facilitate maximal fixation of PCR products. At this stage, if the slides used for printing are not pre-bar coded, then generally it is useful to etch pertinent information into each glass slide with a diamond scribe (while salt residue still defines the border of the array). This information should include slide number, print series, and borders of the printed array. Marking each slide in this manner also serves to distinguish the side of glass that contains DNA.

5. After marking glass slides, denature immobilized PCR products by submerging in a boiling dH_2O bath (turn heating unit off immediately after slides have been submerged) for 2 min.
6. After denaturing DNA, transfer slides to a 95% EtOH bath and fix for 1 min. Spin-dry slides for 5 min at $1000g$ in a benchtop centrifuge. Slides can now be stored in a dust-free microscope slide box at room temperature (see **Note 7**) and are ready for hybridization.

3.2. Total RNA Isolation and Purification

As in all applications that involve the isolation of RNA, care must be taken to avoid introduction of ubiquitous RNAses. Either purchase RNAse/DNAse free reagents (preferred) or treat all solutions (except those containing Tris) thoroughly with diethyl-pyrocarbonate (DEPC) prior to use. Change gloves frequently, and use filter tips and disposable plastics when isolating RNA from parasite material. Generally, it is best to work rapidly during the isolation of RNA (the RNA isolation procedure should be performed in 1 d). The method outlined below produces total RNA free from contaminating carbohydrates and protein and involves the use of both TRIZOL Reagent (Invitrogen) and Qiagen RNeasy affinity columns (Qiagen). RNA isolated in other fashions (although suitable for Northern blotting, RNAse protection, or reverse transcriptase [RT]-PCR assays) generally is not suitable for cDNA microarray applications and can often lead to high background because of biological contaminants. Extremely low amounts of genomic DNA may be isolated during the procedure described here. If this presents a problem (there are reports that Superscript reverse transcriptase II can prime off genomic DNA, leading to false positives; see Genisphere's Web site http://www.genisphere.com). Qiagen provides a protocol for genomic DNA digestion on a column, or alternatively, one can follow any protocol that uses the enzyme RNAse-free DNAse.

3.2.1. TRIZOL Extraction

According to the manufacturer's instructions, add a sufficient volume of TRIZOL to parasite material (1 mL of TRIZOL per 10^7 cells or 1 mL TRIZOL per 100 mg tissue—see **Note 8**). If starting from parasite-infected cell cultures (e.g., *Toxoplasma gondii*-infected fibroblasts or *Plasmodium* sp.-infected red blood cells), then refer to commonly used techniques to enrich for parasite material (if desired). Keep in mind that there is a trade-off between obtaining increased quantities of purified parasite material (enriching away from host material) vs the effect this enrichment has on parasite transcriptional processes. Each researcher will have his or her own protocols for obtaining parasite material (fresh material is best, especially if old parasite material was stored improperly), as well as his or her own favorite method of disrupting/homogenizing samples. For multicellular parasites (helminths, nematodes, or other large metazoans), a homogenizer works quite well for this important application.

1. Using a cleaned homogenizer (clean with 50 mL ddH$_2$O, 50 mL 100% EtOH, and 7 mL TRIZOL between each parasite sample), thoroughly disrupt parasite material (10–15 s). Make sure the tip of the homogenizer stays below the surface of TRIZOL to avoid foaming of solution. After complete homogenization/disruption of sample material, let the sample sit at room temperature for 5 min.
2. Add 0.2 mL chloroform per 1 mL TRIZOL and vigorously shake samples for at least 1 min (see **Note 9**). Let samples incubate at room temperature for 2 min to ensure the complete dissociation of nucleoprotein complexes.
3. Spin samples to separate phases (4°C at 12,000g for 15 min) (see **Note 10**).
4. Carefully transfer aqueous supernatant (containing the RNA) to a new RNAse-free tube without disturbing the interface (containing protein and other cellular debris) and proceed with Qiagen RNeasy affinity purification.

3.2.2. Qiagen Affinity Purification

The second phase of the total RNA purification procedure is to use a Qiagen affinity column to reversibly bind the nucleic acid from the aqueous solution obtained above (see Qiagen RNeasy handbook for detailed principles and procedures of product).

1. Add an equal volume of 70% EtOH to the aqueous supernatant obtained from the TRIZOL extraction (dripwise while vortexing gently to avoid local RNA precipitation) (see **Note 11**).
2. Transfer this solution to a Qiagen affinity column. If the volume of solution is greater than the capacity of the Qiagen column, add the sample in successive aliquots. Spin for 5 min at room temperature (3000–5000g) to facilitate RNA binding.
3. Vortex the collected flow-through (FT) and reapply to Qiagen column. Spin again for 5 min at room temperature (3000–5000g). This step ensures that all RNA in the sample is bound to the column.
4. Discard the FT and stepwise wash the column as indicated in the Qiagen handbook. Volumes of washing buffers will depend on Qiagen column used (midi or maxi column).
5. After the last wash in buffer RPE (supplied in the Qiagen kit) and ensuring that the Qiagen column is completely dry (additional 5 min spin after removal of last FT), carefully transfer the column to a new conical tube (supplied in the Qiagen kit). Elute RNA from the column with three successive equal volumes of RNAse-free ddH$_2$O (also supplied in the kit) (see **Note 12**). All three samples containing the eluted RNA can be collected in the same collection tube.
6. Add 1/10 vol of 3 M sodium acetate (pH 5.2–5.3) and 2.5 vol of 100% EtOH to the RNA sample. Vortex this mixture and incubate on ice for at least 10 min (alternatively, place at –20°C overnight to facilitate precipitation of RNA). Spin pre-

cipitated RNA at 12,000g for 15 min (4°C). You may have to transfer sample to several 1.5-mL microcentrifuge tubes.
7. Wash pellets twice with 75% EtOH by vigorously vortexing sample during each wash. Dry pellets completely in Speed-Vac (*see* **Note 13**).
8. Dissolve RNA pellet(s) from each sample in RNAse-free ddH$_2$O by adding directly to the top of the dried sample and placing at 65°C for 10 min (it may be necessary to incubate longer at 65°C). Do not pipet the sample up and down (RNA may stick to the inside of the pipet tip). After the sample has completely dissolved, combine all aliquots of each sample (if necessary) and transfer to a new microcentrifuge tube. Spin out any insoluble debris (3 min at max speed in microcentrifuge) and quantitate each RNA sample by spectrophotometry (*see* **Note 14**). Aliquot RNA (10-µg aliquots) into RNAse-free microcentrifuge tubes and store samples at –80°C. It is advisable to check the integrity of the RNA sample by electrophoresis through a denaturing agarose gel (*see* **Fig. 1**).

3.3. cDNA Target Labeling With Amino-Allyl dUTP

Although there are alternative cDNA-labeling methods used in cDNA microarray applications (see http://www.nhgri.nih.gov/DIR/Microarray/main.html and http://www.genisphere.com for information on two additional techniques), the protocol listed below describes a procedure that uses amino-allyl dUTP and the fluorescent dyes Cy3 and Cy5. This cDNA-labeling protocol has been slightly modified from one that is available at http://www.microarrays.org (the original protocol was developed at Rosetta Inpharmatics, Kirkland, WA), with the user supplying all the reagents (alternatively, kits that use a similar procedure can be purchased from commercial sources—e.g., Fairplay system from Stratagene). As little as 8 µg of total RNA can be used to generate fluorescently modified cDNA molecules of sufficient quantity and intensity to be used in cDNA microarray hybridizations. The protocol detailed below uses 10 µg of total RNA as the starting input material for each separate cDNA reaction (Cy3 and Cy5). Empirical testing of input RNA is recommended for optimal results (generally, more input RNA leads to generation of cDNA with higher signal intensity).

3.3.1. cDNA Labeling

1. Combine 14.5 µL total RNA (10 µg) with 1 µL oligo-dTV (5 µg/µL) primer in a 0.2-mL RNAse/DNAse-free PCR tube. 10 µg from each sample where gene expression is being compared is necessary for each reaction (e.g., 10 µg from adult female *S. mansoni* RNA and 10 µg from adult male *S. mansoni* RNA).
2. Heat RNA/oligo-dTV mixture at 70°C for 10 min (it is useful to do all incubation steps in a PCR thermal cycler). Immediately place samples on ice and leave for 5 min. This step effectively removes secondary structures associated with total RNA mixtures.

3. Add 12.6 µL of the master mix containing the amino-allyl dUTP to each denatured RNA/oligo-dTV sample (*see* **Note 15**). Place reactions at 42°C to prewarm samples prior to adding 1 µL Superscript Reverse Transcriptase II (SSRTII). Incubate reactions for 1 h, then add another 1-µL aliquot of SSRTII to each and continue incubating for another hour.
4. Remove cDNA reactions from 42°C and add 10 µL 0.5 M EDTA (chelates SSRTII) to each, followed by addition of 10 µL of 1 N NaOH (hydrolyzes any remaining RNA). Place samples at 65°C for 15 min and then immediately on ice. Neutralize sample by adding 25 µL of 1 M Tris-HCl (7.5).
5. To each neutralized reaction, add 450 µL H_2O and transfer contents into separate Microcon-YM30 concentrators (one for each cDNA reaction). Spin at 13,000g in a microcentrifuge for 8 min (*see* **Note 16**). When reservoir volume gets to about 50 µL, add another 400 µL H_2O. Repeat wash with a third 400-µL vol of H_2O (three total washes). These three washes effectively eliminate all unincorporated dNTPs and free Tris (Tris will interfere with Cy dye conjugation reaction).
6. When volume of sample after third wash reaches approx 10 µL, invert the Microcon-YM30 concentrators and place into a new 1.5-mL microcentrifuge tube. Elute washed cDNA samples from filter by spinning at 3000g for 4 min.
7. Dry each cDNA sample (will not see dried pellet) in Speed-Vac (time varies depending on temperature).
8. Resuspend each cDNA pellet in 4.5 µL of DNAse-free ddH_2O.
9. Resuspend an aliquot of dried Cy dye (one cDNA sample will be labeled with Cy3 and the other will be labeled with Cy5) in 4.5 µL of 0.1 M carbonate/bicarbonate buffer (pH 8.5–9.0) (*see* **Note 17**).
10. Mix cDNA with the appropriate Cy dye and incubate in the dark for 1 h at room temperature (Cy dyes will conjugate to the amino-allyl dUTP that has been incorporated during the cDNA reactions).
11. To stop conjugation reaction, add 4.5 µL of 4 M hydroxylamine to each cDNA sample (excess amino groups effectively quench free Cy dye) and incubate in the dark for 15 min.
12. Add 35 µL of 0.1 M sodium acetate (pH 5.2) to each conjugation reaction and combine the two samples in which gene expression will be compared. Transfer combined sample to new 1.5-mL microcentrifuge tube.
13. Add 500 µL PB buffer from QIAGEN PCR clean-up kit to each combined Cy3/Cy5-labeled cDNA sample and apply to QIAquick column. Spin samples for 30–60 s, at maximum speed in microcentrifuge. Reapply FT to column and spin again. Discard FT (all labeled cDNA should be bound to column), and apply 750 µL PE wash buffer (ensure ethanol has been added to PE buffer as per Qiagen instructions) to column.
14. Spin columns again for 30–60 s, at maximum speed in microcentrifuge. Discard FT. As Cy3 and Cy5 dyes can weakly bind to Qiagen column, it is advisable to perform two additional PE washes (three total). After the third wash, spin for 1 min to completely dry column (*see* **Note 18**).

15. Place dried column (containing washed, bound, labeled cDNA) in new 1.5-mL collection tube and add 100 µL elution buffer (EB) to membrane. Let sit at room temperature for 1 min (in dark), then spin for 1 min at 13,000g to elute cDNA. Add same volume of EB and repeat elution (collect both fractions in same tube).
16. Bring the 200 µL eluted, labeled cDNA target to 400 µL with DNase-free H_2O and apply to Microcon-YM30 concentrator. Concentrate sample to approx 20–30 µL using a microcentrifuge (*see* **Note 19**). Add 2.2 µL yeast tRNA (4 mg/mL) and 2.2 µL poly d(A) (8 mg/mL) and dry the remaining solution using a Speed-Vac. Resuspend the pellet in 40 µL prewarmed (48°C) filter-sterilized hybridization buffer containing 40% deionized formamide, 5X SSC, 5X Denhart's reagent, 1 mM Na-pyrophosphate, 50 mM Tris, pH 7.4, 0.1% SDS (*see* **Note 20**).
17. Denature labeled cDNA target for 2 min at 95°C (*see* **Note 20**) and spin for 3 min at maximum speed in microcentrifuge to pellet any insoluble material that may interfere with hybridization. The cDNA target (minus any insoluble pellet) is now ready for hybridization to cDNA microarray.

3.4. cDNA Microarray Preparation, Prehybridization, and Hybridization

cDNA microarrays can be prehybridized for up to 1 h prior to addition of labeled cDNA target. This is an additional cautionary step to prevent nonspecific hybridization events from occurring (may not be needed in all cases).

1. Place slide containing cDNA elements to be used for hybridization in a 50-mL conical tube containing a solution made up of prehybridization buffer (1% BSA/0.1% SDS/5X SSC/50% formamide). Place conical (containing slide) in 42–48°C H_2O bath or hybridization oven, with gentle shaking for 40–50 min.
2. Remove slide from prehybridization conical and briefly wash by immediate immersion and shaking in a Coplin jar containing dH_2O. Repeat wash by transferring slide to another Coplin jar containing an additional aliquot of dH_2O. Finally, submerge slide in a third Coplin jar containing 100% isopropanol.
3. After removal from isopropanol, immediately transfer slides to slide rack and spin dry (do not let air dry) in a benchtop centrifuge (10 min at 1000g). Prehybridized slides now can be stored in a clean 50-mL conical tube. When handling slides, touch only sides of glass and always use powder-free gloves. As an alternative to standard glass coverslips, Lifter-slips (coverslips with Teflon-coated raised edges) are commonly used in cDNA microarray hybridizations (Erie Scientific). Lifter-slips facilitate equal distribution of labeled target cDNA across all elements of cDNA microarray.
4. Place the Lifter-slip directly on top of cDNA microarray glass slide with the Teflon-coated raised edges toward the cDNA. Carefully pipet 37 µL of the spun, denatured, labeled cDNA target (leave 3 µL of the target behind so that any insoluble pellet is not disturbed) between Lifter-slip and glass slide and capillary action will draw the liquid to sufficiently cover a fabricated DNA microarray with dimensions of 25 mm × 60 mm (*see* **Note 21**).

5. Place array into hybridization chamber (make sure chamber has a few drops of ddH$_2$O or 3X SSC in each end to maintain humidity), clamp hybridization lid down tightly, and incubate whole apparatus overnight (14–18 h) at 42–50°C.

3.5. cDNA Microarray Washing

After overnight hybridization, slides must be carefully removed from hybridization chambers and washed to remove unbound target DNAs. The washing procedure listed below is a good place to start, however, optimization of washing times, detergent/salt concentrations, and temperature may be necessary. Three Coplin jars are used in this procedure, but any suitable washing tray may be utilized. If using Coplin jars, only two slides should be processed at one time before changing the wash solutions.

1. Remove cDNA microarray from chamber by carefully picking up one end of the glass slide (coverslip still attached) with forceps and place into Coplin jar 1 containing wash solution one (0.5X SSC/0.1% SDS). Carefully shake slide with forceps until coverslip falls off. Remove coverslip from Coplin jar and continue shaking cDNA microarray slide with forceps for 4 min.
2. Remove cDNA microarray from Coplin jar 1, touch end of slide to paper towel to remove excess wash solution (do not allow slide to dry completely), and transfer slide to Coplin jar 2 (containing wash solution two, 0.5X SSC/0.01% SDS). Again, shake slide with forceps for 4 min.
3. Finally, remove cDNA microarray from Coplin jar 2, quickly touch end of slide to paper towel, and transfer slide to Coplin jar 3 (containing wash solution three, 0.06X SSC). Wash slide by vigorous shaking for 4 min as before.
4. Remove cDNA microarray from Coplin jar 3, immediately submerge in 100% isopropanol, and spin dry using a benchtop centrifuge (5 min at 1000g). You can now store the cDNA microarray in a dark slide box until ready to acquire fluorescent information via a scanner (processed cDNA arrays can be stored for at least a week in the dark before scanning, without losing data).

3.6. Data Acquisition/Image Analysis/Data Mining/Other Issues

Fluorescent data is acquired through the use of a cDNA microarray scanner (review of some commercially available scanners is available *[10]*) and appropriate image analysis software. Data mining and filtering (usually the most time-intensive aspect of cDNA microarray experiments) can be performed using limited software packages (Excel®, Microsoft) or quite statistically thorough software packages such as GeneSpring (Silicon Genetics). The specific details about using such equipment and software are beyond the scope of this chapter. Because of the relatively high cost of scanners and software, anyone performing cDNA microarray experiments should consult appropriate references to ascertain which scanner, image analysis software, and data-mining packages are the most suitable for each laboratory's intended experimental setup. It is also recommended that at least three independent hybridization experiments be performed (analytical replication) with each pair of RNA samples to account for intrinsic variables associated with this gene expression platform *(11)*. Control experiments are also recommended to assess RNA quality, cDNA labeling, and cDNA microarray hybridization conditions. One control experiment includes performing a "self vs self" hybridization (the two RNA samples are identical, but labeled with different dyes). As both RNA species are the same, this control reveals how tightly a cDNA array estimates that the two samples from the same source produce equivalent signals. Deviation away from normalized gene expression ratio of 1 indicates that the array may be failing in some areas (general or specific). If only some portion of the gene expression ratios is far from 1, then their position on the array can be reassessed (background, signal, etc.) before proceeding with more complicated (and valuable) experimental hybridizations. Another control experiment (very good for bimodal comparisons) involves "dye swapping"—switching the Cy dyes used to label the starting cDNA samples. This control experiment provides additional confidence in gene expression between two samples (the same genes should be differentially expressed even when dyes are switched). Finally, independent confirmation of gene expression, using traditional biological techniques (Northern blotting, RT-PCR, real-time PCR, *in situ* hybridization, protection assays, etc.), should be sought on all differentially expressed cDNAs prior to any detailed advanced investigation.

4. Notes

1. Alternative spotting buffers used by other investigators include SSC (1X through 5X final concentration) or 50% DMSO (with this spotting buffer, generally less evaporation of samples in print plates occur during printing run). Some investigators also add betaine (*N,N,N*-trimethylglycine: Sigma) to their spotting solution (typically 3X SSC) at a final concentration of 1.5 M. This additional spotting compound contributes to high spot homogeneity and reduced background, although slides may need to be baked for longer periods of time (overnight) at 80°C to ensure complete fixation of deposited DNA spots *(12)*.
2. Typical PCR conditions for amplifying *S. mansoni* cDNA EST inserts from phage suspension or bacteria solution: initial denaturation at 95°C for 5 min; then 35 cycles of 95°C for 30 s, 59°C for 30 s, 72°C for 4 min; final 72°C extension for 5 min; hold at 4°C.

    Phage suspension was prepared as follows: single plaque cored into 500 µL SM buffer and stored frozen (4 µL of this solution was used as template material for PCR).

    Bacterial suspension was prepared as follows: 10 µL from a 1 mL overnight culture was added to 100 µL dH$_2$O and heated for 5 min at 95°C. This suspension was stored frozen (4 µL of this solution was used as template material for PCR).

Size of amplified product will undoubtedly vary among ESTs. Generally, 500- to 2000-bp ESTs are suitable for arraying (more detailed examination of PCR product size on hybridization kinetics and signal intensity has recently been examined *[13]*).

3. Template DNA generally does not need to be purified prior to PCR amplification. However, in case of particularly difficult PCR amplifications, a template purification step may be necessary. Refer to Sambrook et al. *(14)* for general plasmid and phage DNA isolation protocols or access http://www.nhgri.nih.gov/DIR/Microarray/main.html or http://www.microarrays.org for typical microarray template purification procedures.
4. Centrifugation may replace the use of a vacuum manifold if necessary. In this case, centrifuge 96-well filter plates containing PCR products at 4500g for 10–15 min in substitution of vacuum application.
5. It is critical to shake moderately (not vigorously) to elute DNA samples from the membrane and also to prevent cross-contamination from well to well. Evaluate the rotation speed before elution with a water/dye combination. If you wish to obtain PCR products at a higher concentration, simply elute with less volume. DNA in a range of 0.15–0.25 mg/mL is a good starting concentration for robotic printing (concentration estimates can be made from gel electrophoresis). Alternatively, a precipitation step may be performed using isopropanol, according to protocols at http://www.microarrays.org. Excessively concentrated DNA may clog pins or produce "comet tails" (caused by localized reattachment of excess spot material to the slide surface during postprinting processing), whereas DNA below this concentration leads to weak signal and a decrease in the dynamic range for detecting differentially expressed genes. Refer to Yue et al. *(15)*, http://www.nhgri.nih.gov/DIR/Microarray/main.html, for methods of determining precise starting DNA concentration in source plate (typically a fluorescent assay that can be performed in 96-well plates). Additionally, SYBR GREEN II RNA gel stain (Molecular Probes) can be used as quick test to determine how efficiently the DNA has been deposited onto each microarray prior to hybridization (protocol available at http://www.probes.com/handbook/boxes/0681.html).
6. A review of the specifications and features of some commercially available robotic printers or arrayers is provided by D. Bowtell *(10)*. Additionally, an alternative to printing PCR products onto γ-amino propyl silane-coated slides is to print onto poly-L-lysine-coated slides. Both coating agents essentially perform the same function—provide a hydrophobic and positive charge critical for spotting and positioning of DNA. Methods and reagents to coat Gold Seal brand microscope slides (this particular glass has minimal intrinsic autofluorescence—Erie Scientific) can be found at http://www.nhgri.nih.gov/DIR/Microarray/main.html. Although coating microscope slides in house with poly-L-lysine is more cost-effective than purchasing either poly-L-lysine or γ-amino propyl silane-coated slides, batch-to-batch variation in slide coating often can occur if care is not taken in quality control. There are many varieties of printing tools and pins (split pins are commonly

used), and the pros and cons of each make are beyond the scope of this chapter. Consult an experienced company (e.g., Biorobotics or TeleChem) for details of each tool and pin type.
7. In cases of high background, an extra slide-blocking step can be performed with succinic anhydride (Corning; www.corning.com, provides instructions with their γ-amino propyl silane-coated slides, or alternatively, instructions can be found at http://www.nhgri.nih.gov/DIR/Microarray/main.html). Succinic anhydride converts the residual amines on the surface of the precoated slide into carboxylic moieties (reduces the positive charge of non-DNA-containing areas and, therefore, limits the areas in which cDNA can bind nonspecifically). New techniques *(12)* have improved this blocking procedure by effectively limiting the amount of DNA solubilization that sometimes occurs during succinic anhydride treatment.
8. The correct volume of TRIZOL reagent added to initial parasite sample should be empirically determined for optimal yields of RNA.
9. In some cases (multicellular, multitissue parasites such as adult nematodes and helminths, etc.), a more vigorous shaking step is suggested (3 min instead of just 1 min).
10. If volume of material is too large to spin in a microcentrifuge, use a Sorvall RC-5B (or equivalent) centrifuge and spin at 6,000–10,000g increasing the time to 30 min (Sorvall HB-6 rotor). This centrifugation step will require RNAse-free 15- or 30-mL Corex tubes (treated with 0.1 M HCl in DEPC-H_2O for at least 8 h and subsequently autoclaved).
11. Occasionally, during this step some precipitation of nucleic acid (when starting with approx 1 g of material) is observed. This does not appear to interfere with the subsequent binding of RNA to the Qiagen column as long as the sample is thoroughly vortexed prior to addition to column.
12. The three successive elution volumes are based on the size (midi or maxi) of the Qiagen column used. Suggested elution volumes for each column are indicated in the supplied manufacturer's handbook.
13. As RNA isolated in this manner is free of contaminating carbohydrate molecules (carbohydrates bound to RNA often makes RNA difficult to resuspend), complete Speed-Vac drying of the sample generally does not inhibit the subsequent solubility of each sample.
14. RNA is typically diluted 1/100 in 0.01 M Tris-HCl (7.5) for estimations of purity (OD_{260}:OD_{280} ratio). An OD_{260}:OD_{280} ratio of 2.0–2.3 (in 0.01 M Tris-HCl) generally indicates highly pure RNA free of contamination. For accurate RNA quantitation, measure samples in RNAse-free H_2O.
15. The cDNA master mix in this protocol used a 2:3 ratio of aa-dUTP to dTTP. This ratio works quite well for schistosome RNA. However, empirical determination of the correct ratio for each parasite species is essential for optimal labeling efficiency (this ratio should be a good starting point for optimization experiments).
16. Depending on the microcentrifuge being used as well as fine differences in microcon membrane construction, the time spent washing the cDNA samples may vary. Try to avoid drying the microcon membrane completely, as some difficulty may be encountered when eluting the cDNA (although according to the manu-

facturer's instructions, 10 µL of H₂O can be applied to dried membranes, allowing for recovery of cDNA sample material).
17. It is important not to let the Cy dyes sit in an aqueous buffer prior to conjugation, only rehydrate the dyes directly prior to addition of suspended cDNA sample. Small residual amounts of DMSO do not affect conjugation reaction.
18. Strong color will be observed in the FT following the first wash. This material represents unconjugated dyes. Some color may also be noticed on the Qiagen membrane after spin drying of the samples (after the third wash). This material likely represents trapped Cy3/Cy5-labeled SSRT II (free amino groups on the enzyme capable of being modified by the dyes during the conjugation reaction) and/or labeled cDNA targets. With 10 µg input RNA, one does not consistently observe color on the membrane prior to elution of cDNA or in the eluent.
19. It can be useful to include either human or mouse Cot-1 DNA (1 µg Cot-1 DNA/ 10 µg input RNA) during this concentration step as an additional blocking agent in schistosome cDNA microarray hybridizations. Cot-1 DNA is 50–310 bp in size and enriched for repetitive DNA sequences such as Alu and Kpn (human Cot-1) or B1, B2, and L1 (mouse Cot-1) family members. Including one of these blocking agents prevents undesired hybridization between conserved repetitive DNA sequences in the labeled cDNA and probe DNA by competitive hybridization. Other blocking agents can be added during hybridization, such as yeast tRNA and poly d(A). Yeast tRNA reduces undesired nonspecific DNA hybridization, while poly d(A) that is 40–60 dATP bases in length promotes specific hybridization between the labeled cDNA and probe DNA. Poly d(A) reduces hybridization of the polyA sequences in the probe DNA to the polyT tract in the labeled cDNA.
20. This hybridization buffer is one that works well for schistosome cDNA arrays, although each user should empirically test different components (blocking agents) and formulations (formamide vs aqueous based hybridization buffers) to ensure optimal hybridization kinetics. An aqueous hybridization buffer may also be used consisting of 3.5X SSC/0.275% SDS. In this case, the hybridization is carried out at 65°C overnight instead of 42–50°C. However, dehydration of labeled cDNA sample material between coverslip and cDNA microarray may occasionally occur when using an aqueous hybridization buffer. Using either hybridization buffer, do not place labeled cDNA target on ice after heating at 95°C, as SDS will precipitate.
21. A glass coverslip/Lifter-slip can be cleaned of dust and debris by using compressed air. Ensure that the compressed-air canister does not also spray oil droplets or other potential contaminants onto the glass coverslip. Alternatively, the coverslip/Lifter-slip may be cleaned using EtOH. Care should be taken not to introduce air bubbles between glass coverslips and cDNA microarrays. If air bubbles are observed, then carefully push bubbles to the side of the coverslip with clean forceps prior to hybridization. Small air bubbles will usually dissipate during the hybridization.

Acknowledgments

We thank Drs. Rhian Hayward and Michael Wilson for critically reviewing this chapter. We also thank Drs. David Latto, Tom Freeman, David Johnston, and David Dunne for many helpful discussions concerning cDNA microarray applications for schistosome genomic investigations.

References

1. Adams, M. D., Kelley, J. M., Gocayne, J. D., et al. (1991) Complementary DNA sequencing: expressed sequence tags and human genome project. *Science* **252**, 1651–1656.
2. Hoffmann, K. F., Davis, E. M., Fischer, E. R., et al. (2001) The guanine protein coupled receptor rhodopsin is developmentally regulated in the free-living stages of *Schistosoma mansoni. Mol. Biochem. Parasitol.* **112**, 113–123.
3. Maizels, R. M., Blaxter, M. L., and Scott, A. L. (2001) Immunological genomics of *Brugia malayi*: filarial genes implicated in immune evasion and protective immunity. *Parasite Immunol.* **23**, 327–344.
4. Kappe, S. H., Gardner, M. J., Brown, S. M., et al. (2001) Exploring the transcriptome of the malaria sporozoite stage. *Proc. Natl. Acad. Sci. USA* **7**, 7.
5. Ding, M., Clayton, C., and Soldati, D. (2000) *Toxoplasma gondii* catalase: are there peroxisomes in *Toxoplasma*? *J. Cell Sci.* **113**, 2409–2419.
6. Field, H., Sherwin, T., Smith, A. C., et al. (2000) Cell-cycle and developmental regulation of TbRAB31 localisation, a GTP-locked Rab protein from *Trypanosoma brucei. Mol. Biochem. Parasitol.* **106**, 21–35.
7. Beall, M. J., McGonigle, S., and Pearce, E. J. (2000) Functional conservation of *Schistosoma mansoni* Smads in TGF-beta signaling. *Mol. Biochem. Parasitol.* **111**, 131–142.
8. Cheung, V. G., Morley, M., Aguilar, F., et al. (1999) Making and reading microarrays. *Nat. Genet.* **21**, 15–19.
9. Southern, E. M. (1975) Detection of specific sequences among DNA fragments separated by gel electrophoresis. *J. Mol. Biol.* **98**, 503–517.
10. Bowtell, D. D. (1999) Options available—from start to finish—for obtaining expression data by microarray. *Nat. Genet.* **21**, 25–32.
11. Lee, M. L., Kuo, F. C., Whitmore, G. A., et al. (2000) Importance of replication in microarray gene expression studies: statistical methods and evidence from repetitive cDNA hybridizations. *Proc Natl Acad Sci USA* **97**, 9834–9839.
12. Diehl, F., Grahlmann, S., Beier, M., et al. (2001) Manufacturing DNA microarrays of high spot homogeneity and reduced background signal. *Nucleic Acids Res.* **29**, E38.
13. Stillman, B. A. and Tonkinson, J. L. (2001) Expression microarray hybridization kinetics depend on length of the immobilized DNA but are independent of immobilization substrate. *Anal. Biochem.* **295**, 149–157.
14. Sambrook, J., Fritsch, E. F., and Maniatis, T. (1989) Molecular Cloning: A Laboratory Manual. Cold Spring Harbor Laboratory Press, Cold Spring Harbor, NY.
15. Yue, H., Eastman, P. S., Wang, B. B., et al. (2001) An evaluation of the performance of cDNA microarrays for detecting changes in global mRNA expression. *Nucleic Acids Res.* **29**, E41–E41.

12

DNA Content Analysis on Microarrays

Upinder Singh, Preetam H. Shah, and Ryan C. MacFarlane

Summary

The genome sequencing of protozoan parasites has facilitated the development of powerful postgenomics tools such as DNA microarrays and revolutionized the study of parasite biology. Large-scale genomic comparisons are useful in identifying the extent of genomic variability among related strains and isolates. Identification of deletions between geographically diverse clinical isolates is important in understanding parasite biology and the "fitness" of a given strain in dissemination. Additionally, the development of reliable diagnostic tests or identification of potential vaccine candidates is predicated on the large-scale conservation of the candidate genes. Parasites with variable virulence phenotypes (vaccine strain vs virulent strain) can also be studied for their genomic variability and provide further insights into the potential role of genotypic variability and its relationship to virulence. This chapter outlines the utilization of DNA microarrays to study genomic content.

Key Words: DNA content; genomic abundance; genomic comparisons; genomic deletions; genomic DNA; genotype analysis; microarrays; strain comparisons.

1. Introduction

In recent years, genome sequencing of many protozoan parasites has been undertaken. The availability of sequence data has revolutionized the study of parasite biology and facilitated the development of powerful postgenomics tools such as DNA microarrays. This chapter outlines the utilization of DNA microarrays to study DNA content (presence or absence of a given gene, copy number of genes, etc). Sections of these protocols are identical to, and cross-referenced to, Chapter 11 (cDNA expression microarrays). The basic premise (i.e., use of a microarray to get good-quality data on a genomic scale) is the same; this chapter highlights details specific to the use of microarrays to study gene content. There is a wealth of published information in this arena, including some recent reviews on microarrays *(1–12)* and their uses in parasitology *(13)*.

From: *Methods in Molecular Biology, Vol. 270: Parasite Genomics Protocols*
Edited by: S. E. Melville © Humana Press Inc., Totowa, NJ

Nucleic acid arrays or DNA microarrays are used to monitor genome-wide variation in DNA or cellular RNA content. Although typically thought of as useful for studying RNA abundance, this tool can be modified to look at genomic abundance, as shown recently for *Helicobacter pylori* and *Mycobacterium tuberculosis* *(14,15)*. This approach has also been used to identify gene duplications and deletions in cancerous cells as compared to noncancerous cells *(16)*. In parasite biology, the field of comparative genomics can be used to address numerous issues. The genomic differences between closely related species, various laboratory strains, clinical isolates, and avirulent or vaccine strains can be investigated. The main requirement for these studies is that the genomic sequence of the strain being tested be similar to the strain that was used to generate the microarray, to get adequate "cross-hybridization" on the array. The extent of similarity required to assure cross-hybridization is variable and depends on numerous factors, including the size of the clones spotted on the array, the AT content of the system, and the hybridization and washing protocols. In general, however, if the two strains/species to be studied are ≥75–80% identical, then this approach should be directly applicable. With lower levels of sequence identity between the genomes of interest, slight modifications to the hybridization and washing protocols may need to be implemented to ensure hybridization. The protocols presented here are currently being applied to the study of *Entamoeba* to determine if genomic differences might give insights into the variable virulence phenotypes of the pathogenic *Entamoeba histolytica* vs the commensal *Entamoeba dispar* (P. Shah and U. Singh, unpublished data). Because most genes of these two organisms are highly similar (≥90% identity), microarray-based, genome-wide comparisons are being used to identify candidate virulence genes based on identification of genes that are uniquely present in the *E. histolytica* vs *E. dispar*. A similar approach was used to characterize strains of attenuated *Mycobacterium bovis* strain, BCG. It was found that various BCG strains had acquired deletions in genes that may function to provide immunity against *M. tuberculosis*, providing important insights into why strains of BCG display variable potencies in protecting individuals from infection with *M. tuberculosis* *(15)*.

In parasites that have variable clinical manifestations, and where the clinical variables are most likely related to parasite biology (in contrast to host-specific factors), the ultimate goal of genotype analysis would be the development of rapid assays that can readily distinguish the most virulent strains and allow for strain-appropriate therapy. The development of such assays reflects the best application of bench-to-bedside translation of comparative genomics and takes the technology to the regions endemic for disease. The subsequent development of polymerase chain reaction (PCR) or restriction fragment length polymorphism (RFLP) diagnostic tests, based on clinical isolates from endemic

DNA Content Analysis

developing countries, is useful in identifying genotypes associated with unique epidemiological and clinical outcomes in real situations, giving insights into evolutionary pressures on a given parasite.

Here, we describe how we create DNA microarrays, the cohybridization of genomic DNA derived from different species with a reference DNA, and methods of analysis applied to the resulting data.

2. Materials

2.1. Probe Amplification, Purification, and Printing

For the most part these are identical to those in Chapter 11, and the protocols used are identical.

1. Cloned genomic DNA fragments to be spotted onto the array slides (the "probes," see Chapter 11). For DNA content analysis, these usually comprise a library of DNA fragments derived from genomic DNA in a plasmid vector.
2. Oligonucleotide PCR primers to amplify genomic DNA probes. If the genomic DNA fragments have been cloned into a standard plasmid vector, then these will be standard oligonucleotide primers from a commercial source, such T3/T7 or M13F/M13R.
3. Standard PCR reagents and thermal cycler (capable of handling 96-well PCR plates).
4. Standard 96-well PCR plates.
5. Self-adhering PE foil (Macherey-Nagel 740676).
6. NucleoFast 96-well filter plates (Macherey-Nagel 7435000.4).
7. Vacuum manifold (e.g., Millipore/MultiScreen, Qiagen/QIAvac 96, Promega/Vac-Man 96, BioRad/Aurum).
8. 96-well V-bottom plates (e.g., Nunc 442587).
9. 384-well V-bottom plates (e.g., Genetix X7022).
10. Benchtop centrifuge with rotor for multiwell plates.
11. 3 M sodium acetate.
12. 100% and 70% ethanol. Store in the refrigerator.
13. Speed vacuum apparatus, capable of taking multiwell plates.
14. Bar-coded γ-amino propyl silane-coated slides (Corning 40003).
15. 4X Spotting buffer: 600 mM sodium phosphate, 0.04% SDS. 48.9 mL 600 mM Na_2HPO_4, 1.4 mL 600 mM NaH_2PO_4, 200 µL 10% SDS.
16. Printing tool (and pins).
17. Robotic printer.
18. UV source (e.g., a Stratalinker from Stratagene).
19. 80°C oven.

2.2. Isolation of Genomic DNA From E. histolytica

Here, we describe the isolation of genomic DNA suitable for hybridization to microarrays from *E. histolytica* trophozoites (see **Note 1**).

1. 5×10^7 mid-log-phase trophozoites of *E. histolytica*.
2. Phenol–chloroform premixed with isoamyl alcohol (25:24:1) (Ameresco) and saturated with TE.
3. Chloroform (Fisher Biotech).
4. 3 M sodium acetate, pH 5.2.
5. NET: 100 mM NaCl, 10 mM EDTA, 10 mM Tris-HCl, pH 8.0.
6. TNE: 10 mM Tris-HCl, pH 8.0, 1 mM EDTA, 0.2 M NaCl.
7. RNase: 500-µg/mL stock in H_2O, from bovine pancreas (Roche). Store frozen.
8. Proteinase K: 20-mg/mL stock in H_2O (fungal origin, Invitrogen). Store frozen.
9. Tris-EDTA (TE): 10 mM Tris-HCl, pH 8.0, 1 mM EDTA.

2.3. DNA Labeling, Target Cleanup, Hybridization, and Slide Washing

1. Klenow polymerase and appropriate 10X buffer (Roche Molecular-Boehringer, 1008404).
2. Random nonamer (Operon Technologies) (4 µg/mL).
3. Microcon-YM30 concentrators: 30,000-MW cutoff (Amicon 42410).
4. Deoxynucleotides: 100 mM each (Pharmacia: dATP 272-050, dCTP 272-060, dGTP 272-070,and dTTP 272-080).
5. Cy3- and Cy5-dUTP (Amersham Pharmacia, Cy3 53022, Cy5 55022).
6. Labeling master mix (per sample): 3 µL H_2O, 5 µL 10X Klenow buffer, 5 µL 10X dNTPs, 1.5 µL Cy5 or Cy3, 2 µL Klenow polymerase.
7. Yeast transfer RNA (tRNA): 20 µg/µL (Gibco/BRL 15401-011).
8. Poly-d(A)$_{18}$ synthetic oligonucleotide: 2 µg/µL (New England Biolabs, S131655).
9. 20X SSC: 3 M NaCl, 0.3 M Na citrate.
10. 10% SDS.
11. Hybridization chambers (ArrayIt AHC-1 or equivalent).
12. Coverslips (size depends on cDNA array dimensions).
13. Coplin jars.
14. Slide-swing handle rack (RA Lamb E102).

2.4. Scanning/Software

Numerous different software packages are now available for microarray analysis, including Prediction Analysis of Microarrays (PAM), Significance Analysis of Microarrays (SAM), Genomotyping Analysis by Charlie Kim (GACK). Two of these (PAM and SAM) are written specifically for analysis of expression profiles, but GACK has been written specifically for the comparison of DNA/gene content *(17–19)*.

3. Methods
3.1. Genomic DNA Probe Amplification, Purification, and Printing
3.1.1. Amplification

As described in Chapter 11, **Subheading 3.1.1.**, a 100-µL amplification reaction produces sufficient DNA for the spotting of each DNA probe after 35 ampli-

DNA Content Analysis

fication cycles. Perform the PCR amplification with primers that are complementary to vector sequences flanking the parasite-specific inserts in the starting library (M13R/M13F or T3/T7). (*See* **Note 2**.)

3.1.2. Purification of DNA Probes

Before the parasite-specific DNA probes are suitable for printing, they must be purified away from all contaminating salts, dNTPs, and primers. High-throughput purification of multiple samples via the use of multiscreen FB 96-well filter plates has been outlined in Chapter 11. An alternative method is to precipitate the PCR products with ethanol, the main advantage being reduced cost. Note that the entire process is done in the original source plate, thus minimizing costs as well as avoiding potential problems with errors in plate orientation.

1. Spin PCR products in source plate (10 min at room temperature, 2500g).
2. Add 0.1 vol 3 M sodium acetate and 2 vol cold 100% ethanol. Mix gently and precipitate at –20°C overnight.
3. Spin at 2500g, 4°C, 30 min. Pour off supernatant carefully (it is best to do it in one rapid flicking motion). It is important to avoid contaminating neighboring wells.
4. Add 200 µL cold 70% ethanol.
5. Spin at 2500g, 4°C, 30 min. Pour off supernatant and dry in Speed-Vac. Be careful not to warp plates (avoid high temperature for an extended time or the plates will warp).
6. Resuspend in desired amount of H_2O or 1X SSC.

3.1.3. Printing

Follow the protocol detailed in Chapter 11, including appropriate control DNAs (*see* **Notes 3** and **4**).

3.2. Genomic DNA Isolation and Purification

The entire process depends on the generation of good quality genomic DNA for hybridization to the array (*see* **Note 5**). For any given parasite system, one may have to try numerous genomic DNA isolation methods (*see* **Note 6**).

Listed below is the protocol we use for making genomic DNA from *E. histolytica*. The protocol takes 3 d but yields high-quality DNA suitable for use in our microarrays.

3.2.1. Day 1

1. Usually, we start with 5×10^7 mid-log-phase trophozoites of *E. histolytica* (approx 0.5-mL cell pellet). Resuspend the pellet in a total volume of 8.5 mL NET buffer.
2. Lyse the cells by addition of 80 µL of 20% SDS.
3. Add an equal volume (8.5 mL) of TE saturated phenol–chloroform–isoamyl alcohol (25:24:1) to the cell lysate. Shake the mixture on a nutator for 10 min.

4. Centrifuge in a Beckman JS13.1 rotor at 15,000g for 10 min at 4°C.
5. Remove and keep aqueous phase. Perform two back extractions from the original tube by adding 5 mL of NET buffer each time, then an equal volume of phenol–chloroform–isoamyl alcohol. Spin and harvest aqueous phase as above.
6. Mix all the aqueous collections from all extractions with equal volumes of TE-saturated phenol–chloroform–isoamyl alcohol. Perform the same back extraction steps as above.
7. Mix the total aqueous collection with an equal volume of chloroform, centrifuge, and collect the aqueous phase again.
8. Precipitate DNA overnight at 4°C by adding 0.1 vol of 3 M NaOAc and 2 vol of 100% cold ethanol.

3.2.2. Day 2

9. Centrifuge the tubes in the Beckman JS13.1 rotor at 26,688g for 30 min at 4°C.
10. Wash precipitated DNA with 70% cold ethanol and then resuspend in 5 mL of TNE buffer.
11. Add 15 μL RNase and incubate at 37°C for 30 min.
12. Add 75 μL of proteinase K and incubate at 55°C for 45 min.
13. Extract DNA as described for d 1 with phenol–chloroform–isoamyl alcohol, but perform all back extractions with TNE instead of NET. Precipitate the DNA with ethanol as before.

3.2.3. Day 3

14. Centrifuge the DNA and wash with 70% cold ethanol as before. Air-dry and resuspend in TE (typically 200–300 μL).
15. Check the OD 260:280 ratio and use if ≥1.60. If the DNA solution is not clean enough, perform another phenol–chloroform extraction and reprecipitate with ethanol.

3.3. Labeling of the Target Genomic DNA and the Reference DNA

The protocol detailed below uses 4 μg of genomic DNA as the starting input material for each Cy5 reaction. We routinely label the reference DNA (*see* **Note 7**) with Cy3 (*see* **Note 8**). Generally, more input DNA leads to generation of higher signal intensity on the arrays. Care should be taken to ensure that the experiments are performed using saturating levels (using more reference does not increase signal) of labeled reference DNA. The amount of experimental sample should be adjusted so that its signal is in the linear range (i.e., more DNA gives more signal) (*see* **Note 9**).

1. Prepare total genomic DNA (4 μg) in a total volume of 30.5 μL with TE in an Eppendorf tube. Add 3 μL of random nonamer to each sample.
2. Prepare reference DNA (350 ng) (*see* **Note 10**) in a total volume of 30.5 μL in an Eppendorf tube and add 1.5 μL of each primer (M13F and M13R).

3. Heat DNA sample/nanomer and reference/primer mixtures to 100°C for 1 min. Leave to cool at room temperature for 5 min.
4. Add 16.5 µL of the labeling master mix (containing Cy5 for experimental sample and Cy3 for reference sample) to each denatured sample. Incubate at 37°C for 2 h.
5. After 2 h, centrifuge briefly to pellet any condensation and reincubate at 37°C for a further 2 h (i.e., a total of 4 h).
6. Combine one experimental sample and one reference sample. Add 1 µL yeast tRNA and 1 µL of poly dA18 per pair.
7. Add 500 µL of TE and add to microconcentrator. Spin at 18,000g for 10 min at room temperature in a microcentrifuge. Wash twice, adding 500 µL of TE each time.
8. When the volume of sample after the second wash reaches approx 10 µL, invert the Microcon-YM30 concentrator and place into a new 1.5-mL microcentrifuge tube. Elute washed probe samples from filter by spinning at 18,000g for 10 min.
9. Adjust volume of labeled DNA in solution to 12 µL with TE. Add 2.55 µL 20X SSC and 0.45 µL 10% SDS. Boil for 1 min.
10. Allow sample to cool on benchtop for at least 30 min. Centrifuge briefly and use 14–14.5 µL to hybridize on the array (*see* **Note 11**).

3.4. DNA Microarray Preparation, Prehybridization, Hybridization, and Washing

Follow the protocols provided in Chapter 11, **Subheadings 3.4.** and **3.5.** We routinely hybridize our arrays for 16 h and find that we get the most reproducible results under these conditions. Most protocols recommend hybridizations at 65°C, however, if one is trying to increase hybridization between the DNA of relatively dissimilar species/strains, then lower-stringency conditions can be attempted (i.e., lower temperature). Similarly, the temperature of the washing step may be reduced to lower the stringency. It is advisable to test several different temperatures in these cases. Some laboratories suggest that sealing the coverslip with rubber cement is useful for reducing uneven hybridizations. This is not a routinely used procedure but can be attempted if uneven hybridization (resulting from drying out of the probe) is a problem.

3.5. Data Acquisition/Image Analysis/Data Mining/Other Issues

General comments in Chapter 11, **Subheading 3.6.** apply to these analyses also (*see* **Note 12**). Numerous analytical and statistical tools have been developed for microarray analysis and are now freely available. Most, including SAM and PAM, are written for the analysis of gene expression patterns. However, the program GACK has been written specifically for the analysis of genome abundance using microarrays analysis (*see* **Note 13**).

During data analysis, it is important to realize that the array technology will identify genes as having different genomic abundance between the two strains if the gene is *either* missing in one strain *or* if it is highly divergent in one strain. The technology described here (spotted DNA arrays, one defined hybridization and wash temperature) is currently not sophisticated enough to distinguish genes that are highly divergent from those that are absent.

For DNA content analysis the user has two options in comparing the genomic abundance of a given gene. The first option for comparing strains A vs B is to identify proof-of-concept genes that are known to be present or absent in the comparison strains. Using that information, the user can identify an arbitrary cutoff that appears to distinguish absent genes from ones that are known to be present.

The second option is to use the GACK, a program that was written and developed for genome abundance analysis *(19)*. This program presumes a normal distribution of the data and identifies clones or genes that fall outside the normally predicted distribution. The advantage of this approach is that it is not based on an arbitrary cutoff, but rather depends on the data fitting a normal distribution. Most genomotyping analyses to date use static analysis with constant cutoffs and do not take into account the variations in distributions of signal intensity (for clones on the array) between strains, nor experimental variations. The GACK algorithm uses a dynamic cutoff depending on the shape of the signal–noise ratio curve, which may be different for each array because of divergence at the nucleotide level, variable levels of cross-hybridization, and/or variability resulting from low ratios arising from low intensity values. Additionally, because the output data are graded, the algorithm distinguishes genes that are most likely very divergent or absent from those that may be only slightly divergent.

For use of the GACK algorithm, the results from array data (normally expressed as ratio of red to green signal intensities [R/G]) need to be converted to log(2) scale to validate the assumption of normality. The data cannot be analyzed in linear scale. The converted log(2) data are used as input in GACK. Typically, three different output formats are available for visualization. These include graded, binary, and trinary output. The binary output designates present genes as "1" and divergent genes as "0." The trinary output designates the present genes as "1," divergent genes as "–1," and uncertain genes as "0." We prefer to use the graded output for our analysis, as it assigns a range of values from –0.5 to 0.5 in 0.05 increments. Each rising increment from –0.5 corresponds to lesser likelihood of divergence. Using this output, we can categorize genes ranging from absent/highly divergent, to slightly divergent, to present in both species.

Preliminary GACK data from *E. histolytica* comparative genomics indicates that a graded cutoff of –0.4 distinguishes genes that are present from those that are absent/divergent. This cutoff will be different for each user, depending on the organism/species used. In most cases, the cutoff is close to where the normal curve intercepts the *X* axis, often referred to as the "transition point" (*see* **Note 14**).

DNA Content Analysis

Whichever analytical tool is used, the results of the statistical analysis need to be confirmed by other techniques (*see* **Note 15**). This will help to validate the arbitrary or GACK cutoff used to indicate absence or high divergence of a given gene.

Additional information to be gained from DNA content analysis is the identification of variable copy number of genes or DNA fragments. If a reference sample is used, the signal for a single-copy, vs double-copy, vs multicopy gene can be calculated and may be useful in further characterizing strain and species differences.

4. Notes

1. Isolation of genomic DNA can be accomplished using numerous protocols, and many commercial kits are very suitable. Here, we provide protocols that have worked successfully for *E. histolytica*, in which the main difficulty is the carbohydrate content of the parasite (CTAB method: *E. histolytica* webpage maintained by Graham Clark (http://homepages.lshtm.ac.uk/entamoeba/dnaisoln.htm). In general, we recommend that individuals apply the method of genomic DNA isolation that traditionally is used for their parasite of choice and test the suitability of the DNA for the microarray hybridizations. If the results are not optimal (poor labeling or high background), then modifications to improve the quality of the genomic DNA can be applied, such as cleaning up the DNA via a phenol–chloroform extraction and reprecipitating with ethanol.
2. We have found it easiest to amplify cloned fragments, because this enables the use of a common primer set for all the PCR reactions. We also recommend using the bacterial stocks of the library (without making plasmid preparations for each clone) as the template for PCR, because this is a much cheaper and easier protocol. Aside from the expense of plasmid preparation, concern over the risk of plasmid cross-contamination in high-throughput methods is significant.
3. It is always useful to have some controls printed on the array (yeast tRNA, buffer only, plasmid and phage DNA from which PCR products were amplified, reference regions from the vector). These spots can be very useful when one is trying variable hybridization temperatures and wash conditions, which may be necessary to optimize specific and minimize nonspecific hybridization of probe and target DNAs. In particular, if less stringent hybridization and wash conditions are used to maximize cross-hybridization between more divergent species/strains, then these controls will be even more useful in discriminating and identifying the optimum experimental conditions.
4. Numerous laboratories have made their protocols public, and these provide a valuable resource for newcomers to the field:

Pat Brown:	http://cmgm.stanford.edu/pbrown/mguide/index.html
Joe DeRisi:	http://derisilab.ucsf.edu/
Gary Schoolnik:	http://schoolniklab.stanford.edu/index.html
David Botstein:	http://genome-www.stanford.edu/group/botlab/

5. Genomic DNA that would routinely be acceptable for PCR amplification often is not clean enough to label well with Klenow polymerase. Contaminants such as carbohydrates are the most troublesome, as they inhibit the Klenow labeling reaction and often give a "speckled" background and uneven hybridization. Getting good-quality DNA is especially problematic in carbohydrate-rich parasites such as *E. histolytica*. In this system, we have not had good success with genomic DNA that has an OD 260:280 ratio ≤1.65 and usually aim for samples with a 260:280 ratio of 1.8. Often, we employ an extra phenol–chloroform extraction step to clean up the DNA at the end of any extraction.
6. You may have to determine empirically which method produces the best DNA for the microarray hybridizations. The critical aim is reproducible, even hybridization on the array with minimal background. Controls should be invaluable in distinguishing nonspecific hybridizations and background from specific hybridization. We recommend that the user make genomic DNA as appropriate for the parasite, check the OD 260:280 (ensure that it is ≥1.6), test that it is clean enough to cut with restriction enzymes, and then test it in a microarray experiment.
7. An important technique is the use of a common reference DNA in each hybridization, to which every test DNA sample is individually compared. This approach is especially useful when reagents (genomic DNA from clinical isolates, for example) are limited. The use of a common reference allows an inferred comparison between samples, thus allowing samples processed at different times to be compared to each other. Numerous options are available for use as a reference, for example, genomic DNA or RNA from a parent strain or a common vector sequence from a library. Use of a common vector sequence as a reference is advantageous for numerous reasons. Such a reference is easily and cheaply generated and provides another filter for spot quality, since it should hybridize to all spots relative to the amount of DNA that is spotted. Normalization of the experimental sample signal to the overall signal for the reference probe allows inferred comparison between experiments, and the data can subsequently be analyzed for gene content (presence or absence of a gene) as well as for copy number of a given gene.
8. The arrays we use have a common vector sequence in each spotted clone (probe), thus our reference target DNA is a PCR-amplified DNA fragment that hybridizes with the vector in each clone on the microarray (*see also* **Note 7**). The same primers that were used to generate the PCR products (the probes) for array printing are used for the generation of the reference DNA target, except that an "empty" vector (i.e., with no genomic DNA insert in the multicloning site) is used as template for the PCR. This PCR product (usually approx 150 bp) is then cleaned to remove dNTPs and labeled using the protocol given here.
9. For the genomic DNA sample, empirical testing is recommended to identify the amount of DNA that gives optimal results. We do not routinely use less than 4 µg of genomic DNA, but experiments done with 2 µg have also worked well. If the samples to be used are extremely limiting in quantity, then experiments to find the lowest amount of DNA that give reproducible results, using less limited material, will need to be undertaken.

10. The amount of reference DNA that will give saturating levels of signal needs to be determined. We have found that 350–500 ng of reference DNA works well for our array.
11. Leaving a small amount of probe at the bottom of the tube ensures that you will not add any precipitated SDS or other material to your array. Keep in mind that the size of the printed area will dictate the final amount of probe you will need. For arrays in which you use a 22 mm × 22 mm coverslip, the 14 µL of probe used here is adequate. If the printed size of the array is bigger and necessitates a 22 mm × 40 mm coverslip, then the probe can be resuspended in a final volume of 20 µL and the SSC and SDS increased proportionately.
12. We recommend establishing access to a scanner that you will be able to use throughout the course of your experiments. Its best to avoid switching scanners (even of the same brand) in the middle of a set of experiments, because there may be subtle differences that can affect data reproducibility.
13. The programs can be freely downloaded from the following websites: Significance Analysis of Microarrays (SAM), http://www-stat.stanford.edu/~tibs/SAM/; Prediction Analysis of Microarrays (PAM), http://www-stat.stanford.edu/~tibs/PAM/; Genomotyping Analysis (GACK) by Charlie Kim, http://falkow.stanford.edu/whatwedo/software/programs/GACKman.pdf.
14. For further details please visit http://falkow/stanford.edu/whatwedo/software and http://genomebiology.com/2002/3/11/research/0065.1.
15. Once clones of interest are identified, we highly recommend sequencing the clone and also performing confirmation of gene abundance using a traditional technique such as Southern blotting, prior to any intensive biological experimentation.

References

1. Reinke, V. (2002) Functional exploration of the *C. elegans* genome using DNA microarrays. *Nat. Genet.* **32(Suppl.)**, 541–546.
2. Quackenbush, J. (2002) Microarray data normalization and transformation. *Nat. Genet.* **32(Suppl.)**, 496–501.
3. Churchill, G. A. (2002) Fundamentals of experimental design for cDNA microarrays. *Nat. Genet.* **32(Suppl.)**, 490–495.
4. Holloway, A. J., van Laar, R. K., Tothill, R. W., et al. (2002) Options available—from start to finish for obtaining data from DNA microarrays II. *Nat. Genet.* **32 (Suppl.)**, 481–489.
5. Petricoin, E. F. 3rd, Hackett, J. L., Lesko, L. J., et al. (2002) Medical applications of microarray technologies: a regulatory science perspective. *Nat. Genet.* **32(Suppl.)**, 474–479.
6. MacBeath, G. (2002) Protein microarrays and proteomics. *Nat. Genet.* **32(Suppl.)**, 526–532.
7. Cheung, V. G. and Spielman, R. S. (2002) The genetics of variation in gene expression. *Nat. Genet.* **32(Suppl.)**, 522–525.
8. Gerhold, D. L., Jensen, R. V., and Gullans, S. R. (2002) Better therapeutics through microarrays. *Nat. Genet.* **32(Suppl.)**, 547–551.

9. Chuaqui, R. F., Bonner, R. F., Best, C. J., et al. (2002) Post-analysis follow-up and validation of microarray experiments. *Nat. Genet.* **32(Suppl.),** 509–514.
10. Slonim, D. K. (2002) From patterns to pathways: gene expression data analysis comes of age. *Nat. Genet.* **32(Suppl.),** 502–508.
11. Pollack, J. R. and Iyer, V. R. (2002) Characterizing the physical genome. *Nat. Genet.* **32(Suppl.),** 515–521.
12. Stoeckert , C. J. Jr., Causton, H. C., and Ball, C. A. (2002) Microarray databases: standards and ontologies. *Nat. Genet.* **32(Suppl.),** 469–473.
13. Boothroyd, J. C., Cleary, M. D., and Singh, U. (2003) DNA microarrays in parasitology: strengths and limitations. *Trends Parasitol.* **19(10),** 470–476.
14. Salama, N., Guillemin, K., McDaniel, T. K., et al. (2000) A whole-genome microarray reveals genetic diversity among *Helicobacter pylori* strains. *Proc. Natl. Acad. Sci. USA* **97,** 14,668-14,673.
15. Behr, M. A., Wilson, M. A., Gill, W. P., et al. (1999) Comparative genomics of BCG vaccines by whole-genome DNA microarray. *Science* **284,** 1520–1523.
16. Pollack, J. R., Sorlie, T., Perou, C. M., et al. (2002) Microarray analysis reveals a major direct role of DNA copy number alteration in the transcriptional program of human breast tumors. *Proc. Natl. Acad. Sci. USA* **99,** 12,963–12,968.
17. Tusher, V. G., Tibshirani, R., and Chu, G. (2001) Significance analysis of microarrays applied to the ionizing radiation response. *Proc. Natl. Acad. Sci. USA* **98,** 5116–5121.
18. Tibshirani, R., Hastie, T., Narasimhan, B., et al. (2002) Diagnosis of multiple cancer types by shrunken centroids of gene expression. *Proc. Natl. Acad. Sci. USA* **99,** 6567–6572.
19. Kim, C. C., Joyce, E. A., Chan, K., et al. (2002) Improved analytical methods for microarray-based genome-composition analysis. *Genome Biol.* **3,** RESEARCH 0065.1– RESEARCH0065.17.

13

Typing Single-Nucleotide Polymorphisms in *Toxoplasma gondii* by Allele-Specific Primer Extension and Microarray Detection

Chunlei Su, Christian Hott, Bernard H. Brownstein, and L. David Sibley

Summary

Genotyping is an important tool for epidemiological and population genetic studies in protozoan parasites. The most commonly used method for genotyping is polymerase chain reaction (PCR)-based restriction fragment length polymorphism (RFLP) analysis of single nucleotide polymorphisms (SNPs). However, PCR-RFLP analysis is labor intensive, and only a proportion of the SNPs are recognized by currently available restriction enzymes. Here, we have developed a more efficient microarray-based method to genotype SNPs in the protozoan parasite *Toxoplasma gondii*. This method is sensitive, accurate, and capable of analyzing multiple SNPs simultaneously in a high-throughput format.

Key Words: Allele-specific extension; single-nucleotide polymorphisms; SNP microarrays.

1. Introduction

Rapid genotyping provides an important tool for epidemiological, population, and genetic mapping studies in parasitic organisms. The protozoan parasite *Toxoplasma gondii* has a relatively large (approx 65 Mb) genome that is haploid throughout most of its life cycle. *T. gondii* has a highly clonal population structure consisting of three closely related, yet genetically distinct groups defined as types I, II, and III *(1,2)*. These clonal types are highly similar, with on average only 1–2% divergence at the nucleotide level *(3,4)*. Remarkably, the majority of strains of *T. gondii* possess one of only two alleles at each locus *(5)*. This biallelic pattern combined with the high degree of similarity suggests that a recent meiotic cross gave rise to the majority of existing strains of *T. gondii*.

Table 1
Frequencies of Polymorphisms Between Three Major Genotypes of *T. gondii*

Strain type	Unique SNPs	No. of RFLPs	Frequency of SNPs
Type I (RH strain)	85	47	1 in 272
Type II (ME49 strain)	44	22	1 in 527
Type III (CTG strain)	59	30	1 in 393
Total	188	99	

Based on sequencing of 33 separate loci totaling 23,192 bp.

Amplification of specific DNA segments using polymerase chain reaction (PCR), followed by restriction fragment length polymorphism (RFLP) analysis is the most commonly used method for genotyping of *T. gondii* *(6–10)*. However, this method is labor-intensive and typically requires that each locus be analyzed independently. Analysis of microsatellites has also been applied to *T. gondii*, although this approach is limited by the relative infrequency of short dinucleotide repeats in this organism *(11,12)*.

Despite the close similarity of strains, single-nucleotide polymorphisms (SNPs) are abundant in *T. gondii*. Sequencing of a total of 33 separate loci from three *T. gondii* strains, chosen to represent the three clonal lineages, reveals that SNPs occur approx every 400 bp (**Table 1**). For reasons that are unclear, the frequency of polymorphisms is higher in type I strains than either type II or type III strains. A biallelic pattern is evident at every one of these SNPs, such that whereas one strain contains a unique allele (different nucleotide at that position), the other two strains are identical. Combined with its haploid genome, this simplifies SNP analysis in *T. gondii* in that only two alleles need to be screened at any given locus. However, up to 40% of DNA polymorphisms are not recognized by available restriction enzymes (**Table 1**) and, therefore, a more rapid and robust method of screening SNPs would be advantageous.

SNPs are also abundant in the mouse and human genomes and have been used for mapping and genetic linkage studies *(13)*. Various techniques exist for typing SNPs, including single-strand conformational gel electrophoresis, RFLP analysis, and various PCR-based approaches. To genotype *T. gondii* efficiently using SNPs, we developed a new method to analyze multiple loci by allele-specific primer extension followed by microarray hybridization.

To take advantage of the high-throughput screening capabilities of DNA microarrays, we have used a complementary oligonucleotide tag approach first described by Hirschhorn et al. *(14)*. This method uses pairs of complementary generic oligonucleotides that are optimized for similar thermodynamic properties. One of the members of the pair is synthesized in conjunction with an allele-specific primer to detect the SNP and the other is affixed to a glass slide using an oligo-dT tail. This combination allows the simultaneous analysis of multiple SNP-specific oligonucleotides using one set of standard hybridization conditions. A second method that we incorporated involves the use of single-base mismatch primers to amplify SNPs specifically *(15)*. Although this protocol originally used a primer extension reaction based on RNA samples, we have added a preamplification step to generate double-stranded DNA templates that flank the SNP. Allele-specific extension reactions are conducted on a single-strand PCR reaction using primers that specifically prime only one of the two possible alleles. Finally, we have used the more efficient labeling protocol of amino-allyl dUTP incorporation followed by dye conjugation after the PCR reaction. These improvements have been combined to provide a rapid, high-throughput analysis of genotypes in *T. gondii*. This approach is also applicable to genotyping of various organisms, including other pathogens.

2. Materials

1. Cell lysates of parasite strains to be tested.
2. PCR reagents, including 10X buffer, dNTPs, and AmpliTaq (Roche, Indianapolis, IN).
3. 10-mg/mL Proteinase K (Promega, Madison, WI).
4. PCR primers for preamplifying the region surrounding SNPs (called SNP primers).
5. Allele-specific extension primers (15-nt oligo-dT plus 17–25 bp of allele-specific sequences that define the SNP) (*see* **Note 1**).
6. Shrimp alkaline phosphatase (Roche, Indianapolis, IN).
7. Exonuclease I (Amersham, Piscataway, NJ).
8. Monofunctional *N*-hydroxysuccinamide-ester Cy5 (no. PA23001) and Cy3 (no. PA25001) dyes. (Amersham, Piscataway, NJ). Dissolve dye in 72 µL dimethyl sulfoxide, aliquot 16 × 4.5 µL, dry the aliquots in a vacuum evaporator, and store aliquots at 4°C in a dessicator before use.
9. A solution of 0.05% NP-40, 10% sodium dodecyl sulfate (SDS), 10 mM amino-allyl-dUTP (Sigma, St. Louis, MO).
10. Sonicated herring sperm DNA (Promega, Madison, WI).
11. ArrayHyb LowTemp hybridization buffer (Sigma, St. Louis, MO).
12. PCR purification and Nucleotide Removal Kits (Qiagen, Valencia, CA).
13. Speed-Vac vacuum evaporator (Savant, Farmingdale, NY).
14. 0.1 M carbonate buffer, pH 8.8.
15. 20X SSC, pH 7.0 (3 M sodium chloride, 0.3 M sodium citrate).
16. Array hybridization chamber.
17. Poly-L-lysine (Sigma, St. Louis, MO).
18. Phosphate-buffered saline (PBS).
19. Erie Scientific glass microscope slides and Lifter-slip (Erie Scientific Corporation, Portsmouth, NH).
20. Affymetrix model 417 arrayer and 428 scanner (Affymetrix, Santa Clara, CA).
21. ScanAlyze version 2.35 (Michael Eisen, Stanford University, CA).

3. Methods

The method described here provides rapid, multiplex analysis of SNPs as diagrammed in **Fig. 1**. The strategy is first to amplify a short genomic region, consisting of approx 100 bp, using primers that flank the SNP and thus amplify both alleles. This first PCR step can be multiplexed (referring to the combining of several gene-specific PCR amplifications combined in a single reaction) to allow simultaneous amplification of different loci. In the second step, a primer extension reaction is conducted using two separate samples, one seeded with primers for the A alleles and one seeded with primers for the B alleles. The primers differ by one nucleotide at the 3'-end, which defines the SNP. This reaction also incorporates a modified dUTP that is labeled after amplification with Cy3 (A alleles) or Cy5 (B alleles). The resulting products are cohybridized to complementary oligonucleotides spotted on glass slides. The recognition of these complementary oligonucleotides is based on a unique generic tag attached to the 5'-end of each allele-specific extension primer. This generic oligonucleotide tag does not represent a sequence in the genome of the organism but instead is chosen to provide uniform hybridization to a complementary generic oligonucleotide tag that is cross-linked to a glass slide. The relative hybridization of the A and B allele products to the spotted oligos are quantified using a microarray laser scanner and the resulting ratios are used to determine the genotype of the strain. The advantages of this method are that multiple independent loci can be analyzed simultaneously in a high-throughput manner. The specific methodology is discussed further below.

3.1. Samples

We will illustrate our SNP analysis method by describing the analysis of four *T. gondii* strains (type I, RH, GT1; type II, PTG; and type III, CTG). Parasites were propagated in vitro as described previously *(1)* and were pelleted by centrifugation and resuspended in 1 mL of PBS to a concentration of approximately 10^7 parasites/mL. One hundred microliters of 10X PCR buffer and 10 µL of 10-mg/mL proteinase K were added to the parasites and incubated at 37°C for 1 h, 50°C for 1 h, and 95°C for 15 min. Parasite cell lysates were stored at −20°C before use.

3.2. SNP Markers

We have selected nine independent genetic markers that were previously typed by PCR-RFLP and SNP analysis to illustrate the methods described here. These markers are *ACT1*, L351, cB21, BS226, L358, *SAG1*, *SAG2*, *TUB2*, and *GRA6*, which are dispersed on 8 of the 11 *T. gondii* chromosomes, including Ib, II, III, IV, V, VIII, IX, and X, respectively (**Table 2**).

SNP Typing

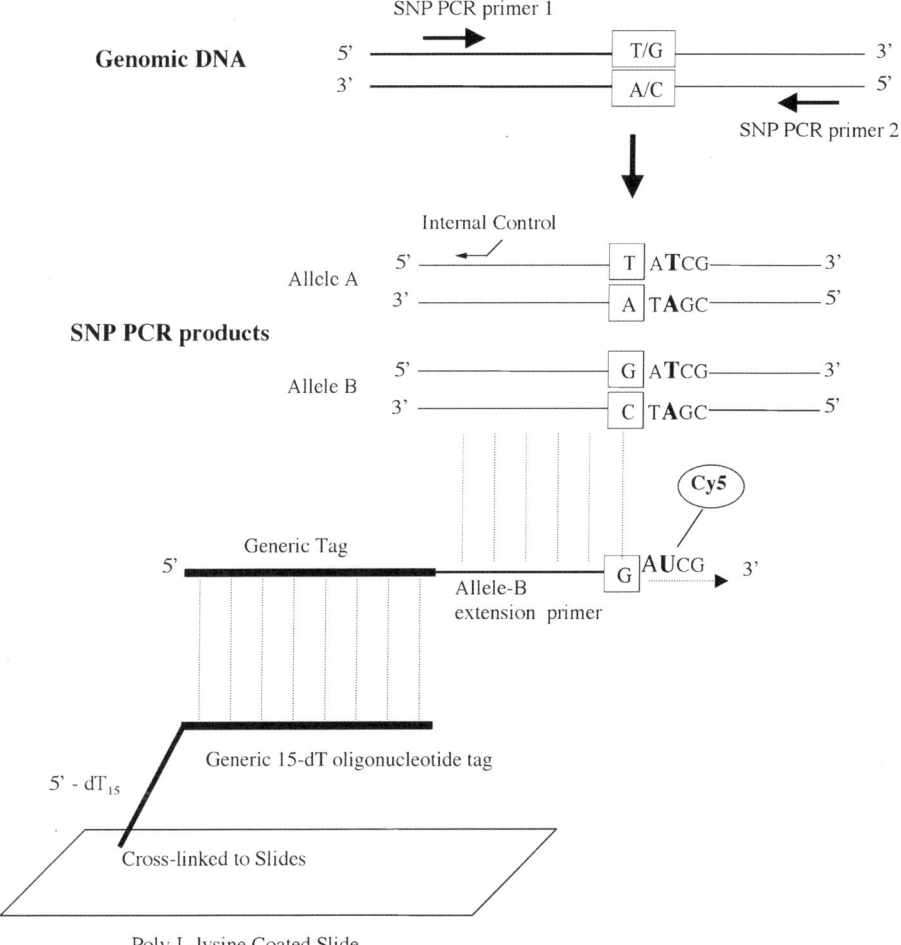

Fig. 1. DNA fragments containing SNPs were amplified by multiplex PCR from genomic DNA. The SNP PCR products served as templates for the allele-specific primer extension reaction. Each extension primer of 17–25 bases contains a 5'-end generic tag complementary to a unique 15-dT oligonucleotide tag that was cross-linked to the slide. The 3'-end of the extension primers was an exact match to one of the two alleles (type A or B allele) for each SNP. The primer extension reaction was carried out in the presence of amino-allyl-dUTP that was later coupled with Cy3 or Cy5. The labeled multiplex primer extension products were pooled and hybridized to the oligonucleotide tag array. Two hybridization patterns were expected, a predominantly green color corresponds to type A alleles and a predominantly red color corresponds to type B alleles (*see* **Fig. 2**).

Table 2
RFLP and SNP Markers Used in the Present Study

Chromosome	Marker	SNP	Strain type	RFLP site
Ib	ACT1	GGT**A**TCCC	I	*Bsm*F1
		GGT**G**TCCC	II, III	GTCCC
II	L351	GTC**G**CTA	I, III	*Mae* III
		GTC**A**CTA	II	GTNAC
III	cB21	ACG**C**GAA	I, III	*Hpy*CH4 IV
		ACG**T**GAA	II	ACGT
IV	Bs226	GTGA**A**CCT	I, II	*Mae* III
		GTGA**C**CCT	III	GTNAC
V	L358	ACA**T**GGG	I, III	*Nla* III
		ACA**C**GGG	II	CATG
VIII	SAG1	TAA**C**TGA	I	*Dde* I
		TAA**G**TGA	II, III	CTNAG
VIII	SAG2	GAT**T**CTG	I, II	*Hinf* I
		GAT**C**CTG	III	GANTC
IX	TUB2	AAT**C**GAG	I, III	*Taq* I
		AAT**G**GAG	II	TCGA
X	GRA6	TTA**A**GCA	I, III	*Mse* I
		TTA**G**GCA	II	TTAA

SNPs are shown as bold and underlined letters for all markers. For each SNP, the top allele is denoted as A and the bottom as B. Strain types include I, II, and III. The listed restriction enzymes are those that recognize restriction site polymorphisms at these locations, as used previously to distinguish different *T. gondii* strain types *(1)*.

3.3. Generic Oligo-dT Tag Array

3.3.1. 15-dT Oligonucleotide Tags

Generic tags are selected from a previously published source *(14)*, from a listing provided at www.genome.wi.mit.edu/publications/SBE-TAGS. The full-length reverse complement sequences of the tags should be synthesized with a 15-dT extension at their 5'-end and spotted onto poly-L-lysine coated slides to create the generic15-dT oligonucleotide tag array *(16)*. The 15-dT stretch is used to facilitate attachment to the slide. In our study of *T. gondii*, the 15-dT oligonucleotides were named to indicate the specific SNP loci that they were designed to detect: ACT1, TTTTTTTTTTTTTTTAGTTCGTGC TTACC GCAGAATGCAG; L351, TTTTTTTTTTTTTTTACAGGATGTGCTCAA CAGACGTT; cB21, TTTTTTTTTTTTTTTACATCAATCTCTCTGACCGT TCCGC; BS226, TTTTTTTTTTTTTTTATGGTGATCAGTCAACCACC AGG; L358, TTTTTTTTTTTTTTTAGACTGCGTGTTGGCTCTGTCACAG; SAG1, TTTTTTTTTTTTTTTGCCTTACATACATCTGTCGGTTGTA; SAG2,

SNP Typing 255

TTTTTTTTTTTTTTTGAGACACCTTATGTTCTATACATGC; TUB2, TTTT
TTTTTTTTTTTGCCACAGATAATATTCACATCGTGT; and GRA6, TTTTT
TTTTTTTTTTCCTCATGTCAACGAAGAACAGAACC.

3.3.2. Positive and Negative Controls

The positive control 15 nucleotide oligo-dT tag, TTTTTTTTTTTTTTT
ACATATCACAACGTGCGTGGAGGC, is also spotted onto the glass slides
and used to monitor the efficiency of array hybridization. The complementary
sequence of this tag should be labeled with Cy3 at its 5'-end (Cy3-GCCT
CCACGCACGTTGTGATATGTA) and spiked in the array hybridization reaction. Two negative control oligonucleotides, referred to as negative control-1
TTTTTTTTTTTTTTTCGCAGGTATCGTATTAATTGATCTG and negative
control-2 TTTTTTTTTTTTTTTACACATACGATTCTGCGAACTTCAA, are used
to monitor nonspecific array hybridization. Additionally, two internal control
oligonucleotides are used to monitor the efficiency of the primer extension reaction as further discussed under **Subheading 3.5.2.** The oligonucleotide tags
used for this step in our *T. gondii* study were LT4-GRA6, TTTTTTTTTTTTTT
TATTGAAGCCTGCCGTCGGAGACTAA, and LT10-cB21, TTTTTTTTTTT
TTTTCACAAGGAGGTCAGACCAGATTGAA, and these were spotted onto
glass slides with the other complementary oligonucleotides.

3.3.3. Array Slide Preparation

Generally, it is prudent to spot different concentrations of the oligonucleotides containing 0.005% NP40 onto poly-L-lysine coated microscope slides
(e.g., 6.25 μM, 25 μM, and 100 μM, as shown in **Fig. 1**) to assess the best level.
After arraying, slides are processed as described by Eisen and Brown *(16)*. In
brief:

1. After spotting, UV-treat the oligonucleotides to cross-link them to the amino groups of the lysines on the slides.
2. Block the slides by submersion in succinic anhydride to convert the remaining free animo groups to amides.
3. Briefly heat the slides by submerging them in boiling water to denature and eliminate the secondary structure of the arrayed oligonucleotides.
4. Air-dry and store in the dark at room temperature. We have used slides for up to 3 mo but find that after this time period the quality deteriorates (*see* **Note 2**).

3.4. Preamplification of SNP Regions

In this step, small fragments of DNA that flank the target SNPs are amplified by multiplex PCR and the products are purified and used as templates to
generate allele-specific probes for the array hybridization steps described in
the next section.

In the *T. gondii* study, PCR primers were designed using PRIMER3.0 (http://www.genome.wi.mit.edu) and used in a multiplex reaction to amplify 70–110-bp fragments containing the nine target SNPs. In the specific case illustrated here, these primers were ACT1-F, CAAGCGCATCGCGTACCG; ACT1-R, ATGATCTAGCTGAGGCGG; L351-F, TGCAAGTGGCAACCTTGAAA; L351-R, TGATTCCTATTAACAGACTA; cB21-F, ACTTGGTGGATCCAGCGA; cB21-R, CCTTCGACGCTGTAGACT; BS226-F, TACACATGCACCTGCGGCGA; BS226-R, CCTCTTGCATCCTTGTCCCC; L358-F, CCGCGAAGAATGGAATGTTC; L358-R, AAGGTTTCGCTTTCTCCGCC; SAG1-F, GCTGACTGGTTGCAACGAGA; SAG1-R, CTTATCACTCGAAGCGTTACCCTGC; SAG2-F, TCTGCGAAGAAAACGAGCGC; SAG2-R, ACTGCAACCCGTGAAACAAC; BTUB-F, TCTGCGTGACAGCTTCGCCA; BTUB-R, TCTCCGCGATTCTCCACAGG; GRA6-F, CAGCAGACAGCGGTGGTGTT; and GRA6-R, TTGCTGTCCACCGCTCGAAC.

Before being used in a multiplex PCR reaction, each SNP PCR primer set is tested individually to ensure that only one DNA fragment of the correct size is amplified from genomic DNA. For the multiplex PCR reaction, the 50-µL mixture should contain 2 µL of *T. gondii* lysate, 200 µM each of the dNTPs, 1.5 mM of MgCl$_2$, 150 nM each of the nine sets of forward and reverse primers, and 1.25 U of AmpliTaq. The reaction is carried out for 30 cycles at 95°C for 30 s, 55°C for 30 s, and 72°C for 1 min. After amplification, the multiplex PCR products are treated with 4.5 U of SAP (Roche) and 25 U of exonuclease I at 37°C for 30 min and 96°C for 15 min. Products are purified by Qiagen PCR Purification Kit following the manufacturer's instruction and eluted in 50 µL of H$_2$O. The purified PCR products are concentrated to 20 µL by vacuum evaporation.

3.5. Allele-Specific Primer Extension

In this step, the purified SNP-containing PCR products created in **Subheading 3.4.** are used to generate allele-specific sequences in a primer extension reaction in the presence of amino-allyl-dUTP. Then, the incorporated amino-allyl-dUTPs are coupled to monofunctional NHS-ester Cy3 or Cy5 dye to label the products prior to use for array hybridization.

3.5.1. Allele-Specific Primers

Each pair of extension primers consists of 17–25 bp that are complementary to one of the two alleles at that locus. Both members of the pair are coupled to the same generic oligonucleotide tag at the 5'-end. The 3'-end of the extension primers is an exact match to one of the two alleles for each SNP. A complementary generic oligonucleotide tag containing 15 nucleotide oligo-dT is cross-

SNP Typing

linked to the slide. In the *T. gondii* study, 12 generic oligonucleotide tags were selected from a pool of previously described oligonucleotide tag sequences and combined with paired extension primers *(14)*. The extension primers were designated as allele-A (type I strains) or -B primers (non-type I strains). For the following extension primers, the first 25 nucleotides are complementary to a generic tag spotted onto the array slides, the underlined 17–25 nucleotides are specific to *T. gondii* sequences, and the two SNP alleles are indicated in parentheses. The extension primers used here were ACT1-(A/B), CTGCATTC TGCGGTAAGCACGAACT AATGATCTAGCTGAGGCG GGA(T/C); L351-(A/B), AACGTCTGTTGAGCACATCCTGTAAAG CTC TTC ACG TGT TGTC (G/A); cB21-(A/B), GCGGAACGGTCAGAGAGATTGATGTTACCGGCCC ATCCATATTC(G/A); BS226-(A/B), CCTGGTGGTTGACTGATCACCATA AACGGCGAAACTCTAGTGA(A/C); L358-(A/B), CTGTGACAGAGCCAA CACGCAGTCT GTTTCGCTTTCTCCGCCC(A/G); SAG1-(A/B), TACAAC CGACAGATGTATGTAAGGCTCAAAGATATTTTGCCAAAATTAA(C/G); SAG2-(A/B), GCATGTATAGAACATAAGGTGTCTACCCGTGA AACAAC ACACAG(A/G); TUB2-(A/B), ACACGATGTGAATATTATCTGTGGCAG CTTCGCCAGTGTAAAT(C/G); and GRA6-(A/B), GGTTCTGTTCTTCGTT GACATGAGGTTCCGAAGGGGTCTGC(T/C).

3.5.2. Internal Controls

Two internal controls LT4-GRA6 (internal control-1) and LT10-cB21 (internal control-2) are included to monitor the efficiency of the primer extension reaction (their location is indicated diagrammatically in **Fig. 1**). These controls generate probes equally well from the SNP PCR products containing either one of the two alleles. The control oligonucleotide LT4-GRA6, TTAGTCTCC GACGGCAGGCTTCAAT CGAGCGGTGGACAGCAA, is designed to generate a product using the GRA6 SNP PCR product as template. The control oligonucleotide LT10-cB21, TTCAATCTGGTCTGACCTCCTTGTGAGAG AAGCGAATAGTCTACAGC, uses the cB21 SNP PCR product as a template to generate a product. For both controls, the oligonucleotides are designed so that the primer extension reactions proceed in the opposite direction from the allele-specific primer extensions (**Fig. 1**), and, therefore, they do not interfere with the genotyping of cB21 or *GRA6*. The products generated by these reactions are detected by hybridization to complementary oligonucleotide tags affixed to glass slides (described in **Subheading 3.2.2.**).

3.5.3. Primer Extension

All extension primers should be tested in the absence of SNP PCR products to check for nonspecific priming. In our study, the primer pair L358-(A/B) showed strong signal from both alleles, and was therefore eliminated from further experiments. Two parallel extension reactions (denoted as Reaction-A and Reaction-B) of 50 µL are prepared, both containing 5 µL of purified primary PCR products, 50 µM dATP, dCTP, and dGTP, 16.7 µM dTTP, 33.3 µM of amino-allyl-dUTP (Sigma), and 1.25 U of AmpliTaq. In Reaction-A, each of the nine Allele-A primers and the two internal controls are added to a final concentration of 150 nM. In Reaction-B, each of the nine Allele-B primers and the two internal controls are added to a final concentration of 150 nM. The extension reactions consists of 35 cycles of 94°C for 30 s, 55°C for 30 s, and 72°C for 1 min. After primer extension, the products are purified by Qiagen Nucleotide Removal Kit following the manufacturer's instruction, eluted in 50 µL of H_2O, dried using vacuum evaporation, and resuspended in 10 µL of 0.1 M carbonate buffer (pH 8.8).

3.5.4. Amino-Allyl Dye Coupling (see **Note 3**).

Amplified PCR products are labeled by direct chemical reaction with derivatized Cy3 or Cy5 to form a covalent adduct with the amino-allyl-dUTP that has been incorporated during PCR, as described at http://cmgm.stanford.edu/pbrown/protocols/aadUTPCouplingProcedure.htm.

1. Mix 4.5 µL of monofunctional N-hydroxysuccinamide-ester Cy3 or Cy5 with 4.5 µL Reaction-A or Reaction-B products, respectively.
2. Incubated at room temperature for 1 h in the dark.
3. Quench the dyes by adding 4.5 µL 4 M hydroxylamine and incubate for 15 min at room temperature in the dark.
4. Add 35 µL 0.1 M NaOAc, pH 5.2, to each reaction to neutralize the pH.
5. Combine the products of the Cy3 and Cy5 reactions.
6. Purify the probes using the Qiagen Nucleotide Removal Kit, elute in 50 µL of H_2O, dry down by vacuum evaporation, and resuspend in 4 µL of H_2O.

3.5.5. Array Hybridization

Labeled PCR products are cohybridized to the array of complementary generic oligonucleotides spotted onto glass slides. The procedure is analogous to hybridizations used in many microarray studies.

1. Denature the 4 µL products obtained above at 96°C for 5 min, snap cool on ice.
2. Add 1 µL of 1 nM positive oligonucleotide control (Cy3-GCCTCCACGCACGT TGTGATATGTA), 1 µL of 5-mg/mL herring sperm DNA, and 12 µL of ArrayHyb LowTemp hybridization buffer.
3. Cover array slide with Erie Scientific Lifter-slip.
4. Load PCR products to slide from the side of the Lifter-slip, allowing it to spread across the slide by capillary action.
5. Incubate slides in a hybridization chamber at 45°C for 16 h.

SNP Typing

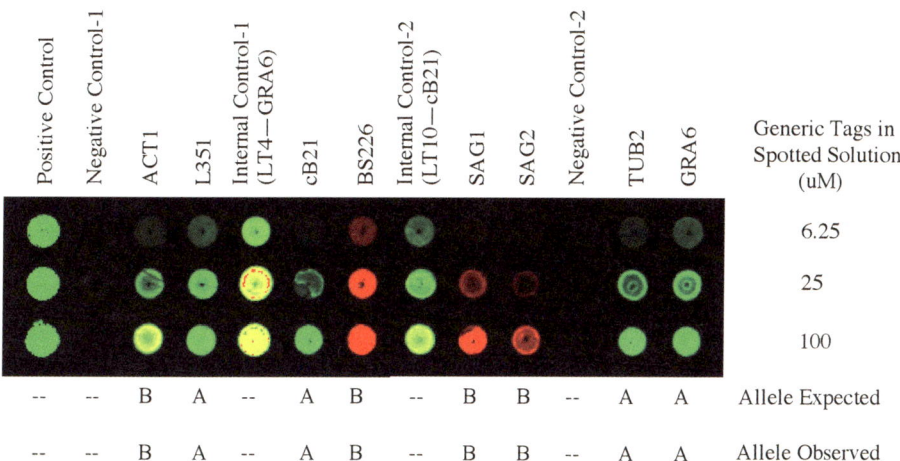

Fig. 2. SNP array for *T. gondii* strain CTG. Green color represents Cy3 that was used for labeling A alleles, and red represents Cy5 that was used for labeling the B allele. Each row was spotted with a different concentration of 15 nucleotide oligo-dT tags, from 6.25 to 100 µM, top to bottom. The positive control labeled with Cy3 shows the expected hybridization to its complementary oligonucleotide. The negative controls show little to no signal and the two internal controls show a combination of Cy3 and Cy5 signals. The SNP alleles were correctly identified for all alleles except ACT1 that did not perform well for all strains (*see* **Table 3**).

6. Wash slides by immersion at room temperature in 2X SSC/0.1% SDS for 2 min, 1X SSC for 2 min, and 0.2X SSC for 5 min.
7. Centrifuge slides at 100g for 5 min to quickly remove residual wash solution.
8. Scan slides using an Affymetrix model 428 laser scanner and store the results as a .tif data file.

3.5.6. Analysis of Array Data

An example of a SNP array detecting polymorphisms in the strain CTG is shown in **Fig. 2**. Analysis of this array was performed using ScanAlyze software. Signal intensities for Cy3 and Cy5 channels were adjusted by subtracting background hybridization based on several randomly chosen areas outside the spots. Additionally, signals were normalized by setting the ratio of Cy3:Cy5 = 1, based on the signal intensities of the two internal controls. The summary of Cy3:Cy5 ratio for SNP typing is shown in **Table 3**. Eight of the nine SNP markers (the exception being ACT1) correctly identified the alleles for the four *T. gondii* strains tested here (*see* **Note 4**). The range of differences in hybridization of the correct allele-A vs correct allele-B ranged from 4-fold (in

Table 3
Summary of Cy3/Cy5 Ratios for Array SNP Analysis

Marker	Toxoplasma strains			
	RH (Type I)	GT1 (Type I)	PTG (Type II)	CTG (Type III)
ACT1	A (2.4)	A (1.2)	B (2.5)	B (0.9)
L351	A (5.9)	A (5.1)	B (0.5)	A (3.4)
cB21	A (3.8)	A (3.4)	B (0.8)	A (6.1)
Bs226	A (2.7)	A (2.2)	A (1.7)	B (0.04)
SAG1	A (16)	A (14)	B (0.7)	B (0.02)
SAG2	A (4.6)	A (2.9)	A (3.2)	B (0.1)
TUB2	A (10.8)	A (8.8)	B (0.7)	A (8.2)
GRA6	A (4.6)	A (3.9)	B (1.0)	A (4.4)

A and B represent the allele types. Numbers in parentheses are Cy3/Cy5 ratios. The A allele was labeled with Cy3, therefore ratios ≥ 1.0 indicate the A allele and ≤ 1.0 indicate the B allele. All of the markers except *ACT1* provide clear results that correctly discriminate between the alleles.

the case of *GRA6*) to almost 50-fold (in the case of Bs226). These results demonstrate that allele-specific primer extension tag arrays provide a powerful method for SNP genotyping in *T. gondii*. The technique described here will be useful in the genetic analysis of multiple strains of many species of pathogenic organisms (*see* **Notes 5** and **6**).

4. Notes

1. The design of allele-specific extension primers is critical to generate specific probes for SNP detection. Because the last nucleotide at the 3'-end of the primer has to be specific and fixed at the SNP site, the options for primer design are limited. The primers with 16–25 bases were selected to hybridize to either the sense strand or the complementary strand of DNA templates. In our experiments, candidate allele-specific extension primers generated from both strands were checked for melting temperature (T_m), self-priming, or dimerization using Oligo Analyzer (http://www.idtdna.com/program/main/home.asp). A T_m of 55–60°C was achieved by adjusting the length of the primers, because their location is determined by the position of the SNP. Primers that were predicted to self-prime or dimerize were eliminated and those that passed this screening were paired with a list of generic tags. Chimeric primers that were predicted to self-prime or dimerize were eliminated.
2. Array slides should be used within 2 mo after preparation. Older slides gave low signal intensity and high background, making the results more variable. In our experience, array slides older than 3 mo were not useable.

3. Direct labeling of allele-specific primer extension products by incorporating Cy3-dCTP and Cy5-dCTP also provides good detection of SNPs using the microarray approach. The main disadvantage is that the background is higher for Cy3-dCTP. For these reasons, we prefer the amino-allyl-dUTP incorporation followed by dye labeling after PCR.
4. In our study of *T. gondii* polymorphisms, seven of nine SNPs were correctly genotyped by microarray. One SNP marker (L358) showed nonspecific signal in the absence of SNP PCR products and was eliminated. Another marker (ACT1) could not distinguish the two alleles among all strain types (*see* **Table 2**), a result that was likely because of nonspecific priming. Redesign of the allele-specific primers could potentially solve these problems. However, given the abundance of SNPs, it may be easier simply to pick another nearby polymorphism to analyze.
5. The advantage of SNP microarray is the capacity to genotype strains on a high-throughput scale compared to, for example, RFLP analysis. Currently, the genetic linkage map of *T. gondii* consists of over 50 RFLP-PCR markers that provide reasonably good coverage over the 11 separate chromosomes *(17)*. Application of microarray SNP genotyping to these markers would enable rapid genetic linkage analysis of complex biological traits. SNP microarray typing will also allow more rapid analysis of the large numbers of new isolates required for epidemiology and population biology studies.
6. It is desirable to establish a cutoff level for interpreting the array hybridization results before genotyping unknown samples. In the experiments presented here, the range of differences in hybridization of the correct allele-A vs correct allele-B was fourfold or higher for all the SNP markers except ACT1. In practice, a three-fold cutoff might be reasonable. For those loci that cannot be distinguished based on array data, another genotyping method such as RFLP is preferred.

Acknowledgments

We are grateful to Michael Grigg and Tovi Lehman, who provided sequence data used in design of SNP primers. This work was partially supported by grants from the National Institutes of Health (AI36629 and AI45806). L. D. Sibley is the recipient of a Scholar Award in Molecular Parasitology from the Burroughs Wellcome Fund.

References

1. Howe, D. K. and Sibley, L. D. (1995) *Toxoplasma gondii* comprises three clonal lineages: correlation of parasite genotype with human disease. *J. Infect. Dis.* **172,** 1561–1566.
2. Sibley, L. D. and Boothroyd, J. C. (1992) Virulent strains of *Toxoplasma gondii* comprise a single clonal lineage. *Nature* **359,** 82–85.
3. Bulow, R. and Boothroyd, J. C. (1991) Protection of mice from fatal *Toxoplasma gondii* infection by immunization with p30 antigen in liposomes. *J. Immunol.* **147,** 3496–3500.

4. Parmley, S. F., Gross, U., Sucharczuk, A., et al. (1994) Two alleles of the gene encoding surface antigen P22 in 25 strains of *Toxoplasma gondii*. *J. Parasitol.* **80,** 293–301.
5. Grigg, M. E., Bonnefoy, S., Hehl, A. B., et al. (2001) Success and virulence in *Toxoplasma* as the result of sexual recombination between two distinct ancestories. *Science* **294,** 161–165.
6. Grigg, M. E. and Boothroyd, J. C. (2001) Rapid identification of virulent type I strains of the protozoan parasite *Toxoplasma gondii* by PCR-restriction fragment length polymorphism analysis of the B1 gene. *J. Clin. Microbiol.* **39,** 98–400.
7. Grigg, M. E., Ganatra, J., Boothroyd, J. C., et al. (2001) Unusual abundance of atypical strains associated with human ocular toxoplasmosis. *J. Infect. Dis.* **184,** 633–639.
8. Honore, S., Couvelard, A., Garin, Y. J., et al. (2000) Genotyping of *Toxoplasma gondii*. *Pathol. Biol.* (Paris) **48,** 541–547.
9. Howe, D. K., Honoré, S., Derouin, F., et al. (1997) Determination of genotypes of *Toxoplasma gondii* strains isolated from patients with toxoplasmosis. *J. Clin. Microbiol.* **35,** 1411–1414.
10. Mondragon, R., Howe, D. K., Dubey, J. P., et al. (1998) Genotypic analysis of *Toxoplasma gondii* isolates in pigs. *J. Parasitol.* **84,** 639–641.
11. Blackstone, C. R., Dubey, J. P., Dotson, E., et al. (2001) High-resolution typing of *Toxoplasma gondii* using microsatellite loci. *J. Parasitol.* **87,** 1472–1475.
12. Costa, J. M., Darde, M., Assouline, B., et al. (1997) Microsatellite in the beta-tubulin gene of *Toxoplasma gondii* as a new genetic marker for use in direct screening of amniotic fluids. *J. Clin. Microbiol.* **35,** 2542–2545.
13. Wang, D. G., Fan, J. B., Siao, C. J., et al. (1998) Large-scale identification, mapping, and genotyping of single-nucleotide polymorphisms in the human genome. *Science* **280,** 1077–1082.
14. Hirschhorn, J. N., Sklar, P., Lindblad-Toh, K., et al. (2000) SBE-TAGS: an array-based method for efficient single-nucleotide polymorphism genotyping. *Proc. Natl. Acad. Sci. USA* **97,** 12,164–12,169.
15. Pastinen, T., Raitio, M., Lindroos, K., et al. (2000) A system for specific, high-throughput genotyping by allele-specific primer extension on microarrays. *Genome Res.* **10,** 1031–1042.
16. Eisen, M. B. and Brown, P. O. (1999) DNA arrays for analysis of gene expression. *Meth. Enzymol.* **303,** 179–205.
17. Su, C., Howe, D. K., Dubey, J. P., et al. (2001) Identification of quantitative trait loci controlling acute virulence in *Toxoplasma gondii*. *Proc. Natl. Acad. Sci. USA* **99,** 10,753–10,758.

14

Transfection of the Human Malaria Parasite *Plasmodium falciparum*

Brendan S. Crabb, Melanie Rug, Tim-Wolf Gilberger,
Jennifer K. Thompson, Tony Triglia, Alexander G. Maier,
and Alan F. Cowman

Summary

Methods to transiently and stably transfect blood stages of the human malaria parasite *Plasmodium falciparum* have been developed and adapted for gene-knockout, allelic replacement, and transgene expression in this organism. These methods are detailed in this chapter, as are approaches used to monitor transfectants during the selection process. The different plasmid vectors that are currently used for gene targeting and transgene expression (including green fluorescent protein expression) are also described.

Key Words: Allelic replacement; DHFR; GFP; knockout; malaria; negative selection; *Plasmodium falciparum*; Rep20; selectable marker; thymidine kinase; transfection; transgene expression.

1. Introduction

Methods to transiently *(1,2)* and stably *(2–5)* transfect blood stages of the human malaria parasite *Plasmodium falciparum* have been developed only relatively recently. These remain inefficient techniques that have not always transferred well to other laboratories. Nevertheless, the technology has been adapted for gene knockout, allelic replacement, and transgene expression in this organism. The methods outlined in this review use electroporation to introduce DNA into parasite-infected erythrocytes. It is important to recognize that some plasmid constructs lead more rapidly than others to the establishment of transfected parasite populations. For example, parasites transfected with plasmids designed to express green fluorescent protein (GFP) grow considerably more slowly than those transfected with other vectors, especially those containing Rep20 sequence *(6)*. In all instances, plasmids are transfected into parasites in the form of undigested circular DNA. These initially replicate episomally as unstably segregating concatamers following transfection and drug selection *(7,8)* but will subsequently integrate into the genome by single- or double-recombination crossover homologous recombination *(9)* if appropriate targeting sequences and selection strategies are used.

Here, we describe a range of vectors that are available for specific purposes, such as negative selection or tagging with GFPs. We then provide protocols for the transfection of malaria parasites in culture, followed by drug selection of transfected cells. Finally, we describe the types of genetic analyses required to determine the nature of the insertion event.

2. Materials

1. Preparation of *Plasmodium falciparum* parasites for transfection: we obtain bags of red blood cells from the Red Cross Blood Service in anticoagulant citrate phosphate dextrose solution. The red blood cells are transferred from the bags to sterile bottles for storage at 4°C and are *not* washed prior to use. Red blood cells are very sensitive to temperature changes; exposing them to repeated temperature alternations will decrease their shelf-life considerably. Red blood cells are used for up to 4 wk after collection.
2. 3.6% w/v $NaHCO_3$ in distilled H_2O.
3. Heat-inactivated human serum.
4. RPMI-HEPES medium: 10.44 g RPMI-1640, 5.96 g 25 mM HEPES, 200 µM hypoxanthine (50 mg), 20-µg/mL gentamicin (10 mL of a 2-mg/mL stock), 960 mL H_2O. Adjust to pH 6.72 with 1 M NaOH. Make up to 1 L with H_2O, filter-sterilize, and aliquot. Store at 4°C. For 100 mL, supplement with 5.8 mL of 3.6% $NaHCO_3$, 5 mL of heat-inactivated human serum, and 5 mL of 5% albumax.
5. Heat-inactivated pooled human serum (HIPHS). Bags of clotted blood are provided by the Red Cross Blood Bank.
   a. Transfer serum to sterile 50-mL tubes.
   b. Centrifuge at 1500g for 10 min to pellet any red blood cells.
   c. If the serum supernatant is clear and yellow, carefully decant into sterile 500-mL bottles. Be careful to avoid the red cell pellet and discard any serum that is cloudy or red.
   d. Include a pool of at least four different donor sera in each bottle to equalize batch variation. (The serum can be stored long-term at −20°C at this stage, then thawed later for heat inactivation.)
   e. Heat-inactivate 500 mL of serum at 56°C for 1 h.
   f. Aliquot into sterile 100-mL bottles and store either at −20°C for long-term storage or at 4°C, where it will be stable for 1 or 2 wk.
6. 5% albumax: Dissolve 5 g of albumax in 100 mL of RPMI-HEPES at 37°C, filter-sterilize, and store at 4°C.
7. CytoMix: 120 mM KCl, 0.15 mM $CaCl_2$, 2 mM EGTA, 5 mM $MgCl_2$, 10 mM K_2HPO_4/KH_2PO_4, pH 7.6, 25 mM HEPES, pH 7.6. For 100 mL: 6 mL 2 M KCl,

7.5 µL 2 M CaCl$_2$, 1 mL 1 M K$_2$HPO$_4$/KH$_2$PO$_4$, pH 7.6 (8.66 mL 1 M K$_2$HPO$_4$ + 1.34 mL 1 M KH$_2$PO$_4$ = 10 mL 1 M phosphate buffer, pH 7.6), 10 mL 250 mM HEPES/20 mM EGTA, pH 7.6 with 10 M KOH, 500 µL 1 M MgCl$_2$; to 90 mL with double-distilled H$_2$O (ddH$_2$O). Adjust pH to 7.6 with 0–350 µL 1 M KOH, add ddH$_2$O to 100 mL, then filter-sterilize. Store in 3 × 33 mL aliquots at 4°C.
8. Cytomix stock buffers: 10 M KOH = 5.61 g/10 mL ddH$_2$O, 250 mM HEPES/20 mM EGTA, 5.96 g HEPES (free acid), 0.76 g EGTA, to 80 mL with ddH$_2$O. Adjust pH to 7.6 with 10 M KOH (approx 1.4 mL). To 100 mL with ddH$_2$O.
9. 200 µM Pyrimethamine (10 mL):
   a. Add 0.012 g to 5 mL 1% glacial acetic acid (in ddH$_2$O). Keep 1% acetic acid at room temperature.
   b. Dilute 200 µL into 10 mL HIPHS (use fresh sterile bottle of medium each time).
   c. Filter-sterilize and store at 4°C. Stable only for 1 mo.
10. WR 99210 (Jacobus Pharmaceuticals).
    a. Dissolve 8.6 mg WR99210 in 1 mL of dimethyl sulfoxide. (This may be stored long-term at −70°C.)
    b. Dilute 1/1000 in RPMI-HEPES.
    c. Filter-sterilize and store at 4°C. Stable for 1 mo at 4°C.
11. Ganciclovir (10 mL): Sigma Ganciclovir (cat. no. G-2536). Dissolve 5.1 mg in 10 mL ddH$_2$O. Filter-sterilize. Store 250-µL aliquots at −70°C. Ensure that precipitates have dissolved after thawing.
12. 5% Sorbitol.
13. TE buffer: 10 mM Tris-HCl pH 7.5, 1 mM EDTA.
14. At least 50 µg of plasmid DNA for transfection.
15. Electroporation equipment (e.g., Genepulser from Bio-Rad) and appropriate cuvets.
16. Petri dishes.
17. Culture incubator.
18. 0.15% saponin: dissolve in RPMI-HEPES.
19. TEN buffer: 40 mM Tris-HCl pH 7.6, 1 mM EDTA pH 8.0, 150 mM NaCl.
20. 0.25 M Tris-HCl, pH 7.6.
21. Dry ice/ethanol bath.
22. Acetyl or buteryl CoA: dissolve 5 mg/mL in ddH$_2$O.
23. ^{14}C-chloramphenicol.
24. Bacterial CAT enzyme (Promega).
25. Ethyl acetate.
26. Thin liquid chromatography (TLC) plate and chamber.
27. Chloroform and methanol.

3. Methods

The methods outlined below emphasise issues we consider important to successfully transfecting *P. falciparum*. We focus particularly on in vitro culturing of blood-stage parasites in preparation for transfection and during drug selection procedures. The approaches we adopt for monitoring and analyzing transfected populations and for constructing the AT-rich *P. falciparum* plasmid vectors are also described.

3.1. Common P. falciparum Transfection Plasmids

Stable transfection of *P. falciparum* has been performed primarily using two types of vectors containing either *Toxoplasma gondii dihydrofolate reductase* (*dhfr*) *(2,3)* or human *dhfr* *(4)* as the gene for selection of transfected parasites. More recently, other genes such as blasticidin, neomycin, and puromycin resistance genes have been used successfully for selection of *P. falciparum* transfectants *(10,11)*. The structure of some commonly used vectors, most of which utilize the human *dhfr* gene as the positive selectable marker, are shown in **Fig. 1** and are described below (*see* **Note 1**).

3.1.1. pHH1

The pHH1 vector *(12)*, its parent vector pHC1 and derivatives *(7,13)* have been useful for gene targeting (*see* **Fig. 2**) and for transgene expression to analyze protein trafficking, merozoite invasion, and drug resistance. This vector allows integration of the plasmid into the genome of *P. falciparum* by single-crossover recombination (*see* **Notes 2** and **3**).

3.1.2. pHTK Vectors

To overcome the problem of persisting episomal plasmid in transfected *P. falciparum*, we developed a new vector (pHTK and derivatives) that utilizes the *thymidine kinase* gene to negatively select against its maintenance *(9)*. This has been very successful, as it allows disruption of genes not previously obtained using pHH1 and also significantly decreases the length of time required to select the *P. falciparum* parasites that have integrated the plasmid. Importantly, this vector allows selection of parasites that have integrated a region of the transfection plasmid by double-crossover recombination. This is an important advance for reasons described above, but also allows more defined deletions and mutations in the *P. falciparum* genome and will also facilitate the production of double mutations and knockouts into the genome (*see* **Note 3**).

Fig. 1. (*Opposite page*) Vector maps for commonly used *P. falciparum* transfection plasmids. In all plasmids the hDHFR cassette is comprised of 1.0 kb of CAM 5'-untranslated region (UTR), 0.56 kb *hdhfr* gene, and 0.6 kb hrp2 3'-UTR. **pHH1** is a single-crossover vector used for targeted disruption and gene replacement or for gene expression. The target sequence is inserted either as a *Xho*I (X)–*Xho*I fragment, or a *Bgl*II (Bg)–*Xho*I fragment if the hsp86 5'-UTR is not required. **pHH2** is similar to pHH1, but contains the gene encoding GFP. The plasmid as shown in this figure contains an acyl carrier protein leader sequence between *Xho*I (X) and *Avr*II (A) sites that can be removed and replaced with the targeting sequence of interest. An ATG needs to

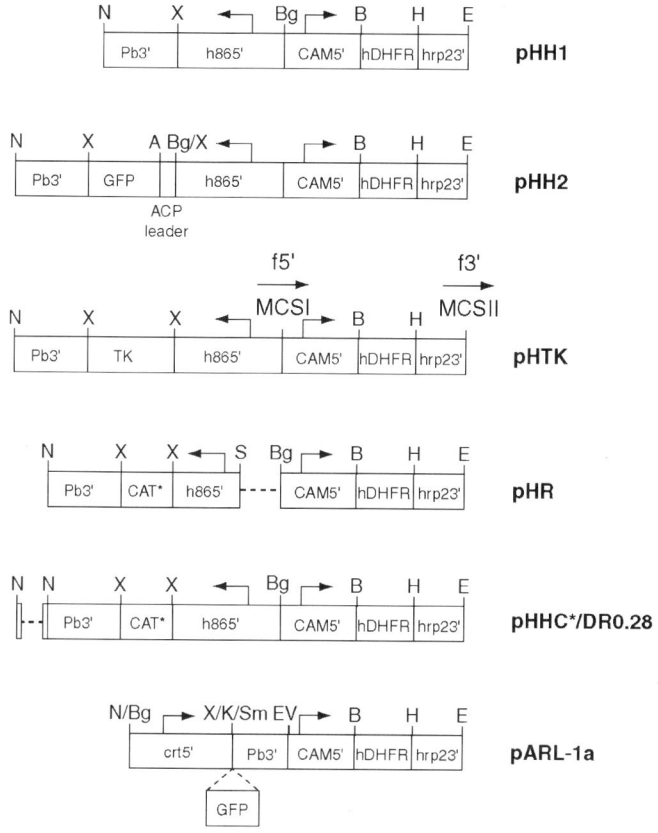

be provided within the target sequence and fused in-frame to GFP. **pHTk** is a double-crossover plasmid used for gene disruption. The 5' and 3' target sequences need to be cloned into the multicloning sites (MCSI and MCSII), respectively, in the same direction as expression of the hDHFR cassette, as shown by the arrows. From 5' to 3' the unique restriction sites within MCSI are *Mlu*I, *Sac*II, *Spe*I, *Bgl*II, *Hinc*II, *Hpa*I, and within MCSII are *Eco*RI, *Cla*I, *Nco*I, *Avr*II, *Bbe*I, *Kas*I, *Nar*I, and *Sfo*I. Two Rep20-containing plasmids have also been included in this figure (*see* **Note 4**), both containing a CAT gene within the *Xho*I site. **pHR** has 509 bp of Rep20 sequence (dashed line) inserted into the *Sac*II (S)–*Bgl*II (Bg) sites of pHH1, whereas **pHHC*/DR0.28** has a 280-bp Rep20 sequence within the *Not*I site of pHH1. Other restriction sites labeled on plasmids include *Hind*III (H), *Bam*HI (B), *Eco*RI (E), and *Avr*II (A). In **pARL-1a** the orientation of the cassettes has been changed and the crt promoter is used to drive expression of transgenes, usually GFP as shown. Sites are as above except for the additions of *Sma*I (Sm), *Kpn* I, and *Eco*RV(EV). Note that both the CAM5' region and the Pb3' are slightly shorter in this construct (850 and 550 bp, respectively) than they are in most other constructs, where they are 1.0 kb and 800 bp, respectively. They remain fully functional regulatory elements when reduced to these smaller sizes.

3.1.3. pHH2

The use of GFP-tagged proteins has been an important application to follow the trafficking pathway of proteins in live *P. falciparum*-infected erythrocytes *(13,14)*. The transfection vector pHH2 allows cloning of sequences into a gene cassette to obtain expression of proteins to GFP. This vector uses the promoter from the *hsp86* gene, which allows a broad expression of the GFP in *Plasmodium* blood stages (*see* **Note 3** and **Fig. 3**).

3.1.4. pARL-1a

The pARL-1a vector is used mainly for expression of GFP-tagged proteins (**Fig. 3**). It uses a tail-to-head orientation of the expression cassettes to avoid the bidirectional influence of the *cam* promoter on the expression of the gene of interest. Additional restriction sites facilitate cloning. In contrast to the *hsp86* promoter driven expression in pHH2, expression in this vector is driven by the *P. falciparum crt* promoter. The distinct *crt* promoter activity has been successful in avoiding cytotoxic levels of GFP expression. For this application the GFP expression cassette of the pHH2 vector is transferred into the *Xho*I site of the pARL-1a vector. The *crt* promoter element can be exchanged to modify the expression profile of the gene of interest.

3.1.5. Rep20 Plasmids

Some of the available plasmids now incorporate stretches of the *P. falciparum* subtelomeric repeat sequences, Rep20 (*see* **Note 4**).

3.2. Preparation of P. falciparum Parasites for Transfection

1. Synchronize *P. falciparum* parasites at approx 1% ring stages using 5% sorbitol 2 d before transfection.
2. It is important to use fresh human erythrocytes incorporating different batches (*see* **Subheading 3.2.**, **step 3**) to ensure that they support growth of parasites during the lengthy initial selection process. The erythrocytes are not washed prior to use.
3. On the day of transfection, parasites should be approx 1–5% parasitaemia (depending on the parasite line). For example, 3D7 parasites generally are transfected at 1% parasitaemia. Parasitaemia can be a higher density (5–10%) for transient transfection.
4. A culture of 5 mL (at 4% hematocrit) will be required for each transfection.

3.3. Preparation of DNA for Transfection

1. Prepare plasmid DNA for transfection by adding 2 vol of ethanol to at least 50 µg of the vector (usually 50–100 µg). Centrifuge at 12,000g at 4°C for 10 min.
2. Carefully remove ethanol and allow the DNA pellet to dry for 5 min in a laminar-flow hood.

3. Resuspend DNA in 15–30 µL of sterile TE. It is essential that the DNA is fully dissolved in the buffer before adding further solutions.
4. Add 370–385 µL of sterile Cytomix to each plasmid DNA pellet for a final volume of 400 µL.

3.4. Electroporation and Plating

1. Centrifuge 5 mL of cultured parasites with erythrocytes at 1500g for 5 min and remove supernatant.
2. Add the Cytomix/plasmid mixture (*see* **Subheading 3.3.**) to the parasitized erythrocyte pellet and pipet up and down gently to mix.
3. Transfer the parasitized erythrocyte/DNA mixture to an electroporation cuvet (0.2-cm gap). We electroporate at 0.310 kV and 950 µF using a Genepulser. The resulting time constant should be between 7 and 12 ms.
4. Immediately add the electroporated sample to a labeled 10-mL Petri dish containing 3–4% erythrocytes in complete RPMI/HEPES medium with 5% serum and 5% albumax.
5. Grow parasites at 37°C in a gas mixture of 5% carbon dioxide, 1% oxygen, and 94% nitrogen. Change medium daily until 5 d posttransfection.

3.5. Positive Drug Selection

1. At 2 d posttransfection, ensure that parasitemia is no higher than 5%.
2. At day 2, add WR99210 to 2.5–10 nM when using the human *dhfr* gene as the select-able marker or 0.2 µM pyrimethamine when using the *Toxoplasma gondii dhfr* gene as the selection system (for other positive selectable markers, *see* **Note 5**).
3. Fresh medium and the appropriate drug are added to cultures daily for an additional 3 d, then every 2–3 d until parasite growth is fully established.
4. Add fresh RBCs (approx 100 µL or 1%) once a week.
5. Parasitized erythrocytes should be detectable in Giemsa-stained smears of erythrocytes after 7–40 d. Some GFP transfectants may take longer.
6. To select parasites with the plasmid vector integrated by homologous recombination, grow the parasites for 3 wk without drug selection, then reapply drug pressure and continue to culture until parasites reappear in the Giemsa-stained blood smears.
7. The parasites should be analyzed by pulsed-field gel electrophoresis (PFGE) and Southern blotting to determine if integration into the relevant gene has been obtained (*see* **Subheadings 3.7.2., 3.7.3.**, and Chapter 19).
8. Continue drug cycling as described in **step 6** until no death is observed after addition of drug.

3.6. Negative Selection Using Thymidine Kinase Vectors

This process is used to select for parasites that have integrated plasmids via double-crossover homologous integration as described (*9*). Vectors such as pHTK are used for this purpose (**Fig. 1**).

1. For selection of transfected parasites using vectors containing the thymidine kinase gene for negative selection, the parasites should be transfected as described above and selected with WR99210.
2. Once drug-resistant parasite populations (meaning parasites containing the episomal plasmid) are established on WR99210, add ganciclovir to 4 µM for 9 d or until parasites are no longer visible. (Note that WR99210 selection *continues* during ganciclovir treatment.) During this period there may be substantial parasite death.
3. Transfer parasites to drug-free medium (i.e., no WR99210 or ganciclovir) and culture until parasites reappear.
4. Reapply WR99210 selection and culture until parasite growth is firmly established.
5. Analyze chromosomes and genomic DNA of the parasites using PFGE and Southern blotting (*see* **Subheading 3.7.2.** and **3.7.3.** and Chapter 19) to determine if integration into the appropriate gene has occurred in these parasites.

3.7. Monitoring Transfectants: Genetic Analysis

It is particularly important to monitor stable transfectants genetically once a drug-resistant population emerges posttransfection and during the drug cycling process. There are three main reasons for this: (a) to ensure that transfected populations do not represent naturally drug-resistant mutants, such as those with a mutation in the endogenous DHFR-TS gene, but are instead transformed with the desired plasmid; (b) to determine if the transfected populations possess episomally replicating and/or integrated copies of the transfection plasmid; and (c) to examine the nature of the integration event (i.e., homologous vs nonhomologous; single- vs double-crossover recombination). A combination of three approaches is used for this analysis.

3.7.1. Amplification by Polymerase Chain Reaction

This approach can be used for all requirements (**Fig. 2**). However, for numerous reasons, we believe that its use is limited and that the technique should be used as a guide only. For example, detection of the transfection plasmid by polymerase chain reaction (PCR) using oligonucleotides specific for a unique sequence (such as a targeting sequence) is confounded by the presence of residual plasmid DNA left over from the original transfection. (However, *see* **Note 6.**) PCR is particularly useful to detect the presence of homologous integration events using a combination of a plasmid-specific oligonucleotide (not specific to the gene targeting sequence) and one directed to genomic sequence located immediately outside the gene targeting fragment found in the plasmid. The presence of such a product (which should be sequenced for confirmation) demonstrates that homologous integration has indeed occurred. Using this approach, however, it is not possible to determine the proportion of the parasites that possess integrated forms of the plasmid.

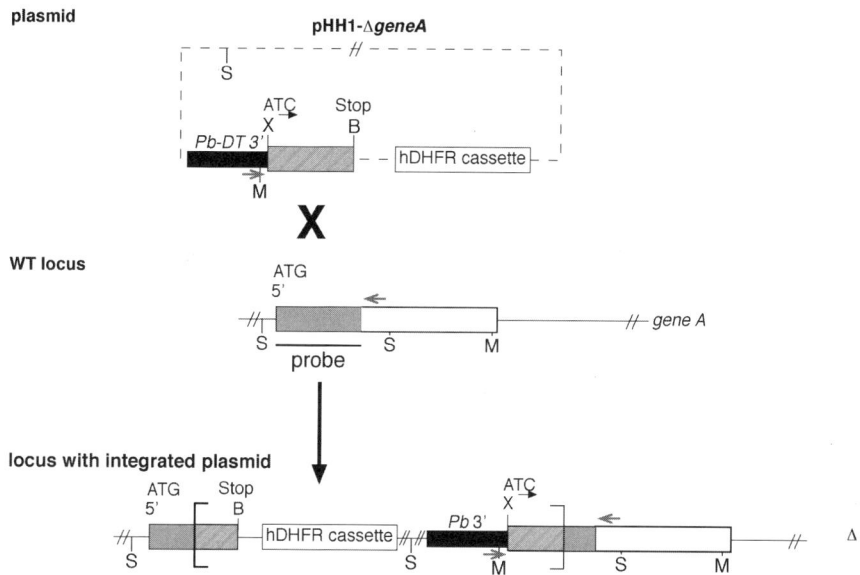

Fig. 2. Schematic representation and strategies for screening a single-crossover homologous integration event designed to generate a truncation of a hypothetical gene (*gene A*). The plasmid construct based on pHH1 is shown (top) with the targeting region of *gene A* represented in gray. The start codon ATG of the endogenous sequence has been changed to an ATC to prevent the reconstitution of the gene after the integration. The wild-type (WT) locus of *gene A* (center) and the locus after the integration event (bottom) are also depicted. The integrated plasmid is shown between the square brackets. Restriction sites: M = *Mfe*I, S = *Sca*I, B = *Bgl*II, X = *Xho*I.

PCR analysis: For the detection of the expected homologous integration event, the primers (arrows) are used. Neither the plasmid nor the wild-type locus should give a specific PCR fragment. Only after integration of the plasmid in the genome should a PCR product be obtained.

Southern blot analysis: In this example, the targeting sequence should be used as a probe for Southern blot hybridisation. After digestion of the DNA from transfected *P. falciparum* cells with *Sca*I and Southern blotting, three possible scenarios may occur:
a. The presence of unintegrated plasmid results in a band equivalent to the size of the linearized plasmid.
b. The presence of wild-type locus yields a band of half the size of *gene A* plus a small portion of its 5' UTR.
c. The integration of the plasmid into the locus of *gene A* results in two bands: one band equivalent to the 5'-UTR, the targeting region, and the *hdhfr* cassette, and the second band consisting of the Pb 3'-end and the 5'-half of *gene A*. A third band the size of the plasmid might also be present, if additional copies of the plasmids have integrated head-to-head (*see* **Note 8**).

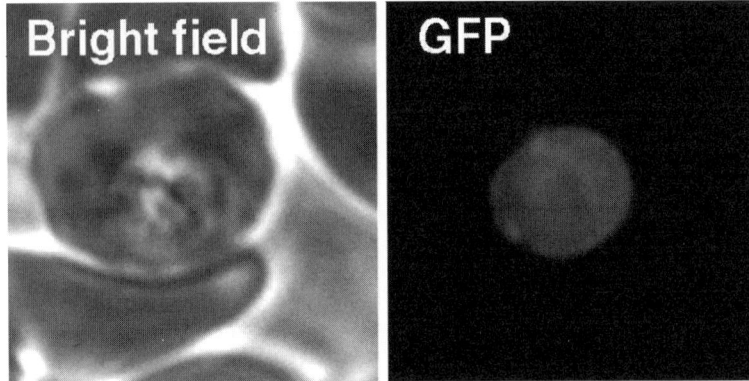

Fig. 3. GFP targeting in *P. falciparum* blood stages. *P. falciparum*-GFP chimeric protein expression at the trophozoite stage of the parasite life cycle. The first image represents a bright field image, the second one the fluorescence signal from the GFP chimeric protein. Trophozoite-stage parasites show parasite cytoplasmic and parasitophorous vacuolar labeling, but with this construct GFP is not exported into the infected erythrocyte cytoplasm. Parasites were transfected with the pARL-1a vector and the chimera expressed from a stably maintained episome under the control of the *crt*-promoter. Fluorescence from GFP was captured in live cells using a Zeiss Axioskop 2 and a Sensi-Cam camera.

3.7.2. Pulsed-Field Gel Electrophoresis

The separation of *P. falciparum* chromosomes by PFGE followed by Southern blotting is a powerful approach to monitor the genotype of transformants *(5,15)*. By hybridizing identical blots with a probe to detect the presence of the transfected plasmid (e.g., plasmid backbone or selectable marker gene) and a probe to detect the targeting sequence present in the plasmid, the progress of episomally replicating to integrated plasmid can be followed. (However, *see* **Note 7**.) A good example of the use of PFGE to analyze transfectants is shown in **ref. *15***.

3.7.3. Southern Blotting of Digested Genomic DNA

To determine if your plasmid has integrated into the intended locus by homologous recombination, genomic DNA (approx 1 µg) is digested by appropriate restriction endonucleases, the fragment separated by agarose gel electrophoresis and Southern blotted to a transfer membrane. The enzymes to be used should be ones that are intended to reveal a distinct difference in the size of the fragments representing wild-type locus, integrated locus, and episomal plasmid when a targeting sequence probe is hybridized to the blot (**Fig. 2**). (However, *see* **Note 8**.)

3.8. Analysis of Transient Transfectants: CAT Assay

Because of the low efficiency of *P. falciparum* transfection, highly sensitive reporter systems are required for use in transient transfection. Two such reporter genes, chloramphenicol acetyl transferase (*CAT*) and luciferase (*LUC*), have been used successfully for promoter analysis following transient transfection. In our laboratories, the *CAT* reporter is most commonly used. To detect CAT expression following transient transfection (performed as in **Subheadings 3.2.–3.4.**), extract parasites at 2–3 d posttransfection and assay for CAT activity as follows:

1. Centrifuge parasites (from 10-mL culture) at approx 1500g for 5 min.
2. Discard supernatant and resuspend pellet in approx 1.5 vol of 0.15% saponin.
3. Place on ice for 10 min.
4. Centrifuge at approx 2800g for 10 min.
5. Carefully remove and discard supernatant.
6. Wash pellet by resuspending in 500 µL TEN buffer and centrifuging at 14,000g for 1–2 min. Repeat one time.
7. Resuspend pellet in 120 µL 0.25 M Tris-HCl (pH 7.6).
8. Freeze/thaw three times in dry ice/ethanol bath.
9. Heat 10 min at 65°C to (to destroy endogenous CAT activity).
10. Centrifuge at 14,000g for 1 min and transfer 114 µL supernatant to fresh tube.

To assay supernatant for CAT activity:

11. Establish substrate mix of 5 µL acetyl (or buteryl) CoA (5 mg/mL in ddH$_2$O) and 0.5 µL ^{14}C-chloramphenicol per sample.
12. Add 5.5 µL substrate mix to 114 µL supernatant.
13. For negative control add substrate mix to 114 µL 0.25 M Tris-HCl (pH 7.6).
14. Positive control is as for point 13, with the addition of 0.5 µL bacterial CAT enzyme prediluted 1/500 in 0.25 M Tris-HCl (pH 7.6).
15. Incubate reactions overnight at 37°C.
16. To extract ^{14}C-chloramphenicol species, add 500 µL ethyl acetate to each tube and vortex for 1 min.
17. Centrifuge at 14,000g for 2 min and remove top phase to a new tube.
18. Evaporate remaining liquid using a vacuum centrifuge or by leaving open tubes in a fume hood.
19. Add 10 µL ethyl acetate to each tube and spot onto TLC plate.
20. Resolve with 97% chloroform/3% methanol in TLC chamber.
21. Allow to dry and expose to film or phosphoimaging plate to detect and quantify unacetylated and acetylated ^{14}C-chloramphenicol spots.

4. Notes

1. Construction of appropriate vectors for stable or transient transfection using some of the available plasmids can be problematic because of instability and poor growth

in *Escherichia coli*. The main reason for this instability appears to be the high AT composition of *P. falciparum* genes and in particular the extragenic region that can be >90% AT. The problems encountered can usually be overcome by testing a number of *E. coli* strains with different genetic backgrounds to identify one that provides a stable vector. The *E. coli* strains PMC103 and XL10-Gold have proven to be very useful for this problem but can provide poor yields of the plasmid. This can be overcome in some instances by transferring the plasmid to another *E. coli* strain with a different genetic background, such as XL1-Blue.

2. Although very useful, this strategy of single-crossover targeting has a major drawback in that it does not allow selection of gene disruptions that are not lethal but are deleterious to parasite growth. This is because of the persistence of episomal plasmid in some parasites despite growth of transfectants in the absence of drug selection. Reapplication of drug pressure selects for parasites that have the plasmid integrated but also for those that contain the plasmid as an episome. If the parasites containing the integrated form of the plasmid grow more slowly, they will be lost in the parasite population and parasites with episomal copies of plasmid will predominate.

3. Subcloning of fragments into the various transfection vectors such as pHH1, pHH2, and pHTK can be very inefficient, and it may be necessary to screen large numbers of *E. coli* colonies to identify those that contain the correct plasmid in an unrearranged state. This can more easily be achieved by screening by PCR, picking a portion of each colony directly into the PCR reaction mixture. To minimize false positives, it is recommended that the PCR be performed with one primer specific for the vector and the other specific for the insert. This facilitates screening of large numbers of colonies to identify those that have the appropriate structure required.

4. It has recently been shown that the inclusion of stretches of the *P. falciparum* subtelomeric repeat sequence Rep20 in transfection plasmids confers improved plasmid maintenance in transfected parasites *(6)*. This occurs because Rep20 sequence allows transfected plasmids to tether to *P. falciparum* chromosomes and as a result plasmids are segregated efficiently between daughter merozoites. The primary advantage of this for transfection technology is that drug-resistant parasite populations are established much more rapidly if Rep20 is included in the transfection plasmid: some 1–2 wk before the appearance of parasites transfected with control plasmids. These plasmids are especially useful for transgenesis experiments (including GFP; *see* **Fig. 3**), but their value for gene targeting approaches has not been evaluated.

5. In addition to the *dhfr*-selectable markers described here, three other positive selectable markers have been used successfully to derive drug-resistant parasite populations. These markers, blasticidin *S* deaminase (BSD) *(10)*, neomycin phosphotransferase II (NEO) *(10)*, and puromycin-*N*-acetyltransferase (PAC) *(11)*, confer resistance to blasticidin S, geneticin (G418), and puromycin, respectively. Both BSD and PAC have shown utility for gene targeting *(11,16)* and are especially useful as second selectable markers for use in parasites lines that already contain the hDHFR marker.

6. This DNA can be destroyed by predigestion of the genomic DNA preparation with *Dpn*I, a restriction enzyme with a frequently occurring recognition sequence that cleaves only methylated (such as that replicated in *E. coli*) and not unmethylated (parasite-replicated) DNA, although *Dpn*I digestion is unlikely to be 100% efficient.
7. However, note that although this approach demonstrates plasmid integration and the chromosome into which this has occurred, it does not reveal the specific nature of the integration event.
8. It should be noted that plasmids that integrate by a single-crossover recombination event often insert a number of head-to-tail plasmid copies into the locus. If this has occurred, a band corresponding to that expected for the episomal plasmid will be observed.

Acknowledgments

We are grateful for the contribution of many past and present members of our laboratories to the development of these methods. We thank the National Health and Medical Research Council of Australia for financial support and the Australian Red Cross Blood Service for the provision of human blood and serum. B. S. Crabb and A. F. Coroman are International Research Scholars of the Howard Hughes Medical Institute.

References

1. Wu, Y., Sifri, C. D., Lei, H. H., et al. (1995) Transfection of *Plasmodium falciparum* within human red blood cells. *Proc. Natl. Acad. Sci. USA* **92,** 973–977.
2. Crabb, B. S. and Cowman, A. F. (1996) Characterization of promoters and stable transfection by homologous and nonhomologous recombination in *Plasmodium falciparum. Proc. Natl. Acad. Sci. USA* **93,** 7289–7294.
3. Wu, Y., Kirkman, L. A., and Wellems, T. E. (1996) Transformation of *Plasmodium falciparum* malaria parasites by homologous integration of plasmids that confer resistance to pyrimethamine. *Proc. Natl. Acad. Sci. USA* **93,** 1130–1134.
4. Fidock, D. A. and Wellems, T. E. (1997) Transformation with human dihydrofolate reductase renders malaria parasites insensitive to WR99210 but does not affect the intrinsic activity of proguanil. *Proc. Natl. Acad. Sci. USA* **94,** 10,931–10,936.
5. Crabb, B. S., Cooke, B. M., Reeder, J. C., et al. (1997) Targeted gene disruption shows that knobs enable malaria-infected red cells to cytoadhere under physiological shear stress. *Cell* **89,** 287–296.
6. O'Donnell, R. A., Freitas-Junior, L. H., Preiser, P. R., et al. (2002) A genetic screen for improved plasmid segregation reveals a role for Rep20 in the interaction of *Plasmodium falciparum* chromosomes. *EMBO J.* **21,** 1231–1239.
7. Crabb, B. S., Triglia, T., Waterkeyn, J. G., et al. (1997) Stable transgene expression in *Plasmodium falciparum. Mol. Biochem. Parasitol.* **90,** 131–144.

8. O'Donnell, R. A., Preiser, P. R., Williamson, D. H., et al. (2001) An alteration in concatameric structure is associated with efficient segregation of plasmids in transfected *Plasmodium falciparum* parasites. *Nucleic Acids Res.* **29,** 716–724.
9. Duraisingh, M. T., Triglia, T., and Cowman, A. F. (2002) Negative selection of *Plasmodium falciparum* reveals targeted gene deletion by double crossover recombination. *Int. J. Parasitol.* **32,** 81–89.
10. Ben Mamoun, C., Gluzman, I. Y., Goyard, S., et al. (1999) A set of independent selectable markers for transfection of the human malaria parasite *Plasmodium falciparum*. *Proc. Natl. Acad. Sci. USA* **96,** 8716–8720.
11. de Koning-Ward, T., Waters, A., and Crabb, B. (2001) Puromycin-*N*-acetyltransferase as a selectable marker for use in *Plasmodium falciparum*. *Mol. Biochem. Parasitol.* **117,** 155–160.
12. Reed, M. B., Saliba, K. J., Caruana, S. R., et al. (2000) Pgh1 modulates sensitivity and resistance to multiple antimalarials in *Plasmodium falciparum*. *Nature* **403,** 906–909.
13. Waller, R. F., Reed, M. B., Cowman, A. F., et al. (2000) Protein trafficking to the plastid of *Plasmodium falciparum* is via the secretory pathway. *EMBO J.* **19,** 1794–1802.
14. Wickham, M. E., Rug, M., Ralph, S. A., et al. (2001) Trafficking and assembly of the cytoadherence complex in *Plasmodium falciparum*-infected human erythrocytes. *EMBO J.* **20,** 5636–5649.
15. Baldi, D. L., Andrews, K. T., Waller, R. F., et al. (2000) RAP1 controls rhoptry targeting of RAP2 in the malaria parasite *Plasmodium falciparum*. *EMBO J.* **19,** 2435–2443.
16. Sidhu, A., Verdier-Pinard, D., and Fidock, D. (2002) Chloroquine resistance in *Plasmodium falciparum* malaria parasites conferred by pfcrt mutations. *Science* **298,** 210–213.

15

A PCR-Based Method for Gene Deletion and Protein Tagging in *Trypanosoma brucei*

George K. Arhin, Shuiyuan Shen, Elisabetta Ullu, and Christian Tschudi

Summary

Sequence information on the *Trypanosoma brucei* genome is rapidly accumulating. As a consequence, there is a need for techniques to analyze gene function systematically. Here, we describe a polymerase chain reaction (PCR)-based method for direct gene deletion and the generation of epitope-tagged fusion proteins. The approach is based on methodologies developed for *Saccharomyces cerevisiae* and involves PCR amplification of a reporter cassette using primers containing flanking sequences specific to the target gene. The PCR product is then transfected directly into procyclic *T. brucei* cells, and homologous recombinants that carry the deleted or tagged target gene are identified.

Key Words: BB2 antibody; epitope-tagging; gene disruption; homologous recombination; *Trypanosoma brucei*; Western blotting.

1. Introduction

The ongoing sequencing of the *Trypanosoma brucei* genome opens new avenues to determine the identity and function of gene products and their role in the pathobiology of the parasite. As a consequence, it has become necessary to establish procedures that will permit the rapid utilization of the available sequence information. Here, we describe a polymerase chain reaction (PCR)-based method for direct gene deletion and the generation of epitope-tagged fusion proteins that will facilitate the identification and functional analysis of *T. brucei* proteins *(1,2)*. Epitope tagging of proteins has become a standard technique to analyze the function, interaction, and subcellular localization of proteins. One major advantage of the PCR targeting method is that it allows synthesis of the tagged protein at its natural expression levels, thus circumventing mislocalization in the cell and assembly into nonphysiological complexes

(2). The approach is based on methodologies developed for *Saccharomyces cerevisiae* *(3)* and entails PCR amplification of a reporter cassette using primers containing flanking sequences specific to the target gene *(2)*. The PCR product is then transfected directly into procyclic *T. brucei* cells, and homologous recombinants that carry the deleted or tagged target gene are identified.

2. Materials

2.1. Module DNA for Transfection Into the Target Genome

These are described in full in **Subheading 3.1.** Dilute DNA to 5 ng/µL.

2.2. Amplification by PCR

1. 10X PCR buffer (provided with enzyme) containing 15 m*M* MgCl$_2$.
2. Deoxynucleotide triphosphates (Roche Diagnostics, cat. no. 1 969 064).
3. Long (115–130-nt) oligonucleotide primers (*see* **Subheading 3.2.1.** and **3.2.2.** for composition requirements of these PCR primers).
4. Short (18- to 21-nt) oligonucleotide primers for secondary PCR (*see* **Note 1**).
5. *Taq* DNA polymerases: Expand High Fidelity polymerase (Roche Diagnostics, cat. no 1 732 641); *Pfu* DNA polymerase (Stratagene, cat. no. 600250-52); *Taq* DNA polymerase (Roche Diagnostics, cat. no. 1 435 094).

2.3. Checking Transformants

2.3.1. Western Blotting

1. Cell wash buffer: 20 m*M* Tris-HCl, pH 7.5, 100 m*M* NaCl, 3 m*M* MgCl$_2$.
2. Sodium dodecyl sulfate-polyacrylamide gel electrophoresis (SDS-PAGE) gels, equipment, and buffers.
3. Protran BA85 pure nitrocellulose transfer and immobilization membrane (Schleicher and Schuell, cat. no. 10402594).
4. Western blot transfer buffer: 25 m*M* Tris, 192 m*M* glycine, 20% v/v methanol.
5. Ponceau S stain, 1:20 dilution in H$_2$O (Sigma, cat. no. P-7767).
6. Blocking and antibody dilution solution: 0.9% NaCl, 10 m*M* Tris-HCl, pH 8.0, 0.5 g/L MgCl$_2$, 3% bovine serum albumin, 10% fetal calf serum; store in aliquots at −20°C.
7. Wash buffers: 1X phosphate-buffered saline (PBS); 1X PBS + 0.3% Tween-20.
8. Monoclonal antibodies: Anti-BB2 *(4)* (*see* **Note 2**), Anti-HA (Roche Diagnostics, cat. no. 1 867 423), Anti-V5 (Invitrogen, cat. no. R960-25), and Anti-Xpress (Invitrogen, cat. no. R910-25).

2.3.2. Genomic DNA Preparation

1. Qiagen DNeasy Tissue Kit (cat. no. 69506).
2. PBS: Dissolve 8.0 g NaCl, 0.2 g KCl, 1.44 g Na$_2$HPO$_4$, and 0.24 g KH$_2$PO$_4$ in 800 mL distilled H$_2$O. Adjust pH to 7.4 and add H$_2$O to 1 L.
3. Ribonuclease A (Sigma, cat. no. R-4875).

3. Methods

The methods below describe the DNA modules used for PCR amplification of tagging cassettes, the detection of the tagged protein, the purification of genomic DNA from *T. brucei* cells, procedures for determining gene deletion, and the correct insertion of the tagging cassette. Please note that no specific modules are needed for gene deletion. The open reading frame (ORF) of drug-resistance markers can be amplified from various available vectors. Finally, the protocol for transfecting *T. brucei* cells with PCR products can be found in Chapter 16.

3.1. DNA Modules for Epitope Tagging

To date, we have assembled 15 DNA modules that specifically introduce tags in frame at either the N- or C-terminal of a *T. brucei* protein (**ref. 2** and **Fig. 1**; *see* **Note 3**). The N-terminal modules contain in a 5'–3' direction (a) a resistance gene that serves as a selectable marker for stable transformants; (b) sequences of the intergenic region separating the translated regions of the *T. brucei* α- and β-tubulin genes (in *T. brucei*, these sequences are required for polyadenylation of the resistance gene and *trans*-splicing of the downstream target gene) *(5)*; (c) an ATG initiation codon; and (d) the sequence of the epitope. The C-terminal modules are composed of similar sequences as the N-terminal modules except that the relative order of the resistance gene and the epitope sequences are reversed (**Fig. 1**). Three different selectable markers are available so far: neomycin/G418 (*NEO*), blasticidin (*BSR*), and phleomycin (*BLE*). Each of these selectable markers was paired with four different epitopes: BB2 *(4,6)*, HA, V5, and Xpress. We chose these particular epitope tags because monoclonal antibodies to these epitopes had little cross-reactivity with *T. brucei* proteins (**ref. 4** and data not shown). Two additional DNA modules for N-terminal tagging pair the *NEO* and *BLE* markers with the tandem affinity purification (TAP) tag *(7)*. All modules were inserted into the multiple cloning site of pBlueScript (Invitrogen) to allow propagation in *Escherichia coli*.

3.2. PCR Amplifications

3.2.1. Amplification of Tagging Cassettes

1. Design and synthesize the required PCR primers. The cassettes shown in **Fig. 1** are amplified by PCR using two primers, Tag-Forward and Tag-Reverse, each between 115 and 130 nucleotides long (*see* **Note 4**). For N-terminal tagging, Tag-Forward is a chimeric primer composed of about 90 nt of the 5'-untranslated region (UTR) of the gene of interest immediately upstream of the ATG initiation codon, plus 24–30 nt annealing to the 5'-end of the cassette. Tag-Reverse is composed of about 90 nt derived from the translated region immediately following the ATG initiation codon of the gene of interest, plus 24–30 nt annealing to the 3'-end of the cassette.

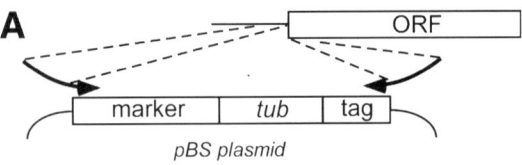

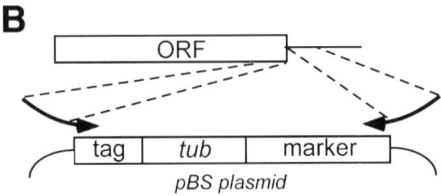

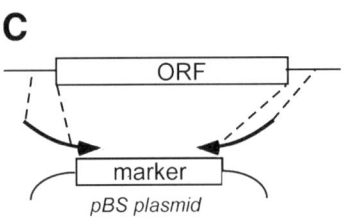

Fig. 1. Strategies for epitope tagging and gene deletion. (**A**) N-terminal epitope tagging. The selectable markers (marker) available so far, *NEO*, *BSR*, and *BLE* confer resistance to G418, blasticidin, and phleomycin, respectively. Epitope tags (tag) available are BB2 (4), TAP (7), HA, V5 (14 amino acid residues, Invitrogen), or Xpress (8 amino acid residues, Invitrogen). (**B**) C-terminal epitope tagging: BB2 and HA are available with a *NEO* resistance marker. tub, α/β-tubulin intergenic region. (**C**) Gene deletion strategy. Oligonucleotide primers (*see* **Subheading 3.2.2.**) are indicated by thick arrows. Drawings are not to scale.

Tag-Reverse is designed in such a way that the epitope-encoding sequence is in frame with the gene of interest. The epitope-encoding sequence is preceded by an ATG initiation codon. Similarly, Tag-Forward for C-terminal tagging is made up of approx 90-nt derived from the 3'-end of the gene of interest (excluding the stop codon, and in frame with the epitope-encoding sequence), plus 24–30 nt annealing to the 5'-end of the module. Tag-Reverse contains about 90 nt of the 3'-UTR immediately following the stop codon, plus 24–30 nt annealing to the 3'-end of the module.

2. Perform the PCR amplification in a 50-µL reaction volume containing 5 ng DNA of the tagging module, 1X PCR buffer (1.5 mM MgCl$_2$), 200 µM of each dNTP, 100 ng of each primer, and 1.75 U of Expand High Fidelity polymerase (Roche Diagnostics; *see* **Note 5**) or 1.25 U of *Pfu* DNA polymerase (Stratagene).

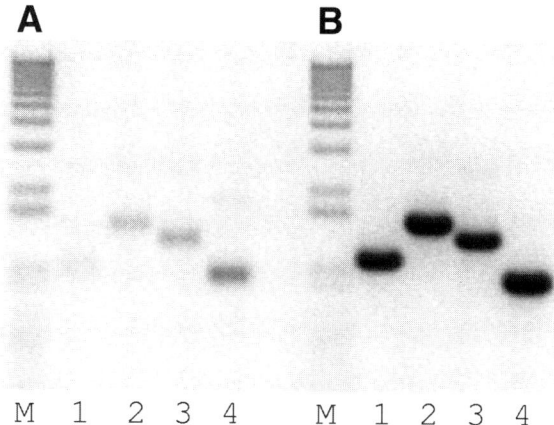

Fig. 2. Four different examples of PCR products from amplification with (**A**) long oligonucleotides (120 nt) and (**B**) short oligonucleotides (24 nt). For the PCR reactions in (**B**), products from (**A**) were diluted 1:100 (*see* **Note 1**). The marker on the left shows the 1 kb plus DNA ladder from GIBCO.

3. We perform PCR in an OMNI-E Thermocycler with a heated lid (Hybaid) using the following conditions (*see* **Note 6**): an initial denaturation at 94°C for 2 min, followed by 30 cycles of amplification at 94°C for 1 min, 55°C for 2 min, and 72°C for 2 min. The PCR amplification reaction is ended with a 10-min extension time at 72°C.
4. Analyze a 5-µL aliquot of the reaction on a 1% agarose gel alongside appropriate molecular-weight markers (**Fig. 2**).
5. Dilute an aliquot of the first PCR reaction (**steps 2–4**) 1:100 and use 1 µL in a second 50-µL PCR reaction (**Fig. 2**), using the short (18- to 21-nt) oligonucleotide primers (*see* **Note 1**).
6. After successful amplification of the tagging cassette, purify the PCR products using the Qiaquick PCR purification kit (Qiagen), following the manufacturer's instructions.

3.2.2. Amplification of Drug-Resistance Genes for Gene Deletion

The generation of a null allele by deleting the ORF is a convenient method for functional analysis of an endogenous gene. In our approach, a gene of interest is inactivated by replacing the ORF with a gene coding for a drug-resistance marker, i.e., the disruption occurs by homologous recombination at the chromosomal locus of the gene.

1. Design and synthesize the required PCR primers. As described above for epitope tagging, our scheme relies on a one-step PCR amplification, using long hybrid primers. The forward primer is composed of about 90 nt immediately upstream of the ORF to be replaced and 24–30 nt annealing to the 5'-end of the drug resistance

gene, whereas the reverse primer is composed of 24–30 nt annealing to the 3'-end of the drug-resistance gene followed by about 90 nt immediately downstream of the ORF to be deleted.
2. Carry out PCR amplifications as described above (**Subheading 3.2.1.**).
3. Following PCR amplification of the tagging module or the drug-resistance gene, use the DNA directly to transfect procyclic *T. brucei* cells (*see* Chapter 16, **Subheading 3.2.**). About 2–5 µg of the PCR fragment are needed for transfecting 5×10^6 cells (*see* **Note 7**).
4. Select stable transformants as described in Chapter 16 with the following final concentrations of drugs: G418, 30 µg/mL; Blasticidin, 10 µg/mL; and Phleomycin, 2.5 µg/mL.

3.3. Checking Transformants

Transformants with integrated tagging cassettes can be identified by Western blot analysis (**Subheading 3.3.1.**), which verifies synthesis of a product of the expected size. However, this test cannot be used in the case of a gene replacement and we verify correct integration by PCR analysis of genomic DNA (**Subheading 3.3.2.**).

3.3.1. Western Blot Analysis

3.3.1.1. Preparation of Whole Cell Extracts

In one simple, fast, but crude way, 1.5 mL of *T. brucei* cells are washed two times with cell wash buffer and resuspended in 50 µL of SDS-PAGE loading dye. The mixture is boiled at 100°C for 5 min. After spinning at 16,000g for 2 min at room temperature, 15 µL of the extract is loaded onto an SDS-PAGE gel.

However, an efficient and preferred method of cell extract preparation for protein analyses is by freeze-thawing, in which cells are subjected to repeated cycles of freezing (in liquid nitrogen or dry ice/ethanol) and thawing at 42°C in a water bath (*see* **Note 8**), as described here:

1. Spin 10 mL of cell culture grown to approx $0.8–1 \times 10^7$ cells/mL for 5 min at 2000g. Resuspend cells in 1 mL cell wash buffer and transfer into a 1.5-mL Eppendorf tube. Spin at 2000g for 1 min, remove supernatant, and repeat wash.
2. Resuspend cells in 100 µL cell wash buffer. Freeze in liquid nitrogen for 30 s. Then transfer Eppendorf tube to a water bath set at 42°C and allow cells to thaw completely.
3. Repeat freeze-thawing six times and at the end spin cells at 16,000g for 10 min at 4°C.
4. Transfer supernatant into a fresh Eppendorf tube. Run 15 µL of the extract on a SDS-PAGE gel.
5. Freeze remainder of extract in liquid nitrogen and store at –80°C. Extracts can be kept frozen for several months.
6. Transfer the gel onto Protran BA 85 pure nitrocellulose transfer and immobilization membrane (Schleicher and Schuell) according to standard Western blotting procedures.

PCR-Mediated Gene Modification in T. brucei

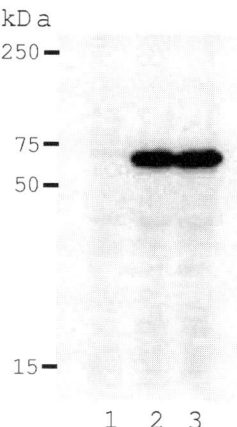

Fig. 3. Western blot analysis of a BB2-tagged methyltransferase. Whole cell extracts prepared from wild-type cells (lane 1); clonal cell lines expressing the N-terminal tagged protein (lane 2), or the C-terminal tagged protein (lane 3) were separated on a 12% SDS-PAGE gel, transferred to a nitrocellulose membrane, and probed with the BB2 monoclonal antibody.

7. Check the efficiency of the transfer by staining the membrane for a couple of minutes with a 1:20 dilution of Ponceau S (Sigma).

3.3.1.2. IMMUNOBLOTTING

Detection of the tagged protein is performed using standard protocols and the manufacturer's recommended concentration of primary and secondary antibody. Protein bands are visualized using an ECL Western Blotting Detection Reagents (Amersham Biosciences; **Fig. 3**).

3.3.2. PCR Analysis of Genomic DNA

3.3.2.1. PURIFICATION OF GENOMIC DNA FROM TRYPANOSOMES

We routinely use the Qiagen DNeasy Kit for the preparation of genomic DNA from *T. brucei* cells (*see* **Note 9**).

1. Spin cells down for 5 min at 2000g and wash cells two times with 1 mL PBS.
2. Resuspend cells in 180 µL PBS, add 20 µL RNase A (10 mg/mL), and incubate at room temperature for 2 min.
3. Follow protocol as outlined by the manufacturer of the kit. Elute DNA in a total volume of 200 µL and use 1 µL for the PCR amplification reaction described below.

3.3.2.2. PCR ANALYSIS

1. Design and synthesize the required PCR primers. Generally, primers for the amplification of genomic DNA are chosen such that they are complementary to sequences

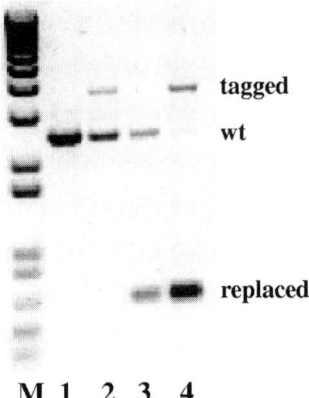

Fig. 4. PCR analysis, using oligonucleotides flanking the targeted polyadenylation factor CPSF73, with genomic DNA from wild-type cells (lane 1), from a clonal cell line where one allele of the gene was tagged with the BB2 epitope at the N-terminus (lane 2), from a clonal cell line where one allele was replaced with the *BSR* resistance gene (lane 3), from a cell population where one allele of the gene was tagged with the BB2 epitope at the N-terminus and the second allele was replaced with the *BSR* resistance gene (lane 4). M, 1 kb plus DNA ladder (Gibco).

outside the region, where homologous recombination occurred. Alternatively, one of the primers can be complementary to sequences within the ORF or to sequences in the tagging cassette.
2. Set up the PCR amplification as described in **Subheading 3.2.1.**, but use regular *Taq* polymerase (Roche Diagnostics; *see* **Note 4**).
3. Perform the reaction with an initial denaturation at 94°C for 2 min, followed by 25 cycles of amplification (94°C for 1 min, annealing at 55°C for 1 min, and extension at 72°C for 3 min). The amplification reaction is ended with a 10-min extension at 72°C.
4. Analyze 10 µL of the PCR reaction on an agarose gel (**Fig. 4**).

The trypanosome genome is diploid. Thus, in cells in which one allele of a gene is tagged, two bands are expected: one corresponding to the wild-type allele and a second representing the tagged allele. Similarly, in cells where one allele was replaced with a gene for a drug-resistant marker, two bands whose sizes correspond to the gene for the resistance marker and the wild-type allele are observed (**Fig. 4**).

3.4. Concluding Remarks

Gene disruption and epitope tagging can be achieved quite easily by homologous recombination in the *T. brucei* genome. The major advantage of the PCR-based method described here is that it allows rapid gene manipulation without requiring plasmid clones of the gene of interest. The modules we have assembled allow for the flexible usage of various markers, i.e., different epitope-tagged proteins can be combined in a single strain. Thus, protein–protein interactions can be studied without the need to generate antibodies.

4. Notes

1. In some cases, the PCR reaction in **Subheading 3.2.1.** (**steps 2–4**) will not be very efficient (**Fig. 2**) or will generate multiple products. For this reason, we also synthesize a second set of short (18 to- 21-nt) primers consisting of the 5'-terminal sequences of each long primer. These short primers are used for a second round of PCR amplification.
2. Monoclonal antibodies to the HA, V5, and Xpress epitopes are commercially available. The BB2 epitope is a 10-amino acid sequence from an immunological major structural protein of the *S. cerevisiae* Ty1 virus-like particle (*4,6*). Monoclonal antibodies against this epitope were developed in Prof. Keith Gull's laboratory and are freely available upon request. The major advantage of this antibody is that it has little to no cross-reactivity with *T. brucei* proteins (*4*).
3. The N- or C-terminus of proteins are typically chosen for epitope tagging, because these sites are likely to be accessible to the antibody and may also be susceptible to modification without affecting function. However, it is possible to insert the tag within the coding sequences, if there is prior knowledge that such an insertion does not disrupt protein structure and/or function.
4. We originally reported that homologous recombination could be achieved with a target sequence as short as 30 nt (*2*). However, in two subsequent experiments using 50 nt of target sequence, we did not recover recombinants. Therefore, we now routinely use 90–100 nt of target sequence with a 100% success rate.
5. To guard against PCR-mediated mutations, it is recommended that high fidelity *Taq* polymerases with proofreading activity are used to amplify the DNA modules for insertion into the *T. brucei* genome. To amplify templates whose termini are GC-rich, the use of Advantage-GC Kit (BD Clontech) is recommended. There is no need to use a *Taq* polymerase with proofreading activity to determine the correct insertion of cassettes into the genome, because this determination is for diagnostic purposes only.
6. Any model of thermocycler can be utilized for the PCR reactions. However, it may be necessary to determine the optimal conditions for PCR amplification.
7. To achieve the 2–5 µg of DNA needed for each transfection, three to four PCR reactions of 50 µL each are assembled.
8. Preparing cell-free extracts by repeated freeze-thawing is convenient and quick for the initial analysis, when multiple cell lines need to be screened. However, this method is not suitable for larger preparations (>20 mL of cells), because cell breakage becomes less efficient and alternative methods, such as sonication, polytron, and french press need to be applied.

9. Any number of methods can be used to prepare genomic DNA from cells after selection. We choose the Qiagen DNeasy Kit, because the yield and quality of the DNA is very high.

Acknowledgments

This study received support from National Institutes of Health Grants AI28798 to E. Ullu and AI43594 to C. Tschudi. C. Tschudi is the recipient of a Burroughs Wellcome Fund New Investigator Award in Molecular Parasitology.

References

1. Gaud, A., Carrington, M., Deshusses, J., et al. (1997) Polymerase chain reaction-based gene disruption in *Trypanosoma brucei*. *Mol. Biochem. Parasitol.* **87,** 113–115.
2. Shen, S., Arhin, G. K., Ullu, E., et al. (2001) In vivo epitope tagging of *Trypanosoma brucei* genes using a one step PCR-based strategy. *Mol. Biochem. Parasitol.* **113,** 171–173.
3. Baudin, A., Ozier-Kalogeropoulos, O., Denouel, A., et al. (1993) A simple and efficient method for direct gene deletion in *Saccharomyces cerevisiae*. *Nucleic Acids Res.* **21(14),** 3329–3330.
4. Bastin, P., Bagherzadeh, Z., Matthews, K. R., et al. (1996) A novel epitope tag system to study protein targeting and organelle biogenesis in *Trypanosoma brucei*. *Mol. Biochem. Parasitol.* **77,** 235–239.
5. Ullu, E., Tschudi, C., and Gunzl, A. (1996) *Trans*-splicing in trypanosomatid protozoa, in *Molecular Biology of Parasitic Protozoa* (Smith, D. F. and Parsons, M., eds.), Oxford University Press, Oxford, UK, pp. 115–133.
6. Brookman, J. L., Stott, A. J., Cheeseman, P. J., et al. (1995) An immunological analysis of Ty1 virus-like particle structure. *Virology* **207,** 59–67.
7. Rigaut, G., Shevchenko, A., Rutz, B., et al. (1999) A generic protein purification method for protein complex characterization and proteome exploration. *Nat. Biotechnol.* **17,** 1030–1032.

16

Analysis of Gene Function in *Trypanosoma brucei* Using RNA Interference

Appolinaire Djikeng, Shuiyuan Shen, Christian Tschudi, and Elisabetta Ullu

Summary

Trypanosoma brucei, a flagellate protozoa of the family Trypanosomatidae, has become one of the model systems for unicellular pathogens to study fundamentally important biological phenomena. The method of choice today to examine gene function in these organisms is RNA interference (RNAi). Messenger RNA (mRNA) degradation is triggered by double-stranded RNA (dsRNA) produced in vivo from transgenes transcribed from opposing tetracycline (tet)-inducible T7 RNA polymerase promoters, or hairpin RNA transcribed from the tet-inducible procyclic acidic repetitive protein promoter. This chapter describes some of the methods we employ for ablation of gene expression by RNAi in *T. brucei* with particular emphasis on transfection and cloning of procyclic cells, induction of dsRNA expression, isolation of RNA, and analysis of dsRNA and target mRNA.

Key Words: Clonal cell line; DNA transfection; hairpin construct; inducible promoter; RNA dot blot; RNAi; *Trypanosoma brucei*.

1. Introduction

Over the last few years, doubled-stranded RNA (dsRNA)-mediated genetic interference or RNA interference (RNAi) has been shown to exist in several organisms representing different levels of the evolutionary tree *(1)*. RNAi is a posttranscriptional mechanism in which expression of dsRNA or delivery of synthetic dsRNA into cells causes specific degradation of the target mRNA. Within a relatively short time of its discovery, RNAi has rapidly been developed as a valuable tool for "reverse" genetic studies. The specific degradation of a mRNA often results in a detectable phenotype and in some cases RNAi phenotypes have been shown to be comparable to loss-of-function phenotypes.

In *Trypanosoma brucei*, RNAi has been used to study several biological processes including, but not limited to, kinetoplast DNA replication, glycosyl phosphatidylinositol biosynthesis, glycosome biosynthesis, RNA editing, and flagellum biogenesis *(2–5)*. Here, we describe methods and reagents available for inducible and heritable RNAi in procyclic *T. brucei* cells.

2. Materials

2.1. Generation of RNAi Constructs

1. pLew79- and pLew100-based vectors for in vivo expression of dsRNA.
2. Oligonucleotide primers.
3. Expand High Fidelity polymerase chain reaction (PCR) system (Roche, cat. no. 1732641).
4. Standard PCR machine.
5. Qiaquick PCR Purification kit (Qiagen, cat. no. 28104).
6. Qiaprep Spin Miniprep kit (Qiagen, cat. no. 27106).
7. Rapid DNA Ligation kit (Roche, cat. no. 1635379).
8. Geneclean II kit (Qbiogene, cat. no. 1001-400).
9. HiSpeed Plasmid Midi kit (Qiagen, cat. no. 12643).

2.2. Transfection of Procyclic T. brucei Cells by Electroporation

1. Procyclic-form *T. brucei* cells (e.g., 29.13.6) that stably express the T7 RNA polymerase and the tetracycline repressor (**ref. 6** and *see* **Note 1**).
2. Cunningham's medium (**ref. 7** and *see* **Note 2**) supplemented with 20% heat-inactivated fetal bovine serum (Gemini Bioproducts, cat. no. 100-106), 2 mM L-glutamine (Gibco, cat. no. 25030-081), penicillin and streptomycin at 100 U/mL and 100 µg/mL, respectively (Gibco, cat. no. 15140-122), and gentamycin at 0.01 mg/mL (Gemini Bioproducts, cat. no. 100-106).
3. CO_2 incubator at 28°C (5% CO_2 in air).
4. Culture flasks and Eppendorf tubes.
5. Cytomix: 120 mM KCl, 0.15 mM $CaCl_2$, 10 mM K_2HPO_4, 25 mM HEPES, 2 mM EGTA, 5 mM $MgCl_2$. Dissolve all the ingredients in 450 mL of HPLC-grade water, adjust to pH 7.6 with KOH, adjust volume to 500 mL, and filter-sterilize. Store at room temperature (stable for several months).
6. Bio-Rad Gene Pulser II.
7. 0.4-cm-gap electroporation cuvet (Bio-Rad, cat. no. 165-2088).
8. Selective drugs: neomycin or G418 (Gibco, cat. no 1181-031) used at 15 µg/mL for maintenance of stable cells lines or at 60 µg/mL for the initial selection of stable transfectants immediately after transfection, hygromycin B (Roche, cat. no. 843555) used at 50 µg/mL, and phleomycin (Sigma, cat. no. P9564) used at 2.5 µg/mL.

2.3. Cloning of T. brucei Cells

1. Multichamber pipet; Finnpipet Varichannel 200-1000 (Labsystems, cat. no. 4347030).

2. 96-Well microtiter plates (Corning, cat. no. 3595).
3. 48-Well microtiter plates (Falcon, cat. no. 35-3078).

2.4. Induction of dsRNA Expression

Tetracycline hydrochloride (Sigma, cat. no. T3383); dissolve the powder in 70% ethanol to make a stock solution of 10 mg/mL. Store in the dark at −20°C. Replenish every week.

2.5. Extraction of Total RNA

1. Cell wash: 20 mM Tris-HCl, pH 7.5, 100 mM NaCl, and 3 mM MgCl$_2$. Filter-sterilize and store at room temperature (stable for several months).
2. RNeasy Tissue kit (Qiagen, cat. no. 69506).
3. RNAse-free DNase I (Roche, cat. no. 776-785).
4. HPLC-grade water (J. T. Baker, cat. no. 7732-18-5).

2.6. RNA Dot Blots

1. Dot blot minifold apparatus (Schleicher & Schuell, cat. no. 44-27510).
2. Protran nitrocellulose membrane (Schleicher & Schuell, cat. no. 10402594).
3. 0.1 N NaOH.
4. 3MM Whatman paper.
5. 20X SSC: 3 M NaCl, 0.3 M Na-citrate.
6. Formamide and formaldehyde.

2.7. Northern Blot Analysis

1. Seakem Gold Agarose (BMA Products, cat. no. 50150).
2. Electrophoresis apparatus with connections to a circulating pump.
3. Hybridization chamber with variable temperature settings from 38 to 60°C.
4. 20X SET: 3 M NaCl, 20 mM EDTA, 0.6 M Tris-HCl, pH to 8.0, filter-sterilize and store at room temperature (stable for several months).
5. Prehybridization solution: 10X Denhardt's, 50% formamide (American Bioanalytical, cat. no. AB00600), 5X SET, 100 µg/mL yeast total RNA, 1% sodium dodecyl sulfate (SDS).
6. 10X MOPS buffer: 0.4 M MOPS, 0.1 M NaOAc, 0.01 M EDTA.
7. Formaldehyde (37% solution, J. T. Baker, cat. no. 50-00-0).
8. 20X SSC: 3 M NaCl, 0.3 M Na-citrate.
9. RNA gel loading solution (Ambion, cat. no. 8552).
10. Hybond™-N membrane (Amersham Pharmacia Biotech, cat. no. RPN303N).
11. Methylene blue solution (ICN, cat. no. 193998); make a working solution of 0.04% in 0.5 M NaCH$_3$COO (pH 5.2) and store at room temperature (stable for several months).
12. DNA fragment to be labeled.
13. [α-^{32}P]dCTP (NEN, cat. no. BLU513H). All necessary precautions should be taken to avoid exposure of self or others to radioactive emission. Dispose of radioactive materials according to local rules.

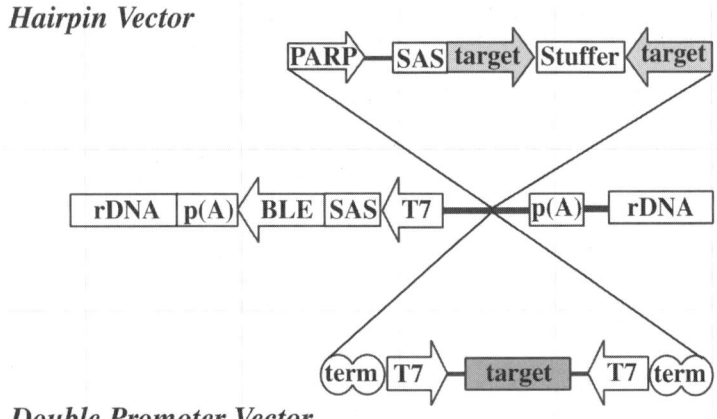

Fig. 1. Vectors for RNAi in *T. brucei*. PARP arrow, tetracycline-inducible PARP promoter; SAS, splice acceptor site; rDNA, ribosomal DNA nontranscribed spacer; p(A), poly(A) addition site; BLE, phleomycin reistance gene; T7 arrow, tetracycline-inducible T7 promoter; term, T7 transcription terminators.

14. Random Prime labeling System (Rediprime™ II, Amersham, cat. no. RPN1633).
15. Bio-spin 6 (P6) chromatography columns (Bio-Rad, cat. no. 732-6002).
16. Hybridization wash solution: 2X SSC, 0.1% SDS.

3. Methods

3.1. Generation of RNAi Constructs

To obtain stable and inducible expression of dsRNA in *T. brucei*, an appropriate construct needs to be inserted into the genome and a stable cell line expressing the dsRNA needs to be selected. At the present time, there are two types of vectors that we and others have constructed for eliciting an RNAi response in *T. brucei* (**Fig. 1**). The first vector uses the tetracycline (tet)-inducible promoter from the procyclic acidic repetitive protein (PARP) genes *(4, 8)* to drive expression of hairpin RNAs (*see* **Notes 3** and **4**). However, the generation of a construct containing two inverted repeats of the gene of interest (*see* **Note 5**) separated by a stuffer fragment requires a two-step cloning procedure *(2)*. The second vector is equipped with opposing tet-inducible T7 RNA polymerase promoters *(2,9)*. In this case, a simple cloning step is required to insert a portion of the gene of interest in between two T7 promoters of pZJM *(2)* for producing dsRNA in vivo (*see* **Note 3**). The hairpin and double-T7 constructs are linearized with Eco RV and Not I, respectively, for integration at the rDNA nontranscribed spacer region of a specific recipient strain of *T. brucei*, named 29.13.6, expressing the tet repressor and T7 RNA polymerase *(6)* (*see* **Note 1**).

3.2. Transfection of Procyclic T. brucei Cells by Electroporation

Procyclic *T. brucei* cultures are maintained in Cunningham's medium *(7)* containing 20% heat-inactivated fetal bovine serum at 28°C in a standard incubator (no CO_2 is required, except for cloning cells in microtiter plates). Each transfection for establishing stable cell lines requires 5×10^7 cells and about 20 μg of DNA.

1. The day before the transfection, dilute the cells in such a way that they reach a density of not more than 10^7 cells/mL the next day.
2. The day of the experiment, first prepare the following:
   a. Aliquot 10 mL of complete Cunningham's medium in the required number of culture flasks (1 per electroporation).
   b. Aliquot 20 μg of linearized DNA into Eppendorf tubes and adjust the volume to 100 μL with H_2O.
   c. Include a no-DNA control tube with 100 μL H_2O.
   d. Prepare and label the required number of 0.4-cm-gap cuvets.
   e. Turn on the Bio-Rad Gene Pulser II and set the capacitance at 25 μF and the voltage at 1.5 kV.
3. Spin down cells (5×10^7 cells per transfection) at 2000*g* for 5 min.
4. Discard the supernatant and resuspend the cell pellet in cytomix in one-half of the original culture volume.
5. Spin again and resuspend the cell pellet in cytomix in one-twentieth of the original culture volume.
6. Aliquot 0.5 mL of cells into each of the Eppendorf tubes containing the plasmid DNA, mix by pipetting up and down, and transfer the mixture into a 0.4-cm-gap cuvet.
7. Close the cuvet cap loosely and zap twice, waiting 10 s between zaps (time constant is normally between 0.4 and 0.5).
8. Transfer the electroporated cells immediately into the flask prepared in **step 2a.** and rinse the cuvet with some medium.
9. Incubate the cells at 28°C for 24 h without drugs. This allows the transfected cells to recover and begin to express the selectable marker.
10. After 24 h, add the appropriate drug (depending on the drug-resistance marker in the vector). At this point, cells can either be cloned (as described in **Subheading 3.3.**) or selected as a population (go to **step 11**).
11. Monitor the cells, and if they grow to approx 10^7 cells/mL, then dilute them to 10^6 cells/mL. Continue selection.
12. Once the control cells die, the drug-selected cells are considered stable and should therefore be tested for expression of dsRNA and also for downregulation of the targeted mRNA.

13. Prepare a frozen stock of the population (*see* **Note 6**). Also prepare clonal cell lines (**Subheading 3.3.**).

3.3. Cloning of Procyclic T. brucei Cells

Although certain RNAi phenotypes are going to be evident in a cell population, others will only become apparent in a clonal cell line (*see* **Note 7**). We clone cells by limiting dilution, which can be applied to a stable cell population or to cells 2 d posttransfection (*see* **Note 8**).

1. Prewarm medium at 28°C for at least 30 min and add the appropriate drug(s) required to maintain selection for transfected cells.
2. Count cells and prepare four sets of dilutions using the media prepared in **step 1**, so as to have one set at 0.3 cell/50 µL, a second set at 2 cells/50 µL, a third set at 4 cells/50 µL, and a fourth set at 10 cells/50 µL. Also dilute one set of wild-type (untransfected) cells so as to have 100 cells/50 µL (although the drug will kill the wild-type cells, they will "condition" the medium). Plan to use 30 wells of a 96-well microtiter plate for one set of dilutions. Add 50 µL of one dilution and 50 µL of wild-type cells to each well.
3. Cover the microtiter plate, seal with Parafilm, and incubate in a humidified incubator at 28°C with 5% CO_2.
4. Monitor the cells microscopically every other day. By 7–11 d there should be noticeable cell growth. For expansion, choose wells from the highest dilution, as these are most likely to contain a single transfected parasite. In a successful cloning, only one of three wells of a dilution should display growth.
5. Add 100 µL of fresh complete medium to each of the chosen wells and continue the incubation. The following day, transfer the 200 µL of cell culture from each chosen well into one well of a 48-well dish and add 500 µL of fresh medium.
6. When cells are dense enough (approx 10^7 cells/mL), transfer them into a culture flask and bring the volume to 3 mL with medium. Cells can now be diluted to large volumes (10–20 mL) and at this point, aliquots of healthy cells should be frozen for long-term storage (*see* **Note 6**).

3.4. Induction of dsRNA Expression and Analysis of RNAi Cells

It is important to start with freshly diluted cells. If cells have been cloned, always plan to analyze several clones, since the level of expression of dsRNA can vary (**Fig. 2**) and this also correlates with differences in the efficiency of degradation of the target mRNA.

1. Dilute cells to 1×10^6 cells/mL and add tetracycline (dissolved in 70% ethanol) to a final concentration of 10 µg/mL. Addition of tetracycline removes suppression of the promoter driving transcription (PARP or T7, *see* **Fig. 1**) and leads to production of dsRNA.
2. Collect a total of 10^8 cells for isolation of total RNA at desired time points (then go to **Subheading 3.5.** for isolation of total RNA). The first time point for RNA

RNAi in T. brucei

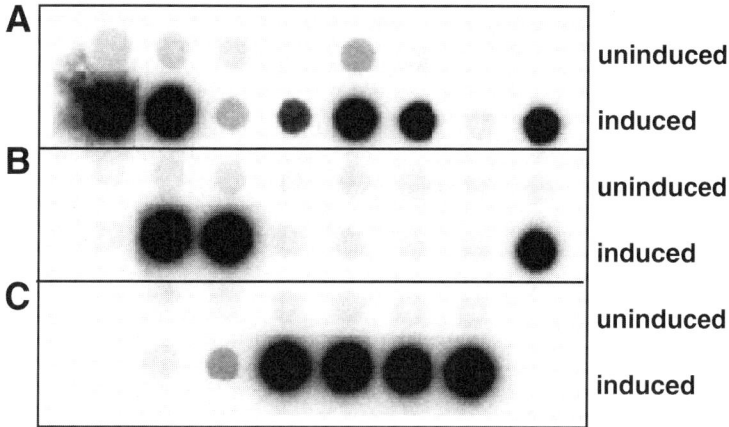

Fig. 2. Induction of dsRNA expression in clonal cells. RNAi cells targeting mRNAs for a *T. brucei* methyltransferase (**A**), capping enzyme I (**B**) and II (**C**) were established. Hairpin constructs expressing dsRNAs were inserted at the rDNA locus and stable cell lines selected and cloned. The expression of dsRNA was induced with tetracycline (at 10 µg/mL) for 24 h. Total RNA was prepared for each cell line, dot blotted on a membrane, and hybridized with a radiolabeled DNA fragment derived from the stuffer sequence.

analysis can be as early as 5 h. Keep the cell concentration at 1×10^6/mL and continue to add tetracycline to culture medium every day.
3. The window for maximum degradation of the target mRNA should be determined for each experiment by analysing aliquots at different time points over the following days. This allows better execution of a downstream assay for functional studies.
4. Also count the cells every day to monitor if RNAi induction affects cell growth.

3.5. Isolation of Total RNA

1. Spin down 10^8 cells in a Falcon tube for 5 min at 2000*g*.
2. Wash cells three times with cell wash buffer and follow the RNeasy Tissue kit for the isolation of total RNA.
3. Use a UV spectrophotometer to estimate the RNA concentration and perform dot blot analysis (**Subheading 3.6.**) or Northern blot analysis (**Subheading 3.7.**).

3.6. RNA Dot Blot Analysis

Total RNA prepared from induced (with tetracycline) and noninduced cloned cells can be used to rapidly determine the expression level of dsRNA by dot blot analysis. The probe can either be a DNA fragment homologous to the dsRNA, or homologous to the stuffer fragment in the case of hairpin constructs.

3.6.1. Assembling the Minifold Dot Blot Apparatus

1. Wash the apparatus with 0.1 N NaOH and rinse well with water.
2. Assemble the apparatus according to the manufacturer's instructions and use 3MM Whatman paper soaked in 20X SSC for at least 1 h.
3. Rinse the chambers with 10X SSC.
4. Turn off the vacuum, and refill the chambers with 10X SSC.

3.6.2. Preparation and Processing of the RNA Samples

1. Aliquot 2–5 µg of total RNA, in H_2O, in a total volume of 10 µL.
2. Add 20 µL formamide, 7 µL formaldehyde, and 2 µL 20X SSC, and incubate at 68°C for 15 min.
3. Cool the samples on ice and add 2 vol of 20X SSC.
4. Turn on the vacuum to remove the liquid and then turn the vacuum off.
5. Load the samples into the chambers, turn on the vacuum, and wash twice with 1 mL of 10X SSC.
6. Keep the vacuum on for an additional 5 min.
7. Remove the nitrocellulose filter, dry at room temperature, and bake the filter at 80°C for 2 h.
8. Set the filter for prehybridization and hybridization as described in **Subheading 3.7.3.**

3.7. Northern Blot Analysis

A successful study of gene function using RNAi depends on the degradation of the corresponding mRNA. Therefore, it is essential to check both the expression of the dsRNA and the degradation of the target mRNA by Northern blot analysis.

3.7.1. Gel Electrophoresis and Transfer

1. Thoroughly clean the gel electrophoresis equipment and rinse it several times with millipore water.
2. Make a 1.2% agarose slurry in 1X MOPS buffer: for a 100-mL gel slurry, weigh 1.2 g of agarose, add MOPS and millipore water to 83 mL. Melt in a microwave oven on medium heat and cool to 60°C.
3. In a chemical hood, add 17 mL formaldehyde to the gel slurry, mix, and pour into the gel casting tray. Let the gel solidify for 30 min.
4. In the meantime, prepare the RNA sample by mixing the following: 1 part of RNA in water and 3 parts of the RNA gel loading solution. Incubate the mixture at 65°C for 15 min, spin briefly.
5. Transfer the gel to the electrophoresis tank containing 1X MOPS buffer and connect to a circulating pump.
6. Load the gel and run at 20–60 V for 6–12 h depending on the desired resolution.
7. After the run, rinse the gel several times with millipore water for 20 min, changing the water every 5 min, and then soak the gel in 20X SSC for 20 min.

RNAi in T. brucei

8. Set the transfer (*see* **Note 9**) with 10X SSC using a Hybond-N membrane.
9. After the transfer, dry the membrane and cross-link RNA onto filter by exposure to UV. The length of exposure to UV depends on the equipment used and should be determined experimentally.
10. To stain the RNA immobilized on the filter, transfer the filter into a solution of 0.04% methylene blue in 0.5 M Na-Acetate at room temperature for 30–120 s.
11. Wash the filter with water to destain and scan or photocopy the filter for documentation.

3.7.2. Probe Labeling

1. Prepare DNA to be used as probe (*see* **Note 10**) by heating 20–25 ng in 45 µL of water to 100°C for 5 min to denature.
2. Place tube on ice for 5 min, then centrifuge briefly to collect condensate.
3. Add the denatured DNA to a rediprime labeling mix, add 5 µL of [α-^{32}P] dCTP, and incubate at 37°C for 60 min.
4. In the meantime, prepare one P-6 column per sample. Spin at 1000g for 2 min and discard the flow-through.
5. Load the entire labeling reaction onto a P-6 column and spin at 1000g for 5 min. Collect the eluate and determine the specific activity by scintillation counting.

3.7.3. Hybridization

1. Place the filter in a clean hybridization container, add enough prehybridization solution to cover the entire filter, and prehybridize at 50°C for at least 30 min.
2. Denature the probe by heating to 100°C for 5 min, then place it on ice for 3 min. Add the probe to the prehybridized filter at a concentration of at least 10^6 cpm/mL of hybridization solution.
3. Allow the hybridization to go overnight and wash the filter with hybridization wash solution at 60°C, until the background is acceptable. Expose filter to a phosphorimager screen or X-ray film.

4. Notes

1. 29.13.6 cells are procyclic-form *T. brucei* that have been genetically modified by insertion into the genome of two cassettes for expression of the T7 RNA polymerase and the tetracycline repressor (*6*). They should be grown in the presence of G418 (at 15 µg/mL) and hygromycin B (at 50 µg/mL), because expression of T7 RNA polymerase and tet repressor were selected with these markers.
2. We routinely culture procyclic-form *T. brucei* cells in suspension at 28°C. We use Cunningham's medium (*7*), which can be prepared in the laboratory, filter-sterilized, and stored in the fridge for several months. Alternatively, we have the medium custom-made by HyClone. Just before use, add fetal bovine serum, Pen/Strep, glutamine, and gentamycin.
3. In the absence of tetracycline, expression from the tet-induced PARP promoter (**Fig. 1**, top) is almost negligible, allowing cloning of toxic products. However, under full induction conditions, high levels of expression can be achieved. On the

other hand, the double-T7 promoter vector (**Fig. 1**, bottom) has the problem that a considerable level of dsRNA is produced even in the absence of tetracycline *(2)*, because binding of the tet repressor to the tet operators does not completely shut off the T7 promoters *(6)*. Thus, in certain instances one should use caution with this particular vector. In addition, the low-level expression can lead to the selection of cells that express dsRNA that target the corresponding mRNA, but have adapted to live normally with low levels of the targeted mRNA. Thus, we use tetracycline-free serum (BD Bioscience, cat. no. 8630-1) whenever we target an essential gene with RNAi. In this instance, we begin to use the tet-free medium immediately after transfection.
4. We had previously used a spliced leader RNA gene fragment and later on we observed that a different stuffer fragment, a piece of the pex11 gene *(2)*, worked better. In fact, RNAi cells generated with constructs containing the pex11 stuffer fragment acquired the phenotype much more rapidly that those generated with constructs made with the SL stuffer (unpublished results).
5. We generally use 100- to 1000-bp DNA fragments for RNAi. If possible, we use only a portion of the gene for RNAi, leaving another region to be used as a probe, since the detection of the dsRNA itself can, in some cases, lead to hybridization background.
6. To prepare glycerol stocks for long-term storage of *T. brucei* cells, do the following: first prepare the freezing medium (20% glycerol in complete Cunningham's media) and filter-sterilize. Then spin down approx 10^8 cells at 2000*g* for 5 min. Remove the supernatant and resuspend the pellet with 1 mL of freezing medium. Transfer the cells into a 1.5-mL screw-cap tube, place the tube between two Styrofoam racks, and transfer directly to –80°C for long-term storage.
7. It is our experience that sometimes the RNAi phenotype is much stronger in clonal cell lines, as compared to populations of stable transfectants. Therefore, we prefer to perform the analysis of the RNAi phenotype in clonal cell lines.
8. Two methods are presently available for the cloning of trypanosome cells. The first is by plating cells on agarose plates *(10)* and the second is by limiting dilution of cells. The cloning method by plating cells onto agarose plates involves several steps and is sometimes problematic.
9. There is a significant variation in methods used for transferring RNA to membranes. The upward capillary method is widely used and requires at least 16 h for maximum transfer of total RNA. We routinely use the downward capillary method. In our hands, this method allows a maximum transfer within 5–6 h.
10. Generally, we use as probe a region of the targeted mRNA not included in the dsRNA construct. This allows the analysis of target mRNA degradation after induction of RNAi. We have also successfully used probes homologous to the stuffer fragment in Northern blot analysis.

Acknowledgments

This study received support from National Institutes of Health Grants AI28798 to E. Ullu. C. Tschudi is the recipient of a Burroughs Wellcome Fund New Investigator Award in Molecular Parasitology.

References

1. Hannon, G. J. (2002) RNA interference. *Nature* **418(6894),** 244–251.
2. Wang, Z., Morris, J. C., Drew, M. E., et al. (2000) Inhibition of *Trypanosoma brucei* gene expression by RNA interference using an integratable vector with opposing T7 promoters. *J. Biol. Chem.* **275,** 40,174–40,179.
3. Wang, Z. and Englund, P. T. (2001) RNA interference of a trypanosome topoisomerase II causes progressive loss of mitochondrial DNA. *EMBO J.* **20,** 4674–4683.
4. Bastin, P., Ellis, K., Kohl, L., et al. (2000) Flagellum ontogeny in trypanosomes studied via an inherited and regulated RNA interference system. *J. Cell Sci.* **113,** 3321–3328.
5. Drozdz, M., Palazzo, S. S., Salavati, R., et al. (2002) TbMP81 is required for RNA editing in *Trypanosoma brucei*. *EMBO J.* **21,** 1791–1799.
6. Wirtz, E., Leal, S., Ochatt, C., et al. (1999) A tightly regulated inducible expression system for conditional gene knock-outs and dominant-negative genetics in *Trypanosoma brucei*. *Mol. Biochem. Parasitol.* **99,** 89–101.
7. Cunningham, I. (1977) New culture medium for maintenance of tsetse tissues and growth of trypanosomatids. *J. Protozool.* **24,** 325–329.
8. Shi, H., Djikeng, A., Mark, T., et al. (2000) Genetic interference in *Trypanosoma brucei* by heritable and inducible double-stranded RNA. *RNA* **6,** 1069–1076.
9. LaCount, D. J., Bruse, S., Hill, K. L., et al. (2000) Double-stranded RNA interference in *Trypanosoma brucei* using head-to-head promoters. *Mol. Biochem. Parasitol.* **111,** 67–76.
10. Carruthers, V. B. and Cross, G. A. (1992) High-efficiency clonal growth of bloodstream- and insect-form *Trypanosoma brucei* on agarose plates. *Proc. Natl. Acad. Sci. USA* **89,** 8818–8821.

17

In Vitro Shuttle Mutagenesis Using Engineered Mariner Transposons

Kelly A. Robinson, Sophie Goyard, and Stephen M. Beverley

Summary

Advances in our understanding of the protozoan parasite *Leishmania* have been facilitated by the development of molecular and genetic tools. One powerful approach for gene identification and analysis is transposon mutagenesis. This can be performed directly in vivo, but often it is more convenient to generate transpositions in vitro for subsequent analysis in vivo, in a process termed "shuttle mutagenesis." The *Drosophila* element *mariner* is well suited for application by either route. Minimal *mariner* elements containing *cis*-acting elements required for transposition have been generated, which can be further modified to suit the needs of the experimenter. Additional genetic markers and/or reporters can be introduced, which are useful for procedures such as insertional mutagenesis, shotgun sequencing, or the generation of protein and transcriptional fusions for subsequent analysis. Active transposase can readily be generated following expression in *Escherichia coli*, and efficiencies of 10^{-3}/target can be obtained, allowing the generation of large transposon insertion libraries suitable for subsequent screening in vivo. This chapter explains the steps necessary to purify active *Mos1* transposase and conduct an in vitro transposition reaction. We also discuss some of the considerations relevant to the design and application of functional *mariner* elements (donor plasmids) relevant to studies in *Leishmania* and other organisms.

Key Words: In vitro transposition; *Mos1* transposon; protozoan parasite; shuttle mutagenesis; Tc1/*mariner* transposon family.

1. Introduction

Protozoan parasites such as *Leishmania* are responsible for numerous illnesses that cause significant mortality and morbidity throughout the world *(1)*. Genetic and genomic tools now available for the study of the parasite promise to increase greatly our understanding of how this parasite survives and causes disease and ultimately lead to improved methods for overcoming this disease by immunization or chemotherapy *(2–5)*.

Two common tasks in parasite genetics are, first, the identification of genes mediating interesting functions, and second, dissection of the role, regulation, and localization of encoded proteins. A powerful tool suitable for this task in many organisms is transposon-based mutagenesis *(6)*. This can be performed directly in vivo, whereby both the transposon and active transposase are introduced or expressed in the parasite, or in vitro, whereby transposition is performed in vitro and the products introduced into the parasite for subsequent analysis (**Fig. 1**). In vivo strategies are especially powerful when incorporated into forward genetic approaches, as mutants generated are simultaneously tagged by the transposon, which can then be used to recover the affected gene. Unfortunately, *Leishmania* is an asexual diploid in the laboratory, and for most loci, recovery of loss-of-function mutations requires at least two genetic events *(7)*. However, this approach is widely used in haploid organisms, or ones in which homozygosity can be readily attained in some manner (for example, by sexual crossing).

In vivo transposition systems can be challenging to set up because of the constraints inherent in engineering transposase expression and controlling transposition. For many purposes, in vitro transposition is more convenient and as powerful. In a process termed "shuttle mutagenesis," transfectable molecular constructs (for example, *Leishmania* DNA cloned in the shuttle *Escherichia coli-Leishmania* vector cLHYG *[8]*) are subjected to transposition in vitro, and then the population of independent insertions is scored for phenotypes following transfection back into *Leishmania* (**Fig. 1**). The transposon insertion library also can be used for rapid and systematic DNA sequencing if necessary, using primers situated within the transposon. We have found this approach especially useful in mapping the active gene within cosmids recovered in various functional genetic screens in *Leishmania*.

Beyond their role as insertional mutagens, transposons can be engineered to contain reporters such as the green fluorescent protein (GFP), β-galactosidase, β-glucosidase, or β-lactamase, or selectable markers such as *NEO*, *HYG*, or *PHLEO*, which mediate resistance to G418/geneticin, hygromycin B, or phleomycin/zeocin, respectively. Following transposition, activation of the reporter or marker can then be used to identify and/or select for transcribed or translated regions of the genome, a procedure commonly referred to as gene/protein "trapping." By studying expression and/or localization of the reporter proteins, one can then conveniently (albeit indirectly) monitor gene expression and protein localization *(9,10)*.

Several transposon systems have been engineered to the point that they are readily incorporated into shuttle transposon mutagenesis strategies; these include Tn7 *(11)*, Tn5 *(12)*, Ty1 *(13)*, Mu *(14)*, and several Tc1/*mariner* family elements *(15–17)*. Relevant factors include the availability, cost, and/or ease of purifica-

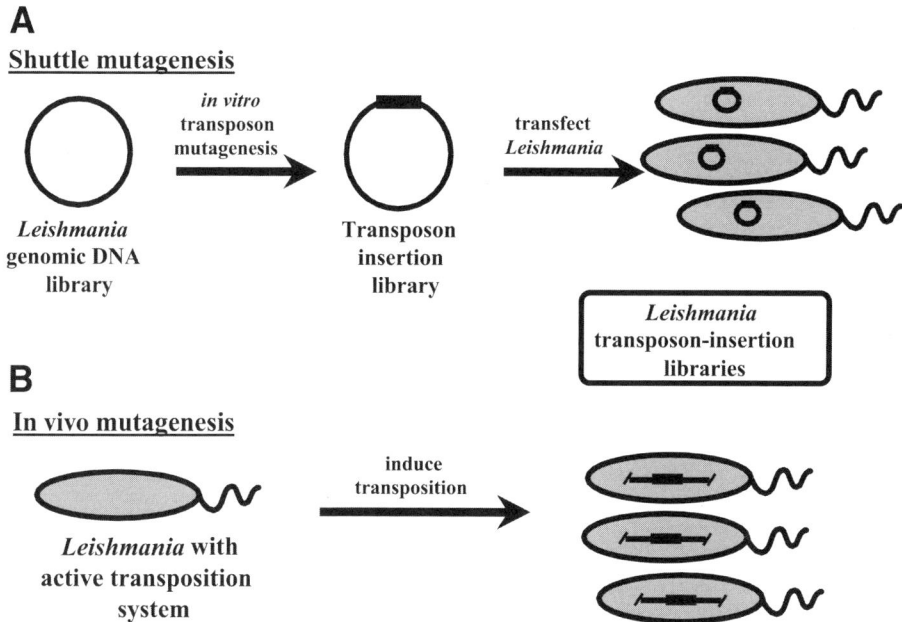

Fig. 1. Transposon mutagenesis strategies. (**A**) Shuttle mutagenesis begins with in vitro transposition into target DNA (plasmid or cosmid) to create an insertion library. The transposon library is then transfected into *Leishmania*, and recovered by selection on the drug-resistance marker, found on the transposon, in various ways. (**B**) In vivo mutagenesis requires the establishment of an active transposition system in the parasite itself. This could be accomplished in several ways; stable expression of transposase and stable introduction of the transposon have been successful *(29)*. Ideally, one would prefer transient introduction of the transposition system, for example, using transient or regulated expression of transposase, and/or transient introduction of donor transposons. Alternatively, one could form a transposase–transposon complex in vivo, and then introduce this for subsequent transposition in vitro. This has not yet been demonstrated in the *mariner* system, but it works well with transposon Tn5 in various eukaryotes *(35,36)*. (Reprinted with permission from **ref. 37**.)

tion of transposase, the randomness with which a given transposon inserts into target DNA, the requirements for specific *cis*-acting elements required within the transposon itself (for example, the size or properties of the flanking inverted repeats [IRs]), the ability of the transposon to carry "cargo" of sufficient size and with the desired properties (such as selectable markers or reporters appropriate for the target organism), and ease of use. We have found the *Drosophila mariner* element *Mos1* to be satisfactory in these respects, and here, we describe the basic elements of the transposon relevant to its application in vitro.

The *Drosophila* element *Mos1* is a member of the *mariner*/Tc1 family, which occurs in most kingdoms of living organisms *(18,19)*. Typically, *mariner*/Tc1 elements are small, encoding only the transposase and *cis*-acting elements required for transposition, such as the flanking IRs *(20,21)*. Transposition occurs through a cut-and-paste mechanism *(21)*, in which recognition of the IRs by the transposase results in excision of the donor element. This is subsequently inserted into a TA dinucleotide of the target molecule, and accompanied by duplication of the TA flanking the insertion site *(22)*. In vitro experiments have shown that the transposition reaction requires only transposase and transposon *cis*-elements, without the need for cellular factors *(15,17,21,23,24)*. For *Drosophila Mos1*, the *cis*-acting elements required for transposition include the 28 bp 5'- and 3'-IRs, along with some internal nucleotides (no greater than 38 and 5 additional internal nucleotides on the 5'- and 3'-sides, respectively; *[17,23]*). Although here we focus on in vitro applications, it is notable that the *mariner* system has been shown to function in vivo in a variety of different organisms including *Leishmania*, insects, and vertebrates *(25–29)*. Most of the *mariner* transposon derivatives described below can be used in vivo as well, in any species.

Our understanding of the mechanism of *Mos1* transposition has lead to the development of a minimal, "empty" transposon donor, pELHY6Δ-0 (**Fig. 2**), into which various transposon "cargos" have been inserted previously *(23)*.

Fig. 2. (*Opposite page*) *Mos1* vectors and the in vitro transposase reaction. (**A**) The "empty" donor plasmid pELHY6Δ-0 contains the minimal *cis*-element (open arrow heads) with the 5'- and 3'-IRs and some internal nucleotides (*see* **Fig. 3A**). The vector contains the *E. coli* OriR6K origin (striped box) for propagation in a λpir+ strain and a *HYG*-selectable marker (internal black line); this particular marker contains a *Leishmania* splice acceptor site (AG) for expression in *Leishmania* and an *E. coli* promoter (black arrow). For specific uses, one can insert various "cargo" within the IRs, at the unique *Msl*I, *Xba*I, or *Sbf*I sites (*see* **Fig. 3A**). In this figure, the donor element is pELHY6Δ-/GEP3/, created by insertion of the /GEP3/ (*see* **Table 1**). For simplicity, only the *E. coli PHLEO*-resistance marker is shown in /GEP3/. (**B**) The basic in vitro transposase reaction contains a donor element, target DNA, and transposase (shaded area). In this example, the transposon target is a cosmid DNA (bottom left in Panel **B**), which contains an *E. coli* ampicillin-resistance marker (Ap) and an OriC origin of replication (open box). After in vitro transposition of the donor plasmid into the TA dinucleotide indicated on the cosmid, the DNA is transformed into a λpir- strain (such as DH10B). Donor plasmids bearing the OriR6K origin of replication cannot replicate in such strains and are selectively lost. Bacterial transformants are plated on LB medium containing Ap/PHLEO to select for transposition. Target DNAs containing transpositions can then be transfected into *Leishmania*, and fusion proteins identified by selecting or screening for PHLEO or GFP expression, respectively.

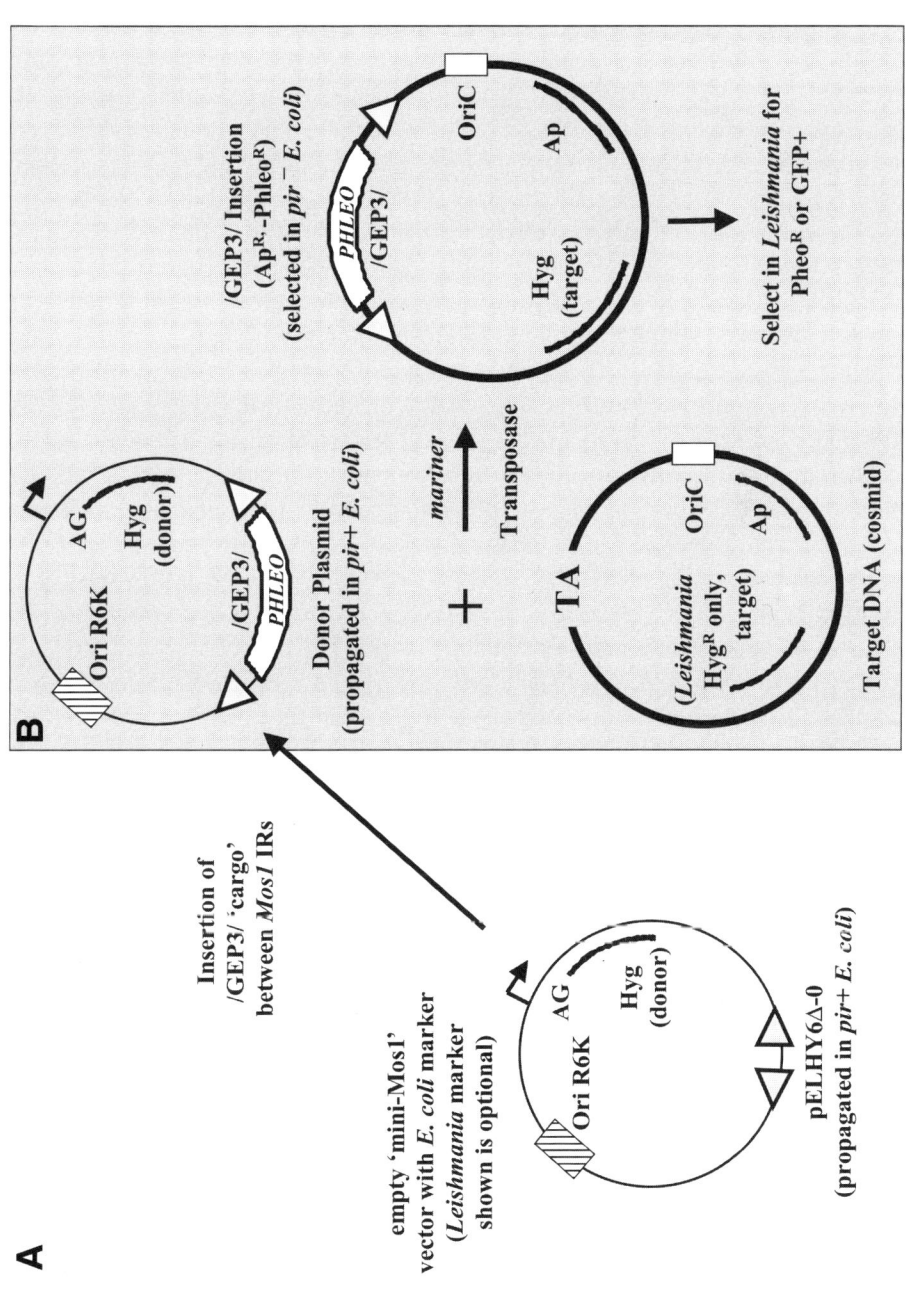

New cargos may be designed and rapidly introduced into this vector as desired by the experimenter. Various transposons have been created previously, and potential applications such as insertional mutagenesis and gene trapping are summarized in **Table 1**. A typical in vitro *Mos1* transposition reaction consists of the donor plasmid (e.g., /GEP3/, **Fig. 2**, **Table 1**), target DNA (e.g., a cLHYG-based cosmid; **Fig. 2**), and purified transposase. The properties and requirements for these elements are discussed below.

The donor plasmid pELHY6Δ-0 contains outside of the minimal transposon a bacterial/*Leishmania*-selectable marker (*HYG* here) and origin of replication (OriR6K) (**Fig. 2**). For propagation in *E. coli*, the OriR6K origin requires the *pir* gene product, which is provided by the use of appropriate *E. coli* host strains when growing this plasmid (often harboring a λ*pir* lysogen). This allows one to select against the donor plasmid following transformation of the in vitro transposition reaction mix into *pir*- *E. coli* (this comprises virtually all common *E. coli* recipients). In this particular donor plasmid, the *HYG* marker is bifunctional, designed to function in *E. coli* because of the inclusion of a bacterial promoter, as well as *Leishmania* because of the presence of a parasite *trans*-splice acceptor site. Thus, this plasmid potentially can be used for in vitro transposition (*E. coli*) as well as in vivo transposition (*Leishmania*). Note that the signals for replication, transcription, and/or mRNA processing differ considerably between *E. coli* and *Leishmania*. Briefly, *E. coli* markers require promoters and plasmids require origins of replication for episomal maintenance; in contrast, in *Leishmania* all that is required is a *trans*-splice acceptor site upstream of the marker open reading frame (ORF), as transcriptional and replication origin requirements are quite relaxed *(30)*. These differences must be taken into account when designing new transposons, and it can also be used to the researcher's advantage in various ways.

As a target plasmid, most common laboratory plasmid, cosmid, and bacterial artificial chromosome (BAC) vectors can be used; the specific requirements are that the target DNA should not contain an OriR6K for replication. The target marker should not be the same as ones borne within the transposon or in its donor background (note that in the cLHYG example shown, the *HYG* gene lacks an *E. coli* promoter and thus does not confer resistance in bacteria). Transposition efficiencies are highest if the target DNA is supercoiled and its "quality" is high *(17)*.

Mos1 transposase is purified and stored as described below. Typically, transposons require Mg^{2+} for activity; however, Mn^{2+} can be used, as this relaxes the requirement for insertion into TA dinucleotides *(17)*. Whereas some *mariner* family transposases show a phenomenon called overproduction inhibition *(15)*, *Mos1* shows simple saturation kinetics, and increasing transposase yields increasing transpositions until a plateau is reached *(17)*.

Table 1
Examples of Mos1 Transposons and Their Properties and Applications

Transposon	Transposon Elements			Translational fusions	Transcriptional fusions	In vivo	In vitro	TIMLI
	Bacterial marker	Eukaryotic marker	Reporter					
pELHY6Δ-/GFP*K	Km	—	—	+	—	—	+	—
pELHY6TK-PG	—	PHLEO	GUS	PHLEO GUS	PHLEO GUS	+	—	—
pELHY6Δ-/GEP3*	PHLEO	PHLEO	GFP	GFP PHLEO	PHLEO	+	+	—
pELHY6Δ-/GEP3//GEP2/	PHLEO	PHLEO	GFP	GFP PHLEO	PHLEO	+	+	—
pELHY6Δ-/NEO*ELSAT	SAT	NEO SAT	—	NEO	SAT	+	+	—
pELHY6Δ-/2x5	Km	—	—	—	—	—	+	+

All transposons listed can be used for insertional mutagenesis and sequencing. Symbols are as described: AG, *Leishmania* splice acceptor; *SAT*, nourseothricin resistance marker; *Km*, kanamycin resistance marker; GFP, modified green fluorescent protein; *PHLEO*, phleomycin/zeocin resistance marker; "/" before or after a gene name indicates a gene lacking a start or stop codon respectively; "*" indicates an in-frame stop codon; black arrows, *E. coli* promoter; open triangle, 5′-IR; gray triangle, 3′-IR; oriC, oriC origin of replication; and BsrGI and SexAI, unique restrictions sites used in TIMLI mutagenesis (described under **Subheading 3.1.**).

Following incubation, the in vitro transposition mix (which contains both donor and target plasmids as well as the desired transpositions) is transformed into *pir- E. coli* and plated on drugs that select for both the transposon (phleomycin in the example of /GEP3/ here) and the target (ampicillin here). Transposition efficiencies can be calculated by comparing platings on ampicillin alone vs ampicillin + phleomycin, and can approach 10^{-3}/target.

The number of individual transpositions required depends on the particular application. Although *mariner* demands TA residues for insertion under standard conditions, these are sufficiently abundant even in the GC-rich *Leishmania* genome to provide plenty of potential target sites, and the requirement for TA can be relaxed if transposition is performed in the presence of Mn^{2+} *(17)*. We have found that, for cosmid targets, several hundred independent insertions usually are sufficient for sequencing and inactivation of most potential target genes. For the recovery of specific gene fusions, larger libraries may be required, because one has the additional constraints of inserting into TAs in the appropriate strand and reading frame. For these purposes, 1000 independent insertions into a cosmid target should suffice. Note that with current in vitro transposition efficiencies and *Leishmania* transfection efficiencies *(31)*, one may contemplate scoring libraries in excess of 10^5 independent insertions.

This chapter describes how to express and purify active *Mos1* transposase and carry out in vitro transposition reactions using donor plasmids from the *Mariner* toolkit.

2. Materials

2.1. Vector

1. A suitable vector from the *mariner* toolkit (*see* **Subheading 3.1.** for details).

2.2. Expression of Mos1 Transposase

1. *E. coli* strain expressing T7 polymerase (BLR [DE3]) from Novagen.
2. Vector expressing His_6-tagged *Mos1* transposase (pET19-Tpase, Beverley lab strain B4289; **ref.** *23*).
3. 1 *M* IPTG (isopropyl-β-D-thio-galactopyranoside) stock solution.
4. Resuspension buffer: 20 m*M* Tris-HCl, pH 7.6, 2 m*M* $MgCl_2$, 25% sucrose, 0.6 m*M* phenylmethylsulfonyl fluoride (PMSF), 1 m*M* benzamidine (BZA) and 1 m*M* dithiothreitol (DTT).
5. Liquid nitrogen or Sonicator.

2.3. Purification of Transposase

1. Lysis buffer: 20 m*M* Tris-HCl, pH 7.6, 4 m*M* EDTA, 200 m*M* NaCl, 1% deoxycholate, 1% nonylphenoxy polyethoxy ethanol (NP-40), 0.6 m*M* PMSF, 1 m*M* BZA, 1 m*M* DTT.

2. DNaseI.
3. 1 M MgCl$_2$.
4. Lysozyme.
5. Buffer A: 20 mM Tris-HCl, pH 8, 500 mM NaCl, 6 M guanidine-HCl, 1% NP-40, 70 mM imidazole.
6. Wash buffer: 50 mM NaH$_2$PO$_4$, 300 mM NaCl, 100 mM imidazole, adjusted to pH 8.0 using NaOH.
7. Elution buffer: 50 mM NaH$_2$PO$_4$, 300 mM NaCl, 500 mM imidazole, adjusted to pH 8.0 using NaOH.
8. Ni-NTA agarose (Qiagen, cat. no. 30210).
9. Purification column (Qiagen, cat. no. 34964).
10. Sodium dodecyl sulfate-polyacrylamide gel electrophoresis (SDS-PAGE) gel.
11. Dialysis slide (Slide-A-Lyzer® dialysis cassette, 10,000-MW cutoff; Pierce ca. no. 66425).
12. Dialysis buffer A: 10% glycerol, 25 mM Tris-HCl, pH 7.6, 50 mM NaCl, 5 mM MgCl$_2$, 2 mM DTT.
13. Dialysis buffer B: 10% glycerol, 25 mM Tris-HCl, pH 7.6, 50 mM NaCl, 5 mM MgCl$_2$, 0.5 mM DTT.
14. 100% Glycerol.

2.4. In Vitro Transposition Assay

1. 10X transposition buffer: 250 mM HEPES, pH 7.9, 10 mM DTT, 50 mM MgCl$_2$, 1 M NaCl.
2. 100% glycerol.
3. 10-mg/mL purified bovine serum albumin (BSA) (New England Biolabs).
4. Donor plasmid (*see* **Subheading 3.1.** for details).
5. Target DNA consists of any plasmid, cosmid, or BAC that contains a different selectable marker than that used on the donor plasmid and that does not contain an OriR6K origin of replication.
6. Transposase (*see* **Subheading 3.2.**)
7. Sterile distilled H$_2$O (sdH$_2$O).
8. Stop buffer: 50 mM Tris-HCl, pH 7.6, 0.5-mg/mL proteinase K, 10 mM EDTA, 250-mg/mL yeast tRNA.
9. 3 M sodium acetate.
10. 100% ethanol.
11. 70% ethanol.
12. Bacterial electroporator.
13. *pir- E. coli* electrocompetent cells, such as DH10B.
14. 25:24:1 Phenol–chloroform–isoamyl alcohol.
15. 10 mM Tris, pH 7.5.
16. Ampicillin, hygromycin, nourseothricin (Dr. Walter Werner; WeBioAge@aol.com) and Zeocin (Invitrogen).
17. Luria Bertani (LB) medium: 10 g Bacto-tryptone, 5 g yeast extract, and 10 g NaCl per liter (with appropriate drug)/LB agar plates: 20 g of agar to 1 L of LB medium.

3. Methods

Here, we describe a range of vectors available for transposition using *mariner* and outline the steps involved in preparing the transposase enzyme and performing an in vitro transposition reaction.

3.1. The mariner *Toolkit*

Table 1 describes some of the *mariner* derivatives that have been developed and used successfully in our laboratory. Various applications and transposons can be envisaged, and the ones below provide some perspective on the factors relevant to their design and utilization. The transposons are described briefly below; many can be used for the recovery and/or selection of gene fusions in *Leishmania* and other organisms, as they lack species-specific regulatory elements (the GEP transposon series, for example). All transposons can be used for primer-island sequencing, and insertional inactivation. Most transposons contain autonomous bacterial selectable markers and can be used in the in vitro system, except for pELHY6TK-PG (thereby restricting it to in vivo applications). Because *Leishmania* uses a polycistronic transcriptional mechanism to generate mRNAs and relies heavily on posttranscriptional regulatory mechanisms to control protein expression, we have given the most attention to transposons that facilitate the recovery of protein fusions.

Transposons /GEP3*, /GEP3/, and /GEP2/ contain a GFP-PHLEO fusion protein (**Table 1**); the linker peptide between the GFP and PHLEO additionally functions as an *E. coli* promoter, and in bacteria this cassette confers phleomycin resistance constitutively (as required for use in the in vitro system). Importantly, the GFP lacks an initiating ATG codon, and thus in eukaryotes GFP-PHLEO expression can only be obtained following insertion of the /GEP transposons into an ORF expressed by the target DNA (**Fig. 3**). Such insertions can be selected for by phleomycin resistance (only in eukaryotes), or screened for by GFP expression. Note that phleomycin resistance can be affected by compartmentalization of the fusion protein; if the PHLEO protein domain is restricted to a compartment such as the glycosome, which is segregated from the nucleus (the site of action of phleomycin), resistance will be abrogated *(32)*.

/GEP3* differs from /GEP3/ and /GEP2/ in that it contains a stop codon following the GFP-PHLEO fusion protein. Thus, the protein fusions recovered bear the N- but not C-terminus of the trapped protein (**Fig. 4A,C**). In /GEP3/ and /GEP2/ the stop codon has been eliminated so that an intact reading frame is maintained across the entire transposon, enabling the recovering of protein fusions that bear both the N- and C-terminus of the trapped protein; this type of transposon is referred to as a "sandwich" transposon (**Fig. 4B,C**). Because *mariner* elements must insert into TAs, which can occur in any reading frame,

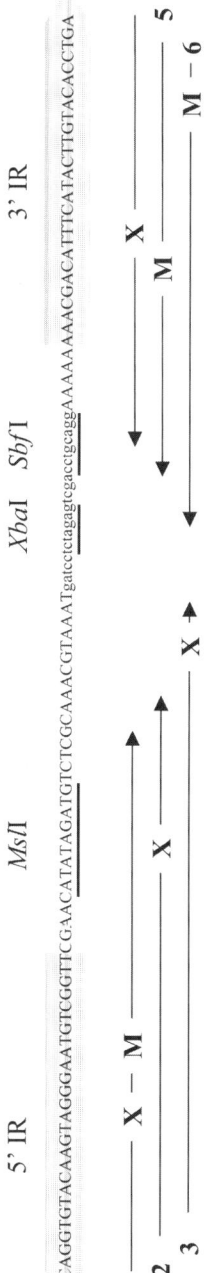
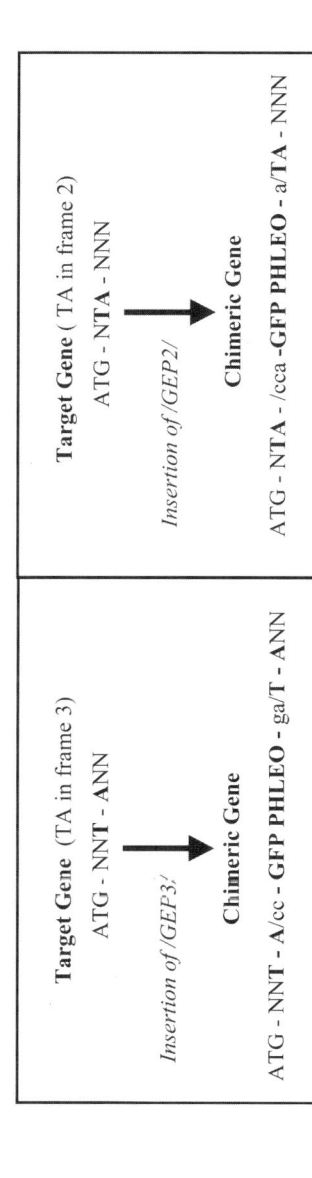

Fig. 3. Nucleotide sequences of *Mos1* cis-element and pELHY6Δ-/GEP3/, pELHY6Δ-/GEP2/ chimeric genes. (**A**) The minimal *Mos1* element used in our work contains essential cis-elements consisting of the 5'- and 3'-IRs (shaded gray block arrows) and the internal 38 and 5 internal nucleotides (nonshaded capital letters). The six potential *Mos1* reading frames, from the flanking DNAs across the IRs, are shown with arrows (labeled 1–6), whereas start and stop codons in each frame are shown by M or X, respectively. Unique restriction sites found within the empty transposon that are suitable for the addition of "cargo" are shown. (**B**) Putative chimeric genes created by insertion of the /GEP3/ and /GEP2/ transposons into target TA dinucleotides in the third or second reading frame, respectively. The "/" symbol represents points of potential fusion of reading frames, in this case with the GEP ORFs and those of the target. (Reprinted with permission from **ref. 23**.)

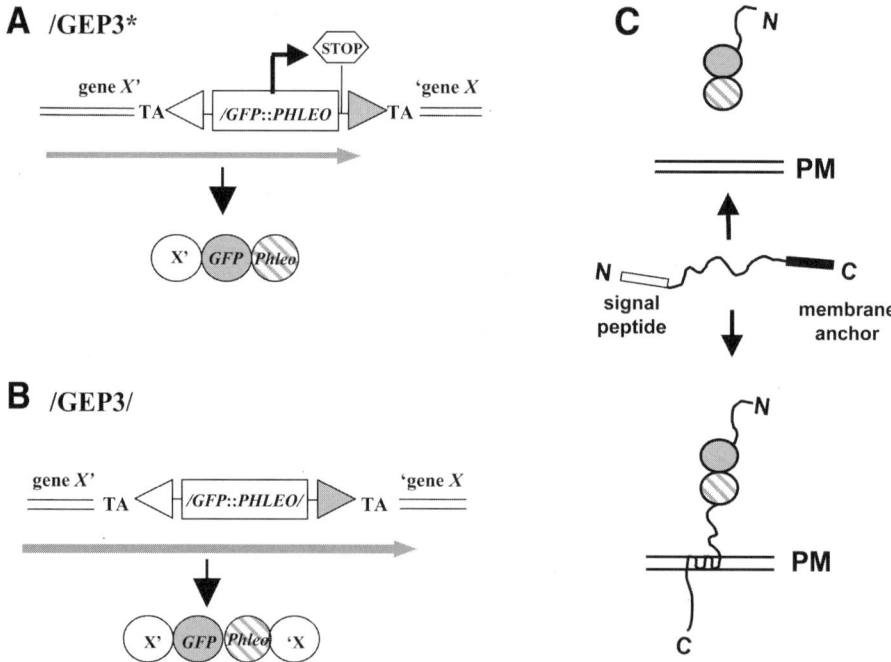

Fig. 4. Diagram of potential translational fusions obtained with transposons /GEP3* and /GEP3/. Both transposons yield translational fusions when inserted in-frame into target ORFs that express a GFP-PHLEO resistance fusion protein domain (**Table 1**). Note that phleomycin resistance can be affected by compartmentalization of the fusion protein; if the PHLEO protein domain is restricted to a compartment such as the glycosome, which is segregated from the nucleus (the site of action of phleomycin), resistance will be abrogated *(32)*. (**A**) The pELHY6Δ-/GEP3* transposon encodes a bifunctional GFP-PHLEO protein with a stop codon after the PHLEO domain. Thus, fusion proteins contain only N-terminal sequence information from the target ORF. (**B**) The pELHY6Δ-/GEP3/ transposon contains an ORF across the entire transposon (both IRs and the bifunctional GFP-PHLEO protein). (**C**) Comparison of the use of "terminator" vs "sandwich" protein fusions. In this example, the properties of fusion proteins generated by /GEP3* and /GEP3/ are compared following in-frame insertion into a typical membrane surface protein, which bears an N-terminal signal peptide and C-terminal membrane anchor. With /GEP3* the C-terminal segment is lost, resulting in secretion of the fusion protein from the cell, whereas with /GEP3/ retention of the C-terminal segment results in the formation of a surface membrane-anchored fusion protein.

/GEP3/ differs from /GEP2/ in which frame can be trapped (because of the sequence of the 5' *mariner* IRs, it is not possible to make a "/GEP1/" transposon for protein trapping; **Fig. 3**). Although all GEP transposons can be used to study translational regulation, the ability of sandwich transposon to retain both

N- and C-terminal sequences, which often contain important protein targeting information (for example, membrane-anchoring domains; **Fig. 4C**), is important for some purposes.

Transposon /NEO*ELSAT (**Table 1**) creates a translational fusion to the *NEO* selectable marker. It should be noted that the ability of the NEO protein to tolerate N-terminal fusions varies considerably amongst different protein targets, in contrast to GFP, PHLEO, and β-galactosidase, which are more permissive. An additional element in this transposon is the selectable marker *SAT* (streptothricin resistance), which contains both a *Leishmania* splice acceptor site and an *E. coli* promoter, allowing for selection for the transposon with *SAT* and protein fusions with *NEO* in *Leishmania*.

Transposon GFP*K (**Table 1**) can be used to generate GFP fusions in a manner similar to the /GEP transposons, as its GFP also lacks an ATG start codon. Additionally, it bears a rare restriction endonuclease (I-*PpoI*), which is helpful in mapping transposon insertion sites in large targets or the genome in vivo.

The transposon carried in pELHY6TK-PG (**Table 1**) contains a PHLEO-GUS translational fusion containing its own start codon. This transposon lacks a constitutive bacterial selectable marker, and contains an *E. coli* oriC replication origin; thus it cannot be used in the in vitro system, and can only be used in vivo. There, the oriC replication origin facilitates the recovery of candidate transpositions back from *Leishmania* into *E. coli* if desired. This transposon donor is carried on pELHY6TK, which is a modified version of pELHY6Δ-0; it additionally bears a conditionally negative selectable marker (herpes virus thymidine kinase) that is active in *Leishmania (33)*.

Transposon /-2x5 (**Table 1**) was designed for transposon-mediated linker-insertional mutagenesis (TIMLI; **ref. *34***). It contains an *E. coli* kanamycin-resistance marker, flanked by a "symmetric" *mariner* element in which the 5'-IR was duplicated. Importantly, this IR contains two sites that occur relatively infrequently in *Leishmania*, *Sex*AI and *Bsr*GI (and most importantly should not occur in targets where they are to be used). In TIMLI mutagenesis, one first generates a large library of transposition events into the target. Then, this transposition pool is collected *en masse*, DNA prepared, digested with *Sex*AI or *Bsr*GI, diluted and self-ligated, and transformed back into *E. coli*. This yields excision of the transposon, leaving behind only an insertion of 12 or 18 nucleotides (encoding 4 or 6 amino acids, respectively). Thus one can generate a library of short-peptide insertions for subsequent functional analysis, such as the mapping of protein domains and activities.

3.2. Expression of **Mos1** *Transposase*

1. Transform BLR (DE3) with plasmid expressing *Mos1* transposase; plate on ampicillin (100 µg/mL).

2. The following day, pick a single colony and resuspend into 5 mL of LB containing 100 μg/mL ampicillin and incubate at 37°C overnight.
3. The next day, use the overnight bacterial culture to inoculate 100 mL of fresh LB medium. Incubate this culture at 37°C to an OD_{600} of 0.6.
4. Induce expression of transposase by adding IPTG to a final concentration of 1 mM. After 5 h of induction, harvest cells by centrifugation at 1303g for 10 min at 4°C.
5. Resuspend bacterial pellet in 0.5 mL of resuspension buffer and flash-freeze by dipping the tube in liquid nitrogen. Store at −80°C (see **Note 1**).

3.3. Purification of Mos1 Transposase

1. Thaw cells at room temperature.
2. Add 1 mg/mL of lysozyme and incubate for 5 min at room temperature. Next, add 1 mL of lysis buffer and incubate at room temperature for 15 min and then for 20 min after adding 60 μg of DNAseI and $MgCl_2$ to 10 mM.
3. All subsequent steps are conducted at 4°C.
4. Typically, most of the expressed transposase is insoluble and forms inclusion bodies. Pellet inclusion bodies at 14,000g in a microcentrifuge and wash three times with 1 mL of 100 mM Tris-HCL, pH 7.6. Resuspend inclusion bodies to a final volume of 4 mL in buffer A. Add 1 mL of 50 % Ni-NTA agarose and gently shake solution for 1 h.
5. Load mixture onto column (Qiagen) and collect flow-through (FT).
6. Wash column twice with 4 mL of wash buffer and collect wash fractions.
7. Elute four times with 0.5 mL of Elution buffer and collect 0.5-mL fractions for analysis later. The eluate contains the transposase protein.
8. Run 20 μL of eluted protein from each of the four tubes on an SDS-PAGE gel. On a 12% SDS-PAGE gel the purified His-tagged *Mos1* transposase runs at approx 50 kDa (see **Note 2**). Also load an aliquot of the noninduced culture, the column FT, and the wash FT.
9. Pool the fractions containing the most transposase. Place this solution of transposase into a dialysis slide, ensuring not to overfill. Conduct dialysis in 1 L of dialysis buffer A for 6–8 h at 4°C. Replace dialysis buffer with 1 L of dialysis buffer B and incubate overnight at 4°C (see **Note 3**).
10. Centrifuge solution at 10,000g for 20 min at 4°C to transfer supernant to a new microcentrifuge tube and discard precipitate.
11. Add glycerol to the transposase in solution to a final concentration of 50%. Store at −20°C.

3.4. In vitro Transposition

1. Set up a standard transposition reaction in a 0.6-mL microcentrifuge tube to a final volume of 20 μL.
   a. 2 μL of 10X transposition buffer.
   b. 2 μL of 100% glycerol (warm glycerol at 65°C to ease pipetting).
   c. 0.5 μL of BSA at 10 mg/mL.

d. 1 μL of donor plasmid (32 fmol).
   e. 6 μL of target plasmid (10 fmol, *see* **Note 4**).
   f. 5 μL of transposase (100 n*M*, *see* **Note 5**).
   g. 3.5 μL of sdH$_2$O.
2. Incubate reaction at 30°C for 1 h to overnight.
3. Add 80 μL of stop buffer and incubate for 30 min at 37°C.
4. Add 100 μL of 25:24:1 phenyl–chloroform–isoamyl alcohol and vortex. Separate phases by centrifugation at 14,000*g* for 15 min. Remove approx 90 μL of the upper layer into a 1-mL microcentrifuge tube. Add 10 μL of 3 *M* sodium acetate and 250 μL of EtOH and incubate at −80°C for at least 1 h.
5. Precipitate DNA by centrifugation at 14,000*g* for 30 min at 4°C. Wash precipitated DNA with 1 mL of 70% EtOH and centrifuge at 14,000*g* for 15 min at 4°C. Resuspend the pellet in 10 μL of 10 m*M* Tris, pH 7.5.
6. Electroporate 2 μL of the purified transposition reaction into DH10B electrocompetent cells (*see* **Note 6**). Add 1 mL of LB medium and incubate for 1 h at 37°C. Plate transfectants onto selective LB plates (*see* **Notes 7** and **8**) and incubate at 37°C overnight.
7. Depending on the purposes, colonies may be picked individually or large pools made for DNA preparations and subsequent analysis *en masse*, for example, following transfection into *Leishmania* or other organisms (*see* **Note 9**; **ref. *31***).

4. Notes

1. Alternatively, one can sonicate cells using a microtip at 40–50% power for 20 bursts. The cells must be kept on ice during sonication. Afterward, one can proceed directly to **Subheading 3.2., step 2**, excluding the addition of lysozyme.
2. The predicted molecular weight of the His-tagged *Mos1* transposase is 43.6 kDa. The altered mobility of the transposase during electrophoresis may result from the presence of the basic histidine residues. Antibodies to the poly-histidine residues specifically recognize the 50-kDa band by Western blot hybridization.
3. The recovery of active, properly folded Tpase is very sensitive to the refolding conditions. Previous work has demonstrated that rapid dilution at low pH or dialysis of detergent-solubilized proteins results in no active enzyme *(17)*. Omission of the column purification step also results in no enzyme activity, likely because of the presence of an unknown inhibitory factor. However, rapid dilution at pH 8.0 or refolding on a column using a linear urea gradient (8–0 *M* urea) has been shown to yield Tpase activity *(17,24)*. Thus, although the refolding protocol described in this article has worked effectively in this laboratory, ultimately the optimal refolding conditions needs to be determined qualitatively by each investigator. We have also been able to purify active transposase from the soluble fraction. Cells are sonicated in lysis buffer and 1 mg/mL of lysozyme and centrifuged at 14,000*g* for 10 min. The supernant is transferred to a 15-mL Falcon tube and the volume is brought up to 4 mL in buffer A. The protocol then continues as described above (**Subheading 3.3., steps 4–11**), eliminating the dialysis step (**Subheading 3.3., steps 9 and 10**). Our experience is that there is a lot of variability in the yield of soluble

active protein and the efficiency of refolding; thus, it is essential that investigators carefully explore these parameters in their own laboratory to find what works best.
4. The quality of DNA is very important to transposition efficiencies. The preparation of donor and recipient DNA containing a high proportion of supercoiled DNA results in high transposition efficiencies. Qiagen midi preparations are generally suitable for this goal. However, when preparing cosmid DNA, one should take special care to avoid shearing the DNA. Transposition efficiency reaches a maximum at around 150 ng of donor plasmid.
5. The concentration of transposase is determined by the micro-BCA method (Pierce). Concentration may also be determined by UV absorbance ($\varepsilon_{280\ nm}$ = 76,989 M^{-1} cm^{-1}); note that these measurements lead to differences in estimation of transposase concentration by a factor of 3 *(17)*. Different batches of purified transposase can have different transposition efficiencies. This difference probably arises from batches of *Mos1* transposase containing different amounts of correctly folded transposase. Therefore, each batch should be tested before conducting large-scale transposition reactions. The transposition efficiency reaches a plateau at a concentration of around 100 nM of transposase and remains at this level at higher concentrations *(17)*.
6. The use of high-efficiency electrocompetent cells works best. Invitrogen GeneHogs® electrocompetent cells can yield 1 × 10^{10} transformants per microgram of pUC vector.
7. Plate 10 μL of a 1/100 dilution of the transformed bacteria onto medium containing the appropriate antibiotic for the resistance marker found on the recipient plasmid (e.g., ampicillin [Ap] at 100 μg/mL, **Fig. 2**). The number of Ap^R colonies multiplied by the dilution factor (in this case by 10,000) is the transformation efficiency. The remaining 990 μL of cells are plated onto medium containing antibiotics for the resistance markers found on the recipient plasmid and the transposon cassette (such as Ap and Phleo in the example shown in **Fig. 2**; 100 μg/mL and 50 μg/mL, respectively). Transposition efficiency is determined by dividing the number of Ap^R,Phleo^R colonies by the transformation efficiency (obtained in previous step). Control transposition efficiencies should range from 10^{-4} to 10^{-3}.
8. One can also estimate the transposition efficiency of a vector such as pELHY6TK-PG, which contains no bacterial-selectable marker. To accomplish this, a negative-selectable marker such as the product of the *ccdB* (control of cell death) gene in placed in the bacterial plasmid, which additionally contains a positive-selectable marker (Kanamycin, Km^R). Examples of this are the pZERO system available from Invitrogen. Transformation of this plasmid into bacteria lacking the gyrase gene results in Km+ colonies. However, cell death occurs when transfected into strains containing gyrase like TOP10. An in vitro transposition reaction is performed with equimolar amounts of this plasmid (Km^R) and a standard plasmid target (Chloramphenicol, Cm^R) in a strain lacking gyrase. The transposition reaction is subsequently transformed into TOP10 bacteria. One half of the transformation is plated onto Lb + chloramphenicol plates (30 μg/mL) and the other half is plated onto Lb + kanamycin plates (50 μg/mL). The number of colonies obtained

when plated on Kanamycin represents transposition events into the *ccdB* gene. The ratio of Km^R/Cm^R represents the transposition efficiency of this transposon.
9. If one is planning transfections into *Leishmania*, it is critical first to determine the sensitivity of your specific strain under the exact circumstances you plan to use. Drug sensitivities vary greatly among different strains and species, in different media, and interactions can occur if two drugs are used simultaneously. First determine the EC_{50} in liquid medium; then, carry out a "mock" transfection followed by inoculation or plating onto media containing drug concentrations ranging upward from three to four times that of the liquid media EC_{50}. Because high drug concentrations inhibit the recovery of *bona fide* transfectants, the goal is to identify the minimal drug concentration that kills untransfected/drug sensitive cells. In *Leishmania*, drug concentrations typically are 15–30 μg/mL for G418/geneticin, 25–40 μg/mL for phleomycin, 15–30 μg/mL for hygromycin B, and 100 μg/mL for nourseothricin, but exceptions are common.

Acknowledgments

The authors thank Lon-Fye Lye for assistance with the transposon protocols and Rosa Reguera for comments on the manuscript. This work is supported by a National Institutes of Health (NIH) Research Fellowship (AI50339) to K. A. Robinson and NIH grant AI 29646 to S. M. Beverley.

References

1. Herwaldt, B. L. (1999) Leishmaniasis. *Lancet* **354,** 1191–1199.
2. Beverley, S. M. (2003) Genetic and genomic approaches to the analysis of *Leishmania* virulence, in *Molecular & Medical Parasitology* (Marr, J. M., Nilsen, T., and Komuniecki, R., eds.), Academic Press, New York, pp. 111–122.
3. Beverley, S. M. (2003) Protozomics: trypanosomatid parasite genetics comes of age. *Nat. Rev. Genet.* **4,** 11–19.
4. Myler, P. J. and Stuart, K. D. (2000) Recent developments from the *Leishmania* genome project. *Curr. Opin. Microbiol.* **3,** 412–416.
5. Swindle, J. and Tait, A. (1996) Trypanosomatid genetics, in *Molecular Biology of Parasitic Protozoa* (Smith, D. F. and Parsons, M., eds.), IRL Press, Oxford, UK, pp. 6–34.
6. Hamer, L., DeZwaan, T. M., Montenegro-Chamorro, M. V., et al. (2001) Recent advances in large-scale transposon mutagenesis. *Curr. Opin. Chem. Biol.* **5,** 67–73.
7. Gueiros-Filho, F. J. and Beverley, S. M. (1996) Selection against the dihydrofolate reductase-thymidylate synthase (*DHFR-TS*) locus as a probe of genetic alterations in *Leishmania major*. *Mol. Cell. Biol.* **16,** 5655–5663.
8. Ryan, K. A., Dasgupta, S., and Beverley, S. M. (1993) Shuttle cosmid vectors for the trypanosomatid parasite *Leishmania*. *Gene* **131,** 145–150.
9. Cecconi, F. and Meyer, B. I. (2000) Gene trap: a way to identify novel genes and unravel their biological function. *FEBS Lett.* **480,** 63–71.

10. Stanford, W. L., Cohn, J. B., and Cordes, S. P. (2001) Gene-trap mutagenesis: past, present and beyond. *Nat. Rev. Genet.* **2,** 756–768.
11. Biery, M. C., Lopata, M., and Craig, N. L. (2000) A minimal system for Tn7 transposition: the transposon-encoded proteins TnsA and TnsB can execute DNA breakage and joining reactions that generate circularized Tn7 species. *J. Mol. Biol.* **297,** 25–37.
12. Goryshin, I. Y. and Reznikoff, W. S. (1998) Tn5 in vitro transposition. *J. Biol. Chem.* **273,** 7367–7374.
13. Merkulov, G. V. and Boeke, J. D. (1998) Libraries of green fluorescent protein fusions generated by transposition in vitro. *Gene* **222,** 213–222.
14. Lavoie, B. D. and Chaconas, G. (1996) Transposition of phage Mu DNA. *Curr. Top. Microbiol. Immunol.* **204,** 83–102.
15. Lampe, D. J., Churchill, M. E., and Robertson, H. M. (1996) A purified *mariner* transposase is sufficient to mediate transposition in vitro. *EMBO J.* **15,** 5470–5479.
16. Fischer, S. E., Wienholds, E., and Plasterk, R. H. (2001) Regulated transposition of a fish transposon in the mouse germ line. *Proc. Natl. Acad. Sci. USA* **98,** 6759–6764.
17. Tosi, L. R. and Beverley, S. M. (2000) *cis* and *trans* factors affecting *Mos1 mariner* evolution and transposition in vitro, and its potential for functional genomics. *Nucleic Acids Res.* **28,** 784–790.
18. Doak, T. G., Doerder, F. P., Jahn, C. L., et al. (1994) A proposed superfamily of transposase genes: transposon-like elements in ciliated protozoa and a common "D35E" motif. *Proc. Natl. Acad. Sci. USA* **91,** 942–946.
19. Robertson, H. M. and Lampe, D. J. (1995) Recent horizontal transfer of a *mariner* transposable element among and between Diptera and Neuroptera. *Mol. Biol. Evol.* **12,** 850–862.
20. Hartl, D. L., Lohe, A. R., and Lozovskaya, E. R. (1997) Modern thoughts on an ancient marinere: function, evolution, regulation. *Annu. Rev. Genet.* **31,** 337–358.
21. Vos, J. C., De Baere, I., and Plasterk, R. H. (1996) Transposase is the only nematode protein required for in vitro transposition of Tc1. *Genes Dev.* **10,** 755–761.
22. Plasterk, R. H. (1996) The Tc1/*mariner* transposon family. *Curr. Topics Microbiol. Immunol.* **204,** 125–143.
23. Goyard, S., Tosi, L. R., Gouzova, J., et al. (2001) New *Mos1 mariner* transposons suitable for the recovery of gene fusions in vivo and in vitro. *Gene* **280,** 97–105.
24. Zhang, L., Dawson, A., and Finnegan, D. J. (2001) DNA-binding activity and subunit interaction of the *mariner* transposase. *Nucleic Acids Res.* **29,** 3566–3575.
25. Coates, C. J., Jasinskiene, N., Miyashiro, L., et al. (1998) *Mariner* transposition and transformation of the yellow fever mosquito, *Aedes aegypti*. *Proc. Natl. Acad. Sci. USA* **95,** 3748–3751.
26. Lohe, A. R. and Hartl, D. L. (1996) Reduced germline mobility of a *mariner* vector containing exogenous DNA: effect of size or site? *Genetics* **143,** 1299–1306.
27. Garza, D., Medhora, M., Koga, A., et al. (1991) Introduction of the transposable element *mariner* into the germline of *Drosophila melanogaster*. *Genetics* **128,** 303–310.

28. Fadool, J. M., Hartl, D. L., and Dowling, J. E. (1998) Transposition of the *mariner* element from *Drosophila mauritiana* in zebrafish. *Proc. Natl. Acad. Sci. USA* **95,** 5182–5186.
29. Gueiros-Filho, F. J. and Beverley, S. M. (1997) *Trans*-kingdom transposition of the *Drosophila* element *mariner* within the protozoan *Leishmania*. *Science* **276,** 1716–1719.
30. Papadopoulou, B., Roy, G., and Ouellette, M. (1994) Autonomous replication of bacterial DNA plasmid oligomers in *Leishmania*. *Mol. Biochem. Parasitol.* **65,** 39–49.
31. Robinson, K. and Beverley, S. M. (2003) Improved methods for transfection and tests of RNA interference (RNAi) activity in *Leishmania*. *Mol. Biochem. Parasitol.* **128,** 217–228.
32. Flaspohler, J. A., Rickoll, W. L., Beverley, S. M., et al. (1997) Functional identification of a *Leishmania* gene related to the peroxin 2 gene reveals common ancestry of glycosomes and peroxisomes. *Mol. Cell. Biol.* **17,** 1093–1101.
33. LeBowitz, J. H., Cruz, A., and Beverley, S. M. (1992) Thymidine kinase as a negative selectable marker in *Leishmania major*. *Mol. Biochem. Parasitol.* **51,** 321–325.
34. Hayes, F. and Hallet, B. (2000) Pentapeptide scanning mutagenesis: encouraging old proteins to execute unusual tricks. *Trends Microbiol.* **8,** 571–577.
35. Shi, H., Wormsley, S., Tschudi, C., et al. (2002) Efficient transposition of preformed synaptic Tn5 complexes in *Trypanosoma brucei*. *Mol. Biochem. Parasitol.* **121,** 141–144.
36. Goryshin, I. Y., Jendrisak, J., Hoffman, L. M., et al. (2000) Insertional transposon mutagenesis by electroporation of released Tn5 transposition complexes. *Nat. Biotechnol.* **18,** 97–100.
37. Beverley, S. M., Akopyants, N. S., Goyard, S., et al. (2002). Putting the *Leishmania* genome to work: functional genomics by transposon trapping and expression profiling. *Philos. Trans. R. Soc. Lond. B: Biol. Sci.* **357,** 47–53.

18

Random Mutagenesis Strategies for Construction of Large and Diverse Clone Libraries of Mutated DNA Fragments

Sudsanguan Chusacultanachai and Yongyuth Yuthavong

Summary

The first important step toward a successful preparation of large and diverse DNA libraries with desired complexity is to select a suitable mutagenesis strategy. This chapter describes three different methods for random mutagenesis, the use of XL1-red cells, error-prone polymerase chain reaction (PCR), and degenerate oligonucleotides-*Pfu* (DOP). These mutagenesis strategies possess different benefits and pitfalls; thus, they are differentially useful for production of DNA libraries with different density and complexity.

The use of XL1-red, an engineered *Escherichia coli* with DNA repair deficiency, is one of the simplest mutagenesis and requires no subcloning step. After plasmid encoding DNA of interest is transformed into the cells, the mutations are simply generated during each round of DNA replication. The mutation frequency of this method is reported to be 1 base change per 2000 nucleotides; however, it can be slightly increased by extending the culture period to allow the accumulation of more mutations. This strategy is suitable for generation of random mutations with low frequency in a large target DNA.

Error-prone PCR is one of the most widely used random mutagenesis. During DNA amplification, misincorporation of nucleotides can be promoted by altering the nucleotide ratio and the concentration of divalent cations in the reaction. We discovered that, by adjusting template concentration, frequency of mutation could be controlled easily and a library with desired mutation rate could be obtained. Additionally, efficiency of subsequent cloning steps to insert the PCR product into plasmid DNA is also a key factor determining size and complexity of the libraries.

DOP mutagenesis is a rapid and effective method for random mutagenesis of small DNA and peptides. This strategy uses two chemically synthesized degenerate oligonucleotides as primers. By controlling the positions and ratios of degenerate nucleotides used during oligonucleotide synthesis, it is possible to control both the position and rate of mutation in degenerated region of the primers. The primers are integrated into newly synthesized plasmid DNA by primer extension reaction using *Pfu* DNA polymerase. After plasmid DNA template encoding wild-type sequence is eliminated from the reaction by *DpnI* digestion, the pool of mutagenized plasmids can then be used directly in screening procedures.

From: *Methods in Molecular Biology, Vol. 270: Parasite Genomics Protocols*
Edited by: S. E. Melville © Humana Press Inc., Totowa, NJ

The different random mutagenesis strategies we describe should have wide applications in the production of libraries of large and diverse DNA libraries and in the generation of mutant proteins for structural and functional studies.

Key Words: Degenerate oligonucleotides; DNA library; DOP mutagenesis; error-prone PCR; in vitro mutagenesis; mutation frequency; *Pfu* DNA polymerase; random mutagenesis; XL1-red.

1. Introduction

Random mutagenesis of DNA is a powerful tool to study gene function and protein structure and is a prerequisite for preparation of randomized DNA libraries, required for many genetic selections. Currently used methods for preparing DNA libraries fall into a few broad categories, each with its own benefits and disadvantages. Therefore, choosing a suitable mutagenesis strategy is a key step for obtaining libraries with the desired complexity for successful selections.

Conventional in vivo random mutagenesis involves the treatment of cells containing a gene of interest, usually in suitable expression plasmids, with chemical mutagens such as 2-aminopurine, 5-bromouracil, ethylmethane sulfonate, or *N*-methyl-*N'*-nitro-*N*-nitrosoguanidine (MNNG) *(1–3)*. These chemicals are DNA-modifying agents, which are highly toxic and require special care in handling. Moreover, these chemical mutagenesis methods produce extremely low frequencies of mutation, and only a certain type of mutation can be generated. For example, MNNG preferentially produces GC-to-AT transitions at only 1 mutation per 10,000 bp *(3)*.

Recently, the use of toxic chemicals for mutagenesis has been superseded by the use of bacterial strains carrying single or multiple mutations in the DNA repair pathways, such as the recently developed XL1-red mutator strain (Stratagene). This mutator *Escherichia coli* carries a combination of a DNA repair deficiency (*mutT*, *mutS*) and a defect in the ε-subunit of DNA polymerase III (*mutD*), which produces mutations during DNA replication. To generate mutations in the gene of interest, the plasmid DNA is simply transformed into these bacterial cells and random mutations are introduced during each cycle of DNA synthesis. The use of such mutator strains is by far one of the simplest mutagenesis strategies, and a very large library of randomly mutated target DNA cloned into a plasmid vector can be obtained. However, this strategy can provide only relatively low mutation frequencies. For example, we found that XL1-red triple mutator strain produced approx 1 mutation per 2000 bp of the target DNA *(4)*, although the mutation frequency was slightly increased by expanding the culture period to longer than the overnight period suggested by the manufacturer.

Among the most popular in vitro mutagenesis strategies is the error-prone polymerase chain reaction (PCR) *(5,6)*, which is based on the creation of mistakes in the incorporation of nucleotides into newly synthesized PCR products.

Construction of Large and Diverse DNA Libraries   321

A commonly used thermostable DNA polymerase in error-prone PCR, *Taq* DNA polymerase, already possesses high intrinsic error rates ($1/10^4$–10^5 bps) because it lacks detectable 3'-5' proofreading exonuclease activity (for the error rate of commercially available thermostable DNA polymerases, see http://micro.nwfsc.noaa.gov/protocols/taq-errors.html). Several conditions are known to enhance the misincorporation by *Taq* DNA polymerase. These include raising the Mg^{2+} concentration to stabilize noncomplementary pairs, the addition of Mn^{2+} to diminish template specificity *(5,7,8)*, use of dNTP analogs *(9)* or alteration of the nucleotide ratio to promote misincorporation *(5,7)*, and use of a high amount of *Taq* DNA polymerase to overcome the mismatch positions *(5)*.

Other factors that influence misincorporation events during PCR include the number of amplification cycles and the initial number of the template copies. Basically, in a reaction containing a lower concentration of template, more rounds of amplification can be achieved and more mutations can be accumulated in the PCR product.

Although error-prone PCR offers the full spectrum of mutations and allows experimental control of the misincorporation level, the PCR products must pass through a series of additional steps. These usually include PCR, restriction digestion, gel purification, ligation, and transformation. In addition to mutations generated during error-prone PCR, these laborious subcloning steps are key steps determining the size and complexity of the mutant library.

In contrast, when the target DNA is quite small (e.g., in the production of short peptide libraries), commonly used methods are based on the integration of oligonucleotide primers carrying random mutations into the gene of interest. During the chemical synthesis cycles used to produce the oligonucleotides, the desired position and rate of mutation can be controlled. The oligonucleotide pools are subsequently incorporated into wild-type genes using a standard subcloning procedure. However, if the integration is accomplished through a more efficient, ligation-free, DNA extension procedure *(10–12)*, as described here, a significantly larger pool size can be obtained.

The strategies described here have been used successfully to construct DNA libraries of the *Plasmodium falciparum* dihydrofolate reductase (pfDHFR) gene for various genetic selection studies. To search for novel drug-resistant mutants of pfDHFR other than naturally existing mutations, random mutagenesis of the entire pfDHFR gene was performed. In nature, single or combinations of four mutations in pfDHFR, including N51I, C59R, S108N, and I164L, are associated with antimalarial antifolate resistance *(13)*. To develop a system comparable to the natural selection process, libraries of pfDHFR mutants containing approx two to three changes in the entire 0.7-kb gene were constructed using error-prone PCR. Libraries were transformed into bacterial host cells and drug-resistant mutants were selected by bacterial complementation. From our selection, both natural and novel drug-resistant mutants of the pfDHFR were obtained. Characterization of those mutants provided useful information for the design and development of antimalarial compounds *(14)*.

This chapter describes three different mutagenesis strategies, XL1-red, error-prone PCR, and degenerate oligonucleotides-*Pfu* (DOP) mutagenesis. Techniques and tips for the control of mutation frequency and producing large and diverse libraries of cloned, mutated target DNA are included.

2. Materials

2.1. Construction of DNA Libraries Using XL1-Red Mutator E. coli

1. XL1-red mutator strain (Stratagene, cat. no. 200129). Keep frozen at −80°C until use.
2. Standard plasmid DNA extraction kit.
3. DNA of interest in a plasmid vector
4. Ice.
5. Falcon tubes: 15 mL.
6. β-Mercaptoethanol.
7. Water bath at 42°C; incubator at 37°C.
8. Luria Bertani (LB) broth medium: 10-g/L tryptone 140, 5-g/L yeast extract, 10-g/L NaCl in distilled H_2O.
9. LB-agar: LB + 15-g/L agar. Add antibiotic appropriate for plasmid vector in use.
10. Sterile toothpicks.
11. Spectrophotometer.
12. Microcentrifuge.

2.2. Construction of a DNA Library by Error-Prone PCR

2.2.1. Error-Prone PCR

1. PCR primers flanking the target DNA and containing appropriate restriction sites for cloning.
2. 10X mutagenesis buffer *(15)*: 500 mM KCl, 100 mM Tris-HCl, pH 8.5, 1% Triton-X100.
3. 10 mM dATP.
4. 10 mM dCTP.
5. 10 mM dTTP.
6. 10 mM dGTP.
7. DNA template encoding the gene of interest (e.g., genomic DNA or a large clone containing entire gene).
8. *Taq* DNA polymerase.
9. A standard thermocycler and mineral oil, if required.
10. Standard agarose, buffer and electrophoresis equipment, and 10-mg/mL ethidium bromide (EtBr) stock.

Construction of Large and Diverse DNA Libraries

11. DNA markers of appropriate size and known concentration.

2.2.2. Desalting and Preparation of the PCR Product for Cloning

1. Nitrocellulose (0.025 μm, White VSWP, 25 mm in diameter (Millipore cat. no. VSWP02500).
2. Sterile Petri dishes, sterile H_2O, and clean forceps.
3. Appropriate restriction endonucleases and buffers.
4. Standard agarose, buffers, 10-mg/mL EtBr, agarose gel electrophoresis equipment.
5. T4 DNA ligase and appropriate 10X buffer.
6. Clean scalpel blade.

2.2.3. Purification of DNA From Agarose Gel

1. QIAquick gel purification kit (Qiagen cat. no. 28706 or equivalent kit).

2.2.4. Ligation

1. Plasmid vector, linearized with appropriate restriction enzymes.
2. Speed vacuum drier, if required.
3. T4 DNA ligase and appropriate 10X buffer.
4. Sterile H_2O.
5. Water bath or incubator at 16 or 25°C.

2.2.5. Electroporation and Harvesting of the DNA Library

1. Electrocompetent *E. coli* cells, e.g., DH5α.
2. An electroporator and electroporation cuvets.
3. High-quality water, e.g., Milli-Q.
4. Standard plasmid DNA extraction kit.
5. LB medium and LB-agar plates with appropriate antibiotic.
6. Standard plasmid DNA extraction kit.

2.3. DOP Mutagenesis

1. Template plasmid DNA carrying gene or DNA segment to be mutated (this should be prepared from an *E. coli* strain that has an intact DNA methyl-transferase system).
2. Top- and bottom-strand degenerate oligonucleotide primers designed for this experiment (OPC-purified) (*see* **Subheading 3.3.1.**).
3. 10 m*M* dNTPs (as above).
4. *Pfu* DNA polymerase or high-fidelity thermostable DNA polymerase and the corresponding 10X buffer.
5. Standard thermocycler and mineral oil, if required.
6. *Dpn*I restriction enzyme.
7. Standard agarose, buffers, 10-mg/mL EtBr, agarose gel electrophoresis equipment.
8. Nitrocellulose (0.025 μm, White VSWP, 25 mm in diameter) (Millipore, cat. no. VSWP02500), sterile Petri dishes, water, and forceps.

9. Electrocompetent *E. coli* cells.
10. LB medium and LB-agar plates with appropriate anitbiotic.
11. Standard plasmid DNA extraction kit.

3. Methods

3.1. Construction of DNA Libraries Using XL1-Red Mutator E. coli: Expanding the Growth Period to Enhance Mutation Frequency

The use of XL1-red mutator cells is one of the simplest and most effective in vivo mutagenesis methods and requires no subcloning step. However, it is important to note that the rapid mutation rate also affects the bacterial chromosome. As a result, the cells grow extremely slowly and a reversal or alteration of mutator phenotype can occur. Hence, during the use of XL1-red cells, growth rate and doubling time must be monitored to ensure that the mutator phenotype is not undergoing change.

The mutation generated from this modified protocol is 1/1400 bp, compared to 1/2600 bp obtained from growing XL1-red cells for a single overnight period only. Therefore, this mutagenesis strategy is suitable for low-frequency mutation of a target gene of 1 kb or larger.

1. Thaw the XL1-red-competent cells on ice.
2. Aliquot 100 µL of the competent cells into a prechilled 15-mL Falcon tube.
3. Add 1.7 µL of β-mercaptoethanol provided with the kit to 100 µL of cells (to get a final concentration of 25 m*M*).
4. Add 50 ng of the template DNA (e.g., a cloned gene of interest in an *E. coli* plasmid vector) to the tube and swirl gently. Incubate the tube for 30 min on ice.
5. Place the tubes containing the cells in a 42°C water bath for 45 s. This subjects the cells to heat shock, which causes transient permeabilization. Immediately return the tube to the ice bucket and incubate the cells on ice for 2 min.
6. Add 0.4 mL LB (or S.O.B. medium; **ref. *16***) and incubate the tube at 37°C for 1 h with vigorous shaking at 225–250 rpm.
7. Plate the entire reaction onto appropriate LB-agar plates containing the appropriate antibiotic, and incubate at 37°C for 24–48 h (*see* **Note 1**). Approximately 100–200 colonies in various sizes should be obtained from one transformation.
8. Use a sterile toothpick to select 200 individual colonies at random from the plate and inoculate each into 5 mL of LB broth containing appropriate antibiotics.
9. Incubate the cells at 37°C with shaking at 225–250 rpm.
10. To ensure the mutator phenotype of the XL1-red cells is retained, do not allow cell density to increase beyond desired point. Cell growth should be monitored every 4–6 h by measuring OD_{600}. The doubling time of the XL1-red cell in LB medium should be 4–6 h at 37°C.
11. When the OD_{600} reaches 0.6, dilute the cells by 1:100 with fresh LB medium (supplemented with appropriate antibiotics).

12. Culture the cells at 37°C with vigorous shaking.
13. Continue monitoring the OD_{600} of the culture and when OD_{600} is approx 0.6, again perform a 1:100 dilution with fresh LB medium containing appropriate antibiotics. Such cycles can be repeated until an abrupt decrease in doubling time from 4–6 h to 1–2 h occurs. This indicates that the mutation efficiency is declining (usually after 6–7 d).
14. Pellet the cells and purify the mutagenized plasmid DNA pool using a plasmid extraction kit or a standard phenol–chloroform extraction and precipitation with ethanol.
15. To assess the mutation frequency, transform the library into DH5α (or any compatible *E. coli* strain) and randomly select 5–10 colonies for DNA sequencing analysis.

3.2. Construction of DNA Libraries by Error-Prone PCR

3.2.1. Error-Prone PCR

We present here a "mild" error-prone PCR protocol. A sufficient yield of PCR products should be obtained (at least 1 µg/ PCR reaction) at approx 1–2% mutation (1 mutation in 100 bp of DNA) (*see* **Note 2**).

1. Prepare 10X mutagenesis buffer *(15)* as described in **Subheading 2.2.1.** Make ali-quots and keep at −20°C until use.
2. Prepare 50-µL reactions as follows: 34 µL sterile H_2O; 1 µL template DNA (e.g., a cloned gene of interest in an *E. coli* plasmid vector) (20 fmol/µL); 1 µL PCR primer 1; 1 µL PCR primer 2; 0.5 µL each of dATP and dGTP; 2.5 µL each of dTTP and dCTP; 5 µL 10X mutagenesis buffer (*see* **Notes 3** and **4** for modified error-prone PCR conditions).
3. Vortex the reaction mixture and spin down briefly.
4. If the thermocycler does not have a heated lid, the reaction should be overlaid with two to three drops of mineral oil.
5. Place each reaction you have prepared in the thermocycler and perform the PCR program as follows: 95°C for 3 min; pause the thermocycler and add 2 µl (5–6 U) of *Taq* polymerase; mix thoroughly by pipetting. Spin the reaction down briefly. Continue 30 cycles of 95°C, 1 min; primer T_m-5°C, 1 min; 72°C, 1 min/kb of target DNA; 72°C for 10 min (*see* **Note 5**).
6. Electrophorese 5 µL of the reaction in an agarose gel and stain with 0.5-µg/mL ethidium bromide to verify that the PCR products are the expected size. The amount of PCR product in 5 µL can be estimated by comparing the fluorescence intensity under UV radiation of the PCR product in the gel to that of the a DNA marker containing a known amount of DNA in each band.

3.2.2. Desalting and Preparation of the PCR Product for Cloning

This simple desalting step is necessary when the salt conditions required in the subsequent enzymatic reactions are not compatible with the error-prone PCR buffer.

1. Pour 15–20 mL sterile H$_2$O into a sterile Petri dish.
2. Use sterile forceps to carefully place VSWP02500 nitrocellulose membrane (white membrane) on the water surface, glossy side up.
3. Pipet the PCR product on the membrane, while the membrane is floating on the water surface. Do not allow the membrane to sink at any time. One membrane can hold up to 100 µL PCR product.
4. Allow to float and desalt for 20–30 min without any disturbance.
5. Use a pipet to gently transfer the desalted PCR product from the membrane to a sterile microfuge tube. Take care not to submerge any part of the membrane during this process. Up to 90% of the PCR product should be recovered from the membrane.
6. Digest the PCR product and plasmid DNA with appropriate restriction endonucleases, using the condition for the best restriction efficiency according to the manufacturer's protocols (*see* **Note 6**).
7. Run the digested product on an agarose gel of appropriate percentage for the expected size of the product.
8. After ethidium bromide staining and visualization under UV light, excise the desired DNA bands from the gel using a sharp blade (precleaned with 70% ethanol).

3.2.3. Purification of DNA From the Agarose Gel

The desired DNA band from the gel slice may be purified using the QIAquick gel extraction kit (see the manufacturer's handbook for detailed principle and protocol). The equivalent protocols or kits for DNA extraction can also be used.

1. Weigh the gel slice and add 3 vol of buffer QG to 1 vol of gel (µL/µg).
2. Incubate at 50–55°C until the gel slice has completely dissolved. Mix occasionally.
3. Add an equal volume of isopropanol to the dissolved gel solution.
4. Apply the sample to the QIAquick column, and centrifuge for 1 min. Discard the flow-through (FT).
5. Wash the column with 0.5 µL Buffer QG and centrifuge for 1 min. Discard the FT.
6. Wash the column with 0.75 µL Buffer PE (in ethanol) and centrifuge for 1 min. Discard the FT.
7. Centrifuge for 2 min to ensure the complete removal of the ethanol in the PE buffer.
8. Place a column into a sterile microfuge tube.
9. Elute the DNA from the column by adding 30 µL of H$_2$O into the center of the column, incubating for 1 min, and centrifuging for 1 min (*see* **Note 7**).
10. Repeat **step 9** to achieve maximum efficiency of the elution. Combine the eluates.
11. The purity and concentration of the eluted DNA fragments can be estimated by agarose gel electrophoresis, as described under **Subheading 3.2.1**.

3.2.3. Ligation Reaction

1. Add an appropriate amount of plasmid DNA, linearized with the appropriate restriction enzyme(s), and mutated PCR products in a single microfuge tube (*see* **Note 8**).
2. Set up a negative control ligation in a separate tube by adding an equal amount of plasmid DNA as in **step 1** but no PCR product.

Construction of Large and Diverse DNA Libraries

3. If the total volume of the DNA exceeds 7 µL, the DNA should be dried under speed vacuum. This drying step also helps to get rid of ethanol carried over from the previous reaction, which greatly inhibits ligation.
4. To each tube, add 2 µL of 5X ligase buffer (*see* **Note 9**); 1 µL of T4 DNA ligase; and up to 10 µL of sterile H$_2$O.
5. Incubate the mixture at 16°C for 16 h. This is particularly important if at least one of the sticky ends is a two-base overhang (for example, AT overhang produced by *Nde*I). However, if both overhangs are four bases or more, the ligation can be performed at 25°C.

3.2.4. Electroporation of Ligation Products and Harvesting the DNA Library

To obtain a large plasmid library of cloned, mutated PCR products from a ligation, a robust transformation method such as electroporation is required. It is recommended that the construction of libraries be performed using highly competent *E. coli* cells (for example, DH5α). The library, in the form of intact plasmid DNA, can be harvested and subsequently transformed into cells required for selection of the desired function. To obtain the best result, media and solutions used during preparation of cells and electroporation should be prepared with Milli-Q water.

1. Thaw the electrocompetent cells (*see* **Note 10**) on ice for about 5–10 min.
2. Add 1–2 µL of ligation mixture into 50-µL electrocompetent cells. Incubate on ice for 2 min.
3. Transfer the cell–DNA mixture into a prechilled cuvet. Make sure that the cell–DNA mixture covers the bottom of the cuvet evenly and is free of large bubbles.
4. Place the cuvet in the chamber in the lock position.
5. If a 1-mm cuvet is used, subject the mixture to electroporation at 25 µf, 200 Ω, and 1.6 kV. For a 2-mm cuvet, the power should be increased to 2.5 kV. A time constant of 3.8–4.5 ms is obtained for high-efficiency electroporation.
6. Immediately take the cuvet out of the chamber and suspend the cell–DNA mixture in 1 mL of LB.
7. Transfer the cells into a sterile tube using a sterile Pasteur pipet.
8. Incubate cells at 37°C for 1 h before plating the reaction on selective medium containing agar and an appropriate antibiotic.
9. Incubate the plates at 37°C overnight. A wide range of colony sizes might be observed from randomly mutagenized pools, especially if the product of the mutated gene is being expressed and is involved in one or more cellular processes.
10. Count or estimate the number of colonies to determine the pool size of the library. Up to several thousand colonies should be obtained from a single transformation, and approx 20,000–30,000 clones can be obtained from one ligation. If a larger pool size is desired, additional ligations and transformations can be performed.

The number of colonies on negative control plates is an indication of the quality of the libraries. The more colonies on the negative control plates, the more wild-type background colonies are present in the libraries (*see* **Note 11**).
11. The colonies of the DNA library can be harvested from plates by adding 3 mL of LB onto the plates and washing off the colonies. Transfer the cells from plate to a new sterile tube.
12. Pellet the cells and purify the plasmid DNA, containing multiple randomly mutated copies of the gene of interest, using the plasmid DNA extraction kit (as described under **Subheading 3.2.3.**) or phenol–chloroform extraction.
13. To examine the type and frequency of mutations, the plasmid in the DNA library should be transformed into new DH5α (or other standard) *E. coli* cells for DNA manipulation.
14. Select 10 colonies at random for DNA sequencing, to assess the type and rate of mutations by comparison of the sequence to that of the wild-type gene.

3.3. DOP Mutagenesis

DOP mutagenesis is modified from Quikchange site-directed mutagenesis (Stratagene). DOP mutagenesis utilizes a pair of oligonucleotide primers, degenerate in the region of interest but flanked by wild-type sequence at both ends *(10)*. The target DNA templates (e.g., a cloned gene of interest in an *E. coli* plasmid vector) are denatured at high temperature, then the primers are allowed to anneal to the target region in the plasmid. In this reaction, thermostable DNA polymerase is added to start primer extension, and the DOP primers carrying random mutations are subsequently incorporated into newly synthesized plasmid DNA. Generally, high-fidelity DNA polymerase, for example, *Pfu* polymerase, is used in the primer extension reaction, to minimize undesired mutations outside the target region. After successful extension cycles, parental plasmid DNA carrying wild-type sequence can be eliminated from the reaction by digestion with *Dpn*I, which recognizes and digests only the methylated DNA that has been produced by replication in *E. coli*. The PCR products are not methylated. The pools of mutant PCR products can be transformed directly into host cells, without additional cloning steps: they are introduced as nicked circular plasmid DNA, and the nicks are fixed inside the cells after transformation.

3.3.1. Primer Design

DOP mutagenesis exploits two complementary primers (top and bottom strands) of similar length (**Fig. 1**). The randomized region (or the mutated region) containing degenerate nucleotides (up to 45 nucleotides) is at the center of the primers, flanked by wild-type sequence at both ends. These wild-type ends are essential for the annealing of the degenerate primers to the target region in the plasmid DNA template. For a successful extension, each of these annealed sections must survive through an extension step (68–72°C). Hence, each end should

Construction of Large and Diverse DNA Libraries

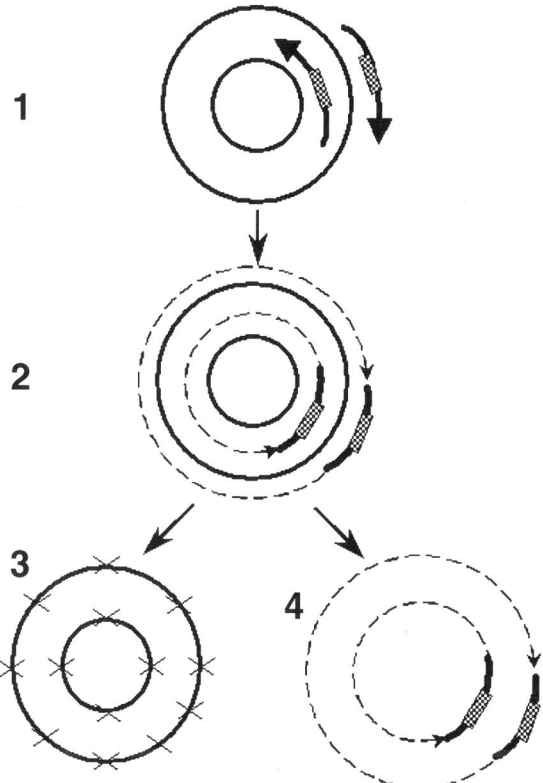

Fig. 1. Primer design and schematic diagram of DOP mutagenesis. DOP primers are a set of complementary primers carrying degenerate oligonucleotides (▦) in the central region. These degenerate nucleotides are surrounded by wild-type sequence (▬) on both ends. After the plasmid DNA templates are denatured at high temperature, DOP primers are annealed to template DNA *(1)*. Random mutagenized sequences in the primers are incorporated into newly synthesized DNA during primer extension reaction *(2)*. The DNA templates carrying wild-type sequence are eliminated from the reaction by *Dpn*I digestion *(3)*. Lastly, DNA libraries can be transformed directly into host cells *(4)*.

have a minimum T_m of 70°C. The length of each entire primer should not exceed 80 nucleotides, which is a limit for the chemical synthesis of high-quality oligonucleotides.

Within the mutated region of the primers, both location and composition of the mutations can easily be controlled during primer synthesis. This unique feature allows the construction of ingenious libraries to maximize the possibility of obtaining mutants with desired phenotype(s). For example, in a selection for zinc finger mutants with altered specificity, four cysteine residues critical for the structural formation of the zinc finger should be intact *(10)*. This can eliminate 160,000 (20^4) misfolded zinc fingers from the libraries prior to the selection process.

3.3.2. DOP Reaction

1. Prepare the reaction as follows:
   a. 39 µL sterile H_2O.
   b. 1 µL Plasmid DNA template (50 ng/µL), prepared from standard *E. coli* cells that have an intact DNA methyltransferase system.
   c. 2 µL dNTP mixture (10 m*M*).
   d. 1 µL Top-strand degenerate primer (125 ng/µL).
   e. 1 µL Bottom-strand degenerate primer (125 ng/µL).
   f. 5 µL 10X *Pfu* buffer.
2. Vortex and spin the reaction briefly.
3. If the thermocycler does not have a hot lid, the reaction should be overlaid with two to three drops of mineral oil.
4. Place the reaction in the thermocycler and start the reaction as follows (*see* **Note 12**):
   a. 95°C for 3 min.
   b. Pause and add 1 µL (3–5 U) of DNA polymerase and mix thoroughly.
   c. Continue 30 cycles of 95°C for 1 min T_m-5°C for 1 min 68–72°C for 2 min/kb (or according to manufacturer's recommendation).
5. Add 1 µL of *Dpn*I restriction enzyme, mix thoroughly and incubate at 37°C for 1–2 h. This should be sufficient to completely digest 50 ng of the wild-type plasmid DNA templates. If a higher amount of DNA template is used, then add more *Dpn*I and incubate the reaction for 10–16 h.
6. Verify that the amplified product is of the expected size by agarose gel electrophoresis. DOP amplification is essentially a linear amplification (i.e., not exponential). One DOP amplification yields approx 300–500 ng of amplified product of similar size to the linear plasmid DNA template.
7. Desalt the reaction using VSWP nitrocellulose membrane (*see* **Subheading 3.2.2.**).
8. Transform 1–2 µL of the product into 50 µL electrocompetent cells (*see* **Subheading 3.2.5.**).
9. Plate 50 µL of the transformed cells on a plate containing appropriate antibiotics. Incubate the plate at 37°C for 16 h.
10. Count the colony on plates to estimate transformation efficiency and pool size. Several thousand colonies can be obtained from a single plate, and up to 100,000 colonies should easily be obtained from a single transformation of 1–2 µL of the DOP product.
11. Harvest the cells and extract plasmid carrying mutagenized gene using a plasmid extraction kit or standard phenol–chloroform extraction methods (*see* **Subheading 3.2.3.**).

4. Notes

1. Generally, tiny colonies are visible after 24–30 h and the plate can be incubated up to 48 h for larger colonies. If the colonies are visible within 16–20 h, it is a sign of reversal phenotype and a very low mutation frequency might be obtained.
2. The error rate produced by the error-prone PCR can be increased by lowering the amount of DNA template. Using 2- and 0.2-fmol DNA template (e.g., a cloned gene of interest in an *E. coli* plasmid vector) should generate approx 5 and 11% mutation frequency, respectively, without any effect on the yield of PCR product.
3. In several error-prone PCR protocols *(5,7,8)*, concentration of dATP was altered by 5–10-fold to promote misincorporation. We found that AT-to-GC transition was preferentially observed when a ratio 1:1:1:0.2 of C:T:G:A was used. However, unbiased mutation was obtained from a reaction containing 1:1:0.2:0.2 of C:T:G:A ratio.
4. The addition of Mn^{2+} is known to enhance the mutation frequency *(5,7)*. However, it also interferes with the PCR reaction, leading to low yield of PCR product. If Mn^{2+} is to be used, the optimum concentration for desired mutation frequency with sufficient PCR yield should be determined experimentally.
5. It is suggested that the mutation rate can also be increased by consecutive rounds of PCR, using product of a previous reaction as a template in a subsequent error-prone PCR *(8)*, and in the instruction manual for Diversify PCR Random Mutagenesis kit, Clontech). In practice, only a small fraction (one-tenth or one-twentieth) of the first round of error-prone PCR is transferred and amplified in the next PCR. This could lead to a clonal bias problem and could hamper the selection of mutants of interest if the transferred fraction contains mutations that are deleterious to the structure or function of the resulting proteins. If this is to be done, the total volume of the second PCR should be increased to ensure that all clones generated from the first error-prone PCR are further amplified. For example, if one-tenth of the first PCR is transferred to the next PCR, 10 reactions of the second PCR should be performed and combined.
6. To maximize the cloning efficiency, plan to use only robust restriction endonucleases known to work well with PCR products (for example, *Eco*RI). Try to avoid end-sensitive enzymes (such as *Hind*III). Those enzymes are less effective at recognizing and digesting the sites near the end of the PCR products, resulting in a low cloning efficiency.
7. According to the manufacturer's instruction, elution buffer (EB) is suggested for the elution of the DNA from the column. However, if the EB pH 8.5 is not compatible with subsequent ligation reaction, sterile H_2O is preferred for the elution of DNA.
8. In a single 10-µL ligation reaction, approx 200 ng of total DNA should be used. The optimum ratio between plasmid DNA and PCR can be varied, depending on the size of the plasmid DNA backbone and PCR insert *(16)*. For 4-kb plasmid and 0.7-kb PCR product, we used 1:5 of plasmid:PCR product.
9. To minimize instability during freezing and thawing cycles, 5X ligase buffer should be divided into aliquots sufficient for several uses and stored at −20°C. If

the buffer has been subjected to many cycles of freezing and thawing, 1 µL of 10 m*M* ATP should be added to the reaction (unless it is a blunt-end ligation).
10. For preparation of electrocompetent cells, *see* http://www.fhcrc.org/labs/hahn/methods/mol_bio_meth/electrocompetent_or consult a standard laboratory manual *(16)*.
11. If a high frequency of background colonies (wild-type) is obtained, the plasmid DNA backbone should be redigested and purified before carrying out another ligation. Additionally, the backbone can be treated with alkaline phosphatase to reduce the background.
12. Do not add an additional extension step, which is usually performed in regular PCR. Once the extension of the entire plasmid is completed, prolonged extension can promote overextension, and duplication of the primer region can occur *(11)*.

References

1. Miller J. H. (1992) Mutagenesis, in *A Short Course in Bacterial Genetics*, Cold Spring Harbor Laboratory, Plainview, NY, pp. 83–128.
2. Beckwith, J. (1991) Strategies for finding mutants. *Meth. Enzymol.* **204,** 3–18.
3. Foster, P. L. (1991) In vivo mutagenesis. *Meth. Enzymol.* **204,** 114–125.
4. Greener, A. and Callahan. M. (1994) XL1-red: highly efficient random mutagenesis strain. *Strategies* **7,** 32–34.
5. Leuang, D. W., Chen, E., and Goeddel, D. V. (1989) A method for random mutagenesis of a defined DNA segment using a modified polymerase chain reaction. *Techniques* **1,** 11–15.
6. Cadwell, R. C. and Joyce, G. F. (1992) Randomization of genes by PCR mutagenesis. *PCR Meth. Appl.* **2,** 28–33.
7. Lin-Goerke, J. L., Robbins, D. J., and Burczak, J. D. (1997) PCR-based mutagenesis using manganese and reduce dNTP. *Biotechniques* **23,** 409–412
8. Shafikhani, S., Siegel, R. A., Ferrari, E., et al. (1997) Generation of large libraries of random mutants in *Bacillus subtilis* by PCR-based plasmid multimerization. *Biotechniques* **23,** 304–306.
9. Spee, J. H., de Vos, W. M., and Kuipers, O. P. (1993) Efficient random mutagenesis method with adjustable mutation frequency by use of PCR and dITP. *Nucleic Acids Res.* **21,** 777–778.
10. Chusacultanachai, S., Glenn, K. A., Rodriguez, A. O., et al. (1999) Analysis of estrogen response element binding by genetically selected steroid receptor DNA binding domain mutants exhibiting altered specificity. *J. Biol. Chem.* **274,** 23,591–23,598.
11. Parikh, A. and Guengerich, F. P. (1998) Random mutagenesis by whole plasmid PCR amplification. *Biotechniques* **24,** 428–431.
12. Miyazaki, K. and Takenouchi M. (2002) Creating random mutagenesis libraries using mega primer of whole plasmid. *Biotechniques* **33,** 1033–1038 (November 2002).
13. Sirawaraporn, W., Sathikul, T., Sirawaraporn, R., et al. (1997) Antifolate-resistant mutants of *Plasmodium falciparum* dihydrofolate reductase. *Proc. Natl. Acad. Sci. USA* **94,** 1124–1129.

14. Chusacultanachai, S., Thiensathit, P., Tarnchompoo, B., et al. (2002). Novel antifolate resistant mutations of *Plasmodium falciparum* dihydrofolate reductase selected in *Escherichia coli*. *Mol. Biochem. Parasitol.* **120**, 61–72.
15. Giver, L., Gershenson, A. Freskgard, P.-O., et al. (1998) Directed evolution of a thermostable esterase. *Proc. Natl. Acad. Sci. USA* **95**, 12,809–12,813.
16. Sambrook, J., Fritsch, E. F., and Maniatis, T. *Molecular Cloning: A Laboratory Manual*, 2nd ed. Cold Spring Harbor Laboratory, Plainview, NY, pp. 1.63–1.67.

19

Separation, Digestion, and Cloning of Intact Parasite Chromosomes Embedded in Agarose

Vanessa Leech, Michael A. Quail, and Sara E. Melville

Summary

The chromosomes of most protozoan parasites cannot be visualized using conventional microscopy because they are too small and do not condense sufficiently at metaphase. Therefore, the development of pulsed field gel electrophoresis allowed the resolution of many parasite karyotypes for the first time. The ability to prepare intact chromosomes in agarose plugs and to isolate individual homologs by electrophoresis has led to many new applications in parasite genomic analysis. This chapter describes the preparation of chromosome plugs from single-celled protozoan parasites, providing numerous tips on how to achieve the highest-quality preparations that will last for years. We also provide detailed protocols for the manipulation of individual excised chromosomes, including restriction mapping and preparation of chromosome shotgun libraries as used in many of the genomic sequencing projects. The protocols provided here underpin several of the advanced methods of genomic analysis and manipulation described in this volume of parasite genomics protocols.

Key Words: Chromosome libraries; chromosome mapping; karyotyping; long-range restriction digestion; PFGE; shotgun cloning.

1. Introduction

In contrast to the chromosomes of metazoan parasites, the chromosomes of many protozoan parasites cannot be visualized using a microscope because they are too small or do not condense sufficiently. The development of pulsed-field gel electrophoresis (PFGE), combined with the release *in situ* of unbroken chromosomes from cells embedded in agarose plugs, allowed definition of the karyotypes of many of these parasites for the first time *(1–6)*. PFGE also revealed the considerable size heterogeneity between apparently homologous chromosomes in different strains and closely related species. Indeed, many analyses have used that size variability to their advantage to determine the number of chromosomes, as we illustrate in **Fig. 1**. One problem of PFGE,

From: *Methods in Molecular Biology, Vol. 270: Parasite Genomics Protocols*
Edited by: S. E. Melville © Humana Press Inc., Totowa, NJ

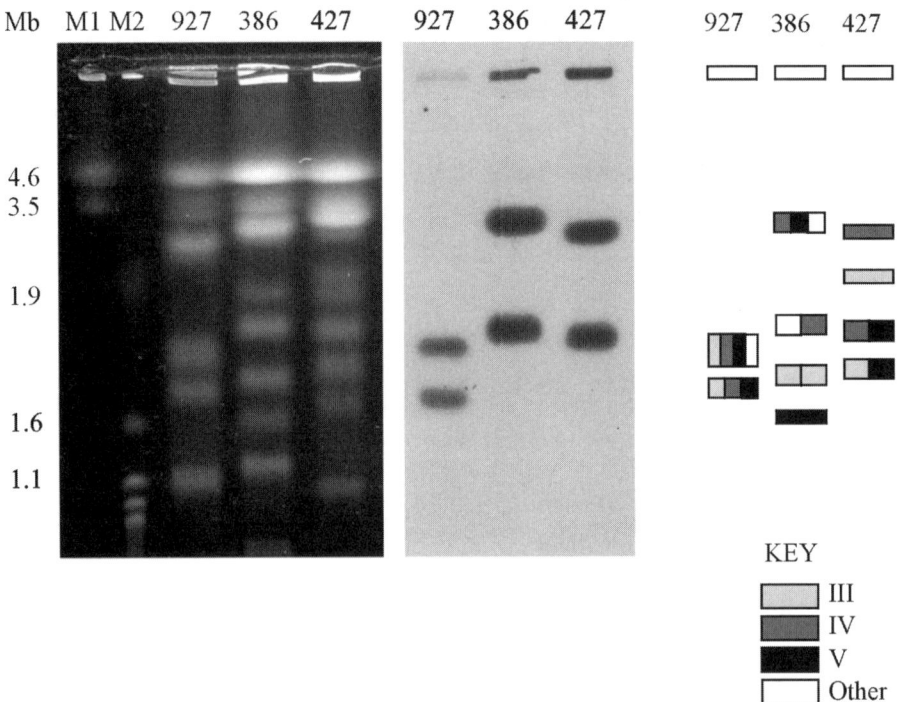

Fig. 1. Resolution of chromosomes III, IV, and V in the *T. brucei* karyotype using PFGE, hybridization of a chromosome-specific marker, and comparison of three karyotypically variable stocks. Note that hybridization of markers specific to these chromosomes to stock 927 only does not determine chromosomal location. CHEF DRIII conditions: 3000–2400 s, 72 h, 1.5 V/cm, 106°, 12°C, followed by 850–450 s, 72 h, 2.4 V/cm, 106°, 12°C. 1% agarose gel in 0.5× TBE. M1, *S. pombe*; M2, *S. cerevisiae*; Mb, Megabases.

compared to visualization of chromosome spreads (*see* Chapter 21), is that nonhomologous chromosomes may be of similar size and, therefore, not resolved in the gel. However, hybridization of single-locus markers to a range of strains with size-variable chromosomes allows the researcher to determine syntenic groups and, hence, the total number of chromosomes.

The separation of individual chromosomes in gels also allows their further manipulation. Early chromosome maps were based on the excision of individual homologs and the use of rare-cutting enzymes to digest the intact chromosome into fragments, again *in situ* in agarose to prevent random breakage (e.g., **refs. 7** and **8**), and these methods remain useful for the analysis of sections of individual chromosomes or, indeed, artificial chromosomes or fragmented chromosomes (*see* Chapter 20).

The ability to elute individual chromosomes from a gel was also exploited by several sequencing teams working with protozoan parasite genomes. The malaria genome is remarkably AT-rich. Large fragments of this genome cloned into vectors that replicate in bacteria (e.g., bacterial artificial chromosomes or cosmids) were found to be highly unstable, enforcing a reliance on yeast artificial chromosomes. To circumvent the many problems inherent in this approach, the sequencing teams developed whole chromosome shotgun (WCS; *see* Chapters 1 and 3), whereby a chromosome is eluted from the gel and shotgun cloned to produce a chromosome-enriched library (usually containing approx 10–13% contamination from other chromosomes). This method has now been used in several sequencing projects (e.g., **refs. 9** and **10**).

Here, we provide basic protocols for the manipulation of chromosome-sized DNA molecules in agarose; their resolution, whole or digested, by PFGE; and their excision and cloning for the production of small-insert chromosomal libraries in plasmid vectors. We base these protocols on our work with African trypanosomes and *Leishmania* species and include additional notes on how to adapt the protocols to work with other protozoan parasites.

2. Materials

2.1. Separation of Intact Chromosomes on an Agarose Gel

2.1.1. Preparation of Intact Chromosomes in Agarose Plugs

1. Parasites: e.g., *Trypanosoma brucei* procyclics in semi-defined medium SDM-79.
2. Trypanosome dilution buffer (TDB), pH 8.0: 118 mM NaCl, 1.2 mM KH$_2$PO$_4$, 30 mM N-tris[hydroxymethyl]methyl-2-aminoethane sulfuric acid (TES), 16 mM Na$_2$HPO$_4$, 5 mM NaHCO$_3$, 5 mM KCl, 10 mM glucose, 1.2 µM MgSO$_4$ or phosphate-buffered saline (*see* **Note 1**).
3. Low-melting-point (LMP) agarose: e.g., SeaPlaque®, GTG®, Flowgen.
4. Sterile 50-mL tubes, e.g., Falcon tubes.
5. Sterile distilled H$_2$O.
6. NDS buffer, pH 9.0 and pH 8.0: 0.5 M EDTA, 10 mM Tris, 1% lauroylsarcosine.
7. 10-mg/mL proteinase K stock.
8. Thoroughly clean plug mold (e.g., Bio-Rad cat. no. 170-3713).
9. Sterile Pasteur pipets.

2.1.2. Pulsed Field Gel Electrophoresis

1. Agarose blocks containing DNA of interest (*see* **Subheading 2.1.1.**).
2. 9-cm-diameter Petri dishes.
3. TE buffer, pH 8.0: 10 mM Tris, 1 mM EDTA.
4. Rotating or rocking platform.
5. TBE or TB(0.1)E or TAE buffer: 0.45 M Tris-borate, pH 8.0, 10 mM EDTA, pH 8.0 (5X TBE) or 0.89 M Tris-borate, pH 8.0, 2 mM EDTA, pH 8.0 (10X TB[0.1]E) or 0.4 M Tris-HCl, 200 mM sodium acetate, 10 mM EDTA, pH 8.2 (10X TAE).

6. CHEF DRII or CHEF DRIII system (Bio-Rad).
7. Low EEO agarose: e.g., Hi-Pure Ultra™, Bio Gene (*see* **Note 2**).
8. Appropriate size markers, e.g., chromosomes of *Saccharomyces cerevisiae*, *Hansenula wingei*, *Schizosaccharomyces pombe* in agarose plugs.
9. 1-L conical flask.
10. 50-mL conical flask.
11. Sterile distilled H_2O.
12. Sterile Pasteur pipets.
13. 10 mg/mL ethidium bromide stock solution.
14. UV transilluminator.
15. Camera/gel imaging system.

2.2. Excision and Restriction Digestion of a Single Chromosome

2.2.1. Excision of a Chromosome From a Pulsed Field Gel

1. Glass plate.
2. Ruler.
3. Ethanol.
4. Black paper/dark surface.
5. Sterile scalpel blades.
6. Containers and electrophoresis buffer from gel tank.
7. 10 mg/mL ethidium bromide stock solution.
8. UV transilluminator.
9. Camera/gel imaging system.
10. Foil/sterile needles.
11. 15-mL Falcon tubes.
12. 0.5 *M* EDTA, pH 8.0.

2.2.2. Digestion of a Chromosome in Agarose and Electrophoresis of the Fragments

1. Excised chromosomal DNA in molecular biology-grade agarose.
2. Sterile scalpel blade.
3. Petri dishes.
4. TE buffer, pH 8.0: 10 m*M* Tris-HCl, 1 m*M* EDTA.
5. Enzymes needed for digestion plus appropriate buffers.
6. Eppendorf tubes.
7. Water bath.
8. Appropriate size markers, e.g., 0.1–200-kb Pulse Marker (Sigma).

2.2.3. Southern Blotting and Hybridization of Single-Locus Markers

1. UV transilluminator.
2. Denaturing solution: 0.5 *M* NaOH, 1 *M* NaCl.
3. Neutralization solution: 0.5 *M* Tris-HCl, 1.5 *M* NaCl.
4. Nylon membrane (e.g., Hybond N+, Amersham Biosciences).

Manipulation of Chromosomes in Agarose 339

5. 3MM paper (Whatman) and paper towels.
6. 10X SSC, pH 7.0: 1.5 M NaCl, 0.15 M Na-citrate.
7. Large tray and glass plates.
8. Labeling and detection reagents (e.g., AlkPhos Direct, Amersham Biosciences).
9. Hybridization oven and bottles.
10. X-ray film (e.g., Kodak X-OMAT AR) and cassette.

2.3. Preparation of a Small-Insert Plasmid Library of Chromosomal DNA

2.3.1. Isolation of Chromosome DNA From LMP Agarose Gel Slices

1. β-Agarase and buffer (NEB).
2. Ultrapure phenol (Invitrogen).
3. Butanol (Sigma).
4. 1 M NaCl.
5. 1 mg/mL glycogen (Sigma) in deionized H_2O.
6. Ethanol (absolute and 70%).
7. TE buffer: 10 mM Tris-HCl, pH 8.0, 1 mM EDTA.

2.3.2. Blunt-End Fragment Preparation Using Sonication and Mung Bean Nuclease

1. 10X MBB buffer: 50% (v/v) glycerol, 10 mM $ZnCl_2$, 0.5 M NaCl, 0.3 M sodium acetate, pH 5.0.
2. Sonicator fitted with a cup horn probe (e.g., model CL4, Misonix, Inc., Farmingdale, NY).
3. DNA molecular-weight markers, e.g., a mixture of lambda digested with *Hind*III and pBR322 digested with *Bst*NI (NEB).
4. Agarose gel electrophoresis equipment and ethidium bromide (EtBr).
5. Mung bean nuclease (Amersham, 156 U/μL).
6. 1 M NaCl.
7. Pellet paint (Novagen).
8. Absolute ethanol.
9. 1X TAE buffer: 40 mM Tris, 20 mM sodium acetate, 1 mM EDTA, pH 8.2
10. Long-wavelength UV transilluminator.
11. TE buffer: 10 mM Tris-HCl, pH 8.0, 1 mM EDTA.
12. Dye solution: 10% (w/v) ficoll 400 (Sigma), 0.1% w/v bromophenol blue in 1X TBE.
13. AgarAce (Promega).

2.3.3. Construction of Shotgun Libraries

1. T4 DNA ligase (Roche, 5 U/μL) and buffer.
2. pUC18 *Sma*I-CIP (40 ng/μL), Qbiogene.
3. 14 mg/mL proteinase K stock (Roche).
4. Electroporation apparatus.
5. Electrocompetent cells, e.g., Electromax DH10B (Invitrogen).

3. Methods

3.1. Separation of Intact Chromosomes on an Agarose Gel

When preparing agarose plugs containing chromosomes from *Trypanosoma* or *Leishmania* species, we begin with a parasite density of approx 1×10^8/mL of culture medium (*see* **Note 3**).

3.1.1. Preparation of Intact Chromosomes in Agarose Plugs

1. Look at organisms under the microscope and determine health and number. Work out the concentration of the cell culture by your usual method (we use a hemocytometer) and the resuspension volume required to achieve the correct concentration (*see* **step 5** below). Prepare an appropriate amount of 1.4% LMP agarose solution and hold at 37°C until required (*see* **Note 4**).
2. Spin down culture at 3000–4000g, at 4–8°C, for 5–10 min only (you do not want to lose too many, but you do not want them packed down such that you damage them when resuspending). Work fast and try to keep them all alive.
3. Resuspend in cool (not too cold) TDB with sterile Pasteur pipet (tip cut off to widen aperture). Resuspend fully but without damaging the cells. This is a wash, so they must resuspend (*see* **Note 1**).
4. Spin down again. Wash again. Do not leave them sitting around.
5. Resuspend cells in TDB at twice the concentration you require in the plugs. (We aim for 5×10^8 cells/mL final concentration, i.e., 5×10^7 per 100 µL plug; therefore, we resuspend to 1×10^9 cells/mL) (*see* **Note 5**).
6. Warm cells briefly in 37°C for 1–2 min, swirling constantly. Add equal amount of 1.4% LMP agarose solution to give 0.7% agarose plugs. Mix quickly, but thoroughly (*see* **Note 5**).
7. Fill plug mold quickly (the pipet tip should not have too narrow a bore, we use a sterile 3-mL plastic Pasteur pipet). Leave to solidify for a few minutes, then place carefully in a dish large enough to accommodate the plug mold but not so large as to waste buffer. Add enough NDS, pH 9.0, to the dish to cover the plug mold and place in the fridge for 10–15 min. This allows the plugs to set completely without sticking to the sides of the plug mold or becoming distorted.
8. Push plugs out of mold with minimal distortion and drop them into NDS, pH 9.0, with 1 mg/mL proteinase K (minimum volume, as the enzyme is expensive). Leave at 50°C for 24 h (*see* **Note 6**).
9. Cool plugs at 4°C (to make them easier to handle and less likely to distort, which leads to breakage of large DNA molecules) and change buffer to NDS, pH 8.0, 1-mg/mL proteinase K. Incubate again for 24 h at 50°C to allow the enzyme to destroy all cellular proteins released into the agarose.
10. Cool plugs. Transfer to fresh NDS, pH 8.0 (sufficient to cover), and store at 4°C. They will keep for many years if they are well prepared and not damaged.
11. Dialyze plugs thoroughly against TE or TBE (electrophoresis buffer) before setting up gel to remove all detergent and reduce EDTA concentration (either over-

Manipulation of Chromosomes in Agarose 341

night, or you may shorten the required time by using a very large volume, movement, and changing buffer several times).

3.1.2. Pulsed Field Gel Electrophoresis

1. Place required number of plugs into 9 cm-diameter Petri dishes and add enough TE buffer, pH 8.0, to cover the plugs. Place on a slowly rotating platform and leave for 30 min–1 h. Replace with fresh TE buffer at least twice more to ensure thorough dialysis (*see* **Note 7**).
2. Prepare 2 L of the required buffer by diluting the stock solution with sterile distilled H_2O (*see* **Note 8**).
3. Place the buffer into the gel tank, switch on the pump and cooling module, and set the cooler to the required temperature. Allow the buffer to equilibrate to the required temperature.
4. Remove 120 mL buffer from gel tank (*see* **Note 9**). Weigh required amount of agarose (*see* **Table 1**) and place in a 1-L conical flask. Add the buffer, weigh the flask plus contents and make a note. Heat carefully to boiling point and check that all the agarose has dissolved. Weigh flask plus contents, and add sterile H_2O to make back up to original weight. Allow to cool to approx 55°C.
5. Cover the teeth of the comb with autoclave tape and place the comb horizontally on a level surface (*see* **Note 10**). Place the plugs in the required order onto the comb. Using a sterile Pasteur pipet, place a few drops of agarose solution onto and between each plug and allow to set for 2–3 min.
6. Position the comb into the mold carefully and pour in the agarose solution, retaining approx 5 mL for use later. Allow the gel to set for 30 min to 1 h.
7. Carefully remove the comb. Melt the retained agarose and use to fill the hole left by the comb. Allow to set for 2–3 min.
8. Transfer the gel to the gel tank, positioning carefully within the support. Ensure that the pump and cooler are switched on and appropriate conditions set, then start run (*see* **Table 1** for suggested conditions). Check that bubbles are visible from the electrodes and make a note of the starting current.
9. When the run is completed, remove the gel from the tank and place into a box containing enough running buffer to cover the gel. Add EtBr to 0.5 µg/mL and allow the gel to stain for approx 1 h (sometimes overnight staining in a lower concentration of EtBr gives a superior gel picture).
10. Place the gel onto a UV transilluminator to visualize the DNA and photograph. It may be necessary to stain the gel for a longer period of time or to destain the gel depending on the concentration of DNA in the gel. Avoid exposing the DNA to UV for long periods of time, as this will break the DNA and allow it to diffuse out of the gel.

3.2. Long-Range Restriction Mapping of Individual Chromosomes

It is absolutely vital that the chromosomal DNA is never allowed to come into contact with ethidium bromide or is exposed to UV light.

Table 1
Examples of PFGE Separation Parameters for *T. brucei* and *L. major* Chromosomes

Size range	Switch time (s)	V/cm	Angle	Run time (h)	Temperature (°C)	Agarose (%)	Buffer
0–200 Kb	3–8	6.0	120	16	13	1.0	1X TB(0.1)E
0.1 Mb–1 Mb	175	2.5	120	120	12	1.0	1X TB(0.1)E
0.5 Mb–2 Mb	400–200	2.5	120	144	12	1.0	1X TB(0.1)E
1 Mb–1.5 Mb	450	2.5	120	96	14	1.0	1X TB(0.1)E
1 Mb–2 Mb	850–450	2.5	120	144	14	1.0	1X TB(0.1)E
1 Mb–3 Mb	1400–700	2.5	120	144	14	1.2	1X TB(0.1)E
3 Mb–6 Mb	3600	1.2	120	168	14	0.8	0.5X TBE
1 Mb–3 Mb	1100–600	2.4	106	72	14	1.0	0.5X TBE
3 Mb–6 Mb	3600–3000	1.5	106	96	14	0.8	1X TB(0.1)E
1 Mb–6 Mb	(a) 3000–2400	1.5	106	72	14	1.0	0.5X TBE
	(b) 850–450	2.4	106	72			

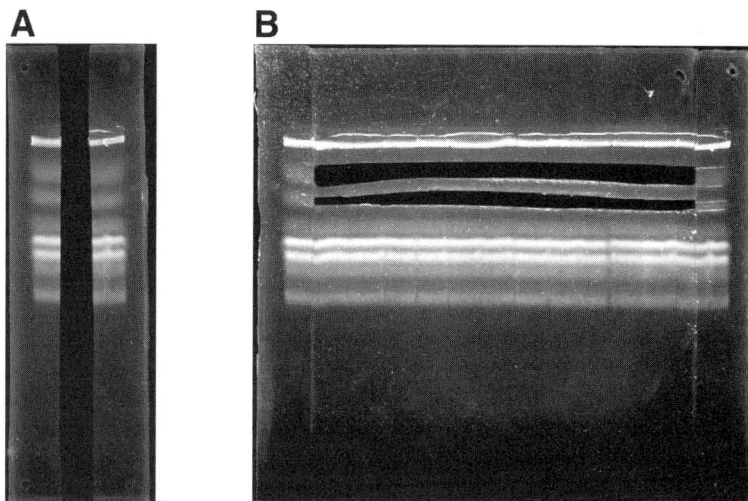

Fig. 2. Excision and digestion of an individual *T. brucei* chromosome. (**A**) The chromosome should be excised from the gel without exposure to EtBr or UV light. This is achieved by staining only the two outside lanes and using these as a guide to excise the remainder of the chromosome. (**B**) The remainder of the gel should be stained with EtBr after excision, to ascertain whether the chromosome has been successfully and accurately excised. In this example, two chromosomal bands have been excised (X and X1 of *T. brucei* stock 92714).

3.2.1. Excision of a Chromosome From a PFG

1. Clean a glass plate and a ruler with ethanol and place on a dark surface, e.g., a piece of black paper. This will make it easier to see where to cut.
2. Using the ruler as a guide, cut a strip off the right and left ends of the gel ensuring that a few millimeters, or a lane, of your sample and/or markers are included in the cut area (*see* **Note 11**).
3. Place these gel strips into a container and add enough buffer from the gel tank to cover the gel plus 0.5 µg/mL ethidium bromide. Allow to stain for approx 1 h. Place the remaining gel into a container with electrophoresis buffer only; *Do not add EtBr.*
4. Place the gel strips onto a UV transilluminator to visualize the DNA and photograph (*see* **Fig. 2A**). Mark the position of the band(s)/region of interest using small cuts, above and below the band, with a sterile scalpel blade. If desired, small pieces of foil or sterile needles can be placed in the cuts to make them more visible and aid alignment.
5. Reassemble the unstained gel plus gel strips on the glass plate and locate the region of interest.
6. Carefully excise the band using a sterile scalpel blade and using the ruler and cut marks as a guide. Place the excised chromosome into a 15-mL Falcon tube and store in 0.5 *M* EDTA, pH 8.0, at 4°C if not used immediately (but *see* **step 8**).

7. Stain the remaining gel pieces in running buffer containing 0.5 µg/mL ethidium bromide, then check that the correct region has been removed by reassembling the remaining gel pieces on the UV transilluminator (*see* **Fig. 2B**).
8. If you intend to extract the chromosomal DNA for preparation of a shotgun clone library, then reset the chromosome slice into a gel prepared with LMP agarose as described under **Subheading 3.1.2.** (This can be difficult to handle and costly, in which case it is possible to produce a composite gel by preparing a standard gel exactly as described above, then excising an area below the chromosomal band and refilling with LMP agarose.) Electrophorese using the same parameters as previously, for approx half the time, then excise the chromosome again as described in **steps 1–7**.

3.2.2. Digestion of a Chromosome in Agarose and Electrophoresis of the Fragments

This method may be used to digest individual chromosomes, or large-insert clones such yeast artificial chromosomes (YACs) excised from a PFG. Do not assume that restriction enzyme sites that are rare in other genomes, such as mammals, will be rare in the parasite genome also. For example, *Not*I cuts the *T. brucei* genome much more frequently because of the higher GC content of this genome. After electrophoresis of the restricted fragments, you will not necessarily see the fragments following EtBr staining (they are just visible in **Fig. 3**). It is not always best to increase the concentration of the DNA in the plugs, as this leads to increased contamination of the excised chromosome, but you should be able to detect the fragment of interest on hybridization of a suitable marker.

1. Cut the gel fragments so that they are approx 6 mm long, 3 mm wide, and 6 mm deep (*see* **Note 12**) and dialyze them in TE buffer, pH 8.0, for several hours as described in **Subheading 3.1.2., step 1**.
2. Place the fragments in 1.5-mL Eppendorf tubes and add sufficient 1X restriction buffer to cover fragment. Incubate for 1 h at room temperature.
3. Remove buffer and replace with 1X restriction buffer containing 50 U enzyme. Place at required temperature for at least 24 h.
4. Dialyze fragments in TE buffer, pH 8.0, for 2 h before setting into an appropriate agarose gel (*see* **Subheading 3.1.2.**).
5. Electrophorese using conditions suitable for expected size (*see* **Table 1**).

3.2.3. Southern Blotting and Hybridization of Single-Locus Markers

Because of the large size of the chromosomal DNA in the PFG, it is necessary to nick the DNA before transferring onto a nylon membrane. This can be achieved by depurination or by UV irradiation. We routinely nick the DNA with UV light at 0.12 J/cm^2, then transfer the DNA onto nylon membrane (Hybond N+, Amersham) using capillary transfer as described by Sambrook *(11)*. We extend the time for transfer to 48 h, with a change of paper towels after 24 h for

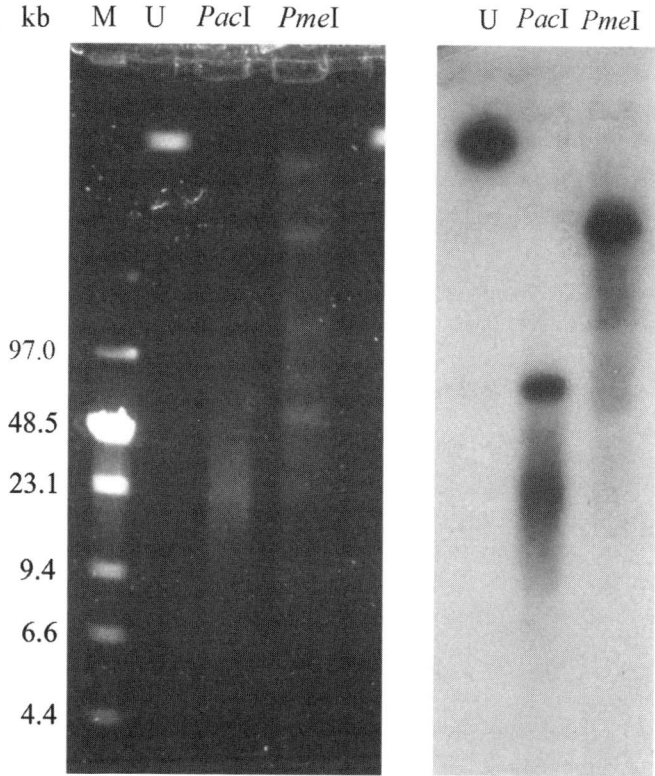

Fig. 3. After digestion, the chromosomal fragments are separated using PFGE parameters appropriate to their size. It is often the case that there is barely sufficient DNA to visualize on the gel after staining with EtBr (*left panel*); however, hybridization of a labeled single-locus DNA fragment (e.g., GPI, as shown here) reveals the position of the fragment in the gel. Its size may then be estimated by comparison to the position of markers of known size (M). Note that nonspecific hybridization is often seen as a result of degradation of DNA fragments. kb, kilobases; U, undigested DNA.

optimal results. After allowing the membrane to dry, cross-link the DNA to the membrane by exposing it to 0.12 J/cm^2 UV light.

Hybridization of DNA probes may be carried out using standard radioactive labeling methods or nonradioactive methods. We routinely use the AlkPhos direct labeling and detection system (Amersham Biosciences), following the manufacturer's instructions, although we increase the temperature of hybridization and washing steps to 65°C instead of 55°C.

You may be able to determine the order of the fragments derived from the chromosome by hybridization of entire BAC or cosmid clones to Southern blots, if such clones do not contain repeated sequence found in multiple fragments

(8). Also, to visualize all fragments derived from restriction of a chromosome, extract a small amount of chromosomal DNA from a gel slice and label and hybridize to the Southern blot *(8)*.

3.3. Preparation of a Small-Insert Plasmid Library of Chromosomal DNA

This method is used by the Sanger Institute to produce chromosome shotgun libraries (**refs. 9** and **10**; http://www.sanger.ac.uk/Teams/Team53/psub/methods.shtml), but it is equally applicable to a large restriction fragment (prepared as in **Subheading 3.2.**) or a large-insert clone, such as a bacterial artificial chromosome (BAC), excised from a LMP agarose gel (*see* **Subheading 3.2.1., step 8**). The yield of DNA is often quite low, 100–500 ng, and so library construction can be difficult.

3.3.1. Isolation of Chromosome DNA From LMP Agarose Gel Slices

DNA of sufficient quality for library preparation can be isolated from agarose gel slices by digesting away the agarose matrix with β-agarase and then extracting with phenol to remove contaminants.

1. Excise the chromosome in LMP agarose, preferably from a gel prepared with TAE buffer (there may be slight variation in migration between gels prepared with TAE and TBE buffers), as described under **Subheading 3.2.1.** (*see* **Note 13**).
2. Soak the gel slice in 1X β-agarase buffer for 30 min on ice. Drain and repeat this wash.
3. Load the agarose plug into the back of a 5-mL syringe barrel fitted with a 23-ga needle. Force into the barrel of another 5-mL syringe. Repeat four times, eventually forcing the agarose into a preweighed sterile tube. Reweigh and calculate volume of agarose by subtraction (*see* **Note 14**).
4. Add one-ninth vol of 10X β-agarase buffer and 8 U of β-agarase per milliliter. Mix and incubate at 37°C for 2 h.
5. Add an equal volume of TE-equilibrated phenol and vortex for 1 min. Incubate on ice for 5 min prior to centrifugation at 3000g for 10 min to separate phases.
6. Take the upper aqueous phase and reduce down to approx 300 µL by successively extracting with an equal volume of butanol (*see* **Note 15**).
7. Remove the aqueous layer to a 1.5-mL microcentrifuge tube and ethanol precipitate by adding 0.1 vol of 1 M NaCl, 10 µL of 1-mg/mL glycogen, and 2.5 vol of ice-cold ethanol. Leave at −20°C overnight, or at −70°C for 30 min.
8. Pellet DNA by centrifugation, wash with 70% ethanol and dry. Resuspend pellet in 21 µL TE, then ascertain yield by running 1 µL of prep, alongside markers of known DNA concentration (we use 50 ng of lambda *Hind*III digest), through a standard agarose minigel (e.g., 0.8% agarose in TBE).

3.3.2. Blunt-End Fragment Preparation
Using Sonication and Mung Bean Nuclease

1. Pool sufficient material gained from several extractions (*see* **Subheading 3.3.1.**) until you have around 10 µg of extracted DNA, and pipet into a sterile 1.5-mL

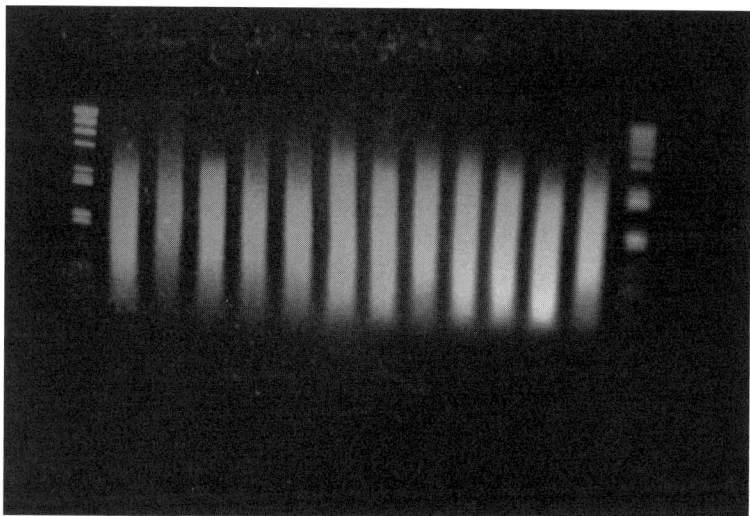

Fig. 4. After sonication of the eluted chromosomal DNA, the size of the resulting fragments is assessed on an agarose gel. You should see a smear of DNA fragments around 500 bp to 4 kb, but very little remaining HMW DNA. This gel picture shows ideally sonicated DNA samples, flanked by DNA molecular-weight markers (a mixture of λ digested with *Hind*III and pBR322 digested with *Bst*NI).

     microcentrifuge tube (*see* **Note 16**). Add 6 µL of 10X MBB buffer and water to give a final volume of 60 µL. Place on ice.
2. Microfuge briefly to settle the tube contents and fragment the DNA by sonication, keeping the sample cold throughout to prevent denaturation of DNA. We use a sonicator fitted with a cup-horn probe on top of which the sample is suspended in a 4°C water bath. Cooling is maintained by siting of the whole apparatus in a cold room. Sonicate at 15% of full power for 2 × 10 s, at a distance of 1 mm from the probe surface. Incubate on ice for 1 min and briefly microfuge to settle contents between sonications (*see* **Note 17**).
3. Run a 1-µL aliquot of each sample alongside a suitable marker on a 0.8% agarose minigel. There are three possible outcomes:
    a. Complete sonication (*see* **Fig. 4**) (no sign of high-molecular-weight [HMW] DNA, smear between ≥4 kb and 500 bp).
    b. Near-complete sonication (smear and faint HMW DNA).
    c. Unsonicated (faint smear and substantial HMW DNA).
   In the case of b., resonicate for 5 s, and check again on a minigel if desired. In the case of c., resonicate again for 2 × 10 s and check again on a minigel before proceeding.

4. Add 0.3 µL of mung bean nuclease to each tube, mix gently, briefly microfuge to settle contents, and incubate at 30°C for 10 min.
5. Add 141 µL of water, 20 µL of 1 M NaCl, 1 µL of pellet paint, and 550 µL cold ethanol. Leave at –20°C overnight or at –70°C for 30 min. Pellet DNA by centrifugation at 13,000g, 4°C for 30 min, wash with 1 mL ice-cold ethanol, and dry.
6. Resuspend DNA pellet in 6.25 µL TE, 0.75 µL 10X TAE, and 2 µL dye solution. Run out through a 0.8% LMP agarose, 1X TAE gel (with 2 µL ethidium bromide per 50 mL gel volume). We run a 15 × 10 cm gel at 50 V, 70 mA, for 2 h.
7. On a long-wave UV transilluminator, cut out 0.6–1 kb, 1–1.4 kb, 1.4–2 kb, and 2–4 kb agarose blocks and place in fresh 1.5-mL microcentrifuge tubes. Melt at 65°C for 5 min or pass through a 26-ga needle to fragment. Equilibrate at 42°C, add 1 U AgarAce (Promega) per 200 µL vol and incubate at 42°C for 20 min (*see* **Note 18**).
8. Extract with TE-equilibrated phenol and ethanol precipitate, adding one-tenth vol 1 M NaCl and 1 µL of pellet paint as described previously.
9. Pellet DNA fragments to be cloned by centrifugation at 13,000g, 4°C for 30 min, wash with 1 mL ice-cold ethanol, and dry. Resuspend in 3 µL of T0.1E if you started the procedure with <1 µg, 5 µL of T0.1E if you started with 1–5 µg, and 10 µL if you started with >5 µg of extracted DNA.

3.3.3. Construction of Shotgun Libraries

1. Set up ligations as follows, in 0.5-mL microcentrifuge tubes: 3 µL insert DNA, 0.3 µL pUC18 *Sma*I-CIP, 0.4 µL 10X ligase buffer, 0.3 µL ligase.
2. Mix by pipetting up and down slowly. Microfuge very briefly to settle contents, then incubate overnight at 12–14°C.
3. Next morning dilute to 50 µL with sterile distilled H$_2$O and add 1 µL of 14-mg/mL proteinase K solution and incubate at 50°C for 1 h to denature the ligase. Thereafter, keep ligations cold (preferably frozen), because there is no buffering in the solution.
4. Test libraries for quality and titer, by electroporation of a 0.2-µL aliquot into high-efficiency electrocompetent cells.

4. Notes

1. Some people use PBS instead of our TDB. We find that TDB works best, probably because it contains ingredients that keep the trypanosomes alive and in good condition until the last moment.
2. Many companies produce PFG-specific agarose, optimized for use with large DNA fragments. These are generally more expensive, and we do not use them routinely.
3. If using parasites that grow intracellularly, e.g., *Toxoplasma*, or parasites grown in laboratory rodents, then it is necessary to completely separate the intact parasites from the host cells before preparing the plugs. The parasites must remain intact until the moment they are set into agarose. When deciding how many parasites per plug you require, take into account variation in genome size—aim for approximately the amount of DNA suggested here, as increasing the amount too much leads to problems in removal of proteins, etc.

4. We have found that estimates of cell numbers by this method vary among laboratory workers. It is best to make plugs at several concentrations in the first instance, e.g., half and/or twice the given concentration also. If there are too few cells in the plugs, then this can make analysis difficult, as you cannot compensate for this subsequently. If there are too many, then you can compensate when setting up gels by cutting up the plugs. However, if cell density is too great, the proteinase K may not remove all proteins, reducing quality and preventing DNA migration into the gel.
5. The cells must be alive until they die slowly *in situ* after addition of agarose. If they die earlier than this, then the cell wall can degrade and the DNA is broken and/or degraded enzymatically. Ideally, the cell wall will be lysed on addition of the plugs to the detergent in NDS. At this point, at least while learning, you could look at them under the microscope and see if they are still wriggling. But don't take long! Just smear a drop onto a slide or a Petri dish and look under any microscope available.
6. The detergent in NDS buffer lyses the cell walls and releases the chromosomes *in situ*. *Trypanosoma* and *Leishmania* parasites have cell walls that are easily and gently disrupted in this way. If the parasites of interest do not lyse so easily, nevertheless a method must be found to lyse them *in situ* in the agarose. If they lyse prior to embedding in agarose—while still in liquid medium—the chromosomes will break
7. Dialysis removes the EDTA and detergent from the storage buffer. The length of time to allow for dialysis is dependent on the number of plugs in each Petri dish; the greater the time and numbers of changes of TE buffer.
8. Use autoclaved deioinized H_2O to avoid introducing fungi and other contaminants.
9. Volumes given are for gels cast using the 14 × 12.7 cm casting stand supplied with the CHEF-DRII or CHEF-DRIII system and will result in a 6-mm-thick gel.
10. Loading plugs onto a comb covered in autoclave tape rather than pushing them into the well prevents distortion of the plugs—and it is easier.
11. It is important to cut a strip from either side of the gel in case the gel has not run straight. Cut enough to get a clear idea of where the bands are. For a wide gel, it may be beneficial to remove a narrow strip from the center of the gel as well.
12. The final size of cut gel fragments is largely dependent on the initial size of the chromosomal band. However, if the gel fragments are too large, the efficiency of digestion may be reduced and a larger volume of enzyme may be needed, increasing the cost.
13. In this instance, we avoid any contact with EtBr during preparation of the chromosomal DNA because this increases cloning efficiency, in contrast to **Subheading 3.2.**, where we avoided EtBr and UV to avoid breakage of the large DNA molecules.
14. This step is designed to allow the β-agarase enzyme access to the agarose matrix. The gel plug may be dissipated either as described here or by incubating at 65°C for 10 min to melt the agarose. In our experience, melting at 65°C is more efficient but can lead to denaturation of A/T-rich DNA sequences and the underrepresentation of such sequences in the resulting library.

15. Note that butanol forms the upper layer in this mixture and that you must retain the lower aqueous layer. If too much butanol is added, all the aqueous layer will be absorbed into it. In such an event, gradually add TE until an aqueous phase can be observed. The DNA should segregate back into the aqueous layer.
16. 10 µg will give the best results, though with care as little as 100 ng for plasmid libraries will be sufficient.
17. An alternative fragmentation method such as nebulization may be used, in which case the fragmented DNA may need to be concentrated by ethanol precipitation before proceeding with the remainder of the protocol. Details of nebulization can be found at the Web site of Bruce Roe's group at the University of Oklahoma: http://www.genome.ou.edu/protocol_book/protocol_partII.html#II, and descriptions of other alternate DNA fragmentation techniques in **ref.** *12*.
18. The desired size range of fragments varies depending on your experiment. However, in general, smaller fragments clone more efficiently. Therefore, usually it is best to clone DNA from several blocks and check the actual size range of the cloned inserts in the resulting libraries.

References

1. Gottesdiener, K., Garcia Anoveros, J., Lee. M.-G., et al. (1990) Chromosome organization of the protozoan *Trypanosoma brucei*. *Mol. Cell Biol.* **10(11),** 6079–6083.
2. LeBlancq, S. M., Korman, S. H., and van der Ploeg, L. H. T. (1991) Frequent re-arrangements of ribosomal RNA-encoding chromosomes in *Giardia lamblia*. *Nucleic Acids Res.* **19(16),** 4405–4412.
3. Henriksson, J., Procel, B., Rydaker, M., et al. (1995) *Trypanosoma cruzi* genome project: chromosome-specific markers reveal conserved linkage groups in spite of extensive chromosomal size variation. *Mol. Biochem. Parasitol.* **73,** 63–74.
4. Wincker, P., Ravel, C., Blaineau, C., et al. (1996) The *Leishmania* genome comprises 36 chromosomes conserved across widely divergent pathogenic species. *Nucleic Acids Res.* **24,** 1699–1694.
5. Melville, S. E., Leech, V., Gerrard, C. S., et al. (1998) The molecular karyotype of the megabase chromosomes of *Trypanosoma brucei* and the assignment of chromosome markers. *Mol. Biochem. Parasitol.* **94,** 155–172.
6. Carlton, J. M. R., Galinski, M. R., Barnwell, J. W., et al. (1999) Karyotype and synteny among the chromosomes of all four species of human malaria parasite. *Mol. Biochem. Parasitol.* **101,** 23–32.
7. Blaineau, C., Bastien, P., Rioux, J. A., et al. (1991) Long-range restriction maps of size-variable homologous chromosomes in *Leishmania infantum*. *Mol. Biochem. Parasitol.* **46(2),** 293–302.
8. Melville, S. E., Gerrard, C. S., and Blackwell, J. M. (1999) Multiple causes of size polymorphism in African trypanosome chromosomes. *Chromosome Res.* **7(3),** 191–203.
9. Bowman, S., Lawson, D., Basham, D., et al. (1999) The complete nucleotide sequence of chromosome 3 of *Plasmodium falciparum*. *Nature* **400,** 532–538.

10. Hall, N., Berriman, M., Lennard, N. J., et al. (2003) The DNA sequence of chromosome I of an African trypanosome: gene content, chromosome organisation, recombination, and polymorphism. *Nucleic Acids Res.* **31(16),** 4864–4873.
11. Sambrook, J., Fritsch, E. F., and Maniatis, T. (1989) *Molecular Cloning. A Laboratory Manual.* Cold Spring Harbor Laboratory Press, Cold Spring Harbor, NY.
12. Quail, M. A. (2003) DNA: mechanical breakage of, in *Encyclopedia of the Human Genome* (Cooper, D. N, ed.), Nature Publishing, London.

20

Chromosome Fragmentation in *Leishmania*

Pascal Dubessay, Christine Blaineau,
Patrick Bastien, and Michel Pagès

Summary

Chromosome fragmentation (CF) constitutes one means of manipulating eukaryotic genomes and provides a powerful tool for examining both the structure and function of chromosomes. During the past 15 yr, CF, which is based on the use of transfection, has been widely used in yeast and mammals to elucidate the functional elements required for normal chromosome maintenance. However, in view of the relatively late development of parasite genome projects, this strategy has only been used recently in parasites. Here, we describe basic methods for CF (except telomere-mediated fragmentation) experiments and analysis in *Leishmania*. Current limitations of this methodology are precisely the lack of knowledge of the nature of centromeres and autonomously replicating sequences in this and other protozoa, the poor understanding of precise recombination mechanisms, as well as the fact that the deletion of unknown genes essential to parasite survival may interfere with recombination events and chromosomal rearrangements. Still, this powerful method has enriched our basic knowledge of chromosomal structure and maintenance.

Key Words: Artificial chromosome; chromosome fragmentation; chromosome stability; *Leishmania*; parasite cultivation; transfection; transfection vector.

1. Introduction

Chromosome fragmentation (CF) constitutes one means of manipulating eukaryotic genomes and provides a powerful tool for examining the structure and function of chromosomes. In particular, it is an essential step toward constructing artificial chromosomes through a better understanding of chromosome functional elements such as centromeres, origins of replication, and telomeres.

During the past 15 yr, CF has been widely used in yeast and mammals to elucidate the functional elements required for normal chromosome maintenance (reviewed in **ref. 1**). It has also proved a useful tool for refining physical mapping of large genomes (in particular human) through yeast artificial chromosome (YAC) fragmentation and the creation of deletion derivatives of large DNA fragments cloned in YACs *(2)*. The different approaches used in CF fall into three broad categories: (a) telomere-mediated chromosome fragmentation, in which telomeres are integrated at different positions along the length of the chromosome, resulting in breakage and production of a truncated chromosome *(3–5)*; (b) large (several hundred kilobases) internal targeted chromosomal deletions *(6)*; and (c) induction of gross recombination events leading to *de novo* minichromosome formation *(7–9)*. The construction of sets of deletion derivatives obviously is very useful for such studies.

These approaches are based on the use of transfection as a means of introducing CF vectors that target a specific site within the genome. Therefore, the construction of appropriate CF vectors as well as a successful transformation system are essential steps toward mastering this technology. After the fragmentation experiment, three main steps are necessary to obtain interpretable results: (a) determining the nature of the chromosomal rearrangements and karyotype changes induced; (b) precisely analyzing the recombination events that occurred, and (c) examining the behavior of the truncated or artificial chromosome in cells cultivated in vitro.

CF has only been used recently in parasites. Early transfection experiments in *Trypanosoma brucei* using small plasmidic constructs produced more or less unstable minichromosomes that integrated into unidentified genomic sequences *(10,11)*, but this was not true CF. Basically, only those parasites in which homologous recombination after transfection is highly efficient and for which genome knowledge is well advanced, e.g., *Leishmania*, *Trypanosoma*, and *Plasmodium*, have benefited from CF developments *(12–16)*. Different model chromosomes have been used, addressing the identification of potential centromeric sequences *(14,16)*, putative replication origins *(15)*, and telomere elongation *(12–14)*.

This chapter describes basic methods for CF experiments and analysis in *Leishmania*. Current limitations of this methodology are the lack of knowledge of the nature of centromeres and autonomously replicating sequences (ARS) in this and other protozoa, the poor understanding of recombination mechanisms, and the fact that the deletion of unknown genes essential to parasite survival may interfere with recombination events and chromosomal rearrangements when selecting the transfectants. The latter factor, combined with the now well-documented genome plasticity in this protozoon, may be responsible for the numerous unexpected and aberrant events obtained when attempting CF.

2. Materials

2.1. Polymerase Chain Reaction

1. 10X polymerase chain reaction (PCR) buffer.
2. 25 mM MgCl$_2$.
3. 5 mM dNTPs.
4. Goldstar *Taq* DNA polymerase (Eurogentec).
5. Standard primers: T7 promoter and SP6 (MWG Biotech).
6. 0.5-mL microfuge tubes.
7. Thermocycler.
8. Mineral oil for PCR.

2.2. Restriction Enzymes

We recommend using high-quality restriction enzymes (e.g., New England Biolabs).

1. Restriction enzymes *Kpn*I, *Xba*I, *Hpa*I, and *Mfe*I and corresponding 10X buffers.
2. Homing endonuclease *Eco57*I and buffer (10X buffer + 0.5 mM S-adenosyl-homo-cysteine) (MBI Fermentas® or Eurogentec).
3. T4 DNA ligase (400 U/mL) and 10X buffer (New England Biolabs).

2.3. Electrophoresis (Analytical and Preparative)

1. High-quality agarose (Seakem GTG) for preparative gels (DNA purification).
2. Agarose (Seakem LE) for analytical gels.
3. 0.5X TBE buffer (stock solution 20X: 1.8 M Tris-borate, 0.04 M EDTA).
4. Ethidium bromide, 0.1 mg/mL (stock solution 10 mg/mL).

2.4. Electroelution

1. Standard double-chamber electroelution block (e.g., CBS Scientific).
2. Dialysis membrane.

2.5. DNA Extraction and Precipitation

1. Phenol.
2. Chloroform/*iso*-amyl alcohol (24:1 v:v).
3. Ethanol, 100% and 70%.
4. 2.5 M ammonium acetate (stock solution 7.5 M).
5. Vacuum apparatus.

2.6. DNA Minipreparation (Minipreps)

1. Solution I: 50 mM glucose, 25 mM Tris-HCl, pH 8.0, 10 mM EDTA, pH 8.0, 50-µg/mL RNase A. Store at +4°C.
2. Solution II: 0.2 N NaOH freshly diluted from a 10 N stock, 1% sodium dodecyl sulfate (SDS); prepare immediately before use.
3. Solution III: 5 M potassium acetate (60 mL), glacial acetic acid (11.5 mL), distilled H$_2$O (28.5 mL). Store at +4°C.

2.7. DNA Maxipreparation (Maxipreps)

1. Affinity columns: e.g., Qiagen Plasmid Maxi Kit (Qiagen).

2.8. Bacterial Culture and Cells

1. Luria Bertani (LB) medium: 10-g/L peptone 140, 5-g/L yeast extract, 10-g/L NaCl.
2. LB-agar: LB + 15-g/L agar.
3. Ampicillin (200-mg/mL stock solution).
4. JM109-competent *Escherichia coli* cells (Promega).

2.9. Sequencing

1. Licor DNA sequencer Long Readir 4200 (Amersham) apparatus.
2. Sequence gels (66 cm): Long Ranger gel solution (BioWhittaker Biomol. Appl.) (4.6 mL), Long Ranger TBE 10X (5 mL), H_2O (32 mL), dimethyl sulfoxide (DMSO) (50 µL), Urea (21 g), TEMED (50 µL), 10% ammonium persulfate (freshly prepared) (350 µL).
3. Labeled primers: T7 IRD700 and Sp6 IRD800 (MWG Biotech).
4. Sequencing kit: Sequitherm EXCEL™ II DNA Sequencing Kit-LC (Amersham).
5. Sequence analysis software: Sequencher (Amersham).

2.10. Parasite Cultivation

1. "Complete" RPMI 1640 medium (all reagents are from Invitrogen Life Tech., and the concentrations given here are final concentrations): RPMI 1640, 20% heat-inactivated fetal calf serum (FCS), 1% v/v nonessential MEM amino acids (100X), 1% v/v MEM amino acids (50X), 1 mM sodium pyruvate (100X), 2-mg/mL dextrose, 25 mM HEPES, pH 7.4, in bidistilled H_2O, 2 mM L-glutamine, 100-U/mL penicillin, 100-µg/mL streptomycin.
2. Selection antibiotics (Euromedex): stock solutions are 50 mg/mL for hygromycin B, puromycin dihydrochloride, and geneticin disulfate, and 20 mg/mL for phleomycin. Aliquot and store at –20°C.
3. Sterile glass bottles.
4. Pipetting aid and sterile pipets.
5. 25-cm^2 and 75-cm^2 sterile culture flasks.
6. 96-well and 24-well microplates.
7. Sterile-flow cabinet.
8. 26°C incubator with a 5% CO_2 concentration and a humidified atmosphere.

2.11. Transfection

1. Electroporation buffer: 21 mM HEPES, pH 7.5, 137 mM NaCl, 5 mM KCl, 0.7 mM Na_2HPO_4, 6 mM glucose. Filter-sterilize and keep at +4°C.
2. Electroporator Easyject Plus (Equibio).
3. Electroporation cuvets (0.4-cm gap).
4. Plasmid DNA resuspended in 50 µL sterile H_2O at a concentration of 80–200 µg/mL.
5. Sterile 50-mL Falcon conical centrifuge tube.

Chromosome Fragmentation in Leishmania

6. Hemocytometer (Thoma cell).
7. Phosphate-buffered saline (PBS), pH 7.2.
8. 1.5-mL microfuge tubes.
9. Micropipet, sterile tips.
10. Ice.
11. SeaPlaque agarose GTG.
12. 10-cm-diameter Petri dishes.
13. 15-mL Falcon polypropylene round-bottomed centrifuge tubes.

2.12. Analysis of Transfectants (Pulse Field Gel Electrophoresis, Southern Blotting, DNA Probe Labeling, and Hybridization)

1. 15-mL conical Falcon centrifuge tubes.
2. Low-melting-temperature agarose (SeaPlaque GTG).
3. PBS.
4. Plug molds (Bio-Rad).
5. Lysis buffer: 0.5 M EDTA, pH 8.0, 1% N-lauroyl-sarcosine. Add 1 mg/mL of proteinase K just before use.
6. 50°C water bath.
7. Complete pulse-field gel electrophoresis (PFGE) apparatus.
8. 1X TE buffer: 10 mM Tris-HCl, pH 8.0, 1 mM EDTA.
9. 0.5X TBE buffer: 45 mM Tris-HCl, pH 8.3, 45 mM boric acid, 1 mM EDTA.
10. Agarose (Seakem GTG).
11. HCl, NaCl, NaOH.
12. Nylon-membranes: Hybond N+ (Amersham).
13. Chromatography paper: Whatman 3MM.
14. Random hexanucleotides pd(N)6.
15. Deoxynucleotides triphosphate dATP, dGTP, dTTP.
16. Radiolabeled α^{32}P-dCTP.
17. DNA polymerase 1 Klenow fragment.
18. 37°C water bath.
19. 20X SSPE: 3 mM NaCl, 200 mM NaH$_2$PO$_4$, 20 mM EDTA, pH 7.5.
20. SDS.
21. 65°C shaking water bath.
22. Hyperfilm (Amersham).
23. Autoradiography cassettes and amplifying screens.
24. Restriction enzymes (NEB) and corresponding buffers.
25. PCR reagents (*see* **Subheading 2.1.**).

3. Methods

3.1. Construction of the Chromosome Fragmentation Vector

Here, we describe the utility of a family of versatile vectors termed pVV (GenBank Accession no. AF315645) that has been successfully used for internal CF. They should also be useful for telomere-mediated CF. Indeed, the telomere-

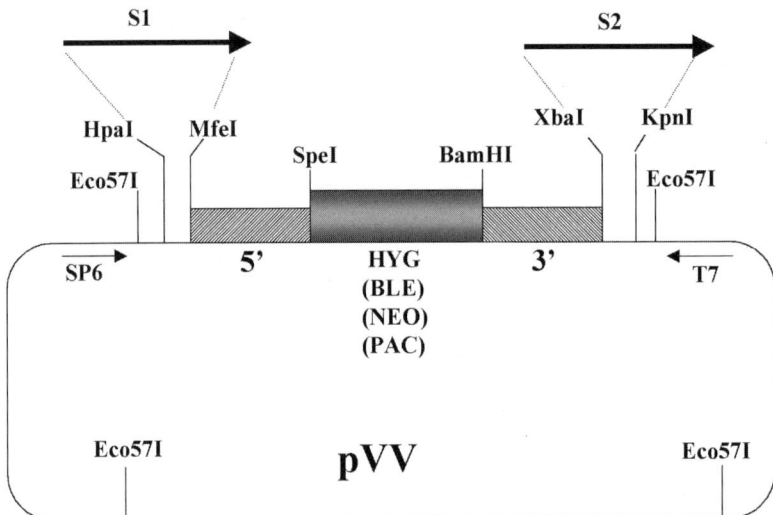

Fig. 1. Schematic representation of the "versatile" transfection vector pVV. See text for description. S1 and S2 = sequences homologous to the targeted chromosomal DNA. *HYG, BLE, NEO*, and *PAC* = resistance marker gene cassette (*see* text for details). 5' and 3' = parts of the 5' and 3' *Leishmania major DHFR-TS* gene flanking regions. SP6 = SP6 RNA polymerase promoter, T7 = T7 RNA polymerase promoter.

mediated CF vector can simply be viewed as a deletion vector in which one of the two sequences homologous to the target genome consists of telomeric repeats. A telomere-mediated CF vector based on telomeric and subtelomeric sequences of *Trypanosoma brucei* has been shown to be functional in *Leishmania* *(14)*.

Vector pVV (*see* **Fig. 1**) was created from the pGEM3Zf bacterial plasmid (Promega). Three essential features were introduced that allow (a) interchange of selection marker cassettes, for sequential targeting; (b) modification of the nature and orientation of homologous sequences; and (c) linearization by restriction within the sequences homologous to the targeted genome.

Four different selectable marker genes can be used, conferring resistance to hygromycin B (*HYG* gene) *(17)*, the aminoglycoside G418 (*NEO*) *(18)*, puromycin (*PAC*) *(19)*, or phleomycin (*BLE*) *(20)*. These resistance gene cassettes can be readily interchanged by *SpeI/Bam*HI double digestion, as shown in **Fig. 1**. 5'- and 3'-DNA sequences (953 and 1738 bp, respectively) have been inserted on both sides of the selectable marker to drive the expression of the drug-resistance gene. They derive from the flanking regions of the *Leishmania major* dihydrofolate reductase-thymidilate synthase (*DHFR-TS*) gene *(21)* (GenBank Accession no. U59231).

Each cassette (5'-region–drug-resistance marker gene– 3'-region) is flanked by two different cloning sites, one containing the restriction sites *Hpa*I and *Mfe*I, and the other *Xba*I and *Kpn*I. These enzymes are rare-cutters in the GC-rich *Leishmania* genome and offer a wide potential for choice of DNA sequences homologous to the targeted locus (here termed S1 and S2, *see* **Fig. 1**). When using telomere-mediated CF, one of the homologous sequences consists of telomeric repeats, whereas the other one defines the targeted locus on the chromosome to be deleted. For example, the vector used by Tamar and Papadopoulou *(14)* included approx 1.5 kb of subtelomeric repeats from *T. brucei* adjacent to the telomeric repeats.

Finally, a restriction site for the homing endonuclease *Eco*57I (CTGAAGN$_{16}$/GACTTCN$_{14}$) has been introduced adjacent to each cloning site. Thus, the vector can be linearized by cutting within the homologous sequences, providing free ends for invasive recombination into the target genome.

3.1.1. Preparation of DNA Fragments Homologous to the Targeted Locus

This step involves preparation of homologous DNA fragments and their insertion into the two cloning sites in the vector. The fragments are PCR-amplified from the target genome using a pair of oligonucleotide primers, each containing a restriction site—e.g., *Hpa*I and *Mfe*I for S1 and *Xba*I and *Kpn*I for S2—such that they can be cloned into the vector in the appropriate orientation.

3.1.1.1. PCR AMPLIFICATION

1. The size of the homologous fragments is generally around 500 bp (*see* **Note 1**). The sequence obviously must be devoid of restriction enzyme recognition sites that will be used in subsequent steps, i.e., *Hpa*I, *Mfe*I, *Xba*I, *Kpn*I, and *Eco*57I.
2. For each homologous sequence, design one pair of complementary oligonucleotide primers of about 20 nucleotides with a T_m at about 60°C.
3. Add an extension of nine nucleotides (nt) in a 5'-position to each primer: three randomly chosen nt + the six nt of the appropriate enzyme recognition site (e.g., 5'-ggggtacc...-3' for *Kpn*I) (*see* **Note 2**).
4. Amplify the homologous fragments from genomic DNA or a genomic library by PCR, in a total reaction volume of 50 µL including 1–5 µL (80 ng) of sample DNA. We use the following optimized concentrations: 5 m*M* MgCl$_2$, 200 µ*M* dNTPs, 50 pmol of each primer, 2 U *Taq* DNA polymerase. We use the following cycling conditions: 95°C for 2 min 30 s, 62°C for 1 min, 30 cycles of 72°C for 1 min then 94°C for 30 s, 62°C for 30 s, and 72°C for 10 min (*see* **Note 3**). Perform two to three replicates of each reaction to have sufficient material for the following steps.

3.1.1.2. PURIFICATION OF THE AMPLIFICATION PRODUCTS

1. Pool the contents of the two or three replicate PCR reactions mixtures and load all into a large slot (made by sealing comb teeth if necessary) of a 1% agarose minigel

(10 × 10 cm) (Seakem GTG agarose in 0.5X TBE buffer + 0.1-mg/L ethidium bromide). Electrophorese for 1 h at 10 V/cm in 0.5X TBE buffer.
2. Visualize the band of interest (at the expected size) under UV light. Using a sharp scalpel blade, cut out a slice of agarose containing the band of interest (*see* **Note 4**).
3. Place a double-chamber electroelution block in an electrophoresis tank. Cover the bottom of each chamber with a dialysis membrane, previously boiled for 10 min in distilled H_2O. Transfer the slice of agarose to the electroelution block. Electroelute at 6 V/cm (120 V) for 1 h in 0.5X TBE buffer.
4. Recover the DNA in 200 µL of buffer in a 1.5-mL microfuge tube; rinse the block with 200 µL of 0.5X TBE.
5. Extract the DNA using a standard phenol–chloroform extraction method (1 vol phenol + 1 vol chloroform). Add ammonium acetate to 2.5 M and 2 vol ethanol to precipitate the DNA. Wash the pellet with 70% ethanol once. Dry the pellet, then dissolve the DNA in 20 µL of distilled H_2O (*see* **Note 5**).
6. Load 1 µL on a 1% agarose minigel (0.5X TBE + 0.1-mg/mL ethidium bromide) along with a DNA size standard of known concentration (e.g. 200 ng of *Pst*I-restricted λDNA). This allows you to estimate of the DNA concentration for each amplified fragment by comparison of fluorescence intensity.

3.1.1.3. ENZYMATIC DIGESTION OF THE PCR PRODUCTS

The PCR-amplified homologous sequences must be subjected to a double digestion (with *Hpa*I-*Mfe*I for S1 and *Kpn*I-*Xba*I for S2) to be inserted into the vector.

1. Digest all of the PCR-amplified DNA fragments (19 µL) in 200 µL of 1X digestion buffer with 1–2 U of enzyme/µg of DNA (*see* **Note 6**). Perform a control digestion of 500 ng of plasmid DNA containing the corresponding restriction sites (e.g., pBluescript) with one-fifteenth of the 200 µL of restriction digestion solution (*see* **Note 7**). Incubate both tubes at 37°C for 1 h 30 min.
2. Load 1–5 µL of the contents of the control tube onto a 1% agarose minigel (0.5X TBE + 0.1-mg/mL ethidium bromide) adjacent to a DNA size standard to verify that the digestion is complete (*see* **Note 8**).
3. Extract the digested fragment using phenol–chloroform as in **step 5** of **Subheading 3.1.1.2.**
4. Dissolve the pellet in 10 µL of distilled H_2O. Load 1 µL on a 1% agarose minigel next to a DNA size standard as in **step 6** of **Subheading 3.1.1.2.**

3.1.2. Insertion of Homologous Fragment S2 Into pVV

3.1.2.1. DOUBLE DIGESTION OF pVV

It is essential to carry out both digestions stepwise and to perform separate control reactions. Partial restriction of the vector DNA must be avoided.

1. Digest 6 µg of pVV DNA with *Kpn*I in 200 µL of 1X enzyme buffer + 1–2 U/µg *Kpn*I of DNA, for 2 h at 37°C.

2. Load 5 µL on a 1% agarose minigel to assess whether the digestion is complete (*see* **Note 8**).
3. Extract the digested product with phenol–chloroform as in **Subheading 3.1.1.2**.
4. Dissolve the DNA pellet in 20 µL of distilled H_2O. Load 1 µL on a 1% agarose minigel to assess DNA concentration (*see* **step 6** of **Subheading 3.1.1.2.**).
5. Proceed with the second digestion using *Xba*I as in **step 1**. Perform a control digestion as in **step 1** of **Subheading 3.1.1.3** (*see* **Note 9**).
6. Load 5 µL of the control digestion on a 1% agarose minigel and verify that the digestion is complete (*see* **Note 9**).
7. Extract the digested vector with phenol–chloroform as in **Subheading 3.1.1.2**.
8. Dissolve the DNA pellet in 10 µL of distilled H_2O. Load 1 µL on a 1% agarose minigel with a DNA size standard (e.g., λ/*Pst*I) as in **Subheading 3.1.1.2**.

3.1.2.2. LIGATION OF S2 INTO PVV

1. Set up a ligation reaction in a microfuge tube containing 1X T4 DNA ligase buffer, 50–75 ng of the double-digested pVV vector, 100–150 ng of the double-digested homologous sequence S2, and 1 µL (400 cohesive end units) of T4 DNA ligase (New England Biolabs) made up to 10 µL with distilled H_2O.
2. Set up an additional control reaction containing the same elements except S2 (*see* **Note 10**).
3. Incubate both tubes overnight at 15–16°C, then store at +4°C.

3.1.2.3. BACTERIAL TRANSFORMATION

We generally transform competent *E. coli* by the "heat-shock" method.

1. Place both ligation reaction tubes on ice.
2. Remove an aliquot of preprepared competent cells (home-made or JM109 Promega) from the –80°C freezer and place immediately on ice for 5–10 min (*see* **Note 11**).
3. Using a sterile pipet tip, quickly transfer 50 µL of cells into both ligation tubes on ice. Gently homogenize tube contents by stirring with pipet tip (but without pipetting).
4. Leave on ice for 30 min, then transfer the tubes to a circulating water bath preheated at 42°C for exactly 1 min.
5. Immediately transfer the tubes to ice and allow the cells to chill for 2 min.
6. Remove the tubes from ice. Using a sterile pipet tip, add 450 µL of LB medium without antibiotic to each tube (*see* **Note 12**). Gently mix by inverting tubes four or five times, then incubate at 37°C for 45–60 min. Invert the tubes every 15 min to mix.
7. From each tube, transfer 50, 100, and 200 µL of the transformed cells onto three 90–100-mm plates containing LB-agar + 70-µg/mL ampicillin. Gently spread over the surface of the agar plate. Leave plates to dry for 10–15 min, then invert and incubate overnight (16 h) at 37°C.

3.1.2.4. SELECTION OF THE RECOMBINANT CLONES

To screen for recombinant clones, we perform both a PCR and the isolation of the clone by transfer of a single colony to culture medium (LB + antibiotic).

1. Estimate the number of recombinant clones by comparing the number of colonies present on the six plates derived from the control and test ligations (*see* **Note 13**).
2. Using a sterile pipet tip, gently remove a well-separated colony from an agar plate. Transfer some of the colony from the tip onto the bottom of an empty PCR tube. Immediately transfer the same pipet tip into a well of a 96-well microplate containing 200 µL of LB medium + 70-µg/mL ampicillin. (We usually screen 10 clones in this manner.)
3. Process one or two clones from the control plate in the same way, to be used as positive controls for the PCR.
4. Incubate the microplate overnight at 37°C.
5. Amplify by PCR to verify insertion of the desired fragment. For this, prepare a PCR mix (1X buffer, 5 mM MgCl$_2$, 200 µM dNTPs, 50 pmol of each primer [*see* **Note 14**], 1 U Goldstar ^{Taq} DNA polymerase, Eurogentec). Transfer 50 µL of the mix into each of the PCR tubes containing the colony smears. Cover the mix with mineral oil. Cycle as follows: 95°C for 2 min 30 s, 62°C for 1 min; 30 cycles at 72°C for 1 min 30 s, 94°C for 30 s, and 62°C for 30 s; finally, 72°C for 10 min.
6. Load 8 µL of each PCR product on a 1% agarose (SeakemGTG) gel (0.5X TBE + 0.1-mg/mL ethidium bromide) adjacent to a DNA size standard (e.g., λ/*Pst*I). Electrophorese at 7 V/cm for 2 h and visualize under UV. The size of the amplicons from nonrecombinant clones is 200 bp, whereas that from recombinant clones is 200 bp + the size of the inserted sequence.

3.1.2.5. DNA MINIPREPARATION AND SEQUENCING

The integrity of the cloned homologous fragment in the pVV vector must be verified by sequencing. For this, the DNA from recombinant clones is purified by a minipreparation method. The protocol given below is adapted from a standard protocol but yields the high-quality DNA necessary for sequencing reactions.

1. Select two to three recombinant clones. Inoculate 3 mL of LB medium + ampicillin (70 µg/mL) with 20 µL of culture from each selected clone in loosely capped 15-mL polypropylene tubes. Incubate overnight at 37°C with vigorous shaking.
2. Centrifuge at 1800*g* for 20 min at room temperature, then discard the supernatant and leave the tubes inverted on a paper towel. Place the tubes at –80°C for 5 min.
3. Remove the tubes and thaw for 2–3 min at room temperature.
4. Add 200 µL of solution I (*see* **Subheading 2.6.**) to each tube. Vortex briefly and gently until the pellet is completely redissolved.
5. Transfer the 200 µL to a 1.5-mL microfuge tube. Add 300 µL of lysis solution II (*see* **Subheading 2.6.**). Mix the contents by inverting the tubes five to six times, and leave 5 min at room temperature.

6. Add 250 µL of ice-cold neutralization solution III (*see* **Subheading 2.6.**). Invert the tubes five to six times, and leave 20 min on ice.
7. Centrifuge at 12,000*g* for 30 min at +4°C, then transfer 600 µL of the supernatant to another microfuge tube and centrifuge again at 12,000*g* at +4°C for 20 min.
8. Transfer 500 µL of the supernatant to another microfuge tube and precipitate with 2.5 times the volume of ethanol at –20°C for 20 min.
9. Centrifuge at 12,000*g* at +4°C for 20 min. Discard the ethanol, wash once with 70% ethanol and resuspend the pellets in 20 µL of distilled H$_2$O. Store at –20°C.

The DNA prepared in this manner should be sequenced (*see* **Note 15**). This step is important, to ensure that the selected clones contain the desired DNA fragment before proceeding to the second stage of cloning. Use the primer T7 IRD800 (MWG, Biotech), located downstream of the *Kpn*I site on pVV, to determine the nucleotide sequence of the cloned S2.

3.1.2.6. PREPARATION OF THE CLONED S2-CONTAINING VECTOR DNA FOR INSERTION OF THE S1 HOMOLOGOUS SEQUENCE

1. Select one of the recombinant clones and perform six separate minipreps of this clone as described under **Subheading 3.1.2.5.**
2. Pool the six minipreps (6 × 20 µL) in a microfuge tube and add 280 µL of distilled H$_2$O.
3. Purify the DNA with phenol–chloroform and ethanol precipitation as in **Subheading 3.1.1.2.** (*see* **Note 16**).
4. Resuspend the DNA pellet in 20 µL of distilled H$_2$O. Visualize 1 µL on a gel as under **Subheading 3.1.1.2.**

3.1.3. Insertion of the Second Homologous Sequence S1 Into pVV

Here, the purified DNA fragment S1 (*see* **Fig. 1**) is introduced into the *Hpa*I and *Mfe*I sites of pVV-S2. Follow the same protocol as used to clone fragment S2 (**Subheading 3.1.2.**), except for a few technical variations at **steps 4** and **6**:

STEP 4. SELECTION OF THE RECOMBINANT CLONES

Use S1-specific PCR primers: standard M13-reverse (5-aacagctatgaccatg-3'), located on pVV upstream of the *Hpa*I site, and another (5'-gtgctactacttagct-3') homologous to the 5'-*DHFRTS* region.

STEP 6. SEQUENCING

Use the labeled primer is Sp6 IRD800 (MWG, Biotech), located upstream of the *Hpa*I site.

3.1.4. Maxipreparation of the CF Vector DNA

Having selected the appropriate clone to be used as the complete vector pVV-S2-S1, you should now perform a maxipreparation of this clone using a commercial kit (we recommend affinity columns such as Maxiprep, QiaGen).

1. Inoculate 3 mL of LB medium containing 70-μg/mL ampicillin with 10 μL of the clone culture that was retained in a 96-well microplate (**Subheading 3.1.2.4.**). Incubate overnight (≥8 h) at 37°C with vigorous shaking.
2. Inoculate 400 mL of LB medium + 70-μg/mL ampicillin with 2 mL of the culture. Incubate overnight at 37°C with vigorous shaking.
3. Extract the vector DNA using a commercial kit according to the manufacturer's instructions.
4. Redissolve the DNA in 150 μL of distilled H_2O (*see* **Note 12**). Estimate the DNA concentration on an agarose gel as under **Subheading 3.1.1.2.** (*see* **Note 17**).

3.1.5. Linearization and Purification of the Transfection Vector pVV-S2-S1

3.1.5.1. Digestion With Eco57I

1. In a 2-mL microfuge tube, digest 60 μg of vector DNA with *Eco57*I (0.4 U/μg of DNA) at 37°C overnight in a final volume of 500 μL.
2. Verify that the digestion is complete by loading 1 μL of the digest on a 1% agarose minigel (*see* **Note 18**).
3. Precipitate the DNA in 2.5 *M* ammonium acetate and 2 vol of ethanol at –20°C for 20 min. Spin at 12,000*g* at +4°C for 30 min. Wash the pellet once with ethanol 70% and dry.
4. Resuspend the DNA precipitate in 400 μL of distilled H_2O.

3.1.5.2. Purification of the Linearized Vector DNA

1. Cast a 0.7% agarose (Seakem GTG) gel (0.5X TBE + 0.1-mg/mL ethidium bromide) containing a single 15-cm-long well. Load the whole digest (400 μL) into the well and migrate at 3 V/cm overnight.
2. Visualize the band of interest (at 4.5–5.5 kb) under UV. Using a sharp scalpel blade, cut out the 15-cm-long slice of agarose containing the band of interest. Place it on another gel plate and directly cast a 0.7% agarose gel (as above) around the slice. Electrophorese at 3 V/cm overnight, followed by 5 V/cm for 4–5 h (*see* **Note 19**).
3. Again, using a sharp scalpel blade, cut out the slice of agarose containing the band of interest (at 4.55–5 kb) and chop up into eight pieces.
4. Electroelute the DNA from the eight agarose pieces in four electroelution blocks (*see* **Subheading 3.1.1.2.**) at 5 V/cm for 4 h.
5. Recover the DNA in 4X 200 μL in a 2-mL Eppendorf tube; rinse the four blocks with the same 200 μL of 0.5X TBE, and pool the 1000 μL.
6. Purify the DNA with phenol–chloroform and precipitate with ethanol as under **Subheading 3.1.1.2.**
7. Resuspend the precipitate in 50 μL of sterile distilled H_2O (*see* **Note 12**). Check the amount and quality of the DNA by loading 1 μL on a 1% agarose gel. The DNA yield is generally between 6 and 10 μg. This amount of linearized vector DNA is necessary for transfection. The DNA is now ready for use.

3.2. Transfection of Leishmania Cells

3.2.1. Parasite Cultivation

The following protocol has been adapted for transfection of *L. major* and *L. donovani* (*see* **Note 20**). We use RPMI 1640 medium enriched with 20% FCS and other complements (*see* **Subheading 2.10.** for "complete" medium) to favor efficient growth (*see* **Notes 21** and **22**).

1. Cultivate the cells in complete medium at 26°C with 5% CO_2 in a humidified atmosphere (*see* **Note 23**). When the selection drug is required (after transfection of the parasites), add it to the medium just before inoculation of the cells.
2. Examine the cells under a microscope on a regular basis (*see* **Note 24**). Passage the cells triweekly with a one-tenth dilution into 25-cm^2 sterile plastic tissue culture flasks (6 mL/flask), just before they reach late log phase (approx 1.5–2×10^7 cells/mL) (*see* **Note 25**).

3.2.2. Sensitivity Test to Selection Drugs

It is necessary to determine the minimal concentration of selection drug at which no survivors are observed in the wild-type population. This test should be performed as a two-step process in liquid culture medium and in agarose plates, because transfected cells will be cultivated on both and the effective concentrations are not the same (*see* **Note 26**).

1. Test the selection antibiotics in liquid medium at increasing concentrations, generally a twofold dilution series from 1 to 100 µg/mL for hygromycin B, 0.5 to 50 µg/mL for phleomycin, 10 to 150 µg/mL for puromycin, and 0.5 to 100 µg/mL for geneticin. Include a drug-free control. For each concentration, inoculate in duplicate 3 mL of medium at 1–2×10^5 *Leishmania* cells per mL in loosely capped 15-mL polypropylene tubes.
2. Count the cells in a haemocytometer daily for 1 wk. The lowest concentration at which no survivors are observed is chosen for the subsequent selection of transfectants in liquid culture.
3. Estimate the approximate EC_{50}. Then test four drug concentrations (from the EC_{50} to several times the EC_{100}) in agarose plates. For this, pour RPMI 1640-agarose plates at the selected drug concentrations, as described in **Subheading 3.2.3.2.** below. Choose the minimal drug concentration yielding no colonies for subsequent selection on agarose plates

3.2.3. Transfection

For highest transfection efficiency, it is essential to use particularly "healthy" parasites, cultivated in optimal conditions (*see* **Note 27**).

3.2.3.1. Electroporation

All steps below must be performed in sterile conditions and as quickly as possible.

1. Two days before transfection (d "–2"), inoculate an adequate volume of medium with promastigotes at 10^6/mL (*see* **Note 28**). For example, if three electroporations (including a control) are foreseen, inoculate 50 mL.
2. On d 0, estimate the number of live cells by counting with a hemocytometer (*see* **Note 29**).
3. Place the electroporation cuvets (0.4-cm gap) containing 50 µL of vector DNA (6–10 µg of purified linearized vector pVV-S2-S1 DNA) on ice.
4. Spin down *Leishmania* cells in a sterile 50-mL Falcon tube at 900g for 10 min at room temperature.
5. Remove the supernatant and resuspend pellet in 10 mL of electroporation buffer by gently pipetting up and down.
6. Centrifuge as above and carefully remove the supernatant, without disturbing pellet, and resuspend in 10 mL EP buffer.
7. Repeat centrifugation as above, then gently remove the supernatant and resuspend the cells at a concentration of 10^8/mL in EP buffer.
8. Incubate the cells on ice for 10 min.
9. Set up the electroporator (Easyject plus™, Equibio). We routinely use the following conditions: one pulse with 2 kV/cm, 2310 Ω, and 25 µFarrads. With these settings, we get a 57.7-ms pulse time (*see* **Note 30**).
10. In each electroporation cuvet, mix 800 µL of the *Leishmania* cell suspension (8 × 10^7 cells) with the 50 µL of ice-cold vector DNA (= test sample [TS]).
11. Prepare a "mock cuvet" containing the same amount of cells but no vector DNA (= control without DNA [CWD]). This will serve as a control for the mortality of nontransfected cells in the presence of the selection drug.
12. Carefully dry the outside walls of the cuvet before placing it in the chamber. Zap (subject to electric field) immediately, then place the cuvets on ice for 10 min.
13. Transfer the cells from each cuvet into a 50-mL tube containing 10 mL of drug-free complete medium.
14. Also process a control electroporation with 800 µL of nonelectroporated cells (NEC).
15. Incubate all tubes for 24 h in the 26°C CO_2 incubator until you commence drug selection on d 1 (*see* **Note 31**).

3.2.3.2. Preparation of Selective Plates (*see* **Note 32**)

Two methods may be used to isolate the transfectant cells: either in liquid culture or on RPMI medium-agarose plates. The latter obviously has the advantage of isolating clones in one step, whereas the former requires subsequent cloning in liquid medium (*see* **Note 33**). We describe the second method below. All steps must be performed in sterile conditions.

1. Melt and sterilize 6% (10X) LMT agarose (SeaPlaque GTG) in double-distilled H_2O by autoclaving (120°C for 20 min). Place the flask in 65°C water bath until use.
2. Prepare complete medium and keep at 37°C.
3. In a sterile-flow cabinet, pour medium into the 10X agarose (9:1 v/v) and mix gently while maintaining the flask at 42°C until completely homogenized. (You need about 25 mL/plate).

Chromosome Fragmentation in Leishmania

4. Immediately pipet 25 mL onto one or two plates for use as drug-free controls.
5. Add the selective drug at appropriate concentration (*see* **Subheading 3.2.2.**) to the agarose-medium mix. Rapidly homogenize and pipet 25 mL of the mix onto each plate. Carefully avoid air bubbles. Do two to three plates for each electroporation cuvet (TS and CWD).
6. Leave the plates uncovered in the hood for 1 h 30 min to solidify (*see* **Note 34**).

3.2.3.3. PLATING OF *LEISHMANIA* CELLS AND DRUG SELECTION

1. Vigorously shake the cultured *Leishmania* cells to break down cell aggregates. Count the cells in the control tubes NEC and CWD to assess cell viability after electroporation.
2. For each sample (TS and CWD), transfer 50 µL of cells to 1 mL of drug-free medium in a 24-well microplate to test viability of electroporated cells in liquid medium (= viability control [VC]).
3. For each sample, transfer 1 mL of cells to a 25-cm^2 culture flask containing 8 mL of complete medium with the selection drug at the appropriate concentration (**Subheading 3.2.2.**). We routinely use 30-µg/mL hygromycin B for *L. major* and 50-µg/mL hygromycin B for *L. donovani*, 8-µg/mL geneticin for *L. major* and 50-µg/mL geneticin for *L. donovani*, 10-µg/mL phleomycin and 100-µg/mL puromycin. Incubate in the 26°C CO_2 incubator with the cap loose. This selective liquid culture (SLC) is maintained as a safeguard in case plating efficiency is too low.
4. Centrifuge the remaining 9 mL at 900*g* for 10 min at +4°C, then gently remove the supernatant with a pipet, carefully avoiding the cell pellet.
5. Resuspend the pellet in 600 µL of sterile medium by flicking the tube.
6. Transfer 10 µL of this suspension to 5 mL of medium, mix, and transfer 100 µL to 1 mL of medium. Plate 200 µL of this dilution on a nonselective plate to evaluate plating efficiency.
7. Divide the remaining 600 µL of TS and CWD samples between three plates each by depositing 200 µL of cells onto the center of the plate and spreading them by gently tilting the plate (*see* **Note 35**). Leave uncovered for a few minutes, then place the closed plates in the 26°C CO_2 incubator. It is not necessary to wrap the plates in plastic film or Parafilm™.

3.2.3.4. ISOLATION OF TRANSFECTANTS

1. On the drug-free plates, colonies begin to emerge after 4–7 d. Count these to assess plating efficiency.
2. On the CWD plate, many aggregates containing live cells are visible during the first 4 d but then disappear progressively. On d 10, no survivors should be observed.
3. On the selective plates, colonies start to emerge after 7–14 d (*see* **Note 36**). They should be transferred to liquid culture as soon as possible (*see* **Note 36**). Identify reasonably sized and well-separated colonies. Using a sterile pipet tip, transfer each selected colony to 200 µL of drug-containing complete medium in a 96-well microplate. Rinse the tip by pipetting up and down. Change tip for each colony. Maintain sterility at all times.

4. Wrap the plates with Parafilm and store at room temperature: growth will slow down, and the plates may prove useful at a later stage.
5. After 1–2 d, transfer the contents of each well to separate wells of a 24-well microplate containing 1 mL of drug-containing complete medium.
6. After 1–2 d more, transfer each well to 2 mL of selective medium in a 25-cm^2 flask with the cap loose. Scale up the cultures by adding 3 mL of selective medium 1–2 d later and again 4 mL after 1–2 d more.
7. When cells reach log phase, remove 9 mL of culture for extraction of DNA. Add fresh selective medium to the flasks, and maintain the cultures (*see* **Subheading 3.2.1.**) until the DNA has been analyzed (*see* **Subheadings 3.3.2., 3.3.3., and 3.3.4.**).

3.2.3.5. Cultivation of the Control Flasks in Liquid Medium

These flasks are used only to assess the success of different stages of the process. Check cell viability in the VC drug-free cultures for a few days after electroporation and discard when they reach stationary phase. In the SLC cultures, the cell density should decrease considerably after 3–7 d. In the control CWD, no survivors should be observed after 10 d, showing that the drug concentration was sufficient to kill all nontransfectant cells. In the TS preparations, a low proportion of cells survive and multiply. This culture may be discarded when numerous colonies have been isolated from selective plates.

3.3. Primary Analysis of Transfectants

Transfectants with the desired CF event are identified by analyzing the molecular karyotype by PFGE. This necessitates preparation of chromosomal DNA. To screen a high number of clones simultaneously, we harvest the cells from a small volume of cell suspension. This is usually performed 3 wk after the electroporation.

3.3.1. Chromosomal DNA Preparation

1. Prepare 2% LMT agarose (SeaPlaque GTG) in PBS and keep in water bath at 42°C.
2. Remove 9 mL of cell suspension containing 1.2–1.6 × 10^8 parasites in late log phase. This will be enough to prepare two DNA plugs.
3. Spin down the cells at 900g at +4°C for 10 min and resuspend the cell pellet in PBS buffer. Repeat this process twice to wash the pellet.
4. Prior to the final resuspension, count the cells using a hemocytometer. Resuspend the cells in PBS at a concentration of 1.6 × 10^9/mL (in about 0.1 mL).
5. Add an equal volume of LMT agarose kept at 42°C. Mix quickly and pour immediately into the plug molds (100 µL per plug). Leave to solidify 15 min at +4°C.
6. Remove the agarose plugs from the molds using a fine spatula, and drop the plugs into labeled microfuge tubes containing 1 mL of lysis buffer with proteinase K (1 mg/mL).
7. Incubate overnight at 50°C, then store at +4°C until required.

3.3.2. Pulse-Field Gel Electrophoresis

1. Wash the agarose plugs 4 times in 1X TE buffer for 15 min with gentle shaking.
2. Prepare a 1.5% agarose (Seakem GTG) gel (*see* **Note 37**) in 0.5X TBE, then place the plugs in the wells and the gel in the tank in freshly prepared 0.5X TBE (*see* **Note 38**).
3. Electrophorese using appropriate parameters for the chromosome size class to be separated (*see* **Note 39**).
4. Stain the gel in 0.5X TBE containing 0.5-µg/mL ethidium bromide for 1 h. Photograph the gel under UV light, with a ruler placed on the gel to aid karyotype analysis.

3.3.3. Southern Blotting

Depurinate the gel in 0.4 N HCl for 15 min twice and then transfer onto a nylon membrane in denaturing conditions following a standard alkaline Southern blotting method (*see* **Note 40**).

3.3.4. Hybridization

The verification of the deletion of the targeted region and of its replacement by the drug-resistance gene is done by hybridization to a Southern blot of DNA probes homologous to the resistance gene and to chromosomal segments that are within and flanking the targeted chromosomal locus (*see* **Note 41**). This step allows you to select the clones of interest and discard the rest.

1. Label 30–40 ng of purified DNA probe by random priming labeling and hybridize to the DNA on the nylon filter (*see* **Note 40**).
2. Wash the filters at high stringency (at 65°C in 0.1X SSPE + 0.1% SDS).
3. Nylon blots may be stripped and rehybridized 5–10 times.

3.4. Cloning of the Transfectants by Limiting Dilution

This method has been described in several publications (for a review, *see* **ref. 23**).

1. Perform cloning of parasites at the beginning of log-phase growth, for example, 1–2 d after a one-tenth dilution of the cells (*see* **Note 42**).
2. Transfer the cell sample into a sterile 15-mL conical tube and allow the cell aggregates to sediment for 1 h.
3. Transfer 1 mL from the higher phase (which contains the most motile cells) to another tube containing medium without FCS. Homogenize vigorously by pipetting up and down to ensure that no more cell aggregates remain.
4. Count the cells precisely with a hemocytometer (mean of two countings at two dilutions).
5. Prepare tubes with one-tenth serial dilutions of the cells in medium without FCS. The last dilution should be 2.5 cells/mL in 40 mL of complete medium containing the selection drug. Mix by pipetting.

6. Pipet 0.2 mL of the final dilution into the wells of two 96-well microplates. This should result in a mean distribution of 0.5 cells per well (*see* **Note 43**). Continually swirl the tube while distributing the cells. Incubate in the 26°C CO_2 incubator.
7. From d 5, examine the plates using an inverted microscope. The clones will grow by forming rosettes. Therefore, discount any well containing more than one rosette. Using the 0.5 cell/well concentration, the cloning efficiency usually is between 50 and 90%, i.e., about 25–45 clones will grow on each of the two microplates.
8. After 2–3 wk, when dense growth is observed, transfer each sample thought to contain a single cloned parasite population to 1 mL of drug-containing complete medium in a 24-multiwell plate.
9. After 1–2 d, transfer the contents of each well separately to 2 mL of complete medium with selective drug in a 25-cm^2 flask. After 1–2 d, add a further 3–4 mL of medium. Because the limiting dilution method cannot guarantee the purity of each subclone (particularly with *Leishmania*), it is recommended to perform a second subcloning using the same protocol.

3.5. Further Analysis of the Recombination Events

The recombination events that have occurred in each selected clone must be characterized and the rearranged chromosomal region should be analyzed further. This is performed by enzymatic restriction of total chromosomal DNA, as well as PCR and sequencing of the junction regions. The correct or, conversely, incorrect insertion of the vector into the chromosomal locus, as well as aberrant events, can be detected in this way.

Increase the culture volumes of the selected transfected *Leishmania* clones up to 100 mL. From these cultures, prepare chromosomal DNA as in **Subheading 3.3.1.** This will yield about 20 agarose plugs, which can be kept at +4°C for >1 yr.

3.5.1. Enzymatic Digestion of Chromosomal DNA

We digest the chromosomal DNA directly in the agarose plugs prepared for PFGE.

1. Place half an agarose plug (representing 3–4 × 10^7 cells, i.e., 2–3 μg of total DNA) in a 50-mL polypropylene tube.
2. Wash in 20–40 mL of sterile cold 1X TE for 30 h at +4°C, changing the buffer six times during this period (*see* **Note 44**).
3. Transfer to a 15-mL round-bottomed polypropylene tube. Equilibrate in 1 mL of 1X digestion buffer for 2 h.
4. Remove the buffer and replace with 200 μL of new buffer or as small a volume as possible (just enough to cover the plug). Add 80 U of enzyme per tube (it is essential to use high-quality restriction enzymes) and incubate overnight at the recommended temperature.
5. Separate the restriction fragments by electrophoresis. For fragments <10 kb, load the plugs onto a 0.7–0.9 % agarose gel (Seakem GTG). For fragments >10 kb, it will be

necessary to perform PFGE (test PFGE conditions prior to the experiment to determine the best conditions for the size of fragments to be resolved). Include appropriate DNA size standards such as lambda DNA restricted with *Pst*I, *Hind*III, and/or *Eco*RI.
6. After electrophoresis, proceed to Southern blotting and hybridization as described in **Subheadings 3.3.3.** and **3.3.4.**

3.5.2. Analysis of the Integration Locus by PCR and Sequencing

This step involves PCR amplification of both junctions between the targeted chromosome and the inserted vector, and determination of the nucleotide sequence of the amplified fragments.

1. For each junction, design a pair of oligonucleotide primers, one homologous to the inserted vector sequence (in the 3'- or 5'-*DHFRTS* region, depending on the junction tested) and the other to the chromosome sequence immediately upstream or downstream (depending on the junction tested) of the insertion locus.
2. Melt an agarose plug containing the DNA of the clone of interest in a water bath at 65°C for 10 min.
3. Extract the DNA with phenol–chloroform and precipitate as in **Subheading 3.1.1.2.**
4. Resuspend the DNA pellet in 50 µL of distilled H_2O. Estimate the DNA concentration on a 1% agarose gel as in **Subheading 3.1.1.2.**
5. PCR-amplify each of the junctions in two separate tubes using 80 ng of template DNA and the parameters described in **Subheading 3.1.1.1.**, but with annealing temperatures appropriate to the T_m of the primers.
6. Load the PCR products on a 1% agarose gel. After electrophoresis, excise and purify both PCR fragments as in **Subheading 3.1.1.2.**
7. Clone the PCR products into a TA-cloning type vector (e.g., pGEMT-Easy, Promega) according to the manufacturer's instructions.
8. Sequence the products as in **Subheading 3.1.2.6.**

3.5.3. Analysis of Chromosomal Stability

To determine whether the chromosome fragmentation has affected mitotic stability of the truncated chromosome, the parasite clone is cultivated in the absence of drug pressure for 100–150 generations. Samples of cells are collected at intervals and the persistence of the fragmented chromosome is assessed both by PFGE and drug susceptibility testing (for further details, see **ref. 15**) as follows:

1. At various time intervals, collect three samples of each culture in mid-log phase and process in three ways:
   a. Transfer the first sample into both drug-containing and drug-free medium at a concentration of 2×10^5 cells/mL. Count promastigotes daily with a hemocytometer from days 2 through 7. The percentage of drug-resistant cells in a given line is estimated by comparing the cell concentrations in both cultures in the mid-log phase (days 2–3).

b. Clone the second sample by limiting dilution as in **Subheading 3.4.** Select 20–30 subclones. For each subclone, (1) assess the drug-resistant phenotype by double culture as in **Subheading 3.5.3.1.**, and (2) test for the presence of the truncated chromosome by PFGE as in **Subheadings 3.3.1. and 3.3.2.**
c. Assess the presence of the truncated chromosome in the population by PFGE analysis of the third sample (*see* **Subheadings 3.3.1. and 3.3.2.**).

4. Notes

1. The size of the homologous sequences is usually between 200 and 1000 bp. It may exceed this size but should not be <200 bp. Below this threshold, the efficiency of integration into the *Leishmania* genome is low or null *(22)*.
2. It is necessary to add 2–3 nt upstream of the restriction sites, so they are not located right at the ends of the PCR products. Most restriction enzymes will only recognize their cutting site if it is flanked by a minimal number of nt.
3. The cycling conditions, and particularly the annealing temperatures, may be optimized according to the theoretical T_m of the primers and the thermocycler, but note that the annealing temperature of the primers should not be modified becuase of addition of restriction sites.
4. Minimize the time of exposure to avoid radiation damage to the DNA. Also, remember that UV light is dangerous, particularly to the eyes. Wear a full safety protective mask and gloves for this procedure.
5. It is possible to use a commercial DNA extraction kit to purify the PCR products. However, it is not practical because several columns must be used to process the contents of the two to three PCR tubes, and the elution volume must be kept to a minimum.
6. We use New England Biolabs® enzymes with buffer 2 for *Kpn*I-*Xba*I and buffer 4 for *Hpa*I-*Mfe*I.
7. It is essential to ensure that purified amplification product is completely restricted for high-efficiency insertion into the vector. As restriction sites have been added to the S2 and S1 homologous fragments, double digestion will yield a fragment 12 bp smaller than the undigested PCR product. This size difference is too small to be visible on a simple agarose gel. By adding one-fifteenth of the DNA to be digested to the plasmid DNA control digestion, it is possible to monitor the progress of the sample DNA digestion by following that of the control DNA. To do this, use a plasmid in which the double digestion will yield a fragment detectable on an agarose gel (>100 bp). Otherwise, you should include one control tube for each of both enzymes. Note that this procedure tests the purity of the reaction medium and not merely the enzyme efficiency. An alternative, more lengthy procedure is to clone the PCR product into a TA-cloning type vector (e.g., pGEMT-Easy, Promega), carry out a double digestion on the recombinant plasmid, and assess the quality of the digestion by the presence of both the insert and plasmid bands on an agarose gel.
8. If the digestion is not complete, incubate for a further 1 h 30 min. Be careful to avoid overdigestion, which may alter the ends of the homologous sequences and considerably lower cloning efficiency. Ideally, follow the manufacturer's instructions.

9. The *Xba*I and *Kpn*I sites are too close on the vector to be able to verify the second digestion on a gel, hence the need for a control as described in **Subheading 3.1.1.3.** It is essential that the digestion be complete to avoid recircularization of the vector during the subsequent ligation step. Recircularization leads to considerable background (false positive clones) during selection of transfectants, making the latter much more cumbersome.
10. It is recommended to include a control consisting of the vector alone to check for the recircularization of the vector, and thus the quality of the double digestion, as pVV is lacking the α-complementation system used for identifying recombinant clones after transformation (blue/white colonies). By comparing the number of colonies obtained after transformation with the test and control ligations, respectively, one can estimate the percentage of recombinant clones (*see* **Subheading 3.1.2.4.**)
11. For optimal results, it is critical that the cells be kept at < +4°C during the whole procedure until the "heat shock."
12. It is preferable to carry out these steps in sterile conditions, preferably in a flow cabinet.
13. The yield may be considered satisfactory if the ratio of clones obtained from the control ligation compared to that obtained from the test ligation is lower than 0.25. This restricts the number of clones to be screened to isolate four to five recombinant clones.
14. PCR amplification aims to detect insertion of the fragment into the cloning site of the vector. The primers are located on pVV on each side of the *Kpn*I and *Xba*I sites: downstream of the *Kpn*I site (5'-gtaaaacgacggccag-3') and in the 3'-*DHFRTS* region (5'-cctcggcgtctcgtcatca-3') (*see* **Fig. 1**). Note that different primers are used to verify insertion of S1 and S2 sequences (*see* **Subheading 3.1.3.**).
15. We use a Licor® DNA sequencer Long Readir 4200 (Amersham). Sequencing reactions are performed using the Sequitherm EXCEL™ II DNA Sequencing Kit-LC (Epicentre Tech.) and read using the Sequencher program (Amersham). Instructions for performing sequence reactions may be found in any of the standard kits.
16. Most often, such miniprep DNA is not completely devoid of impurities. It is essential to purify to achieve complete restriction digestion—a crucial step for the construction of the vector.
17. We recommend digesting 1 µL of the purified DNA with a restriction enzyme that linearizes the vector, and loading several dilutions of this digest adjacent to varying concentrations of a DNA size standard (e.g., λ/*Pst*I). The DNA yield obtained from 400 mL of culture generally approximates 400 µg (i.e., 1 mg/mL).
18. Four *Eco*57I restriction sites are present in pVV-S1-S2. Thus, *Eco*57I digestion will yield three fragments of 800, 900, and 1200 bp and one fragment of 4500–5500 bp, depending on the size of fragments S1 and S2. The larger fragment represents the linearized vector DNA that will be introduced into the *Leishmania* cells.
19. This double purification in agarose allows a considerable reduction in the presence of circular forms of the vector, which constitute a severe limitation in transfection success.
20. The in vitro cultivation of *Leishmania* promastigotes is generally straightforward in various culture media (reviewed in **ref. 23**). However, variations in growth

efficiency are observed among different species and strains and different media. When a strain is to be cultivated in a new culture medium, it should be adapted progressively by frequent passaging at higher cell densities. Transfected cells, perhaps because they are exposed to drug selection, are more difficult to grow than wild-type cells. If the medium is not adequate (e.g., not rich enough), the plating and cloning efficiencies after transfection may be considerably reduced.

21. Culture media should be prepared in sterile conditions and stored at +4°C. Avoid storing medium more than 3–4 d. Sterility should be tested by incubating an aliquot at 37°C for a few days, followed by microscopic examination to detect microorganisms, in particular fungi, which are slower to grow and more difficult to detect.
22. Different batches of FCS may influence parasite growth as well as transfection and cloning efficiencies. It may be necessary to test several batches before a suitable batch can be purchased.
23. If no CO_2 incubator is available, use a hermetically sealed container with a CO_2 source such as Alka-Seltzer or a candle jar.
24. The culture may be observed directly on an inverted microscope without manipulation. This allows a good estimate of the cell density, morphology, and motility, and therefore vitality.
25. Regular passaging is necessary to avoid overcrowding and decline of cultures. On the other hand, be cautious not to overdilute the cells, especially transfectant clones. For species that grow faster than *L. major* and *L. donovani*, passage more frequently or with a greater dilution.
26. This test is critical. Indeed, a too high or too low drug concentration may lead to loss of transfectants or to false positives, respectively. Moreover, the test must be performed for each strain, because the lethal dose of antibiotics varies among strains. Aim to test the drug sensitivity of each strain on a regular basis, as progressive inactivation of drugs may be induced by aging or light and heat sensitivity of stock solutions. We recommend to aliquot and store stocks at –20°C.
27. After thawing, culture cells for 2–3 wk before transfection. Also, do not perform transfection with cells that have been cultured for a long time.
28. These concentrations may have to be modified depending on the species and the culture medium used. We recommend establishing a growth curve for the model in use.
29. The cell concentration is critical for best transfection efficiency. Cells should be taken in mid- to late log phase, and concentration should not exceed 8×10^6/mL (for *L. major*) or 10^7/mL (for *L. donovani*).
30. These conditions are indicative. It is necessary to test electroporation conditions for each type of parasite and apparatus. Our conditions yield 30–50% of survivors after electroporation, shown to be adequate for efficient transfection.
31. It is important to leave the transfected cells without drug selection for 24 h, to allow expression of the introduced resistance gene.
32. Careful preparation of the plates is critical. We recommend preparing both culture medium and agarose plates on the day of use (day 1).
33. If drug selection can be obtained only in liquid medium, divide contents of each cuvet into four samples immediately after electroporation, and transfer to 1 mL of

drug-free medium in a 24-well microplate for 24 h. On the next day (d 1), transfer the contents of each well to 8 mL of complete medium containing the selection antibiotics in 25-cm² culture flask. Allow to replicate, then isolate clones by the limiting dilution method (*see* **Subheading 3.4.**).
34. It is important to let the plates dry sufficiently before plating, to avoid movement of the motile *Leishmania* cells between colonies.
35. Be very careful at this stage, as the agarose is semisolid, hence particularly soft. We do not recommend using a solid spreader for this step.
36. Using our conditions, the transfection efficiency varies from 0.2 to 1×10^{-4} transfectants/electroporated cells for *L. major* and *L. donovani* and an efficient transfection should yield about 100–150 colonies. If the colonies are not picked soon enough, they will rapidly merge into a continuum, making selection of individual colonies impossible.
37. It is important to use high-quality agarose for long migrations such as PFGE and for enzymatic DNA digestion and purification.
38. To analyze a large number of transfectant clones simultaneously, it is useful to have access to several PFGE apparatuses.
39. PFGE migration conditions vary considerably among apparatuses. We use a homemade apparatus and migrate at 7.5 V/cm for 1–3 d, depending on the chromosome size class. The pulse times for different size classes can be found in Wincker et al. *(24)* for our apparatus and on the *Leishmania* Genome Network website, www.ebi.ac.uk/parasites/leish.html, for the Bio-Rad CHEF apparatus.
40. Use standard protocols *(25)*.
41. Despite all precautions taken in the purification steps, recircularization as well as multimerization of the transfection vector may occur during or after electroporation. This will give rise to false positives, i.e., drug-resistant clones that have not integrated the resistance gene. These can only be detected by hybridization analysis of Southern-blotted PFGE gels.
42. Promastigotes have a tendency to grow as "rosettes," increasing with cell density, and it is important to minimize this in the sample to be cloned. Avoid using late growth phases in which density is higher.
43. Cloning efficiency varies depending on the strain and the drug used. In some experiments, the 0.5 cell/well dilution may not yield any clones. For these "fastidious" strains, we recommend inoculating one additional plate with 1 (and sometimes 2) cells/well.
44. This step aims to remove all traces of Sarkosyl-EDTA and proteinase K. To avoid nucleases in the reaction buffers, use double-distilled H_2O and sterile tubes. Wash at +4°C.

References

1. Larin, Z. and Mejía, J. E. (2002) Advances in human artificial chromosome technology. *Trends Genet.* **18,** 313–319.
2. Pavan, W. J., Hieter, P., Sears, D., et al. (1991) High-efficiency yeast chromosome fragmentation vectors. *Gene* **106,** 125–127.

3. Vollrath, D., Davis, R. W., Connelly, C., et al. (1988) Physical mapping of large DNA by chromosome fragmentation. *Proc. Natl. Acad. Sci. USA* **85,** 6027–6031.
4. Farr, C. J., Stevanovic, M., Thomson, E. J., et al. (1992) Telomere-associated chromosome fragmentation: applications in genome manipulation and analysis. *Nat. Genet.* **2,** 275–282.
5. Barnett, M. A., Buckle, V. J., Evans, E. P., et al. (1993) Telomere directed fragmentation of mammalian chromosomes. *Nucleic Acids Res.* **21,** 27–36.
6. Surosky, R. T., Newlon, C. S., and Tye, B. K. (1986) The mitotic stability of deletion derivatives of chromosome III in yeast. *Proc. Natl. Acad. Sci. USA* **83,** 414–418.
7. Praznovszky, T., Kereső, J., Tubak, V., et al. (1991) De novo chromosome formation in rodent cells. *Proc. Natl. Acad. Sci. USA* **88,** 11,042–11,046.
8. Harrington, J. J., Van Bokkelen, G., Mays, R. W., et al. (1997) Formation of *de novo* centromeres and construction of first-generation human artificial microchromosomes. *Nat. Genet.* **15,** 345–355.
9. Csonka, E., Cserpan, I., Fodor, K., et al. (2000) Novel generation of human satellite DNA-based artificial chromosomes in mammalian cells. *J. Cell Sci.* **113,** 3207–3216.
10. Lee, M. G., E., Y., and Axelrod, N. (1995) Construction of trypanosome artificial mini-chromosomes. *Nucleic Acids Res.* **23,** 4893–4899.
11. Patnaik, P. K., Axelrod, N., Van der Ploeg, L. H. T., et al. (1996) Artificial linear mini-chromosomes for *Trypanosoma brucei*. *Nucleic Acids Res.* **24,** 668–675.
12. Horn, D., Spence, C., and Ingram, A. K. (2000) Telomere maintenance and length regulation in *Trypanosoma brucei*. *EMBO J.* **19,** 2332–2339.
13. Pace, T., Scotti, R., Janse, C. J., et al. (2000) Targeted terminal deletions as a tool for functional genomics studies in *Plasmodium*. *Genome Res.* **10,** 1414–1420.
14. Tamar, S. and Papadopoulou, B. (2001) A telomere-mediated chromosome fragmentation approach to assess mitotic stability and ploidy alterations of *Leishmania* chromosomes. *J. Biol. Chem.* **276,** 11,662–11,673.
15. Dubessay, P., Ravel, C., Bastien, P., et al. (2001). Effect of large targeted deletions on the mitotic stability of an extra chromosome mediating drug resistance in *Leishmania*. *Nucleic Acids Res.* **29,** 3231–3240.
16. Dubessay, P., Ravel, C., Bastien, P., et al. (2002) Mitotic stability of a CDS-free version of *Leishmania* major chromosome 1 generated by targeted chromosome fragmentation. *Gene* **289,** 151–159.
17. Cruz, A, Coburn, C. M., and Beverley, S. M. (1991) Double targeted gene replacement for creating null mutants. *Proc. Natl. Acad. Sci. USA* **88,** 7170–7174.
18. Laban, A., Tobin, J. F., Curotto de Lafaille, M. A., et al. (1990) Stable expression of the bacterial neo^r gene in *Leishmania enriettii*. *Nature* **343,** 572–574.
19. Freedman, D. J. and Beverley, S. M. (1993) Two more independent selectable markers for stable transfection of *Leishmania*. *Mol. Biochem. Parasitol.* **62,** 37–44.
20. Jefferies, D., Tebabi, P., Le Ray, D., et al. (1993) The *ble* gene as a new selectable marker for *Trypanosoma brucei*: fly transmission of stable procyclic transformants to produce antibiotic-resistant bloodstream forms. *Nucleic Acids Res.* **21,** 191–195.

21. Ryan, K. A., Dasgupta, S., and Beverley, S. M. (1993) Shuttle cosmid vectors for the trypanosomatid parasite *Leishmania*. *Gene* **131,** 145–150.
22. Papadopoulou, B., Roy, G., and Dumas, C. (1997) Parameters controlling the rate of gene targeting frequency in the protozoan parasite *Leishmania*. *Nucleic Acids Res.* **25,** 4278–4286.
23. Jaffe, C. L., Grimaldi, G., and McMahon-Pratt, D. (1984) The cultivation and cloning of *Leishmania*, in *Genes and Antigens of Parasites. A Laboratory Manual, 2nd ed.* (Morel, C. M., ed.), Fundaçao Oswaldo Cruz & UNDP/World Bank/WHO, Rio de Janeiro, Brazil, pp. 47–91.
24. Wincker, P., Ravel, C., Blaineau, C., et al. (1996) The *Leishmania* genome comprises 36 chromosomes conserved across widely divergent human pathogenic species. *Nucleic Acids Res.* **24,** 1688–1694.
25. Sambrook, J., Fritsch, E. F., and Maniatis, T. (1989) Molecular Cloning. A Laboratory Manual. Cold Spring Harbor Laboratory Press, Cold Spring Harbor, NY.

21

FISH Mapping for Helminth Genome

Hirohisa Hirai and Yuriko Hirai

Summary

Basic techniques for fluorescence *in situ* hybridization (FISH) mapping that have been used in genome projects on schistosomes and filariae are introduced. The chapter shows techniques specific for bacterial artificial chromosome (BAC) and yeast artificial chromosome (YAC) clones and includes experiences of chromosome preparation, DNA labeling, hybridization, microscopy, and localization of BAC clones.

Key Words: BAC clones; chromosome mapping; fluorescence *in situ* hybridization (FISH); parasite genome; schistosomes.

1. Introduction

Fluorescence *in situ* hybridization (FISH), established in the 1980s, is a good tool for chromosome mapping of eukaryote genomes (e.g., **ref. 1**). During the past decade, many genome projects have been initiated, as yeast artificial chromosome (YAC) (e.g., **refs. 2** and **3**) and bacterial artificial chromosome (BAC) (e.g., **refs. 4** and **5**) cloning systems made it feasible to clone whole genomes of various eukaryotes. In schistosomes, YAC (**6**) and BAC (**ref. 7**, http://www.chori.org/bacpac/schis103.htm) clones have been developed for whole genome analyses. To take advantage of such large fragments (average 350 and 100 kb, respectively), chromosome localization techniques have been developed to map YAC and BAC DNA clones. For example, a FISH technique was established for mapping genes to the chromosome of schistosomes (**8**). Additionally, *in situ* polymerase chain reaction (PCR) on chromosomes (PRINS) has been applied to map repetitive sequences such as telomere repeats (**9**). As chromosome mapping techniques were first established for humans and mammals, several steps were modified for application to schistosome or filarial chromosomes.

The steps of the FISH and PRINS techniques, from chromosome preparation to a description of results showing the location of cloned genomic DNAs, will be illustrated in this chapter.

2. Materials

2.1. Chromosome Preparation

1. Snails infected with sporocysts of *Schistosoma mansoni*.
2. Filaria adult worms.
3. Glass slides (Matsunami, S-2111).
4. Ethanol (99.5%).
5. Glacial acetic acid.
6. Colchicine (Sigma, C-9754).
7. 1% sodium citrate.
8. Pasteur pipets with a tip of 2 mm diameter.
9. Two needles for insect dissection.
10. 5-mL syringe with a 18-ga needle.
11. Polypropylene 15-mL conical tube.
12. Cavity culture glass slide (Matsunami) and coverslips.
13. Ice.
14. Kimwipes.
15. 37°C incubator.
16. 1% sodium citrate or 0.56% potassium chloride.
17. Fixative F-1: 60% 1/1 ethanol/acetic acid.
18. Fixative F-2: 1/1 ethanol/acetic acid.
19. Fixative F-3: glacial acetic acid.

2.2. Labeling for Probes

1. Double-distilled water (ddH_2O).
2. Ice.
3. BioPrime DNA labeling system (Invitrogen, 18094-011).
4. 99.5% and 70% ethanol (store at −20°C).
5. Formamide (Intergen, S-4117).
6. BioNick labeling system (Invitrogen, 18247-015).
7. 0.5 M EDTA.
8. 3 M sodium acetate.
9. 20X SSC: 3 M NaCl, 0.3 M sodium citrate.
10. TE: 10 mM Tris-HCl, pH 8.0, 1 mM EDTA.
11. 5 M sodium chloride.
12. Salmon testes DNA, 9.5 mg/mL (Sigma, D-7656).
13. Schistosome genomic DNA digested with *Sau*3a.
14. 30% dextran sulfate (Pharmacia).
15. Etachinmate (Takara) (a reagent for precipitating small probes).

2.3. Hybridization and Detection of Signals

1. Sodium hydroxide pellets.
2. Parafilm.
3. BI buffer: 0.1 M sodium bicarbonate, 0.1% Igepal (Sigma, I-3021).
4. Nonfat milk (Carnation).
5. FITC-avidin DCS (Vector, A-2011).
6. *p*-Phenylendiamine (Sigma, P-6001).
7. DAPI (4',6-diamine-2'-phenylindole dihydrochloride) (Sigma, D-9542).
8. Propidium iodide (PI) (Sigma, P-4170).
9. *In situ* PCR (PRINS) machine (Hybaid, HB-TR3-CMFB).
10. Suitable oligonucleotides (e.g., [CCCTAA]$_7$ for telomere localization, as described in **Subheading 3.3.3.**).
11. Dig-PRINS labeling kit (formerly Boehringer Mannheim, 1695932): 500 µM each of dATP, dCTP, and dGTP and 50 µM of Dig-11-dUTP.
12. *Taq* DNA polymerase (Gibco-BRL) and appropriate 10× buffer.
13. Rubber cement.
14. Stop solution: 50 mM NaCl, 50 mM EDTA.
15. Fluorescence microscope (Zeiss, Axioplan 2).
16. IPLab spectrum imaging software (Scanalytics, Inc.).
17. Photoshop graphic software 6.0 (Adobe Systems).
18. Canvas graphic software 7.0 (Deneva).
19. Personal computer (Apple).
20. PXL 1400 CCD or Cool SNAP HQ camera system (Photometrics).

3. Methods

The following sections describe chromosome preparation for schistosomes and filarial parasites (**Subheading 3.1.**), labeling of probes (**Subheading 3.2.**), *in situ* hybridization, posthybridization wash, detection of signals, PRINS (**Subheading 3.3.**), microscopy and computation (**Subheading 3.4.**), and the description of *in situ* locations of probes (**Subheading 3.5.**).

3.1. Chromosome Preparation

This step is critical for good FISH results. If the quality of chromosome preparations is not adequate, then it will affect the hybridization reactions. Below, we present our methods, which have consistently provided good-quality chromosome preparations from schistosomes and filarial parasites (*see* **Note 1**).

3.1.1. Schistosomes or Other Trematodes

This technique of chromosome preparation (**Fig. 1**), which results in chromosomes with suitable length for FISH analysis in a short time *(10)*, was modified from the original method *(11)*. First, one needs *S. mansoni*-infected snails with sporocysts that just started shedding cercariae, generally 25–35 d postinfection.

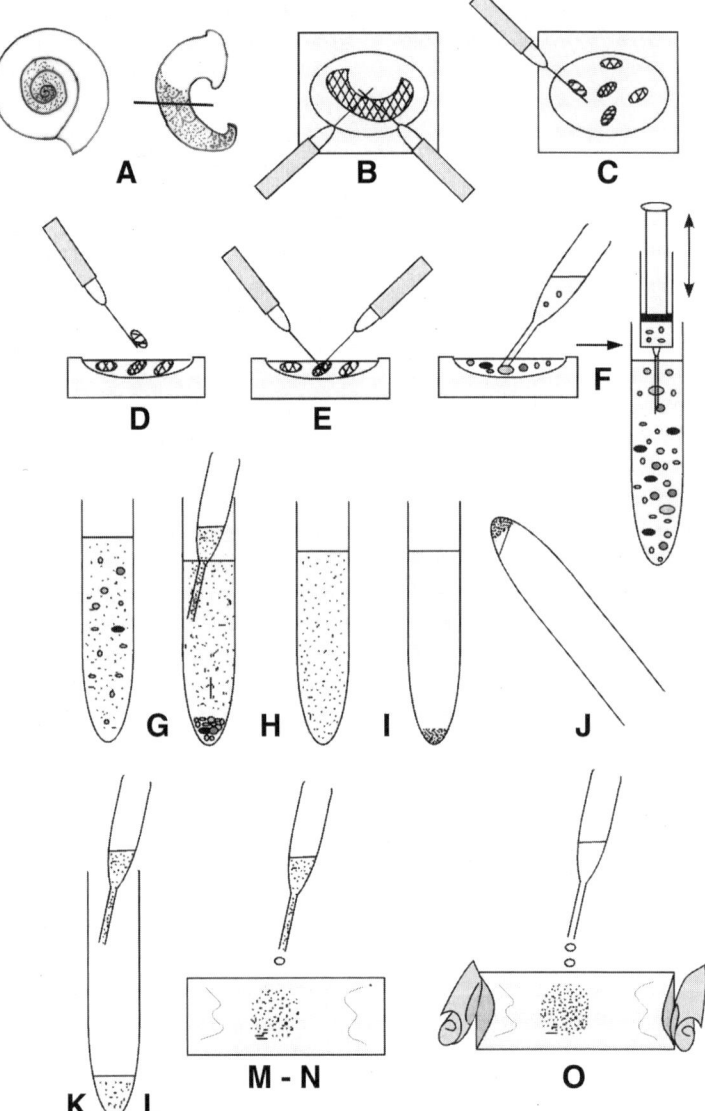

Fig. 1. Chromosome preparation technique using intramolluscan stages for schistosomes or other trematodes. See text for details.

1. Gently crush and dissect infected snails to remove only hepatic tissues containing sporocysts, using two dissection needles (**Fig. 1A**).
2. Put the infected tissue into 0.005% colchicine hypotonic solution (in distilled H_2O), cut into several blocks for sporocysts to be well exposed to the hypotonic solution, and remove unnecessary (nonsporocyst) tissues (**Fig. 1B**).

FISH Mapping 383

3. Place at room temperature for 20–30 min to make hypotonic treatment and to arrest cell division at metaphase (**Fig. 1C**).
4. Transfer the tissues into fixative (3:1ethanol:acetic acid) in a cavity culture glass slide (**Fig. 1D**).
5. Tease the tissues with two dissection needles into pieces as small as possible (**Fig. 1E**).
6. Transfer the small tissues to a 15-mL polypropylene conical tube together with fresh fixative (**Fig. 1F**).
7. Adjust the final volume to 10 mL and triturate the cells from the tissues by extrusion five times with a 5-mL syringe set with an 18-ga needle (**Fig. 1F**).
8. Let mixture stand at room temperature for 10 min to obtain a cell suspension (**Fig. 1G**).
9. Transfer supernatant only to a new conical tube (**Fig. 1H**).
10. Centrifuge at 780–970g for 5 min (**Fig. 1I**).
11. Discard supernatant (**Fig. 1J**) and repeat **steps 10** and **11** once more to wash the cells, but do not repeat more than twice, otherwise cells with good chromosome condition may be lost.
12. Add a small volume of fixative (usually several hundred microliters depending on cell number; check cell number by examining a slide preparation with a phase-contrast microscope) to make cell suspension with a Pasteur pipet (**Fig 1K**).
13. Chill the suspension on ice for at least 10 min or, if not used immediately, store at –20°C until needed for slide preparation (**Fig. 1L**).
14. Gently apply 1 drop of the chilled cell suspension onto a dried glass slide precleaned with ethanol (99.5%) (**Fig. 1M**).
15. Place the slide on a horizontal plate. As the fixative evaporates, add two drops of chilled fixative (**Fig. 1N**). Remove surplus fixative from both edges with Kimwipe (**Fig. 1O**). Repeat this once more.
16. After removing surplus fixative from the edges, place the slide on a horizontal plate until dry.
17. Finally, store the preparations in a 37°C incubator for a few days before using (*see* **Notes 2** and **3**).

3.1.2. Filarial Parasites or Other Nematodes

Because nematodes do not have an alloiogenesis stage as trematodes do, only adult female and male worms are useful for preparing chromosome slides. That is, only testes and ovaries and/or eggs *in utero* are available to obtain chromosomes. A technique that was originally established for insects is appropriate for nematode chromosome preparation *(12)* (*see* **Fig. 2**).

1. Dissect adult worm and isolate testes or ovaries from each sex (**Fig. 2A**).
2. Put the tissues into 0.005% colchicine hypotonic solution in 1% sodium citrate or 0.56% potassium chloride for 20–30 min (**Fig. 2B**).
3. Transfer tissues with two dissection needles (or a Pasteur pipet) to a dried glass slide precleaned with ethanol (99.5%) (**Fig. 2C**).

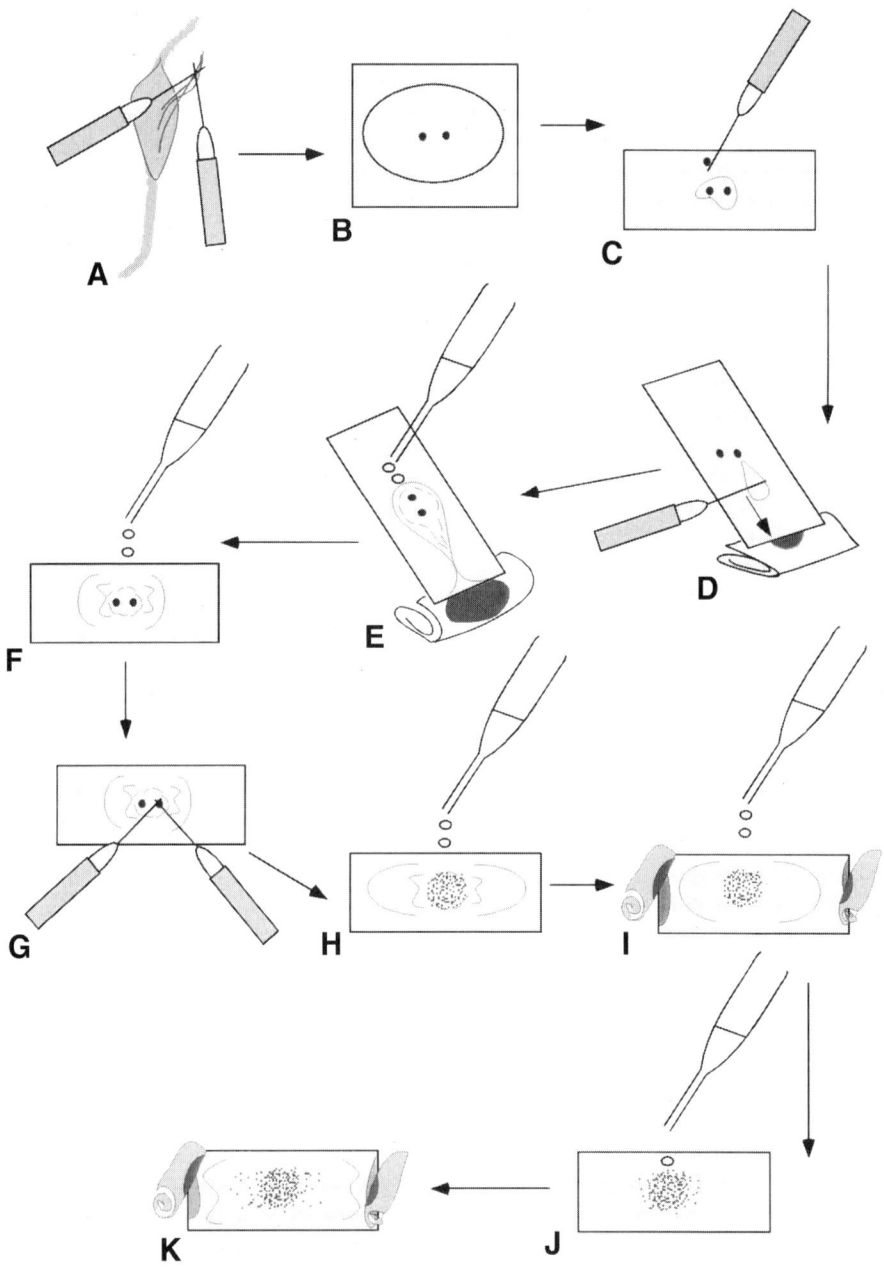

Fig. 2. Chromosome preparation technique using adult worms for filarial parasites or other nematodes. See text for details.

FISH Mapping 385

4. Remove surplus hypotonic solution around the tissues with a needle (**Fig. 2D**), wash the tissues with Fixative F-1, then saturate tissues and surface of glass slide by dropping F-1 with a Pasteur pipet (**Fig. 2E**).
5. Remove saturated fixative using Kimwipes and by reclining the glass slide. Completely remove fixative containing hypotonic solution from around tissue with a needle (**Fig. 2E**).
6. Add two drops of F-1 on the tissue (**Fig. 2F**).
7. Tease the tissues as small as possible with two needles under a dissecting microscope (**Fig. 2G**).
8. After teasing the tissue, add two drops of F-2 (**Fig. 2H**).
9. Remove surplus fixative from both edges of the slide with Kimwipes and carefully observe drying (**Fig. 2I**).
10. Add two drops of F-3 just before the fixative evaporates (**Fig. 2J**).
11. Remove surplus fixative and allow to dry horizontally (**Fig. 2K**).
12. Place the dried preparations in a 37°C incubator for a few days (*see* **Note 1**). The preparations dried in the incubator are put in a slide container and can be preserved for a couple of years in a –80°C freezer.

3.2. Labeling for Probes

A labeling system is chosen according to the size of insert DNA and the type of vector. YAC and BAC clones should be labeled with a random priming system, plasmid and cosmid DNAs with a nick translation system.

3.2.1. Labeling for YAC and BAC Clones

The BioPrime DNA labeling system (Invitrogen), which contains biotin-14-dCTP as a hapten and the large fragment of DNA polymerase I (Klenow enzyme), is suitable for labeling human YAC clones *(13)*. We have confirmed the usefulness of this kit for labeling YAC *(6)* and BAC *(7)* clones of schistosomes. Here, we outline the labeling technique for BAC clones, which is similar to the technique for YAC clones.

1. Denature 1 µg of BAC clone DNA in 10 µL of ddH$_2$O by boiling for 5 min, followed by chilling in ice for a few minutes.
2. Prepare a total 50-µL reaction mixture with components of the BioPrime labeling system (10 µL denatured DNA solution, 5 µL 10X dNTP, 20 µL 2.5X buffer containing primer, 14 µL ddH$_2$O, 1 µL Klenow enzyme [use 2 µL Klenow enzyme for YAC clones]).
3. Incubate the mixture at 37°C for 180 min.
4. Stop the reaction by heating at 65°C for 10 min.
5. Prepare a total 740-µL mixture for ethanol precipitation (50 µL labeled DNA solution, 147 µL ddH$_2$O, 74 µL 5 *M* NaCl, 2 µL ssDNA [9.5 µg/µL], 2 µL [0.3 µg] *Sau*3A digested schistosome genomic DNA, 465 µL cold [–20°C] 99.5% ethanol).
6. Keep the mixture at 4°C for 60 min.

7. Spin down at 21,500g at 4°C for 5 min (see **Note 4**).
8. Discard supernatant and dry briefly in a 37°C incubator for 5 min.
9. Add 40 µL formamide to the DNA pellet.

3.2.2. Labeling for Small Fragment DNA Clones

Employing a nick translation system rather than the random prime system is better for DNA clones with smaller fragments. We use the BioNick labeling system (Invitrogen). This kit uses biotin-14-dATP as a hapten.

1. Make a total 50 µL reaction mixture (1 µg probe DNA, adequate volume of ddH$_2$O depending on volume of DNA solution, 5 µL 10X dNTP mix, 5 µL enzyme mix).
2. Briefly mix and place in a 16°C incubator for 1 h.
3. Stop the reaction by adding 5 µL of 0.5 M EDTA for 5 min at room temperature.
4. Add 45 µL ddH$_2$O to give a total volume of 100 µL.
5. Precipitate the mixture by adding 3.3 µL of 3 M sodium acetate, 1 µL salmon testes DNA, 1 µL Ethachinmate, and finally 200 µL 99.5% ethanol at room temperature.
6. Completely mix, then spin down the mixture at 21,500g at room temperature (see **Note 4**).
7. Discard supernatant and dry in a 37°C incubator for 5 min.
8. Add 40 µL formamide to the DNA pellet.

3.3. Hybridization and Detection of Signals

The methods for hybridization are the same as for other eukaryotes except for denaturation of chromosomal DNA. In schistosomes, an alkaline denaturing method *(8)* is used so that C-band information, which is useful for identifying each chromosome, can be obtained. However, the alkaline method erases the chromosomes on a slide preparation of filarial parasites, and therefore, a heat method *(1)* is used for genome mapping of filarial parasites.

3.3.1. In Situ *Hybridization*

1. Prepare hybridization buffer with 15 µL 20X SSC and 45 µL 30% dextran sulfate.
2. Prepare hybridization mixture: in the case of large probes, with 30 µL hybridization buffer and 20 µL labeled probe DNA saturated in 40 µL formamide; in the case of small probes, with 15 µL hybridization buffer and 10 µL labeled probe.
3. Denature the probe DNA at 72°C for 10 min.
4. Denature the chromosome DNA: in the case of schistosomes, with 0.05 M NaOH (pH 12.5) in 2X SSC for 4.5 min, followed by dehydration with 70% and 99.5% ethanol for 5 min each and air-dry; in the case of filarial chromosome DNA, use a heat treatment with 70% formamide in 2X SSC at 72°C for 2 min, followed by dehydration with 70% and 99.5% ethanol. Chill on ice.
5. Apply each volume of the denatured DNA probe mix onto a slide containing denatured chromosome and cover the solution with Parafilm (cut to size) or an adequate coverslip.
6. Incubate the slide in a moist chamber at 37°C for 12–16 h.

FISH Mapping

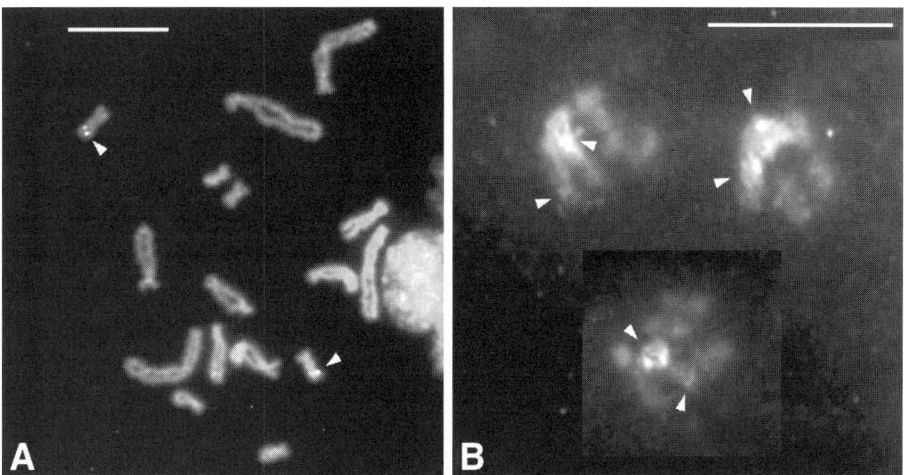

Fig. 3. FISH images of *Schistosoma mansoni* chromosomes hybridized with a BAC clone (**A**) and of *Brugia malayi* chromosomes (meiotic metaphase-I) hybridized with a plasmid clone (**B**). Scale is 10 μm. White arrowhead indicates location of probe hybridization and fluorescence.

3.3.2. Posthybridization Wash and Detection of Signals *(see also **Note 5**)*

1. Gently wash the hybridized slide preparation with 40% formamide in 2X SSC at 42°C, then 2X SSC (no formamide) at 42°C, and then 2X SSC at room temperature for 10 min each.
2. Immerse the washed slide in BI buffer for 5 min.
3. Block the surface with 5% nonfat milk in BI buffer for 10 min.
4. Apply 50 μL detection solution made of 5% milk BI buffer and 2 μg FITC-avidin onto the area of hybridized DNA.
5. Keep the slide in a moist chamber at 37°C for 60 min.
6. Wash the slide with BI buffer on a shaker for 10 min. Repeat wash.
7. Apply 20 μL antifade solution containing the fluorochromes, PI and DAPI, at 300 ng/mL each.
8. Cover with a 24 × 36 mm² coverslip and remove the surplus solution with a Kimwipe (*see* **Note 6**).

The preparation is now ready to observe signals with a fluorescence microscope. The hybridized slide preparation exhibiting signals (*see* **Fig. 3A,B**) can be kept in a refrigerator for a couple of months without experiencing fading.

3.3.3. PRINS Reaction

The PRINS technique has improved the rapidity, simplicity, and sensitivity of chromosome localization of specific DNAs *(14,15)*. It was developed to locate

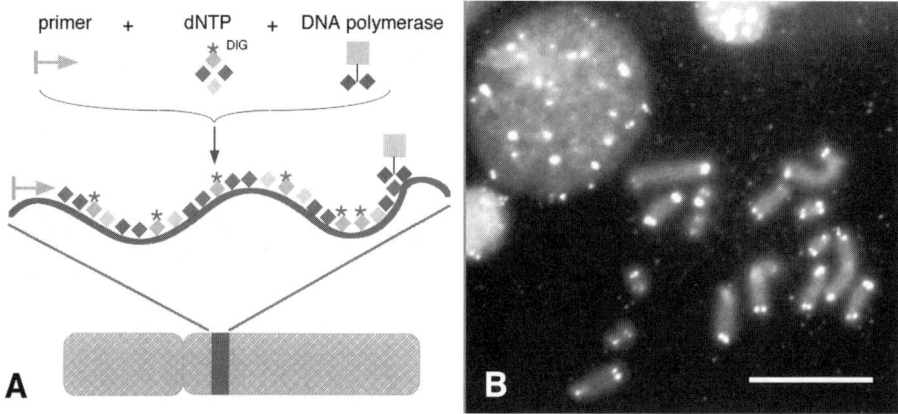

Fig. 4. PRINS technique. (**A**) Outline of the technique (modified from **ref. *16***). (**B**) PRINS location of telomere sequence in *Schistosoma japonicum*. Scale is 10 µm.

DNA sequences on chromosomes using oligonucleotide primers (*see* **Fig. 4A**). In **Fig. 4B**, the chromosome location for the telomere sequence of a schistosome, $(TTAGGG)_n$ (*9,17*), is shown as an example.

1. Prepare a 30 µL-mixture consisting of 100 pMol oligonucleotide primer $(CCCTAA)_7$, labeling, reaction buffer, and 2.5 U of *Taq* DNA polymerase.
2. Apply the mixture onto a slide preparation containing chromosomes and cover with a clean 22 × 40 mm^2 coverslip. Seal with rubber cement (Elmer).
3. After the rubber cement dries, place the slide onto an *in situ* block in a PCR machine.
4. Run PCR program: 93°C for 5 min to denature chromosomal DNA and 61°C for 30 min to anneal the primer and to extend the chain.

Stop the reaction with a stop solution (50 m*M* NaCl, 50 m*M* EDTA). Post-PCR wash and signal detection are the same as described for FISH mapping, but the carrier to detect reacted regions must be selected according to the hapten used for labeling DNA. Also, the PCR conditions are different for each primer or template DNA used (*see* **Note 7**).

3.4. Microscopy and Computation

To observe and record location of signals from probes hybridized *in situ*, a fluorescence microscope and a software for image analyses are required. For obtaining images, we have used an epifluorescence microscope mounted with a charge-coupled device (CCD) camera and, for controlling images, a personal computer running the IPLab imaging software. Because the CCD camera system is fitted with an autowheel filter set regulated by the associated software, images can be reconstructed on a monitor without image aberration by using three different fluors: DAPI, PI, and FITC.

This system is very useful for saving image data in a short time and identifying the accurate location of signals. Because the image of DAPI or PI alone shows locations of the constitutive heterochromatin (C-band), comparison with FITC image signals enables us to identify the chromosome by number and to locate the *in situ* position of a probe. C-bands are a good landmark for identifying each chromosome in schistosomes, together with centromere position and chromosome size *(8–10)*. Unfortunately, G-bands, which are the best tool for mapping mammal genes, are not useful for metazoan parasites such as schistosomes. However, the appearance of C-bands in a single image is very useful and makes it easy to locate the probes. In particular, interstitial C-bands in chromosome arms are good markers for chromosome mapping (*see* **Fig. 5A,B**).

3.5. Description of Signal Location

As G-bands are not available for schistosomes, C-band and centromere position are used as landmarks to identify chromosomes. However, these characteristics are not enough to read accurately the position of signals on schistosome chromosomes. Therefore, to locate DNA probes exactly, we use the distance of the signal position from the tip of the short arm as shown in **Fig. 6**. Because the IPLab imaging software has a function to measure the distance on the monitor, it can be done on a FITC image on the computer monitor. The value obtained as the pixel number can be transformed into the relative length (percent) by dividing it with the total pixel number (total length) of each chromosome. Using relative length, the location of each probe can be indicated on each schematic standard chromosome.

Before measuring the position of the signal, a schematic standard chromosome map needs to be constructed for each chromosome showing the size of each chromosome and location of C-bands. **Figure 7** is a schematic of a standard chromosome diagram for *S. mansoni*. The size of each chromosome allows an estimate of the DNA content of each chromosome (*see* **Table 1**), assuming the haploid genome size of *S. mansoni* is 270 Mbp *(18)*.

For filarial parasites, the standardization of chromosomes is more difficult because their chromosomes are holocentric, which means they have no primary constriction on the chromosome. Therefore, there are no markers to distinguish one chromosome from the other. The genome size of the filarial parasite (*Brugia malayi*) is smaller (100 Mbp) *(19)* than that of *S. mansoni*. Sex chromosomes of *B. malayi* are larger than the autosomes: the largest X-chromosome has a C-band at each end of the chromosome and the medium-sized Y-chromosome is

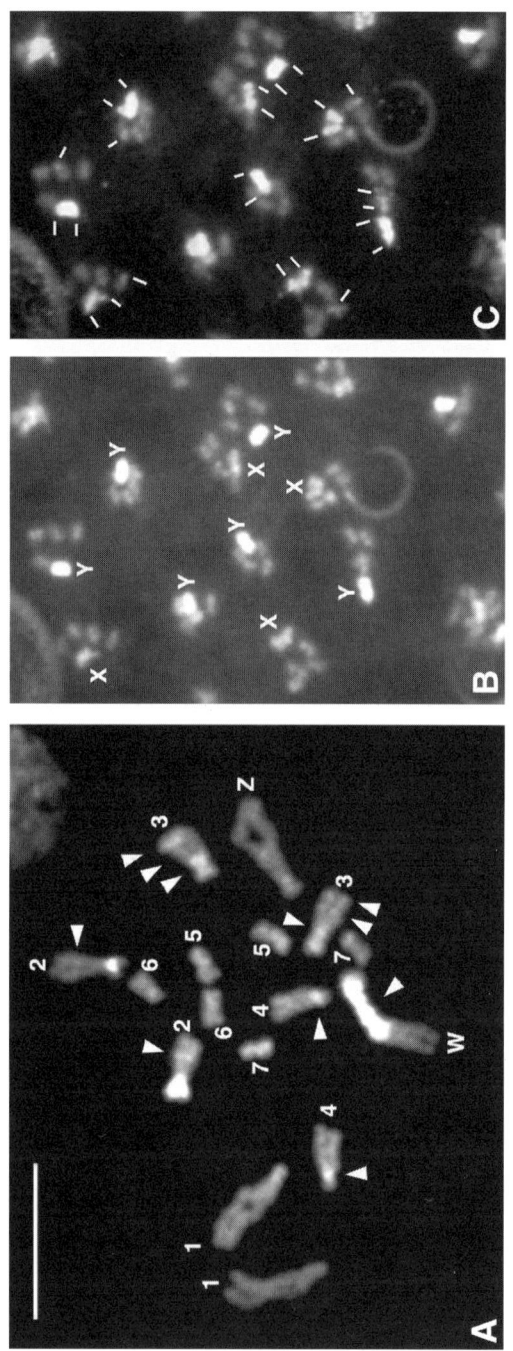

Fig. 5. C-band pattern of *Schistosoma mansoni* (**A**) and of *Brugia malayi* (meiotic metaphase-II) (**B**) chromosomes. (**C**) Relationship of C-band pattern and a repetitive sequence in *B. malayi*. Arrow head indicates C-band region in *S. mansoni* chromosomes. Autosomes are numbered (1–7), and male (ZZ) and female (ZW) sex chromosomes are identified in schistosomes. In contrast, filarial parasites have male (XY) and female (XX) sex chromosomes. In sex chromosome nomenclature, ZW and XY are used for female and male heterogametic systems, respectively. White small bars indicate chromosome regions hybridized with a repetitive sequence of *B. malayi*. Scale is 10 μm.

FISH Mapping 391

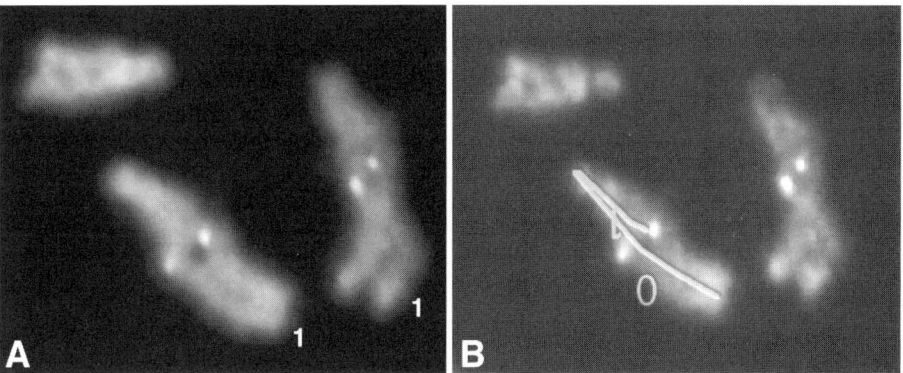

Fig. 6. Hybridization signal in chromosome 1 of *S. mansoni* (**A**). Measurement of signal location with imaging software (**B**). The distance of the signal position from the tip of the short arm is obtained and then divided by the total length of the chromosome.

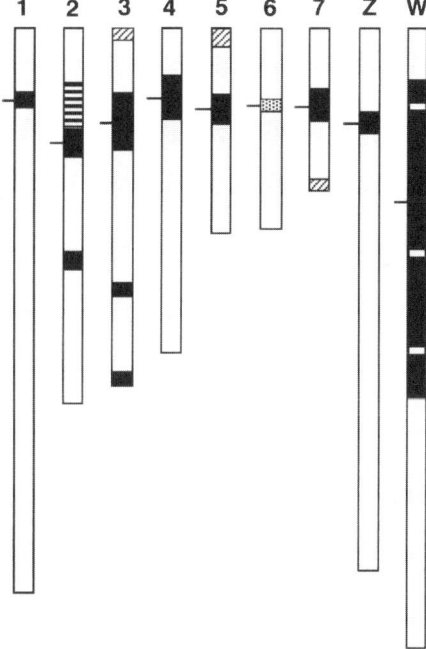

Fig. 7. Schematic of *S. mansoni* chromosomes based on ratios of measurements of relative length to total length of karyotype. White region is euchromatic DNA, black virgule, and dotted regions are constitutive heterochromatic (C-band) regions. Horizontal lines in chromosome 2 represent a nucleolar organizer region (NOR). Small bar to left side of each chromosome indicates centromeric region.

Table 1
Relative Length and DNA Content in Each Chromosome of *S. mansoni*

Chromosome	Relative length (%)	DNA content (Mbp)
1	20.8	56.16
2	13.7	36.99
3	13.6	36.72
4	11.7	31.59
5	7.2	19.44
6	7.1	19.17
7	5.7	15.39
Z	20.2	54.54
W	23.1	62.37
Total	100	270

totally heterochromatic. Chromosome 1 has a C-band at one end *(20)*. **Figure 5B** shows the chromosomal features in meiotic second meta-phase (M-II) chromosomes of a male filarial parasite. The chromosome spread has only one X-chromosome consisting of two sister chromatids and another one possesses only a Y-chromosome, consisting of two sister chromatids. The location of repetitive DNA in the heterochromatic regions of chromosomes 1, X and Y is shown in **Figs. 3B** and **5C**. So far the other chromosomes have no specific identifying characteristics.

4. Notes

1. The best conditions for chromosome preparation are 60–70% humidity and 20°C.
2. As the chromosome preparation is most critical to obtain good results from FISH, it is important to obtain samples with good cell division, not too young or too old. Humidity of 60–70% is best for obtaining elongated straight chromosomes. Slide preparations with too many cells are prone to exhibit more false positives, so the cell suspension should be adjusted to optimize the number of cells on a slide.
3. The fixed cell samples should be gently changed to new fixative if they have been stored in a freezer for a long time after preparation, and both the cell suspension and the fixative for droppings on the slide should be chilled on ice before making chromosome spreads. Additionally, one has to suitably adjust the timing of the second and third dropping of fixative by carefully watching the surface of the slide, being careful not to drop too early or too late.
4. For ethanol precipitation after DNA labeling, the concentration of salt and temperature of centrifugation is determined in terms of the size of the clone. These factors seem to affect the recovery of the labeled DNA.

5. Drying the slide preparation after hybridization appears to make the signal dull. Do not block for too long with nonfat milk BI buffer, as this results in a signal remote from the region hybridized with the probe, causing the signal image to be out of focus. The dried milk has to be nonfat.
6. If antifade solution mounted on the slide is contaminated by lens immersion oil, it makes the image dull and out of focus. If room temperature is higher than 25°C, a device to cool the CCD camera from outside is required to obtain good images over a long period and to prevent damage of the imaging analyzer.
7. The cycle PCR reaction may also be applied to chromosomes, and single-copy genes may be detected *in situ* by a combined method of PCR and FISH techniques. This technique is very useful to identify each chromosome after FISH location using appropriate repetitive sequences, particularly where chromosomes lack good diagnostic markers, such as those that are small or holocentric.

Acknowledgments

The authors thank Dr. P. T. LoVerde for his long-time collaboration, valuable comments, and revision of English in the manuscript. We also thank Dr. J. Foster for the opportunity to map a clone of *Brugia malayi*. This work was supported by Japan Society of Promotion of Science Grant 13557021 (to H. Hirai) and National Institutes of Health grant 1-U01-AI48828-01 (Subrecipient H. Hirai).

References

1. Pinkel, D., Straume, T., and Gray, J. W. (1986) Cytogenetic analysis using quantitative, high sensitivity, fluorescence hybridization. *Proc. Natl. Acad. Sci. USA* **83,** 2934–2938.
2. Imai, T. and Olson, M. V. (1990) Second-generation approach to the construction of yeast artificial chromosome libraries. *Genomics* **8,** 297–303.
3. Bellanné-Chantelot, C., Lacroix, B., Ougen, P., et al. (1992) Mapping the whole human genome by fingerprinting yeast artificial chromosomes. *Cell* **70,** 1059–1068.
4. Shizuya, H., Birrcn, B., Kim, U.-J., et al. (1992) Cloning and stable maintenance of 300-kilobase pair fragments of human DNA in *Escherichia coli* using an F-factor-based vector. *Proc. Natl. Acad. Sci. USA* **89,** 8794–8797.
5. Kim, U.-J., Birreb, B. W., Slepak, T., et al. (1996) Construction of a human bacterial artificial chromosome library. *Genomics* **34,** 213–218.
6. Tanaka, M., Hirai, H., LoVerde, P. T., et al. (1995) Yeast artificial chromosome (YAC)-based genome mapping of *Schistosoma mansoni*. *Mol. Biol. Parasitol.* **69,** 41–51.
7. Le Paslier, M.-C., Pierce, R. J., Merlin, F., et al. (2000) Construction and characterization of a *Schistosoma mansoni* bacterial artificial chromosome library. *Genomics* **65,** 87–94.
8. Hirai, H. and LoVerde, P. T. (1995) FISH techniques for constructing physical maps on schistosome chromosomes. *Parasitol. Today* **11,** 310–314.

9. Hirai, H., Taguchi, T., Saitoh, Y., et al. (2000) Chromosomal differentiation of the *Schistosoma japonicum* complex. *Intl. J. Parasitol.* **30,** 441–452.
10. Hirai, H., Spotila, L. D., and LoVerde, P. T. (1989) *Schistosoma mansoni*: chromosome localization of DNA repeat elements by *in situ* hybridization using biotinylated DNA probes. *Exp. Parasitol.* **69,** 175–188.
11. Short R. B. and Grossman, A. I. (1981) Conventional Giemsa and C-banded karyotypes of *Schistosoma mansoni* and *S. rodohaini*. *J. Parasitol.* **67,** 661–671.
12. Imai, H. T., Taylor, R. W., Crosland, M. W. J., et al. (1988) Modes of spontaneous chromosomal mutation and karyotype evolution in ants with reference to the minimum interaction hypothesis. *Jpn. J. Genet.* **63,** 159–185.
13. Qin, S., Zhang, C. M., Isaacs, S., et al. (1993) A chromosome 11 YAC library. *Genomics* **16,** 580–585.
14. Koch, J. E., Kolvraa, S., Petersen, K. B., et al. (1989) Oligonucleotide-priming methods for the chromosome-specific labeling of alpha satellite DNA *in situ*. *Chromosoma* **98,** 259–265.
15. Gosden, J., Hanratty, D., Starling, J., et al. (1991) Oligonucleotide-primed *in situ* DNA synthesis (PRINS): a method for chromosome mapping, banding, and investigation of sequence organization. *Cytogenet. Cell Genet.* **57,** 100–1004.
16. Hindkjær, J., Koch, J., Mogensen, J., et al. (1994) Primed *in situ* (PRINS) labeling of DNA, in *Methods in Molecular Biology, Vol. 33:* In Situ *Hybridization Protocols* (Choo, K. H. A., ed.), Humana, Totowa, NJ.
17. Hirai, H. and LoVerde, P. T. (1996) Identification of the telomere on *Schistosoma mansoni* chromosomes by FISH. *J. Parasitol.* **82,** 511–512.
18. Simpson, A. J. G., Sher, A., and McCutchan, T. F. (1982) The genome of *Schistosoma mansoni*: isolation of DNA, its size, base, and repetitive sequences. *Mol. Biochem. Parasitol.* **6,** 125–137.
19. Maina, C. V., Grandea, A. G., III, Tuyen, L. T. K., et al. (1987) *Dilofilaria immitis*: genome complexity and characterization of a structural gene, in *Molecular Paradigm for Eradicating Parasitic Helminths* (MacInnis, A. J., ed.), Liss, New York, pp. 193–204.
20. Sakaguchi, Y., Tada, I., Ash, L. R., et al. (1983) Karyotypes of *Brugia pahanagi* and *Brugia malayi* (Nematoda: Filarioidea). *J. Parasitol.* **69,** 1090–1093.

22

Fiber-FISH: Fluorescence *In Situ* Hybridization on Stretched DNA

Klaus Ersfeld

Summary

High-resolution fluorescence *in situ* hybridization (FISH) on deproteinized, stretched DNA prepared by *in situ* extraction of whole cells immobilized on microscope glass slides allows the visualization of individual genes or other small DNA elements on chromosomes with a resolution of approx 1000 bp. Applications of fiber-FISH range from the determination of numbers of repetitive genes to establishing the physical order of cloned DNA fragments along continuous sections of individual chromosomes. Particularly in organisms with relatively small and gene dense genomes, such as protozoan parasites, fiber-FISH can easily be used as a complementary technique to classical in vitro mapping approaches.

Key Words: Digital microscopy; DNA fibers; genome mapping; repetitive DNA; stretched DNA.

1. Introduction

Fiber-FISH is defined as the application of fluorescence *in situ* hybridization (FISH) techniques on extended DNA molecules. In most cases, the target DNA is prepared by deproteinizing chromosomal chromatin by extraction with detergents or high salt concentration *(1–9)*. This treatment removes histones from chromatin and leads to a complete unwinding of the DNA to the level of a linear double helix of approx 340 nm/1000 bp. In many protocols, chromatin extraction is done *in situ* by treating whole cells attached to microscope slides. Alternatively, "naked DNA" prepared by other techniques, such as the isolation of chromosomes after pulse-field gel electrophoresis, has been used successfully *(10,11)*. Fiber-FISH techniques are employed mainly as an additional or complementary approach to map chromosomal regions physically. This may involve the ordering of a series of genomic clones (cosmids, bacterial artificial chromosomes, etc.) on a piece of continuous DNA, or, at a higher level of resolution, the visualization of the location and order of individual genes. Fiber-FISH has also been used to study DNA replication in yeast artificial chromosomes *(6)*. The resolution of Fiber-FISH can be less than 1 kb. Because 1 kb corresponds to ca. 340 nm on a completely relaxed DNA double helix, the optical resolution of a light microscope can be the limiting factor of this technique (for review, *see* **ref. 12**). An advantage over traditional mapping techniques, such as restriction analysis, is the immediate visual information obtained. This can be particularly useful when characterizing "difficult" areas on chromosomes, such as highly repetitive regions *(8,9,13,14)*. Here, the number of repeats can often be deduced directly from the microscopic analysis of a single slide.

The vast majority of published Fiber-FISH protocols describe the application of these techniques on mammalian cells. Although in principle these protocols should be applicable to DNA from other organisms such as protozoan parasites, modifications may be necessary to obtain comparable results. An essential step of the procedure is the successful extraction of nuclei and chromatin to release and spread DNA onto a slide. Because the protein compositions of these cellular components are poorly characterized in lower eukaryotes, the outcome of such biochemical manipulations is not necessarily predictable. Here, I describe two different techniques to extract chromatin *in situ*. The first protocol uses hypotonic lysis and extraction with sodium dodecyl sulfate (SDS) *(7)* and the second mainly extraction was with high salt concentration *(1,9)* to release proteins from chromatin. As far as I am aware, these are the only published protocols that have also been applied to parasites, namely, *Trypanosoma brucei*.

1.1. Fiber-FISH Protocol 1

This protocol originally was developed using metaphase-arrested cultured lymphoblasts and cultured fibroblast cells. However, the same protocol has been used on trypanosomes to map the order of genomic P1-clones (S. Melville, personal communication) (**Fig. 1**). The technique is based on the initial lysis of cells in a hypotonic buffer and a solution containing Triton X-100. Chromatin extraction is subsequently achieved by further extraction with the detergent SDS. Some details (cell numbers, *g*-force during centrifugation) have to be adapted for different types of cells. The protocol described under **Subheading 3.1.** has been used with trypanosomes (*T. brucei*).

1.2. Fiber-FISH Protocol 2

The second technique described here originally was developed to study the dynamics of DNA replication in the nucleus. ^{3}H-thymidine pulse-labeled cells were extracted with high concentration of salt *(15)*. This treatment led to the

Fiber-FISH

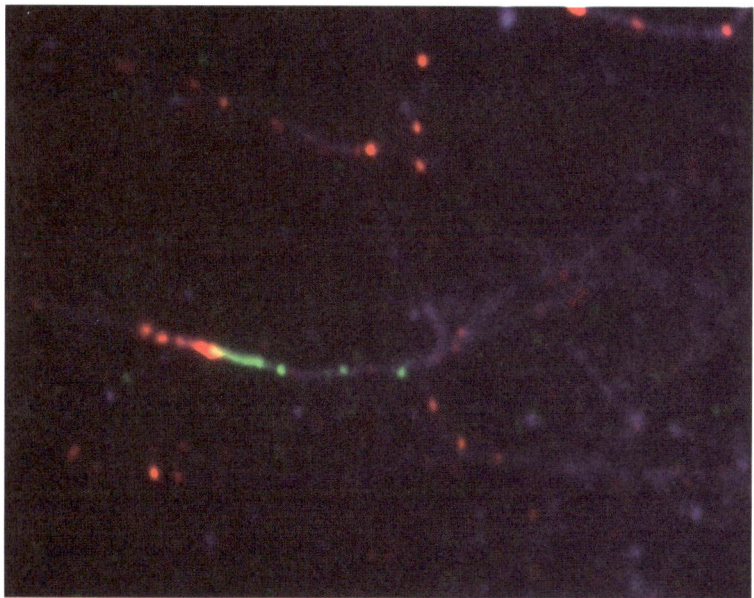

Fig. 1. Fiber-FISH on chromosomal DNA from *T. brucei* using Protocol 1. Two P1 plasmid clones with an average insert size of 60 kb were hybridized to spread DNA and simultaneously detected using a FITC conjugate (green signal) or a rhodamine-conjugate (red signal). The signals reveal an overlap of both probes (yellow area) on a chromosome. (Image courtesy of S. Melville, University of Cambridge.) Sequencing and in vitro mapping experiments confirmed the overlapping location of these two clones (located on chromosome I).

formation of DNA halos around the remnants of the nucleus and *in situ* autoradiography was used to trace the location of newly replicated DNA within these halos. Much later, this approach of extracting chromatin and releasing it from its nuclear compartment was the basis of the first published Fiber-FISH protocol *(1)*. The protocol initially was used for Fiber-FISH on mammalian tissue culture cells, but we subsequently used a slightly modified protocol successfully to study the organization of tandemly repeated genes in *T. brucei* *(9)* (**Fig. 2**). Under **Subheading 3.2.**, I describe the protocol as used with trypanosomes.

2. Materials

2.1. Fiber-FISH Protocol 1

1. Live parasites (e.g., growing in culture medium, *see* Chapters 14–16, 19, or 20).
2. Clean microscope slides.

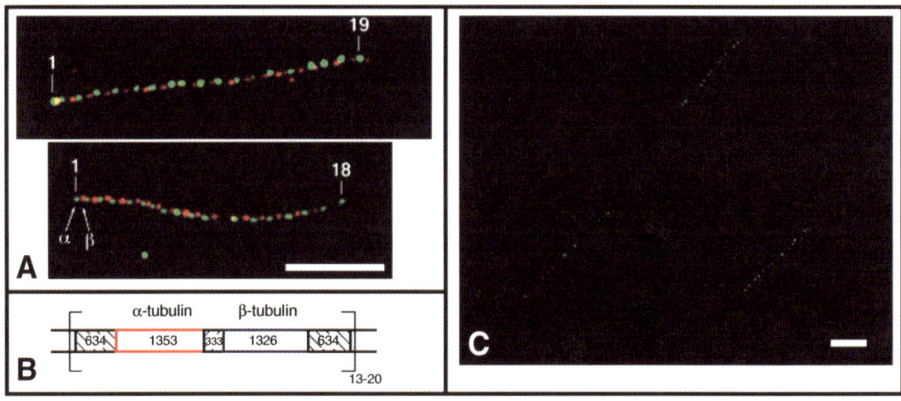

Fig. 2. Detection of repetitive gene organisation by Fiber-FISH. DNA from *T. brucei* was prepared for Fiber-FISH according to Protocol 2. Two PCR fragments, representing the coding regions for α- and β-tubulins, were labeled with digoxigenin and biotin, respectively. The probe mix was hybridized to the spread DNA and detected with a sheep-antidigoxigenin antibody and an antisheep FITC conjugate and with Cy3-streptavidin. This example demonstrates the power of resolution of Fiber-FISH. The open reading frames of α- and β-tubulin are each approx 1300 bp in length, separated by intergenic regions of ca. 300 bp (between α and β) and 600 bp (between β and α). The signals that detect the alternating α-/β-genes (α is green and β is red) are clearly separated (1000 bp of naked DNA correspond to approx 300 nm). The number of signals is comparable with the number of copies of the tandemly arrayed α-/β-tubulin genes described in the literature in this trypanosome stock. (**A**), Composite image of the merged green and red channels. (**B**), The organization of the α-/β—tubulin gene cluster as determined by sequencing. (**C**), A field of view showing only the green channel (α-tubulin). The microscope used was a Leica DMR equipped with a 100-W mercury light source and a ×100 Planapo objective (NA = 1.4). Images were taken using a Princeton Instrument Pentamax cooled CCD camera (Roper Scientific), controlled by IP-Lab software (Signal Analytics) running on a Macintosh computer. Bar, 10 μm. (Image reproduced from **ref. 6** with permission of Springer-Verlag.)

3. Coplin jars, 50-mL vol, to hold 5 slides (*see* **Note 1**).
4. Hypotonic buffer: 75 m*M* KCl, 0.2 m*M* spermine, 0.5 m*M* spermidine. Pass through a 0.22-μm filter. Prepare fresh for each experiment.
5. Polyamine buffer: 15 m*M* Tris-HCl, 0.2 m*M* spermine, 0.5 m*M* spermidine, 2 m*M* EDTA, 0.5 m*M* EGTA, 80 m*M* KCl, 20 m*M* NaCl, 14 m*M* β-mercaptoethanol, 0.25% (v/v) Triton X-100, pH 7.2. Pass through a 0.22-μm filter. Prepare fresh for each experiment. Keep on ice immediately before use.
6. Glass slides: clean with ethanol before use.

Fiber-FISH 399

7. Ice.
8. Detergent buffer: 10 mM Tris-HCl, 100 mM EDTA, 0.5% (w/v) SDS.
9. Methanol and acetic acid (mix 3 vol to 1 vol).
10. Denaturation solution: 1.5 M NaCl, 0.5 M NaOH.
11. Neutralization solution: 0.5 M Tris-Hcl, 3 M NaCl, pH 7.2.
12. 2X SSC: 0.3 M NaCl, 0.03 M trisodium citrate, pH 7.0.

2.2. Fiber-FISH Protocol 2

1. Microscope slides coated with organosilane.
2. Coplin jars, 50-mL vol, to hold 5 slides (*see* **Note 1**).
3. 2 N hydrochloric acid (HCl).
4. Double-distilled water (ddH$_2$O).
5. Acetone.
6. Phosphate-buffered saline (PBS).
7. Extraction buffer 1: 0.5% (v/v) Nonidet P-40, 10 mM MgCl$_2$, 0.5 M CaCl$_2$, 25 mM Tris-HCl, pH 8.0, 1 mM phenylmethylsulfonyl fluoride.
8. Extraction buffer 2: 2 M NaCl, 0.2 mM MgCl$_2$, 25 mM Tris-HCl, pH 8.0.
9. Propidium iodide.
10. Wash buffer 1: 0.2 M NaCl, 0.2 mM MgCl$_2$, 25 mM Tris-HCl, pH 8.0.
11. Wash buffer 2: 0.2 mM MgCl$_2$, 25 mM Tris-HCl, pH 8.0.
12. Deionized formaldehyde in PBS.
13. Ethanol: 70, 90, and 100%. Prepare dilutions with distilled H$_2$O.

3. Methods

3.1. Fiber-FISH Protocol 1

1. Culture cells to a density of approx 10^7/mL.
2. Centrifuge cells at 2000g for 2 min, and discard supernatant.
3. Resuspend trypanosomes in 10 mL of hypotonic buffer and allow to swell for 15 min at room temperature.
4. Centrifuge at 2000g for 5 min, then ensure that the supernatant is completely removed.
5. Resuspend pellet in 200 µL of freshly prepared polyamine buffer. Leave for 10 min on ice.
6. Vortex the suspension for 12 s.
7. Remove cellular and nuclear debris by a 1-min centrifugation at 100g. The supernatant contains the chromosomal material.
8. Pipet a 10-µL drop of this suspension on to an ethanol-cleaned glass slide near one end or close to the frosted section.
9. Add an equal volume of detergent buffer, gently mixing using the pipet. This step causes the extraction of remaining chromatin-associated proteins, e.g., histones.
10. Spread the liquid in a line across the width of the slide, then place the slide at an angle of 30° and allow the liquid to flow to the bottom of the slide. Drain off excess liquid.

11. Fix the preparation in 3:1 methanol–acetic acid for 2 min, air-dry and the dehydrate by submerging in 70% ethanol, then 80% ethanol, then 100% ethanol for 90 s each. Air-dry again.
12. Prior to hybridization, denature the DNA by incubating the slide for 3 min in 1.5 M NaCl, 0.5 M NaOH.
13. Neutralize the DNA by incubation in 0.5 M Tris-HCl, 3 M NaCl, pH 7.2, for 3 min.
14. Wash in 2X SSC for 3 min.
15. Perform FISH on the immobilized DNA fibers using standard methods, as described elsewhere (*see* **Notes 2** and **3** and Chapter 21). **Figure 1** shows the result of FISH using probes prepared from two overlapping clones of genomic DNA from chromosome I of *T. brucei* using this protocol.

3.2. Fiber-FISH Protocol 2

For cells grown in suspension, such as trypanosomes, the following method is used to attach the cells to microscope slides.

1. Coat slides with organosilane to facilitate attachment:
   a. Wash standard microscope slides for 5 min in 2 N HCl.
   b. Wash slides with ddH$_2$O and acetone for 1 min each.
   c. Incubate slides with a 2% solution of aminopropyltriethoxysilane (Sigma) in acetone for 1 min.
   d. Rinse briefly with acetone and air-dry. Slides can be stored in a box for up to 2 wk.
2. Collect trypanosomes grown in suspension by low-speed centrifugation as under **Subheading 3.1.**
3. Resuspend the pellet in PBS.
4. Pipet 50 µL of cell suspension (containing about 50,000 cells) onto a silanized slide and allow to settle for 5 min at room temperature.
5. Briefly wash slides in PBS to remove any unbound cells.
6. Dip slides for 45 s each in ice-cold extraction buffer 1.
7. Dip slides in extraction buffer 2 containing 40-µg/mL propidium iodide.
8. Wash slides once with wash buffer 1, once with wash buffer 2, and twice with H$_2$O for 1 min each.
9. Treat preparations with 3.6% deionized formaldehyde in PBS for 5 min at room temperature, then wash twice in PBS.
10. Dehydrate the preparations by submerging in 70, 90, then 100% ethanol at –20°C for 5 min each and finally air-dry.
11. FISH is done according to standard methods described elsewhere (*see* **Notes 2** and **3**, and Chapter 21). **Figure 2** shows the result of FISH using tubulin gene probes from *T. brucei* chromosome I using this protocol.

4. Notes

1. Unless stated otherwise, all extraction and washing steps are done in Coplin jars filled with 50 mL of the appropriate solution.

2. Procedure for the preparation of nonradioactive DNA probes (e.g., digoxigenin- or biotin-labeled) and protocols for the *in situ* hybridization procedure are identical to those described elsewhere and therefore will not be discussed here (for example, **refs.** *16,17*, and Chapters 11, 21).
3. Fiber-FISH usually requires the simultaneous application of differentially labeled DNA probes and their simultaneous detection, as shown in **Figs. 1** and **2**, and Chapters 11, 12). Therefore, to make full use of the molecular resolution of Fiber-FISH, it is highly recommended to use a microscope system with high NA objectives and equipped with a high-resolution, high-sensitivity CCD camera.

Acknowledgments

The work in my group is funded by The Wellcome Trust.

References

1. Wiegant, J., Kalle, W., Mullenders, L., et al. (1992) High-resolution *in situ* hybridization using DNA halo preparations. *Hum. Mol. Genet.* **1,** 587–591.
2. Parra, I. and Windle, B. (1993) High resolution visual mapping of stretched DNA by fluorescent hybridization. *Nat. Genet.* **5,** 17–21.
3. Fidlerova, H., Senger, G., Kost, M., et al. (1994) Two simple procedures for releasing chromatin from routinely fixed cells for fluorescence in situ hybridization. *Cytogenet. Cell Genet.* **65,** 203–205.
4. Tocharoentanaphol, C., Cremer, M., Schrock, E., et al. (1994) Multicolor fluorescence *in situ* hybridization on metaphase chromosomes and interphase halo-preparations using cosmid and YAC clones for the simultaneous high-resolution mapping of deletions in the dystrophin gene. *Hum. Genet.* **93,** 229–235.
5. Florijn, R. J., Bonden, L. A. J., Vrolijk, H., et al. (1995) High-resolution DNA fiber-FISH for genomic DNA mapping and color bar-coding of large genes. *Hum. Mol. Genet.* **4,** 831–836.
6. Rosenberg, C., Florijn, R. J., Van De Rijke, F. M., et al. (1995) High resolution DNA Fiber-fish on yeast artificial chromosomes: direct visualization of DNA replication. *Nat. Genet.* **10,** 477–479.
7. Mann, S. M., Burkin, D. J., Grin, D. K., et al. (1997) A fast, novel approach for DNA Fiber-fluorescence *in situ* hybridization analysis. *Chromosome Res.* **5,** 145–147.
8. Shiels, C., Coutelle, C., and Huxley, C. (1997) Analysis of ribosomal and alphoid repetitive DNA by fiber-FISH. *Cytogenet. Cell Genet.* **76,** 20–22.
9. Ersfeld, K., Asbeck, K., and Gull, K. (1998) Direct visualisation of individual gene organisation in *Trypanosoma brucei* by high-resolution *in situ* hybridisation. *Chromosoma* **107,** 237–240.
10. Heiskanen, M., Karhu, R., Hellsten, E., et al. (1994) High-resolution mapping using fluorescence *in situ* hybridization to extended DNA fibers prepared from agarose-embedded cells. *Biotechniques* **17,** 928–933.
11. Weier, H. U. G., Wang, M., Mullikin, J. C., et al. (1995) Quantitative DNA fiber mapping. *Hum. Mol. Genet.* **4,** 1903–1910.

12. Heiskanen, M., Peltonen, L., and Palotie, A. (1996) Visual mapping by high resolution FISH. *Trends Genet.* **12,** 379–382.
13. Erdel, M., Hubalek, M., Lingenhel, A., et al. (1999) Counting the repetitive kringle-IV repeats in the gene encoding human apolipoprotein(a) by Fiber-FISH. *Nat. Genet.* **21,** 357–358.
14. Tsuchiya, D. and Taga, M. (2001) Application of Fiber-FISH (fluorescence *in situ* hybridization) to filamentous fungi: visualization of the rRNA gene cluster of the ascomycete *Cochliobolus heterostrophus*. *Microbiology-SGM* **147,** 1183–1187.
15. Vogelstein, B., Pardoll, D. M., and Coffey, D. S. (1980) Supercoiled loops and eucaryotic DNA replication. *Cell* **22,** 79–85.
16. Eckwall, K. and Partridge, J. F. (1999) Fission yeast chromosome analysis: fluorescence *in situ* hybridization (FISH) and chromatin immunoprecipitation (CHIP), in *Chromosome Structural Analysis: A Practical Approach* (Bickmore, W.A., ed.), Oxford University Press, Oxford, UK, pp. 39–57.
17. Ersfeld, K. and Stone, E. (2000) The simultaneous detection of proteins and DNA in cells, in *Protein Localization by Fluorescence Microscopy: A Practical Approach* (Allan, V., ed.), Oxford University Press, Oxford, UK, pp. 51–66.

23

Yeast Two-Hybrid Assay for Studying Protein–Protein Interactions

Ahmed Osman

Summary

Protein–protein interactions occur in a wide variety of biological processes and essentially control the cell fate from division to death. Today, the identification of proteins that interact with a protein of interest is a focus of intensive research and is an essential element of the rapidly growing field of proteomics. Yeast two-hybrid assays represent a versatile tool to study protein interactions in vivo. GAL4-based assay, for example, uses yeast transcription factor GAL4 for detection of protein interactions by transcriptional activation. Some transcription factors (such as GAL4) possess a characteristic phenomenon that the transactivation function can be restored when the factor's DNA-binding domain (DBD) and its transcription-activation domain (AD) are brought together by two interacting, heterologous proteins. GAL4-yeast two-hybrid assay uses two expression vectors, one uses GAL4-DBD and the other uses GAL4-AD. DNA sequences encoding the two proteins of interest (or a protein and a complementary DNA library) can be cloned in the GAL4-DBD and GAL4-AD vectors to form the bait and the target of the interaction trap, respectively. A selection of host cells with different reporter genes and different growth selection markers provides a means to detect and confirm protein–protein interactions and highlight the flexibility of these assays to fit different applications. This chapter presents an outline for the GAL4-based yeast two-hybrid system with a detailed description of the vectors, host cells, and methods for detection and verifying protein interactions.

Key Words: ADE2; Auxotrophic markers; GAL4; GAL4-AD; GAL4-BD; HIS3; LacZ; LacZ assays; LEU2; MEL1; pADH; TRP1; yeast plasmid preparation; yeast transformation; yeast two-hybrid.

1. Introduction

The rapid progress in the field of genome sequencing, which has provided full sequence information for some organisms and large regions of the genomes of others, indeed paves the road for the era of proteomics. Studying protein interactions illustrates how cells communicate in a network and how an individual cell takes different paths ranging from division to death. This helps us to understand the mechanistic basis of different biological processes.

Two-hybrid assays have emerged as a powerful tool enabling the study of protein function in the context of other interacting proteins. Different versions of the two-hybrid system have been developed, which are appropriate for various strategies and use different test organisms. This chapter discusses the GAL4-based yeast two-hybrid system and its use to identify an interacting partner(s) of a protein of interest, or to evaluate the interaction pattern of two known proteins.

The first yeast two-hybrid system was introduced in the late 1980s by Dr. Stanley Fields and coworkers *(1)*. The system is based on the phenomenon that some transcription factors divide the protein sequence into two separate domains, a DNA-binding domain (DBD or BD) that binds to the promoter sequence of responsive genes and a transcription-activation domain (AD) that is responsible for transcription activation. Neither domain is able to activate transcription separate from the other, and bringing both domains within close vicinity (even in a *trans*-configuration, i.e., separated by foreign, interacting protein sequences) usually is enough to reconstitute an actively functioning transcription factor that is able to activate reporter gene(s) transcription. GAL4 is a yeast transcription factor that is involved in the regulation of galactose metabolism *(2)*.

Some of the terms that will be used frequently in this chapter are defined here. Yeast selection marker, also referred to as auxotrophic marker, describes the nutritional requirements that are necessary for growth and need to be depleted from growth media to select for a yeast transformant of interest. Upstream activation sequence (UAS) describes the responsive element in the promoter region, to which GAL4-DBD binds to bring the fused peptides upstream of the reporter gene(s). A reporter gene is a gene that is either integrated into the genome of the yeast host strain (such as β-galactosidase reporter gene; *LacZ*) downstream of GAL4 responsive UAS (UAS_{GAL4}) or whose promoter region has been modified and replaced by UAS_{GAL4} (e.g., imidazole-glycerolphosphate [IGP]-dehydratase reporter; *HIS3*). Its transcription is activated upon binding and reconstitution of the hybrid GAL4 transcription factor, providing a measure to determine positive interaction (**Fig. 1**).

2. Materials

2.1. Selection and Preparation of Recombinant Vectors and Host Strains

1. Purified yeast two-hybrid vector DNAs: for example, pGBK-T7 and pGAD-T7 (BD Biosciences Clontech) or pBD-GAL4cam and pAD-GAL4-2.1 (Stratagene) (*see* **Subheading 3.1.1.** and **Fig. 1**).

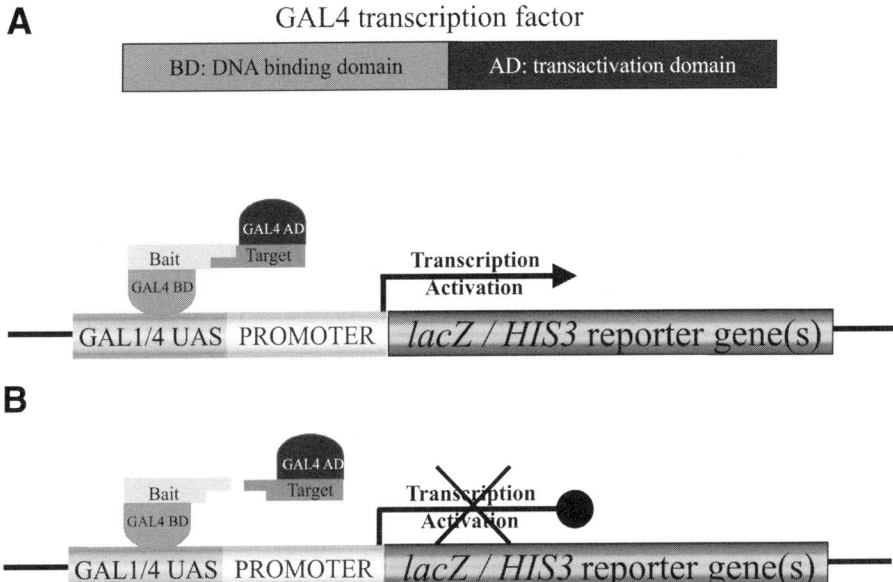

Fig. 1. Mechanism of the yeast two-hybrid assay. GAL4 transcription factor consists of two domains: DNA binding and transcription activation. Panel **A** shows the interaction of a bait construct (GAL4 BD fusion) with a target protein (GAL4-AD fusion), resulting in the reconstitution of the GAL4 transcription factor and the induction of reporter gene(s) transcription activation. In absence of such interaction (Panel **B**), reporter gene(s) transcription is not activated.

2. Standard cDNA library preparation kit or restriction enzymes and DNA ligase and appropriate buffers for cloning of genes of interest (*see* **Subheading 3.1.3.**).
3. Standard plasmid DNA preparation kits or protocols.

2.2. Yeast Transformation

1. YPAD (yeast extract, peptone, adenine, dextrose) medium: to make 1 L, dissolve 20 g bacto peptone (Difco) and 10 g yeast extract (Difco), in up to 940 mL distilled H$_2$O. Adjust pH to 5.8, sterilize by autoclaving for 20 min at 121°C. Cool autoclaved medium to 55°C, then add: 50 mL 40% dextrose solution (autoclave-sterilized), 10 mL adenine hemisulfate (Sigma) (4 mg/mL, filter-sterilized). Make up to 1 L.
2. YPAD plates: add 15–20 g/L bacteriological agar to YPAD medium and autoclave. Add L-amino acids according to requirements (*see* **step 14** below and **Table 1**).
3. Spectrophotometer.
4. Sterile Falcon tubes and centrifuge.
5. Sterile distilled H$_2$O.

Table 1
Selective Media

Host strain	Plasmid(s) pGAL4/pSV40	p53/pLamin C	pSV40 + p53	pSV40 + pLamin C	BD (bait)	AD (library or target)	BD + AD
AH109	-Leu	-Trp	-Leu/-Trp	-Leu/-Trp	-Trp	-Leu	-Leu/-Trp
	-Leu/-His/-Ade/3-AT[a]	-Trp/-His/-Ade/3-AT	-Leu/-Trp/-His/-Ade/3-AT	-Leu/-Trp/-His/-Ade/3-AT	-Trp/-His/-Ade/3-AT	-Leu/-His/-Ade/3-AT	-Leu/-Trp/-His/-Ade/3-AT
PJ69-4A	-Leu	-Trp	-Leu/-Trp	-Leu/-Trp	-Trp	-Leu	-Leu/-Trp
	-Leu/-His/-Ade/3-AT[a]	-Trp/-His/-Ade/3-AT	-Leu/-Trp/-His/-Ade/3-AT	-Leu/-Trp/-His/-Ade/3-AT	-Trp/-His/-Ade/3-AT	-Leu/-His/-Ade/3-AT[a]	-Leu/-Trp/-His/-Ade/3-AT
Y187[c]	-Leu	-Trp	-Leu/-Trp	-Leu/-Trp	-Trp	-Leu	-Leu/-Trp
Y190	-Leu	-Trp	-Leu/-Trp	-Leu/-Trp	-Trp	-Leu	-Leu/-Trp
	-Leu/-His/3-AT[b]	-Trp/-His/3-AT	-Leu/-Trp/-His/3-AT	-Leu/-Trp/-His/3-AT	-Trp/-His/3-AT	-Leu/-His/3-AT	-Leu/-Trp/-His/3-AT
YRG2	-Leu	-Trp	-Leu/-Trp	-Leu/-Trp	-Trp	-Leu	-Leu/-Trp
	-Leu/-His	-Trp/-His	-Leu/-Trp/-His	-Leu/-Trp/-His	-Trp/-His	-Leu/-His	-Leu/-Trp/-His

A list of selective media (SD) to be used with different combinations of host strains and test or control plasmids. For each strain, the top row lists the selective media required for control plate(s) by which transformation efficiency is checked.

[a] 3-AT is used at 2.5 mM final concentration.
[b] 3-AT is used at 30–35 mM final concentration.
[c] Control plates and test plates are the same for this strain.

6. 1 M lithium acetate (LiOAc; Sigma), pH 7.5. Sterilize by autoclaving.
7. 50% polyethylene glycol (PEG-3350; Sigma): 50 g PEG-3350 in up to 100 mL distilled H_2O. Sterilize by autoclaving.
8. 10X TE buffer (pH 7.5): 100 mM Tris-HCl, pH 7.5, 10 mM EDTA, pH 8.0. Sterilize by autoclaving.
9. Salmon sperm DNA (ssDNA; Sigma): 1 g ssDNA in 100 mL distilled H_2O. Heat and stir to dissolve, sonicate or randomly shear through an 18-ga needle, extract once with TE-equilibrated phenol, once with phenol/chloroform/isoamyl alcohol (Chisam) (25:24:1), and once with chloroform. Ethanol-precipitate, resuspend in 1X TE buffer, pH 8.0, at a final concentration of 2 mg/mL. Store as 1-mL aliquots at −20°C. Boil for 5 min before use.
10. Transformation mix: 500 µL PEG-3350 (50%), 80 µL LiOAc (1 M), 125 µL ssDNA (2-mg/mL), 100 µL 1X TE (pH 7.5). Add 10 µg of each plasmid DNA.
11. Dimethyl sulfoxide (DMSO).
12. 42°C water bath.
13. 10X yeast nitrogen base without amino acids (YNB): 67 g YNB without amino acids (Difco), dissolve in up to 1 L distilled H_2O. Adjust pH to 5.8, autoclave or double-filter-sterilize. Dispense in 100-mL aliquots, and store protected from the light at 4°C.
14. 20X universal amino acid dropout solution (UDO): Dissolve the following L-amino acids (Sigma) in 1 L distilled H_2O: 600 mg isoleucine (Ile), 3000 mg valine (Val), 400 mg arginine-HCl (Arg), 600 mg lysine-HCl (Lys), 1000 mg phenylalanine (Phe), 4000 mg threonine (Thr), 600 mg tyrosine (Tyr), 2000 mg glutamic acid (Glu), 2000 mg aspartic acid (Asp), and 800 mg serine (Ser). Filter-sterilize, and store in 100-mL aliquots at 4°C for up to 1 yr. Warm if necessary before use to dissolve precipitated components.
15. 200X individual dropout (DO) solutions (dissolve each component in 100 mL distilled H_2O): 400 mg adenine hemisulfate* (Ade), 400 mg histidine HCl·H_2O (His), 2000 mg leucine (Leu), 400 mg methionine (Met), 400 mg tryptophan (Trp), 400 mg uracil* (Ura). Filter-sterilize each component, store at 4°C for up to 1 yr. The components marked with (*) will need warming before use.
16. 1 M 3-amino-1,2,4-triazole (3-AT) solution in distilled H_2O: 0.85 g/10 mL. Filter-sterilize, store at −20°C. Note that this substance is heat-labile and needs to be added to medium that has been cooled down to ≤50°C.
17. Synthetic dextrose (SD) minimal medium for plates (200 mL): 3–4 g agar, 155 mL distilled H_2O. Autoclave, keep in 65°C water bath until ready to pour, then add the following: 10 mL 40% dextrose solution, 20 mL 10X YNB, 10 mL 20X UDO, 1 mL each of 200X individual DO component(s)*, 0.5 mL (2.5 mM) 1 M 3-AT solution for AH109 and PJ69-4A or 7 mL (35 mM) for Y190. *One or more individual DO components are omitted from the medium to prepare the specific SD minimal medium lacking this/these component(s), e.g., SD-His is prepared as above by excluding the 200X His DO solution.
18. 30°C incubator.

For alternative protocol (*see* **Subheading 3.2.2.**):

19. LiOAc/TE solution (0.1 *M*/1X, prepared fresh): 1 mL LiOAc (1 *M*, pH 7.5) solution, 1 mL 10X TE buffer, pH 7.5, 9 mL sterile distilled H_2O.
20. PEG-3350/LiOAc/TE solution (40%/0.1 *M*/1X, prepared fresh): 8 mL PEG-3350 (50%), 1 mL LiOAc (1 *M*, pH 7.5) solution, 1 mL 10X TE buffer, pH 7.5.

2.3. LacZ Assays

1. Liquid and solid selective media as above.
2. Reinforced nitrocellulose membrane (e.g., Nytran, S & S).
3. Liquid nitrogen.
4. Whatman filter paper.
5. 10X Z-buffer (100 mL): 16.1 g sodium dibasic phosphate ($Na_2HPO_4 \cdot 7H_2O$), 5.5 g sodium monobasic phosphate ($NaH_2PO_4 \cdot H_2O$), 0.75 g potassium chloride (KCl), 0.25 g magnesium sulfate ($MgSO_4 \cdot 7H_2O$), make up to 100 mL with distilled H_2O. Adjust pH to 7.0, autoclave or filter-sterilize, store at room temperature.
6. 50-mg/mL X-Gal prepared in N,N-dimethyl-formamide and stored at −80°C in 1-mL aliquots.
7. Working Z-buffer (freshly prepared, make approx 2 mL for each 87-mm filter to be assayed): 1 mL 10X Z-buffer, 9.0 mL distilled H_2O, 27 µL β-mercaptoethanol, 200 µL X-Gal (50-mg/mL).
8. Humid chamber and 30°C incubator.

For alternative protocol (*see* **Subheading 3.3.2.**):

9. ONPG substrate: 5 mg/mL *o*-nitrophenyl β-D-galactopyranoside (OPNG) in 1X Z-buffer containing 27 µL β-mercaptoethanol/10 mL. Prepare fresh. Note that this substance takes about 2 h to dissolve.
10. 1 *M* sodium carbonate ($NaCO_3$).

2.4. Yeast Plasmid Preparation

1. YPAD and selective media as above.
2. Glycerol.
3. Resuspension buffer: 1% Triton X-100, 1% sodium dodecyl sulfate (SDS), 10X TE, pH 8.0.
4. Acid-washed glass beads, 450–600 µm in diameter (Sigma).
5. Zymolase (Zymo research) or Lyticase (Clontech) or β-glucuronidase (5000 U, Sigma).
6. Phenol/Chisam (*see* above).
7. Ethanol and 3 *M* sodium acetate (NaOAc), pH 5.2.
8. DNA miniprep column purification kit, such as Wizard (Promega), Qiaprep Spin (Qiagen), or Chroma Spin (Clontech); or glass milk (GeneClean III, Bio 101).
9. Standard PCR amplification reagents, thermocycler, and vector primers (*see* **Note 1**).
10. *Escherichia coli*-competent cells: e.g., TOP10 (Invitrogen), DH10B (Invitrogen), JM109 (Promega), XL1-BlueMRF' (Stratagene).

Table 2
Control Plasmids

Control plasmid	Commercial source	Insert description (anticipated results)	Selection marker (yeast)	Selection marker (bacteria)
pGAL4	Stratagene	Wild-type, full-length GAL4 (standalone positive control)	*LEU2*	Ampicillin
PSV40	Stratagene	Large T-antigen of SV40 virus, aa. 84–708 (negative control by itself, positive control along with p53) *(21)*	*LEU2*	Ampicillin
P53	Stratagene	Murine p53, aa. 72–390 (negative control by itself, positive control along with pSV40) *(21)*	*TRP1*	Ampicillin
pLamin C	Stratagene	Human lamin C, aa. 67–230 (negative control by itself, negative control along with pSV40) *(22)*	*TRP1*	Ampicillin
pCL1	Clontech	Wild-type, full-length GAL4 (standalone positive control)	*LEU2*	Ampicillin
pGADT7-T	Clontech	Large T-antigen of SV40 virus (negative control by itself, positive control along with p53) *(21)*	*LEU2*	Ampicillin
pGBKT7-53	Clontech	Murine p53 (negative control by itself, positive control along with pSV40) *(21)*	*TRP1*	Kanamycin
pGBKT7-Lam	Clontech	Human lamin C (negative control by itself, negative control along with pSV40) *(22)*	*TRP1*	Kanamycin

A list of control plasmids, their commercial sources, insert descriptions, anticipated results after transformation in yeast and selection markers in yeast and bacteria.

2.5. Control Plasmids

Using control plasmids, along with the test constructs, provides positive and negative control data required to assess how well the system works. A list of control plasmids with insert descriptions, growth conditions in yeast and bacteria, anticipated results, and commercial sources is presented in **Table 2**.

3. Methods

3.1. Selection and Preparation of Recombinant Vectors and Host Strains

3.1.1. Choice of Vectors

The system consists of two GAL4 two-hybrid vectors and a yeast host strain. The first vector (BD vector) harbors the sequence of GAL4-DBD with the

TRP1 gene (phosphoribosyl-anthranilate isomerase) as a selection marker *(3)*. Yeast cells containing this plasmid grow in medium lacking tryptophan (SD minus Trp). Downstream of GAL4-DBD, the sequence encoding the protein of interest should be cloned in-frame with GAL4-DBD sequence to produce a fusion protein of the test gene preceded by GAL4-DBD. This plasmid is termed the "bait vector." The second vector, GAL4-AD, is under selection of the *LEU2* gene, which encodes β-isopropylmalate dehydrogenase and complements leucine (Leu) auxotrophy of a *leu2* mutant *(4)*. The AD vector is used to produce a fusion protein with a target protein sequence (or cDNA library in the case of genome-wide searches). Expression of both proteins is driven by a constitutive yeast gene expression promoter, the alcohol dehydrogenase I (*pADH1*) (*see* **Note 2**) *(5)*, to assure high-level expression of the fusion proteins in vivo. Both BD and AD vectors contain antibiotic-resistance genes, preferably different from each other, as selection markers in bacteria. When it is necessary to isolate plasmids from potential positive yeast colonies, having the AD and BD vectors with different bacterial selection markers helps to select for the desired AD plasmid from a prepared mixture of both plasmids. The vectors pGBK-T7 and pGAD-T7 (BD Biosciences Clontech) and pBD-GAL4cam and pAD-GAL4-2.1 (Stratagene) are examples of BD and AD vectors, respectively, that use *pADH1* as a gene expression-driving promoter.

3.1.2. Choice of Host Strain

The genome of the yeast host strains are modified in a way that allows the selection of transformed yeast cells and detection of interacting proteins. For example, *LEU2* and *TRP1* genes (used for selection) are mutated and the promoter of *HIS3* gene (used for detecting the interaction) is modified. GAL4 and another related regulatory protein, GAL80 *(6)*, are also mutated to prevent interference from endogenous yeast genes/proteins. Reporter genes commonly used in yeast host strains are *HIS3* and *LacZ*. The *HIS3* gene encodes IGP-dehydratase, an enzyme that is involved in histidine biosynthesis. Upon activation by a reconstituted GAL4 (positive interaction), expression of the *HIS3* reporter gene enables the transformed yeast cell to grow on selective medium lacking histidine. Similarly, the activated *LacZ* reporter translates into an active and functional β-galactosidase enzyme that cleaves the X-Gal substrate, producing blue color as an indication for a positive interaction. In certain host strains, the *HIS3* reporter may exhibit leaky expression in absence of induction with GAL4 because of the promoter composition *(7)*. The effect of this leaky expression can be minimized by adding 3-amino-1,2,4 triazole (3-AT), a metabolic inhibitor for the IGP-dehydratase enzyme, to the selective medium lacking histidine to neutralize the basal expression of *HIS3* reporter. Other reporter

Table 3
Yeast Host Strains and Their Nutritional Requirements

Yeast Host strain[a]	-Ade	-His	-His 2.5 mM 3-AT	-His/ 25 mM 3-AT	-Leu	-Met	-Trp	-Ura	Reporter gene(s)
AH109 (Clontech)	–	±	–	–	–	±	–	+++	ADE2, HIS3, LacZ, MEL1
PJ69-4A (23)	–	+	±	–	–	±	–	–	ADE2, HIS3, LacZ[b], MEL1
Y187 (Clontech)	–	–	–	–	–	+	–	–	LacZ, MEL1
Y190	–	+++	+	±	–	+++	–	+++	HIS3, LacZ, MEL1
YRG2 (Stratagene)	–	–	–	–	–	+++	–	+++	HIS3, LacZ

A list of some yeast strains used in the yeast two-hybrid assay and their nutritional requirements as determined by comparing the growth rate (after up to 7 d incubation at 30°C) on selective media to that on nonselective, rich medium, assigned with (+++), which represents the optimal growth conditions.

[a]For long-term storage, prepare glycerol stocks of the yeast strain by adding sterile glycerol to overnight, saturated culture to a final concentration of 20–25%. Store in 1-mL aliquots at –80°C. Revive by scraping off the surface of the frozen stock with a sterile loop and streak onto YPAD plate. Incubate at 30°C for 2–3 d.

[b]The use of this strain is not recommended for LacZ qualitative assay.

genes that are used less frequently are *ADE2* and *MEL1*. *ADE2* enables yeast cells with positively interacting proteins to grow on adenine (Ade)-lacking medium. *MEL1* is a gene that is normally regulated by GAL4. Transcription activation of *MEL1* gene by GAL4 leads to production of the α-galactosidase enzyme, capable of cleaving X-α-Gal substrate producing a blue color, another parameter that helps to identify a positive protein–protein interaction. Yeast host strains may contain one or more reporter genes. **Table 3** shows a list of some host strains available commercially, with the nutritional markers (auxotrophic markers) used for selection of transformed yeast cells and/or detecting a positive interaction, and the reporter gene used. Nutritional requirements were determined by plating different host strains onto selective media lacking one of the following markers: Ade, His, Leu, Met, Trp, and Ura.

3.1.3. Cloning cDNA or Gene of Interest Into Selected Yeast Two-Hybrid Vectors

Standard cloning strategies described in any molecular biology reference book *(8)* would suffice for cloning either a cDNA insert, a mutant construct of a gene of interest, or constructing a cDNA library. The cDNA insert should be cloned in both AD and BD vectors to check for homodimerization and to be ready to switch the system if necessary, as explained below (*see* **Note 3**).

cDNA libraries are, in general, made by priming poly-A^+ RNA with oligo-dT to exclude ribosomal RNA from the cDNA clones generated. Different kits are available commercially to construct either phage or plasmid libraries. A good representative library contains at least 100-fold more original clones than the number of genes in the organism. For example, the genome of *Schistosoma mansoni*, the parasite under study in our laboratory, contains approx 20,000 genes *(9,10)*, and a cDNA library with 2.5×10^6 original clones be a good representation for the transcriptome of this organism (*see also* **Note 4**).

Vectors are prepared using plasmid preparation techniques such as CsCl gradient centrifugation or any commercially available plasmid prep kit. Midi or large-scale DNA extraction kits usually yield satisfactory quality and quantity for yeast transformations. Plasmid DNA of the cDNA library to be used for screening should be prepared from an inoculum that is representative for the number of original clones (*see* **Note 5**).

3.2. Yeast Transformation

Most yeast transformation protocols are based on the use of the alkali cation method developed by Ito et al. and later modified by Schiestl and Gietz and by Hill et al. *(11–13)*.

Two general transformation protocols that are used successfully in our laboratory to transform yeast are given here. The choice of which protocol to follow depends on the application. For library transformation, high efficiency transformation is very important to screen as many clones as possible, whereas for evaluating protein–protein interactions, transformation efficiency is not as important because few colonies will provide the required information.

3.2.1. Preparation of Competent Yeast Cells and Library Transformation (Protocol 1)

1. Inoculate 5 mL of YPAD medium with a single colony from a fresh YPAD plate (less than 4 wk old). Shake at 30°C, overnight.
2. Dilute 0.1 mL of the overnight culture 10X with fresh YPAD and determine the A_{600} reading of the overnight culture (usually around 3.0).
3. Inoculate 30 mL YPAD medium with the overnight culture to an A_{600} between 0.2–0.3. Shake for an extra 4–6 h at 30°C until A_{600} reaches 1.2–1.35.

4. Harvest the cells in a sterile Falcon tube or equivalent by centrifugation at 1000g, at 15°C for 5 min.
5. Pour off the supernatant and gently resuspend the cells in 15 mL of sterile distilled H_2O. Harvest as before.
6. Repeat the above washing step.
7. Resuspend cell pellet in 200 µL (final volume should be approx 300–350 µL).
8. Aliquot 100 µL/tube (in microcentrifuge tubes), spin briefly for few seconds, and pipet off the water. Cells are now ready for transformation.
9. To each tube, add the following to the top of the cell pellet: 500 µL PEG-3350 (50%), 80 µL LiOAc (1 M), 125 µL ssDNA (2-mg/mL). Finally, add 100 µL 1X TE, pH 7.5, containing 10 µg of each plasmid.
10. Fully resuspend the cell pellet into this transformation mix by homogenizing the pellet with a blunt pipet tip (heat-sealed), then by inversion several times. Incubate at 30°C for 45 min, mixing every 10–15 min by inversion.
11. Add 50 µL DMSO, mix, and heat-shock in 42°C water bath for 18 min, mixing every 5 min.
12. Centrifuge at 2000g for 2 min. Aspirate off the supernatant and resuspend the pellet into 1 mL of SD-Leu/-Trp/-His medium.
13. Shake for 30 min at 30°C, spin down as above, and resuspend into 1 mL of sterile distilled H_2O.
14. Spread 200–250 µL/150-mm plate of selective medium, based on the host strain used (SD-Leu/-Trp/-His for YRG2; SD-Leu/-Trp/-His/-Ade/2.5 mM 3-AT for either PJ69-4A or AH109; SD-Leu/-Trp/-His/35 mM 3-AT for Y190 or SD-Leu/-Trp for Y187).
15. Spread 50 µL of the cells onto a control plate of SD-Leu/-Trp to determine co-transformation efficiency of both plasmids (an average of 3000–5000 colonies on the control plate is usually obtained, which means a total transformation efficiency of $1.5-3.0 \times 10^5$).
16. Incubate the plates at 30°C for 5–7 d.

A set of control plasmid transformations should be performed in parallel to the library transformation (*see* **Subheading 2.3.** and **Table 2**). The protocol outlined below can be used for this purpose.

3.2.2. Preparation of Competent Yeast Cells and AD/BD Co-Transformation (Protocol 2)

1. Inoculate 5 mL of YPAD with a single yeast colony, shake at 30°C for overnight. Take A_{600} reading as in Protocol 1 (usually around 3.0).
2. Use about 1.5 mL of the overnight culture to inoculate 25 mL of YPAD medium to an A_{600} of approx 0.15–0.2. Incubate for an extra 3–5 h to reach an A_{600} of 0.8–1.0.
3. Pellet the cells and wash with distilled H_2O as outlined in **Subheading 3.2.1.**
4. Resuspend in 1 mL of 0.1 M LiOAc/1X TE, pH 7.5, and transfer the cell suspension to a microcentrifuge tube.

5. Spin down at 2000g for 2 min in a microcentrifuge, discard the supernatant, and resuspend in 0.5 mL of LiOAc/TE solution.
6. Add 50 μL of cell suspension to a microcentrifuge tube containing 0.1–1.0 μg of the plasmid(s) to be transformed in a volume up to 10 μL plus 15 μL of ssDNA (30 μg).
7. Add 500 μL of PEG-3350/LiOAc/TE solution.
8. Mix the cells gently with a blunt-end tip and then by inversion several times.
9. Incubate at 30°C for 30 min, mixing every 10 min, heat shock at 42°C for 15 min, mixing every 5 min.
10. Centrifuge the cells at 2000g for 10–15 s, remove the supernatant, and resuspend cell pellet in 100 μL of sterile distilled H_2O.
11. Plate the cells onto the appropriate selective medium (50 μL/90-mm plate). Make sure to have a control plate to check transformation efficiency (i.e., SD-Leu/-Trp for AD/BD double transformations, SD-Leu or SD-Trp for single transformations with *LEU2* and *TRP1* selection markers, respectively).
12. Incubate the plates at 30°C for 5–7 d. It usually takes about 2–3 d for the transformed yeast colonies to become visible.

3.3. LacZ Assays

Two alternative methods for the assay of activation of the *LacZ* reporter gene are given here.

3.3.1. Colony-(Filter-)Lift Assay (Protocol 1)

The LacZ filter-lift assay is a qualitative yet sensitive assay to test *LacZ* reporter gene activation *(14)*. Positive interaction is indicated by a blue colony in which GAL4 has been reconstituted and has activated the *LacZ* reporter gene. The data obtained from this assay are confirmatory for the results of growth selection on minimal media, which examines the *HIS3* reporter gene for Y190 and YRG2 or *HIS3* and *ADE2* reporters in case of AH109 and PJ69-4A (*see* **Note 6**).

1. Grow yeast colonies on selective media for 5–7 d (until 1–3 mm in diameter).
2. Make two master plates of the grown colonies on the appropriate selective media. This helps to reduce number of filters, save reagents, and keep a backup copy of grown colonies, especially useful in the case of library screening (*see* **Note 7**). If the interaction of two known proteins is being examined, then one master plate is usually enough.
3. Carefully overlay a reinforced nitrocellulose membrane (e.g., Nytran; S & S) on the top of the plate. Use a glass spreader to make good contact of the membrane with the plate surface. Poke holes asymmetrically to orient the filter to the plate.
4. Peel the filter off the plate and put it in liquid nitrogen (in a regular ice bucket or a Styrofoam box) with the colonies side facing up. Leave the filter for about 1 min, take it out of the liquid nitrogen, and let it thaw on clean 3-mm Whatmann filter paper.

5. Repeat the above freeze–thaw cycle.
6. Overlay the membrane, with the colonies side up, onto a Whatmann filter paper #1, presoaked with approx 2 mL of working Z-buffer.
7. Incubate at 30°C for 2 h to overnight in a humid chamber, but it may take up to 3 d for detecting weak interactions (you can place a tray of distilled H_2O in the 30°C incubator to provide humidity).
8. Identify positive (blue) colonies and proceed to yeast plasmid preparation (**Subheading 3.4.**).

3.3.2. Liquid β-Gal Assay (15)(Protocol 2)

This assay is relatively less sensitive than the filter-lift assay, but it is used as a quantitative assay to determine the relative strength of protein–protein interactions when compared to each other or to a standard interaction (*see* **Note 8**).

1. Grow yeast colonies on selective media for 5–7 d (1–3 mm in diameter).
2. Pick a single colony and resuspend in 5 mL of appropriate SD medium, grow at 30°C overnight. (It is important to use SD medium rather than YPAD to maintain the selective pressure on the yeast cells.)
3. Dilute 2 mL of the overnight culture with 8 mL of YPAD and shake at 30°C for 3–5 h. Meanwhile, prepare the substrate ONPG substrate. This must be prepared fresh and, as it takes about 2 h to dissolve, one must allow enough time to have it ready when needed.
4. Take A_{600} reading (usually approx 0.5–0.8). Dispense 1.5 mL of the culture into each of three microcentrifuge tubes (samples are processed in triplicate).
5. Centrifuge in a microcentrifuge at maximum speed for 30 s. Remove the supernatant. Resuspend (by vortexing or pipetting up and down) in 1.0 mL of 1X Z-buffer. Centrifuge as before and discard the supernatant.
6. Resuspend the cell pellet into 300 µL of 1X Z-buffer. Dilute 100 µL into 1 mL and recheck the A_{600} reading. Record the reading of the original culture (remember that you have twofold dilutions of the original culture). Transfer 100 µL of cell suspension to another tube. Place the tubes in liquid nitrogen (approx 1 min). Thaw the cells in a 37°C heat block (approx 1–2 min).
7. Repeat freezing/thawing cycles two more times.
8. Prepare blank tubes (three tubes) with 100 µL 1X Z-buffer.
9. Add 700 µL 1X Z-buffer/β-mercaptoethanol to each tube.
10. Add 200 µL substrate solution. Record zero time (substrate addition). Incubate at 30°C. After yellow color develops, add 0.4 mL of 1 M Na_2CO_3 and record elapsed time in minutes.
11. Centrifuge the reactions at maximum speed for 10 min and take the A_{420} readings for samples relative to the blank.
12. Calculate β-galactosidase activity in units, where 1 U is the amount of enzyme required to hydrolyze 1 µmole of ONPG substrate to *o*-nitrophenol and D-galactose per min per cell *(16)*. β-galactosidase units = 1000 × A_{420}/(time × vol of culture × A_{600}) (*see* **Note 9**).

3.4. Yeast Plasmid Preparation

Several methods were developed for plasmid preparation from yeast. They all use similar principles with slightly different approaches. The cell wall of yeast cells is disrupted, the DNA solution recovered, and then purified and concentrated.

The yeast cell wall can be disrupted by the use of acid-washed glass beads (450–600 μm in diameter, Sigma) coupled with phenol extraction *(17)* or by different enzymes that digest the cell wall components, such as Glucanex *(18)*, Zymolase, β-glucuronidase, or lyticase. After the cell wall lysis step, protocols for DNA purification are similar to bacterial plasmid preps, which involve alkaline treatment followed by a neutralization step. DNA can then either be precipitated or purified through a binding column/matrix and then eluted in the desired volume. The protocol outlined below summarizes some of the different approaches used for yeast plasmid preparations.

1. Grow overnight cultures of yeast colonies of interest in 3–5 mL YPAD or appropriate selective medium. Make glycerol stocks for future use, store them at –80°C.
2. Transfer 1.5 mL overnight culture to a microcentrifuge tube. Spin down at 2000g for 2 min, discard the supernatant, and resuspend the cell pellet in 0.25 mL resuspension buffer.
3. Add 0.25 mL phenol/Chisam and 0.3 g glass beads, vortex vigorously for 2 min, spin in a microcentrifuge at high speed for 3 min, and transfer the top aqueous layer to a new tube. Alternatively, add 2 μL of Zymolase (10 U), 10 μL of Lyticase (50 U), or 5 μL β-glucuronidase (5000 U) and incubate at 37°C for 60 min.
4. Extract with 0.25 mL of phenol/Chisam.
5. Transfer the aqueous layer to a new tube. Ethanol-precipitate the DNA by adding 0.1 vol of 3 M NaOAc, pH 5.2, and 2.5 vol of absolute ethanol. Leave at –80°C for 1 h, then spin down in a microcentrifuge at 4°C, for 15 min.
6. Resuspend the pellet in 200 μL 1X TE, pH 8.0.
7. Further purify the prepared DNA through any miniprep column format, or simply by using 10 μL glass milk, then elute the DNA in 30 μL 1X TE, pH 8.0.
8. Confirm that you have obtained the AD and BD plasmid DNAs by performing polymerase chain reaction (PCR) reactions using vector or insert-specific primers and 1 μL of the DNA solution to amplify the cloned gene(s).
9. Use 2 μL to transform *E. coli* competent cells (*see* **Note 10**).
10. Spread the transformed bacteria onto the appropriate selective medium, i.e., Luria Bertani (LB)-ampicillin to select for either AD vector pAD-GAL4-2.1 or pGAD-T7, LB-kanamycin for the BD plasmid pGBK-T7, or LB-chloramphenicol for the BD plasmid pBD-GAL4cam.
11. Select more than 20 colonies from the AD-transformed LB-Amp plate (in the case of library screening) to perform colony PCR with AD-fwd and AD-rev primers (*see* **Note 1**). In the case of multiple-size PCR products, different recovered plas-

mids must be checked individually by retransformation into yeast to identify the plasmid(s) responsible for reporter gene transactivation and consequently the potential interacting partner.

3.5. In Vitro Verification of the In Vivo Protein–Protein Interaction

After isolating and identifying a cDNA or gene sequence that codes for a protein that is interacting in vivo in the two-hybrid assay with a protein of interest, this interaction should be verified by testing the interaction of the isolated proteins in vitro also. Both BD and AD vectors, pGBK-T7 and pGAD-T7, contain a T7 promoter located in the junction between the GAL4 domain and the fused protein sequence. This promoter drives the transcription of the downstream sequences, catalyzed by T7 RNA polymerase. In addition, these vectors contain different epitope tags (hemagglutinin, HA in pGAD-T7; and c-Myc in pGBK-T7) for immunodetection and/or precipitation. As an alternative option, if not using these vectors, the coding sequence of the cDNA of interest can be recloned into a vector suited for coupled in vitro transcription/translation reactions. Such vectors usually contain a promoter to drive transcription reactions catalyzed by any of the viral RNA polymerases (T7, T3, or Sp6). In vitro transcribed RNA can then be translated in cell-free translation machinery such as a rabbit reticulocyte lysate. Examples of such vectors are the pCRII vectors (Invitrogen) and the pGEM series (Promega), in which T7 and Sp6 promoters flank the sequence of interest, and the pCITE vector family (Novagen). pCITE vectors possess another useful feature, which is enhanced translation efficiencies because of presence of viral sequence that enhances translation of cap-less messages (Cap-Independent Translation Enhancer).

This in vitro assay involves incorporating ^{35}S-methionine (or another suitable label) into one of the two interacting proteins during translation, while producing the second protein in a nonlabeled form (the latter may be a bacterially expressed protein). Protein interactions can be detected by precipitating the radiolabeled protein with a reagent that is specific for the nonlabeled protein. Then the specific interaction can be determined as percentage of the precipitated product compared to the input amount of the labeled protein. The specific reagent can be an affinity resin or an antibody reagent directed to the encoded protein or to the epitope tag, which can then be precipitated with protein-A or protein-G resin. Background (nonspecific) interactions between the labeled protein and the reagent used for precipitation should be considered and subtracted from the specific values. Another approach for detecting the in vitro interaction pattern between two proteins is co-precipitation. In this approach, the first protein is precipitated (after a binding reaction with the second protein) with a reagent specific to the first protein, subjected to SDS-polyacrylamide gel electrophoresis, and then subjected to Western blotting using a specific antibody reagent directed against the second protein. The reagents used to precipitate the first protein or detecting the second one should be highly specific and should not cross-recognize the other species. Again, potential background interactions should be considered.

Cell-free protein translation lysates (such as TNT lysates from Promega, STP-3 from Novagen, PROTEINscript from Ambion) are used to translate the in vitro transcribed RNA into protein, either labeled or nonlabeled, following the manufacturer's instructions. Bacterial expression systems, utilizing different tag formats (glutathione-*S*-transferase [GST]-tagged, including the pGEX series from Amersham-Pharmacia Biotech and pET-42 vectors from Novagen; 6× His-tagged, such as the pET series from Novagen; and maltose-binding protein [MBP] tag such as pMAL vectors [*see* **Note 11**]) from New England Biolabs are used to produce protein bound to an affinity matrix, which can then be used to precipitate either labeled or nonlabeled interacting protein. In addition to the two proteins, protein–protein binding reactions should contain a buffering system (Tris, phosphate, HEPES, etc.), and other specific component(s) necessary for stabilizing protein interactions, such as metal ions, nucleotides, co-enzymes, etc., and may contain a detergent. Incubation time of the reaction varies from 1 h to overnight, depending on the studied protein(s). Long incubations should be allowed to proceed at 4°C. Several washing steps should be included to decrease the nonspecific interactions. Reactions are then size-separated by electrophoresis and specific interactions can be detected using suitable methods based on the system used.

4. Notes

1. One pair of primers (forward and reverse primers) is designed for each of the AD and BD vectors, flanking the multiple cloning sites. These primers can be used for sequencing cloned cDNA inserts from both strands and/or PCR amplification to verify insert recombination into the vector. The primers work with any version of GAL4-based vectors.
   AD-fwd primer: 5'-CAG GGA TGT TTA ATA CCA CTA C-3'
   AD-rev primer: 5'- GCA CAG TTG AAG TGA ACT TGC-3'
   BD-fwd primer: 5'- GCG ACA TCA TCA TCG GAA GAG-3'
   BD-rev primer: 5'- AAG AAA TTC GCC CGG AAT TAG C-3'
2. Older versions of yeast two-hybrid vectors used a truncated alcohol dehydrogenase promoter (*pADH1*), which weakly drives gene expression. A drawback for these earlier systems was that such weak interactions could not be detected.
3. The bait plasmid (BD) should be checked before use to make sure that it does not activate transcription of the reporter gene(s) in the absence of an interacting partner fused to the GAL4-AD: for example, genes encoding proteins containing transcription activation domain(s), which, when fused to GAL4-DBD, bind to GAL4-

responsive elements in the promoter region of the reporter genes and activate the transcription of the reporter gene(s) with the intrinsic transactivation function independent from GAL4-AD. Examples of such proteins are schistosome homolog of human TGF-β signal transducer Smad2 (SmSmad2) *(19)* and *S. mansoni* nuclear receptor SmRXR1 *(20)*. In such cases, the bait construct is not valid for library screening in this form and different modifications should be tried, including cloning individual domains or deleting or mutating potential regions to obtain a protein that lacks the transactivation feature without disrupting its protein–protein interacting capacity. However, the bait gene can be cloned in the AD plasmid (called target, in this case) and used to check the interaction pattern of its encoded protein with proteins encoded by other known genes cloned in BD plasmid, as long as the BD constructs fail to activate transcription by themselves.

4. These are useful links to protocols of kits from different manufacturers used in the construction of cDNA libraries, yeast transformation, reporter gene(s) assays and plasmid preparation: yeast two-hybrid system HybriZap 2.1 (Stratagene), http://www.stratagene.com/displayProduct.asp?productId=256; yeast protocol handbook (Clontech), http://www.clontech.com/techinfo/manuals/PDF/PT3024-1.pdf; yeast protocol handbook Addendum (Clontech), http://www.clontech.com/techinfo/manuals/PDF/PT3024-4.pdf; matchmaker GAL4 yeast two-hybrid system 3 (Clontech), http://www.clontech.com/techinfo/manuals/PDF/PT3247-1.pdf; yeast plasmid isolation kit (Clontech), http://www.clontech.com/techinfo/manuals/PDF/PT3049-1.pdf; yeast transformation kit (Clontech), http://www.clontech.com/techinfo/manuals/PDF/PT1172-1.pdf; yeast plasmid isolation kit (Zymo Research), http://www.zymoresearch.com/pdf/d2001.PDF; yeast transformation kit (Zymo Research), http://www.zymoresearch.com/pdf/t2001.PDF.

5. In our laboratory, a phage cDNA library (HybriZap 2.1; Stratagene) of 2×10^6 original clones was amplified to 8×10^{10} clones/mL, of which 1 mL was in vivo excised into a phagemid library following the manufacturer's instructions. The phagemid library was transfected into bacteria and plated onto 10 (150 mm) LB/Amp plates. Colonies were then scraped off plates, resuspended into 100 mL of LB/Amp medium, used to inoculate 6 L of LB/Amp, and incubated overnight with shaking at 37°C. Large-scale centrifugation through cesium chloride (CsCl) gradients was performed from this culture and the prepared DNA, representing a *S. mansoni* cDNA library, was used to transform yeast. DNA recovered from this preparation was almost 10^4-fold the original number of clones and 10^6-fold greater than the predicted number of genes in the organism.

6. Yeast host strain PJ69-4A is not recommended for the LacZ-filter assay, because permeabilization procedures used in this assay (such as freeze–thaw cycles or chloroform treatment) induce the *GAL7*, promoter resulting in blue color regardless of GAL4 domain interaction status (James P., University of Wisconsin; personal communication). However, this strain works well in the liquid β-Gal assay, utilizing ONPG substrate, as described above.

7. For comparison purposes, it is sometimes useful to prepare a master plate of selective medium to which all or some individual dropout solutions were added, to

allow plating colonies grown on different selective media. For example, a master plate of SD + Leu, + Trp, + His, + Ade, can be used to plate colonies transformed with positive and negative control plasmids as well as bait plasmid, target plasmid, and a mixture of both.

8. Samples such as pGAL4 (positive control) or strong interacting proteins (such as p53/pSV40) may produce the yellow color much faster than weakly interacting proteins. In this case, a smaller number of cells (as determined by the volume of culture) can be used, and compensated for, to achieve a relatively slower color development rate, which will be comparable to the experimental samples.
9. For example, a sample with initial A_{600} reading of 0.5, 1.5-mL aliquots processed as above (i.e., 0.1 mL processed is equivalent to 0.5 mL original culture) and giving an average A_{420} reading of 0.3 developed in 2 h will have β-galactosidase activity as follows: β-galactosidase units = 1000 × 0.3/(120 min × 0.5 mL × 0.5 A_{600}) = 10 units.
10. Because of the low quality of the plasmid DNA extracted from yeast cells (usually because of contamination with yeast genomic DNA), use of electrocompetent *E. coli* is preferred to use of chemically competent cells, as they give higher transformation efficiencies.
11. Unlike GST-tagged and His-tagged proteins, MBP fusion proteins should not be eluted off the amylose resin if the resin is going to be the specific precipitating reagent. Alternatively, the eluted fusion protein can be precipitated using a specific antibody reagent (directed against the tag or the recombinant protein) and either protein A or protein G resins (i.e., by immunoprecipitation).

Acknowledgments

I thank Dr. Philip T. LoVerde and Dr. Edward G. Niles (SUNY at Buffalo, School of Medicine) for reviewing this chapter and for their helpful suggestions. I also thank Dr. Philip James (University of Wisconsin, Medical School) for providing the yeast host strain PJ69-4A and for his useful comments on using this strain.

A. Osman received financial support from the UNDP/World bank/WHO Special Programme for Research and Training in Tropical Diseases (TDR; A20357) and the National Institutes of Health (NIH; AI46762).

References

1. Fields, S. and Song, O. (1989) A novel genetic system to detect protein–protein interactions. *Nature* **340(6230),** 245–246.
2. Rine, J. (1991) Gene overexpression in studies of *Saccharomyces cervisiae*, in *Guide to Yeast Genetics and Molecular Biology, Vol. 194* (Guthrie C., ed.), Academic Press, New York, pp. 239–250.
3. Kingsman, A. J., Clarke, L., Mortimer, R. K., et al. (1979) Replication in *Saccharomyces cerevisiae* of plasmid pBR313 carrying DNA from the yeast trpl region. *Gene* **7(2),** 141–152.

4. Storms, R. K., Holowachuck, E. W., and Friesen, J. D. (1981) Genetic complementation of the *Saccharomyces cerevisiae* leu2 gene by the *Escherichia coli* leuB gene. *Mol. Cell Biol.* **1(9)**, 836–842.
5. Castanon, M. J., Spevak, W., Adolf, G. R., et al. (1988) Cloning of human lysozyme gene and expression in the yeast *Saccharomyces cerevisiae. Gene* **66(2)**, 223–234.
6. Hashimoto, H., Kikuchi, Y., Nogi, Y., et al. (1983) Regulation of expression of the galactose gene cluster in *Saccharomyces cerevisiae*. Isolation and characterization of the regulatory gene GAL4. *Mol. Gen. Genet.* **191(1)**, 31–38.
7. Flick, J. S. and Johnston, M. (1990) Two systems of glucose repression of the GAL1 promoter in *Saccharomyces cerevisiae. Mol. Cell Biol.* **10(9)**, 4757–4769.
8. Sambrook, J. and Russell, D. W. (2001) *Molecular Cloning: A Laboratory Manual, 3rd ed*. Cold Spring Harbor Laboratory, Plainview, NY.
9. Franco, G. R., Valadao, A. F., Azevedo, V., et al. (2000) The *Schistosoma* gene discovery program: state of the art. *Int. J. Parasitol.* **30(4)**, 453–463.
10. Oliveira, G. and Johnston, D. A. (2001) Mining the schistosome DNA sequence database. *Trends Parasitol.* **17(10)**, 501–503.
11. Ito, H., Fukuda, Y., Murata, K., et al. (1983) Transformation of intact yeast cells treated with alkali cations. *J. Bacteriol.* **153(1)**, 163–168.
12. Schiestl, R. H. and Gietz, R. D. (1989) High efficiency transformation of intact yeast cells using single stranded nucleic acids as a carrier. *Curr. Genet.* **16(5–6)**, 339–346.
13. Hill, J., Donald, K. A., Griffiths, D. E., et al. (1991) DMSO-enhanced whole cell yeast transformation. *Nucleic Acids Res.* **19(20)**, 5791.
14. Breeden, L. and Nasmyth, K. (1985) Regulation of the yeast HO gene. *Cold Spring Harb. Symp. Quant. Biol.* **50**, 643–650.
15. Giacomini, A., Corich, V., Ollero, F. J., et al. (1992) Experimental conditions may affect reproducibility of the beta-galactosidase assay. *FEMS Microbiol. Lett.* **79(1–3)**, 87–90.
16. Miller, A. L., Frost, R. G., and O'Brien, J. S. (1977) Purified human liver acid beta-D-galactosidases possessing activity towards G(M1)-ganglioside and lactosylceramide. *Biochem. J.* **165(3)**, 591–594.
17. Hoffman, C. S. and Winston, F. (1987) A ten-minute DNA preparation from yeast efficiently releases autonomous plasmids for transformation of *Escherichia coli. Gene* **57(2–3)**, 267–272.
18. De Sampaio, G., Bourdineaud, J. P., and Lauquin, G. J. (1999) A constitutive role for GPI anchors in *Saccharomyces cerevisiae*: cell wall targeting. *Mol. Microbiol.* **34(2)**, 247–256.
19. Osman, A., Niles, E. G., and LoVerde, P. T. (2001) Identification and characterization of a Smad2 homologue from *Schistosoma mansoni*, a transforming growth factor-beta signal transducer. *J. Biol. Chem.* **276(13)**, 10,072–10,082.
20. Freebern, W. J., Osman, A., Niles, E. G., et al. (1999) Identification of a cDNA encoding a retinoid X receptor homologue from *Schistosoma mansoni*. Evidence for a role in female-specific gene expression. *J. Biol. Chem.* **274(8)**, 4577–4585.

21. Li, B. and Fields, S. (1993) Identification of mutations in p53 that affect its binding to SV40 large T antigen by using the yeast two-hybrid system. *FASEB J.* **7(10),** 957–963.
22. Bartel, P., Chien, C. T., Sternglanz, R., et al. (1993) Elimination of false positives that arise in using the two-hybrid system. *Biotechniques* **14(6),** 920–924.
23. James, P., Halladay, J., and Craig, E. A. (1996) Genomic libraries and a host strain designed for highly efficient two-hybrid selection in yeast. *Genetics* **144(4),** 1425–1436.

24

From Genomes to Vaccines for Leishmaniasis

Carmel B. Stober

Summary

A total of 2183 clones derived from four life cycle stage-specific, spliced-leader cDNA libraries of *Leishmania major* LV39 Neal strain were randomly picked and sequenced to generate expressed sequence tags (ESTs). Then 1094 unique genes were identified, with 18.2% having BLAST hits with known genes/proteins and 81.8% failing to match genes currently deposited in public databases. Approximately 250 unique genes were obtained from a lesion-derived amastigote complementary DNA (cDNA) library, the form of the parasite that is infective to the mammalian host. Polymerase chain reaction (PCR)-amplified ESTs were spotted onto glass slides, and DNA microarray used to identify a further approx 100 unique cDNAs highly expressed in amastigotes.

One hundred unique, randomly selected amastigote-expressed genes were PCR amplified to exclude the 5'-spliced leader, and the full-length genes subcloned into the TOPO-TA cloning vector. The genes were sequence-verified, excised using restriction enzyme digestion, and cloned upstream of the eukaryotic cytomegalovirus promoter into the expression vector pcDNA3. Expression plasmids were sequence-verified and large-scale, endotoxin-free plasmids prepared. Then 100 µg of each expression plasmid (DNA vaccine) was delivered subcutaneously to the rump of susceptible BALB/c mice, the mice boosted 4 wk later and then challenged 2 wk post-boost with 2×10^6 *L. major* LV39 parasites to the hind footpad. Infection was monitored on a weekly basis by measuring footpad depth with digital calipers. Protection was scored by comparing the footpad depth of mice receiving empty vector DNA to those immunized with DNA containing *L. major* amastigote-expressed genes.

Key Words: ESTs; cloning; DNA vaccination.

1. Introduction

Expressed sequence tags (ESTs) represent short stretches of sequence-verified complementary DNAs (cDNAs) and have proved an efficient method of gene identification in protozoan parasites such as *Plasmodium falciarum* (*1*), *Toxoplasma gondii* (*2*), and *Leishmania major* (*3*) (*see* Chapters 4 and 5). The presence of a 5'-spliced-leader sequence on most trypanosomatid messenger

RNAs (mRNAs) *(4)* combined with the 3'-polyadenylated tail enables the construction of spliced-leader cDNA libraries containing directionally cloned, full-length genes. This is in contrast to human and murine cDNA libraries, which often consist of truncated gene products. In our laboratory, we possess four spliced-leader cDNA libraries from *L. major* LV39 Neal strain, representing mRNA derived from different life-cycle stages of this protozoan parasite.

L. major parasites exist as two distinct physiological forms according to whether they reside within the gut of the insect vector or the macrophage phagolysosome of the mammalian host, termed promastigotes and amastigotes, respectively. Promastigotes can be further differentiated biochemically corresponding to the maturation stages in the invertebrate host gut *(5,6)*. The progressive differentiation of noninfective, procyclic promastigotes to infective, metacyclic promastigotes (a process termed "metacyclogenesis") can be induced in vitro. Day 3 promastigotes in culture are noninfective, procyclic promastigotes in the early logarithmic phase of growth, d 7 promastigotes are a heterogeneous population of parasites in late logarithmic-early stationary phase, whereas d 10 promastigotes are mostly stationary-phase, infective metacyclic parasites. Spliced-leader cDNA libraries of d 3, 7, and 10 *L. major* promastigotes were constructed using Lambda ZAP II (Stratagene) as the cloning vector, as described by Ajioka et al. *(7)*. A lesion-derived amastigote cDNA library was generated using a parallel approach in the laboratory of Debbie Smith and colleagues (Imperial College, London, UK).

In our laboratory, more than 2000 *L. major* ESTs derived from the four spliced-leader cDNA libraries described above were bioinformatically analyzed and pooled according to sequence similarity *(3,8)* (*see also* Chapter 5). From 2183 cDNAs, 1094 unique genes were identified, which included approx 250 amastigote-expressed genes. Of these, 18.2% of the ESTs were genes (EMBL) or proteins (Swissprot) with previous BLAST hits, whereas the remaining 81.7% were novel genes with no significant matches to other genes/proteins. This information is available at http://www.ebi.ac.uk/parasites/leish.html. Polymerase chain reaction (PCR)-amplified ESTs were subsequently spotted onto glass slides to be used in microarray experiments comparing different life-cycle stages of *Leishmania* (*see also* Chapter 11). Approximately 200 amastigote-expressed genes were reproducibly identified using microarray analysis (unpublished data). Using the two resources available in our laboratory (amastigote-expressed ESTs and amastigote-specific genes identified using microarray), 100 amastigote-expressed genes were randomly selected and subsequently cloned into an expression vector for delivery to BALB/c mice as DNA vaccines.

DNA vaccines comprise a foreign gene of interest cloned into a bacterial plasmid engineered for optimal expression in eukaryotic cells. The bacterial plasmid contains a strong promoter, such as the viral cytomegalovirus promoter, upstream of the inserted foreign gene. Bacterial DNA also provides the advantage of possessing unmethylated cytidine–phosphate–guanosine (CpG) dinucleotide sequences that are able to stimulate the mammalian innate immune response directly. DNA vaccines are delivered to the host by the intramuscular, subcutaneous, or intradermal routes and are taken up by host cells. The foreign gene is expressed by host cells, processed, and presented to the host immune system. The mechanism by which antigen presentation occurs following DNA vaccination has not been clearly defined, but numerous studies have demonstrated that both CD4+ and CD8+ cellular as well as humoral immune responses are elicited. Therefore, DNA vaccination was selected as the mode of immunization for these studies because: (a) generation of a DNA vaccine was possible without the need to predict the open reading frame of the *L. major* gene of interest or express the protein in vitro, (b) no adjuvant was required because of the inherent adjuvanticity provided by CpG motifs contained within the backbone of the bacterial plasmid, and (c) both cellular and humoral immune responses are generated, which is an important factor for immune responses to intracellular infections.

Numerous studies have examined the use of DNA vaccines in the murine model of leishmaniasis, and promising vaccines include the *Leishmania* homolog of the receptor for activated C kinase *(9)*, *L. major* stress-inducible protein 1 (LmSTI1) *(10)*, and thiol-specific antioxidant *(10)*. LmSTI1 was identified as a protective antigen by screening sera from *L. major*-infected mice using a *L. major* amastigote cDNA expression library *(11)*. Expression library immunization (ELI), in which entire libraries or sublibraries are delivered to the host as DNA vaccines, have been tested in numerous disease model systems, including *Mycoplasma pulmonis* *(12)*, *Plasmodium chabaudi* *(13)*, *Trypanosoma cruzi* *(14)*, and *Leishmania* *(15,16)*. The study of Piedrafita et al. *(15)* screened 10^5 clones to obtain a protective sublibrary of 1000 clones, and Melby et al. *(16)* commenced their screen with 30,000 clones and concluded with 40 protective vaccines. Therefore, ELI provides a mechanism of screening for vaccines, but it is likely that protective antigens contained in large pools will be missed when using this approach.

Therefore, our strategy was to identify protective vaccine candidates in the murine model of cutaneous leishmaniasis using 100 *L. major* amastigote-expressed genes, each cloned into the expression vector pcDNA3. Preliminary studies in our laboratory showed that protective antigens could be identified when examining pools of four vaccines *(8)*. However, protection mediated by one of our DNA vaccines was masked when combined with three other nonprotective antigens (unpublished observation). Therefore, in contrast to the studies of Piedrafita *(15)* and Melby *(16)*, our approach has been to test individual vaccine candidates and compare the performance of these antigens alone compared to when they were coadministered with nine other DNA vaccines. Our experimental approach for obtaining DNA vaccines is as follows: Putative amastigote-expressed genes are amplified from phage stocks of the spliced-leader *L. major* cDNA libraries, subcloned into a TOPO TA cloning vector, then excised and cloned into the expression vector pcDNA3. Following DNA sequence verification, endotoxin-free, large-scale plasmid preparations are generated for subsequent delivery to the mammalian host as a DNA vaccine. These procedures are described here.

2. Materials

2.1. cDNA Amplification and Purification

1. Oligonucleotide primers to amplify cDNA (*see* **Note 1**).
2. cDNA template: Single phage plaques stored at 4°C in 800 µL SM buffer (10 mM NaCl, 10 mM MgSO$_4$, 50 mM Tris-HCl, pH 7.5, 0.01% gelatin) with 100 µL chloroform).
3. dNTPs (Bioline).
4. BIO-X-ACT DNA polymerase (Bioline) (*see* **Note 2**).
5. PCR machine.
6. Standard 96-well PCR plates.
7. Agarose gel equipment, agarose prepared in 1X TBE buffer (0.09 M Tris-borate, 0.002 M EDTA), ethidium bromide (10-mg/mL stock), and 6X xylene cyanol loading buffer (0.25% xylene cyanol FF, 30% glycerol in water) *(17)*. (Ethidium bromide is mutagenic and toxic, so gloves should be worn at all times when handling this reagent.)
8. QIAquick PCR purification kit (Qiagen).
9. Ethanol.
10. Microcentrifuge and 1.5-mL microfuge tubes.

2.2. TOPO TA Cloning and Analysis of Positive Clones

1. TA Cloning® kit with pCR® 2.1 vector (box 1) and One shot® TOP10 chemically competent *Escherichia coli* (box 2) (Invitrogen).
2. 5-Bromo-4-chloro-3-indoyl β-D-galactopyranoside (X-gal) made up to 40 mg/mL in dimethylformamide and stored at −20°C in the dark (Sigma).
3. Luria Bertani (LB) medium (1% bacto-tryptone, 0.5% bacto-yeast extract, 172 mM NaCl, adjusted to pH 7.0 *[17]*), containing 50-µg/mL ampicillin.
4. LB agar plates (LB medium with 1.5% bacto-agar) with 50-µg/mL ampicillin.
5. 42°C water bath.
6. 37°C shaking incubator.
7. QIAprep® spin miniprep kit (Qiagen).
8. Ethanol.
9. Microcentrifuge and 1.5-mL microfuge tubes.
10. *Eco*RI restriction endonuclease, corresponding 10X buffer, and 100X bovine serum albumin (BSA).

2.3. Sequence Confirmation and Mini or Maxi Preparation of Plasmid DNA

1. QIAprep spin miniprep kit (Qiagen).
2. Ethanol.
3. Microcentrifuge and 1.5 mL-microfuge tubes.
4. Restriction endonucleases, corresponding 10X buffers, and 100X BSA.
5. Agarose gel equipment, agarose, ethidium bromide, and xylene cyanol loading buffer (17).
6. DNA sequencing reagents.
7. Endofree plasmid maxi, mega, or giga kit (Qiagen).
8. Endotoxin-free, DNAase-free plasticware.
9. Endotoxin-free saline or phosphate-buffered saline (PBS) (Sigma-Aldrich).

2.4. Cloning of cDNA to Eukaryotic Expression Vector

1. QIAquick gel extraction kit (Qiagen).
2. Ethanol.
3. Agarose gel equipment, agarose, ethidium bromide, and xylene cyanol loading buffer (17).
4. Eukaryotic expression vector such as pcDNA3 (Invitrogen).
5. Restriction endonucleases, corresponding 10X buffers, and 100X BSA.
6. T4 DNA ligase with buffer (Roche Diagnostics or similar).
7. Chemically transformation-competent *E. coli* strain DH5α or XL1-blue (17).
8. 14-mL polypropylene snap-cap tubes (BD Falcon).
9. As for **Subheading 2.2., step 3.**
10. As for **Subheading 2.2., step 4.**
11. Microcentrifuge and 1.5-mL microfuge tubes.

2.5. Administration of DNA Vaccine

1. BD MICRO-FINE 0.5-mL insulin syringe (0.33 × 13 mm—29 ga) (Becton Dickinson).
2. Experimental animals.

3. Methods

3.1. cDNA Amplification by the PCR and Purification of PCR Products

3.1.1. Amplification

1. Use 1–10 µL of phage suspension as template for PCR in a total reaction volume of 50 µL. Standard components of the PCR reaction are as follows (final concentration): 1.5–2 mM MgCl$_2$, 5 µL 10X reaction buffer (OptiBuffer□), 250 µM of each dNTP, 100 ng of forward and reverse primers (*see* **Note 1**), 2U BIO-X-ACT□ DNA polymerase, 1–10 µL phage suspension, made up to a total volume of 50 µL. Overlay with mineral oil if the PCR machine does not have a heated lid.

2. Standard PCR reaction conditions will vary according to the insert size and primer characteristics, but typical reaction conditions are as follows: initial denaturation at 94°C for 5 min, 30 cycles of (94°C for 1 min, 60°C for 1 min, 72°C for 2 min), 72°C for 5 min. The tubes then remain at 4°C until removed.
3. Electrophorese 5–10 µL of the PCR product plus 1–2 µL xylene cyanol loading buffer on a 0.8% agarose gel containing 1-µg/mL ethidium bromide to verify size and purity of the PCR reaction. View under UV illumination, ensuring that the user is shielded from the light source.
4. The PCR product can be stored at –20°C prior to purification.

3.1.2. Purification

Purification of PCR products can be achieved using a QIAquick PCR purification kit with spin columns (or similar) to remove contaminating primers, nucleotides, polymerase, and salts. All of the reagents required are contained within the kit (add ethanol to buffer PE as instructed) and are stored at room temperature. The protocol, essentially following the manufacturer's instructions, is as follows:

1. Add 5 vol of buffer PB to 1 vol of the PCR reaction and mix. Removal of mineral oil or kerosene is not required.
2. Apply the sample to a QIAquick spin column placed in a 2-mL collection tube and centrifuge for 30–60 s.
3. Discard the flow-through (FT).
4. Add 0.75 mL of buffer PE (with ethanol added) to the QIAquick column and centrifuge for 30-60 seconds.
5. Discard the FT and recentrifuge for 1 min at maximum speed.
6. Place the QIAquick column within a clean 1.5-mL microcentrifuge tube.
7. To elute the DNA, add between 30 and 50 µL of heated buffer EB (10 mM Tris-HCl, pH 8.5) to the center of the QIAquick membrane, incubate for 1–5 min (*see* **Note 3**) and centrifuge at maximum speed for 1 min.
8. The purifed PCR product can be stored at –20°C until required.

3.2. TOPO TA Cloning and Analysis of Positive Clones

TOPO TA cloning provides a rapid, one-step cloning strategy for the direct insertion of *Taq* or BIO-X-ACT polymerase-amplified PCR products into a plasmid vector. The plasmid vector pCR 2.1-TOPO (**Fig. 1**) is supplied linearized with single 3'-thymidine (T) overhangs enabling PCR inserts with single deoxyadenosine (A) overhangs to ligate efficiently into the vector. The TOPO TA cloning kit contains all reagents required unless otherwise stated. Box 1, containing TOPO TA cloning reagents, should be stored at –20°C, and Box 2, containing One shot TOP10 chemically competent *E. coli*, requires storage at –80°C. All reagents, with the exception of the pCR 2.1-TOPO vector and TOP10 *E. coli*, should be equilibrated to room temperature before use. The TOPO TA cloning reaction is essentially performed as described here.

From Genomes to Vaccines

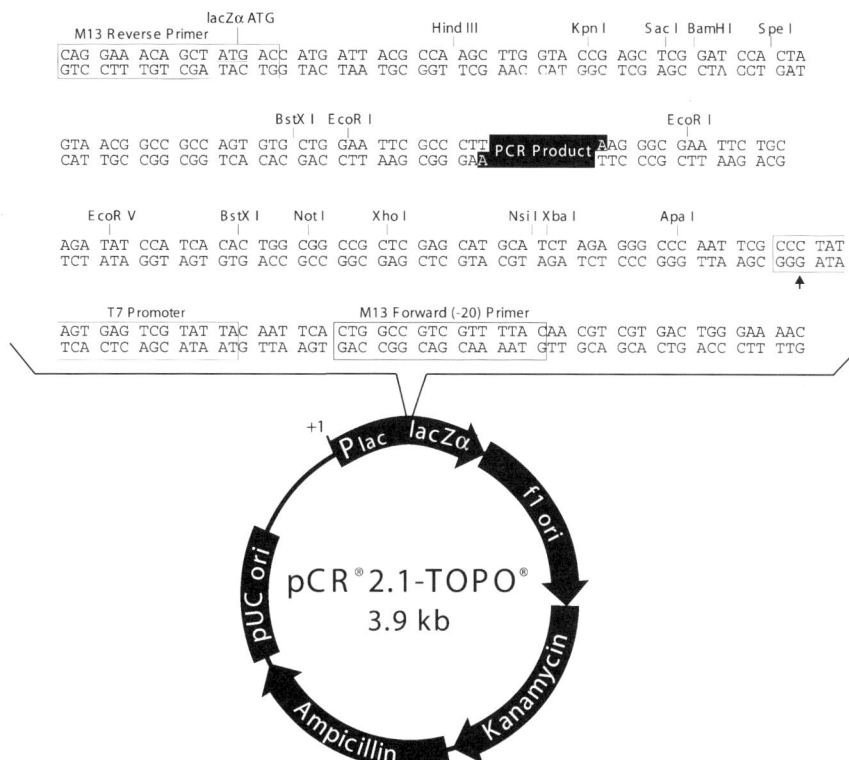

Comments for pCR® 2.1-TOPO®
3931 nucleotides

LacZα fragment: bases 1-547
M13 reverse priming site: bases 205-221
Multiple cloning site: bases 234-357
T7 promoter/priming site: bases 364-383
M13 Forward (-20) priming site: bases 391-406
f1 origin: bases 548-985
Kanamycin resistance ORF: bases 1319-2113
Ampicillin resistance ORF: bases 2131-2991
pUC origin: bases 3136-3809

Fig. 1. Vector map of TOPO TA cloning vector, pCR2.1-TOPO, taken from the Invitrogen Web pages (http://www.invitrogen.com).

3.2.1. TOPO TA Cloning

1. To 0.5–4 μL of purified PCR product, add 1 μL of salt solution provided in kit, sterile water to a total volume of 5 μL, and 1 μL of pCR 2.1-TOPO vector. Mix the reaction gently and incubate for 5 min at room temperature. Transfer to ice or store the TOPO cloning reaction at −20°C overnight.

2. Add 2 µL of the TOPO cloning reaction to a vial of One shot TOP10 chemically competent *E. coli* (thawed on ice) and mix gently without pipetting up and down.
3. Incubate on ice for 5–30 min.
4. Heat-shock the cells for 30 s by placing the tubes in a 42°C water bath without shaking.
5. Add 250 µL of room-temperature SOC medium (supplied with One shot *E. coli*), cap the tube tightly, and shake the tube horizontally (200 rpm) at 37°C for 1 h.
6. Prewarm two selective LB agar plates containing 50-µg/mL ampicillin per TA cloning reaction to 37°C for 30 min. Spread 40 µL of 40-mg/mL X-gal (a chromogenic substrate) on each LB plate and incubate at 37°C until ready for use.
7. Add 20 µL of the transformation mixture plus 20 µL of SOC medium to one LB plate plus X-gal, and 50 µL of transformation to the second plate. In each case, spread evenly across the plate. Incubate the plates overnight at 37°C and store at 4°C until required. The blue color of the colonies will develop further following cold storage.

3.2.2. Analysis of Positive TA Colonies

Colonies obtained following TOPO TA cloning that contain plasmid with cDNA inserts will either be white or light blue, but not dark blue, in color. For each cloning reaction, between 5 and 10 colonies should be selected to determine the presence of inserts of the correct size. A QIAprep spin miniprep kit can be used for this application or, alternatively, minpreps can be performed manually using alkaline lysis as described by Sambrook et al. *(17)*. All kit reagents should be stored at room temperature:

1. Pick between 5 and 10 white or light blue colonies into 3 mL of LB medium containing 50-µg/mL ampicillin and incubate overnight at 37°C in a shaking incubator (200–250 rpm).
2. Transfer 1 mL of the culture to a 1.5-mL microfuge tube, centrifuge, and resuspend pelleted bacterial cells in 250 µL of buffer P1.
3. Add 250 µL of buffer P2, gently invert the tube four to six times, and incubate for a maximum of 5 min.
4. Add 350 µL of buffer N3, invert the tube four to six times, and centrifuge on high speed for 10 min.
5. Apply the supernatant to a QIAprep spin column within a 2-mL collection tube. Centrifuge for 30–60 s and discard the FT.
6. Wash the column by adding 0.75 mL of buffer PE (with ethanol added) and centrifuge for 30–60 s. Discard the FT and centrifuge for an additional 1 min to remove residual wash buffer.
7. Transfer QIAprep column to a clean 1.5-mL microfuge tube and elute with 50 µL of heated buffer EB (10 m*M* Tris-Cl, pH 8.5) to the center of the QIAprep membrane, incubate for 5 min (*see* **Note 3**), and centrifuge at maximum speed for 1 min.
8. Use 10 µL of the plasmid for restriction enzyme digestion with EcoRI as follows: 10 µL of eluted plasmid, 1.5 µL of 10X buffer, 1.5 µL of 10X BSA, 2 µL of water.

From Genomes to Vaccines 431

Digest at 37°C for a minimum of 1 h. (An *Eco*RI site is present on either side of the insert site of the PCR product [**Fig. 1**], therefore this digestion will rapidly identify the presence of inserts of the correct size.)

9. Add 3 µL of 6X xylene cyanol loading buffer, terminate the reaction by heat inactivation (5 min at 65°C), and electrophorese the digested plasmid on a 0.8% agarose gel containing 1-µg/mL ethidium bromide. View the gel for the presence of inserts under UV illumination.
10. Plasmids containing inserts of the correct size should be further analyzed by DNA sequencing using vector primers such as the M13 reverse primer or the T7 promotor primer (**Fig. 1**). This will determine the orientation of the insert. Approximately 200–300 ng of a purified plasmid is required per sequencing reaction.
11. For plasmids containing correct inserts, combine 850 µL of the overnight culture from **step 1** with 150 µL of sterile glycerol and transfer to a cryovial for storage at –80°C.
12. The plasmids can be stored at –20°C for extended periods.

3.3. Cloning of cDNA to a Eukaryotic Expression Vector

3.3.1. Gel Extraction of Digested Insert From pCR2.1-TOPO Vector

1. On determining which of the TA plasmids contain inserts, digest 30 µL of the plasmid with appropriate restriction endonucleases (*see* **Note 4**) as follows: 1 µL of each restriction enzyme, 4 µL 10X buffer, 4 µL 10X BSA. Incubate at 37°C for a minimum of 1 h and terminate the reaction by the addition of 8 µL xylene cyanol loading buffer and heat inactivation.
2. Electrophorese the digested plasmid on 0.8% agarose gel stained with ethidium bromide *(17)* and view under UV illumination wearing a protective mask. Excise the insert from the agarose gel using a sharp scalpel and transfer to a preweighed 1.5-mL microfuge tube.

The following steps employ the QIAquick gel extraction kit.

3. Determine the weight of the agarose gel piece, add 3 vol of buffer QG and incubate at 50°C for 10 min with occasional agitation.
4. Add 1 vol of isopropanol equal to the volume of gel to the sample and mix.
5. Apply the sample to a QIAquick column contained within a 2-mL collection tube and centrifuge for 1 min. Discard the FT and repeat for any remaining sample (the QIAquick columns hold a maximum volume of 0.75 mL).
6. Repeat **steps 4–7** of **Subheading 3.1.2.**
7. The eluted excised insert can be stored at –20°C (short-term) or –80°C (long-term).

3.3.2. Gel Extraction of Eukaryotic Expression Vector

The expression vector to be used for DNA vaccine delivery can be digested and gel-extracted simultaneously with the TA plasmid, as equivalent enzymes will be used for this reaction. The eukaryotic expression vector employed in

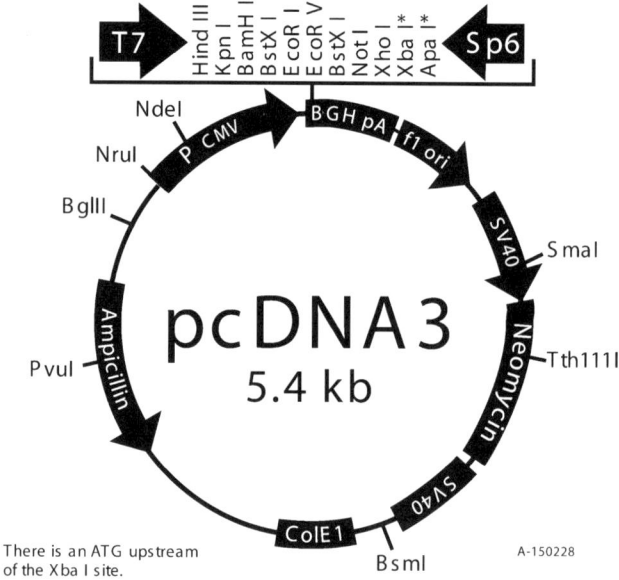

Fig. 2. Vector map of the eukaryotic expression vector, pcDNA3, taken from the Invitrogen Web pages (http://www.invitrogen.com).

our laboratory is pcDNA3 (Invitrogen) (**Fig. 2**), which has been modified by removing the neomycin resistance gene. Approximately 10 ng of the digested plasmid is required per ligation reaction.

1. Digest an appropriate amount of pcDNA3 (e.g., 1 µg) in a 40-µL reaction containing 1 µL of each restriction enzyme, 4 µL 10X buffer, 4 µL 10X BSA, as described in **step 1**, **Subheading 3.3.1.**
2. Repeat **steps 2–6** in **Subheading 3.3.1.**
3. The eluted, digested plasmid can be stored at −20°C (short-term) or −80°C (long-term) in small aliquots to avoid freeze–thawing.

3.3.3. Quantitation of Excised Insert and Digested Plasmid

The insert and plasmid DNA can be quantified using either UV spectrophotometry or gel electrophoresis as described by Sambrook et al. *(17)*. Readings should be taken at 260 and 280 nm to determine the amount of contaminating protein. Alternatively, *Hind*III digest of lambda DNA (NEB) (2–23 kbp) or Hyperladder™I (Bioline) (200–10,000 bp) can be used as standards on an agarose gel stained with ethidium bromide, and the relative fluorescence of the standards and the prepared DNA can be compared.

3.3.4. Ligation of Insert Into pcDNA3 and Transformation Into Chemically Competent E. coli

1. The ligation reaction should be performed in a total reaction volume of 10 µL. Add 10 ng of digested plasmid to both ×3 or ×10 molar ratio of insert plus 1 µL of 10X buffer and make up to a total volume of 9 µL with water. Heat at 45°C for 5 min.
2. Transfer to ice and add 1 µL (1 U) of T4 DNA ligase and gently mix.
3. Incubate at 4°C for 16 h.
4. Transfer 100–150 µL of chemically competent *E. coli* strain DH5α or XL1-blue (thawed on ice) (prepared as described by Sambrook et al. *(17)* or purchased from a commercial source) to a 14-mL snap-cap tube.
5. Add the 10 µL ligation reaction to the competent cells in the 14-mL snap-cap tube and incubate for 30 min on ice.
6. Heat-shock the competent cells plus ligation reaction components at 42°C for 1 min, transfer to ice for 2 min, and add 800 µL of LB medium (without ampicillin).
7. Incubate at 37°C for 45 min in a shaking incubator (200–250 rpm).
8. Plate 100 µL of the transformation mixture onto LB agar plates containing 50 µg/mL ampicillin by spreading evenly across the plate.
9. Transfer the remaining mixture to a 1.5-mL microfuge tube, centrifuge to pellet the bacterial cells, remove the supernatant, and resuspend the pellet in 150 µL of LB medium. Plate out the resuspended bacterial cells.
10. Incubate the LB plates at 37°C overnight and store at 4°C until required.

3.4. Mini or Maxi Preparation of Plasmid DNA and Sequence Confirmation

3.4.1. Minipreparation of Plasmid DNA and DNA Sequencing

Minipreps are performed using the QIAprep spin miniprep kit as described in **Subheading 3.2.2.**

1. Pick between 5 and 10 colonies per ligation reaction into 3 mL of LB medium containing 50-µg/mL ampicillin and incubate overnight at 37°C in a shaking incubator (200–250 rpm).
2. Repeat the **steps 2–10** as described under **Subheading 3.2.2.**

3. Plasmids containing inserts of the correct size should be further analyzed by DNA sequencing using vector primers such as the T7 promotor primer (forward primer in pcDNA3) or the SP6 promotor primer (reverse primer in pcDNA3). Approximately 200–300 ng of a purified plasmid is required per sequencing reaction.
4. Once the integrity of the plasmids is confirmed, storage of the transformed bacteria as glycerol stocks at −80°C is highly recommended. Combine 850 µL of the overnight culture from **step 1** with 150 µL of sterile glycerol and transfer to a cryovial for storage at −80°C.

3.4.2. EndoFree Plasmid Maxipreparations

Endotoxins, also termed lipopolysaccharides or LPS, are cell membrane components of Gram-negative bacteria (e.g., *E. coli*) that are major contaminants when purifying plasmid DNA from bacterial cell lysates using conventional methods. Bacterial LPS is a powerful activator of host immune responses interacting with receptors expressed on antigen-presenting cells such as dendritic cells and macrophages. The removal of contaminating endotoxin prior to the in vivo administration of plasmid DNA is therefore essential. DNA purified using EndoFree plasmid kits contain negligible amounts of endotoxin (<0.1 EU/µg plasmid DNA).

Endotoxin-free plasmids are obtained using EndoFree plasmid maxi, mega, or giga kits (Qiagen) according to whether a yield of 0.5, 2.5, or 10 mg of plasmid DNA is required. The protocol described follows the manufacturer's instructions for Endofree plasmid maxi kits unless otherwise stated. Kit reagents are stored at room temperature with the exception of buffer P1 following the addition of RNase A, which is stored at 4°C.

1. Streak out the glycerol stock of recombinant *E.coli* containing the DNA vaccine of interest onto an LB agar plate containing 50-µg/mL ampicillin and incubate overnight at 37°C.
2. Pick a single colony and inoculate a starter culture of 2–5 mL of LB medium containing 50-µg/mL ampicillin. Incubate for approx 8 h at 37°C with vigorous shaking (200–300 rpm).
3. Dilute the starter culture 1/500 into 100–200 mL of LB medium containing 50-µg/mL ampicillin and incubate at 37°C for 12–16 h with vigorous shaking (200–300 rpm).
4. Harvest the bacterial cells by centrifugation at $6000g$ for 15 min at 4°C.
5. Resuspend the bacterial pellet in 10 mL of buffer P1 (with RNase A added) and transfer to a 50-mL centrifuge tube.
6. Add 10 mL buffer P2, gently mix by inversion four to six times, and incubate at room temperature for 5 min.
7. Add 10 mL chilled buffer P3, mix by gentle inversion four to six times, and transfer to the barrel of a capped QIAfilter cartridge. Incubate at room temperature for 10 min.

8. Remove the cartridge cap, insert the plunger, and plunge the cell lysate into a clean 50-mL tube. Add 2.5 mL of endotoxin removal buffer and incubate on ice for 30 min.

To reduce contamination with endotoxin, the following steps are performed in a flow cabinet using endotoxin-free reagents and plasticware.

9. Equilibrate a QIAGEN-tip 500 by applying 10 mL buffer QBT, and allow to empty by gravity flow. Apply the filtered lysate from the previous step and allow to enter the resin by gravity flow. Wash the column twice with 30 mL of buffer QC, discarding the FT.
10. Elute the DNA with 15 mL of buffer QN to a fresh 50-mL endotoxin-free centrifuge tube. The eluate can be stored at 4°C overnight at this stage.
11. Add 10.5 mL room-temperature isopropanol to the eluted DNA. Mix and centrifuge at 5000g for 60 min at 4°C.
12. Wash the DNA pellet with 5 mL endotoxin-free, room-temperature 70% ethanol (ethanol is added to endotoxin-free water supplied with the kit), and centrifuge at 5000g for 60 min at 4°C.
13. Decant the supernatant and air-dry the pellet for 5–10 min. Redissolve the DNA in approx 300 µL of endotoxin-free saline or PBS. Transfer to 4°C overnight to ensure solubilization of the plasmid.
14. Quantitate the plasmid DNA using UV spectrophotometry and adjust to a concentration of 1 or 2 µg/µL. Transfer to endotoxin-free 1.5-mL microfuge tubes for storage at –20°C until required. The preparation is now ready for use as a DNA vaccine.

3.5. Administration of DNA Vaccines

The dose, site of administration, and number of immunizations required for an individual DNA vaccine should be determined for each application and are outside the scope of this chapter. In our laboratory, BALB/c mice are primed with 100 µg of DNA vaccine delivered in a 50–100-µL volume subcutaneously to the shaven rump using a MICRO-FINE insulin syringe. The mice are boosted 4 wk later using the same immunization regimen. Two weeks postboost, the animals are challenged with 2 million stationary-phase *L. major* substrain LV39 parasites subcutaneously into the hind footpad. The development of infection is subsequently monitored by measuring footpad depth on a weekly basis with digital callipers. Vaccine efficacy is determined by the ability of the DNA vaccine to generate a *Leishmania*-specific immune response resulting in a significant decrease in footpad swelling, lesion development and ulceration following *L. major* challenge when compared to mice immunized with empty vector plasmid DNA. The mechanism by which DNA vaccination inhibits the clinical outcome in murine cutaneous leishmaniasis is thought to involve the cellular arm of the host immune response and both CD4+ and CD8+ T-cells. Further information on the use of genetic immunization in animal and human model systems may be found at http://www.dnavaccine.com (*see* **Note 5**).

4. Notes

1. On constructing the cDNA library, spliced-leader-*Not*I and oligo dT-*Xho*I oligonucleotides were initially utilized to amplify *L. major* cDNA for subsequent ligation into the λZAPII vector. Vector primers were then employed for PCR amplification and DNA sequencing of selected plaques. However, for amplification of the insert of interest from bacteriophage plaques for DNA vaccine generation, an EST-specific 5' forward oligonucleotide designed to anneal downstream of the spliced-leader sequence was used in combination with the vector primer, M13-20 (GTAAAACGACGGCCAGT). This avoided cross-contamination and generated a full-length cDNA clone excluding the spliced leader. The forward primer included a restriction enzyme site, either *Bam*HI or *Eco*RI, at the 5' end.
2. A high-fidelity DNA polymerase that leaves an "A" overhang, such that primer extension products are suitable for effective integration into TA cloning vectors, is required.
3. Heating of buffer EB on full power in a microwave for 15 s prior to application to the column and an incubation time of 1–5 min prior to elution increases the yield from the QIAquick or QIAprep columns.
4. For our application, *Bam*HI or *Eco*RI (contained within the forward primer used for PCR amplification from the library) is used in combination with *Xho*I (the enzyme site originally used to create the spliced-leader library). For undertaking double digestions, New England Biolabs (NEB) provide a comprehensive list of appropriate buffer conditions. An alternative strategy is to use restriction enzyme sites contained within the pCR2.1-TOPO vector. However, the orientation of the insert will need to be determined by DNA sequencing.
5. Invitrogen now provides GATEWAY™ cloning technology. This system provides an alternative strategy to that described here, whereby the gene of interest is cloned into an "entry vector." The insert can be subsequently cloned into an appropriate "destination vector" using site-specific recombination rather than restriction endonucleases and ligase. Further information may be found at http://www.invitrogen.com.

References

1. Chakrabarti, D., Reddy, G. R., Dame, J. B., et al. (1994) Analysis of expressed sequence tags from *Plasmodium falciparum*. *Mol. Biochem. Parasitol.* **66,** 97–104.
2. Kiew-Lian, W., Blackwell, J. M., and Ajioka, J. W. (1996) *Toxoplasma gondii* expressed sequence tags: insight into tachyzoite gene expression. *Mol. Biochem. Parasitol.* **75,** 179–186.
3. Levick, M. P., Blackwell, J. M., Connor, V., et al. (1996) An expressed sequence tag analysis of a full length, spliced-leader cDNA library from *Leishmania major* promastigotes. *Mol. Biochem. Parasitol.* **76,** 345–348.

4. Perry, K. L., Watkins, K. P., and Agabian, N. (1987) Trypanosome mRNAs have unusual "cap 4" structures acquired by addition of a spliced leader. *Proc. Natl. Acad. Sci. USA* **84,** 8190–8194.
5. Beverley, S. M. and Turco, S. J. (1998) Lipophosphoglycan (LPG) and the identification of virulence genes in the protozoan parasite, *Leishmania. Trends Microbiol.* **6,** 35–40.
6. Sacks, D. L. and Perkins, P. (1984) Identification of an infective stage of *Leishmania* promastigotes. *Science* **223,** 1417–1419.
7. Ajioka, J. W., Melville, S. E., Coulson, R. M. R., et al. (1995) Techniques associated with protozoan genome analysis, in *Methods in Molecular Genetics, Vol. 6* (Adolph, K. W., ed.), Academic Press, London, pp. 30–47.
8. Almeida, R., Norrish, A., Levick, M., et al. (2002) From genomes to vaccines: *Leishmania* as a model. *Phil. Trans. R. Soc. Lond. B* **357,** 5–11.
9. Gurunathan, S., Sacks, D. L., Brown, D. R., et al. (1997) Vaccination with DNA encoding the immunodominant LACK parasite antigen confers protective immunity to mice infected with *Leishmania major. J. Exp. Med.* **186,** 1137–1147.
10. Mendez, S., Gurunathan, S., Kamhawi, S., et al. (2001) The potency and durability of DNA- and protein-based vaccines against *Leishmania major* evaluated using low-dose, intradermal challenge. *J. Immunol.* **166,** 5122–5128.
11. Webb, J. R., Kaufmann, D., Campos-Neto, A., et al. (1996) Molecular cloning of a novel protein antigen of *Leishmania major* that elicits a potent immune response in experimental murine leishmaniasis. *J. Immunol.* **157,** 5034–5041.
12. Barry, M. A., Lai, W. C., and Johnston, S. A. (1995) Protection against mycoplasma infection using expression-library immunisation. *Nature* **377,** 632–635.
13. Smooker, P. M., Setiady, Y. Y., Rainczuk, A., et al. (2000) Expression library immunization protects mice against a challenge with virulent rodent malaria. *Vaccine* **23,** 2533–2540.
14. Alberti, E., Acosta, A., Sarmiento, M. E., et al. (1998) Specific cellular and humoral immune response in Balb/c mice immunised with an expression genomic library of *Trypanosoma cruzi. Vaccine* **16,** 608–612.
15. Piedrafita, D., Xu, D., Hunter, D., et al. (1999) Protective immune responses induced by vaccination with an expression genomic library of *Leishmania major. J. Immunol.* **163,** 1467–1472.
16. Melby, P. C., Ogden, G. B., Flores, H. A., et al. (2000) Identification of vaccine candidates for experimental visceral leishmaniasis by immunization with sequential fractions of a cDNA expression library. *Infect. Immun.* **68,** 5595–5602.
17. Sambrook, J., Fritsch, E. F., and Maniatis, T. (1989) *Molecular Cloning, A Laboratory Manual,* 2nd ed. Cold Spring Harbor Laboratory Press, New York.

Index

A

AFLP, *see* Amplified restriction fragment length polymorphism
Allele-specific primer extension, *see* Single nucleotide polymorphisms
Amplified restriction fragment length polymorphism (AFLP),
 adapter preparation and ligation, 180, 181, 183, 184
 applications, 173, 174, 177
 DNA preparation, 179, 180, 183
 DNA recovery from gels, 183
 materials, 177, 178
 overview of steps, 174, 177
 polyacrylamide gel electrophoresis, 182, 184
 preamplification, 181
 primer end-labeling, 181, 184
 restriction endonuclease digestion of genomic DNA, 180, 183
Annotation,
 automation, 18, 42
 base composition in gene identification, 22, 23
 expressed sequence tags, 94
 Gene Ontology project, *see* Gene Ontology
 gene function prediction,
  annotating without sequence similarity, 34, 35
  domain homology, 30–33
 orthologs and paralogs, 33, 34
 sequence comparison,
  algorithms, 27–29
  database searching, 29, 30
 gene-finding algorithms, 20, 21
 gene model refinement, 24
 gene prediction, 18, 19
 metabolism reconstruction, 35, 36
 open reading frames, 19, 20
 pseudogenes, 24
 RNA genes, 24, 25

B

BAC, *see* Bacterial artificial chromosome
Bacterial artificial chromosome (BAC),
 continuous sequence generation, gap filling, 8, 9
 shotgun library construction, 7
 shotgun sequencing and assembly, 7, 8
 fluorescence *in situ* hybridization, *see* Fluorescence *in situ* hybridization
 walking, 9, 10
BLAST,
 expressed sequence tag similarity analysis,
  commands, 111, 112, 124
  local database building, 109–111, 123, 124
  multiple searches, 112, 113

439

overview, 108, 109
parsing of outputs, 113, 116, 118
gene function prediction, 27–29, 33
ortholog identification,
BLAST searches, 134, 135
materials, 131, 132
overview, 128, 129, 131
reciprocal best match retrieval, 135, 144, 145
sequence retrieval and processing, 134
BLOSUM, gene function prediction, 27, 28

C

CAP3, consensus sequence prediction for expressed sequence tag clusters, 107
CF, *see* Chromosome fragmentation
Chromosome fragmentation (CF),
applications, 353, 354
Leishmania transfection,
cell plating and drug selection, 367, 375
control flask cultivation, 368
electroporation, 365, 366, 374
parasite cultivation, 365, 373, 374
selective plate preparation, 366, 367, 374, 375
sensitivity test to selection drugs, 365, 374
transfectant isolation, 367, 368, 375
limitations in parasite studies, 354
materials, 355–357
transfectant analysis,
chromosomal DNA,
digestion, 370, 371, 375
preparation, 368
chromosomal stability analysis, 371, 372
integration locus polymerase chain reaction and sequencing, 371
pulsed-field gel electrophoresis, 369, 379
Southern blot, 369, 375
transfectant cloning by limited dilution, 369, 370, 375
vector construction,
homologous DNA fragment preparation,
amplification product purification, 359, 360, 372
amplification product restriction digestion, 360, 372
polymerase chain reaction, 359, 372
homologous S1 fragment insertion into pVV, 363
homologous S2 fragment insertion into pVV,
double digestion, 360, 361, 373
bacterial transformation, 361, 373
ligation, 361
plasmid minipreparation and sequencing, 362, 363, 373
recombinant clone selection, 362, 373
linearization,

Index

*Eco*57I digestion, 364, 373
   purification, 364, 373
   maxipreparation, 363, 364, 373
   pVV features, 357–359
Chromosome,
   mapping, *see also* Fluorescence *in situ* hybridization; *In situ* polymerase chain reaction on chromosomes
      digestion in gel and electrophoresis of fragments, 344, 349
      excision from gels, 343, 344, 349
      Southern blotting and hybridization of single-locus markers, 344–346
      preparation of intact chromosomes in agarose plugs, 340, 341, 348, 349
      pulsed-field gel electrophoresis, *see* Pulsed-field gel electrophoresis
      shotgun cloning, *see* Whole-chromosome shotgun
CLOBB, clustering expressed sequence tags into putative genes, 103, 105, 106, 123
Clone-by-clone strategy,
   bacterial artificial chromosome continuous sequence generation,
      gap filling, 8, 9
      shotgun library construction, 7
      shotgun sequencing and assembly, 7, 8
   bacterial artificial chromosome walking, 9, 10
   map-as-you-go approach, 3
   minimal tiling path, 3
   paired end-sequence marker generation, 3–6
   physical map construction, 2, 3
   validation, 10
CLUSTAL W,
   alignment for positive selection scanning, 137, 138, 147, 148
   gene function prediction, 31

D

Databases, *see also specific databases*,
   expressed sequence tags, relational databases, 118, 119
   submission, 97, 98
   mining, 68, 70, 71,
   parasite resources,
      data consolidation, 58, 59
      goals, 58
      prospects, 71
Decoder, protein prediction from consensus sequences, 121
Degenerate oligonucleotides-*Pfu* (DOP) mutagenesis, *Plasmodium falciparum* dihydrofolate reductase gene,
   materials, 323, 324
   primer design, 328–330
   principles, 328
   reaction conditions, 330
DIANA-EST, protein prediction from consensus sequences, 121
Dihydrofolate reductase, *see Plasmodium falciparum*
DNA microarray,
   amino-allyl dUTP labeling of complementary DNA

targets, 228–230, 234, 235
applications, 219, 220, 237, 238
complementary DNA probes,
  amplification, 224, 232, 233
  printing, 225, 226, 233, 234
  purification, 224, 225, 233
gene content analysis in genome,
  genomic DNA isolation and
    purification, 241, 242, 246
  genomic probe,
    amplification, 240, 241, 245
    printing, 241, 245
    purification, 241
  imaging and interpretation, 243–245, 247
  labeling of target and reference DNA, 242, 243, 246, 247
  materials, 239, 240, 245
  microarray hybridization and washing, 243
  rationale, 238, 239
hybridization, 230, 231, 235
imaging and interpretation, 231, 232
materials, 222–224, 232
prehybridization, 230
principles, 220, 222
RNA preparation,
  Qiagen affinity purification, 227, 228, 234
  TRIZOL extraction, 226, 227, 234
single nucleotide polymorphism microarrays, *see* Single nucleotide polymorphisms
washing, 231
DNA vaccine,
  leishmaniasis,
    administration, 435, 436

expressed sequence tag analysis, 424
materials, 426, 427, 436
murine model studies, 425, 426
plasmid preparation,
  maxipreparation, 434, 435
  minipreparation, 433, 434
polymerase chain reaction of complementary DNA,
  amplification, 427, 428, 436
  expression vector preparation, 431–433, 436
  positive TA colony analysis, 430, 431, 436
  product purification, 428, 436
  TOPO TA cloning, 428–430
mechanism of action, 425
Domains, homology searching, 30–33
DOP mutagenesis, *see* Degenerate oligonucleotides-*Pfu* mutagenesis

E

EBI, databases, 50, 58
EC numbers, *see* Enzyme Commission numbers
Electroporation,
  Leishmania transfection, 365, 366, 374
  Plasmodium falciparum transfection, 269
  Trypanosoma brucei transfection, 291, 292, 296
EMBL, content, 46
Enzyme Commission (EC) numbers, assignment, 35, 36

Epitope tagging, *see Trypanosoma brucei*
EST, *see* Expressed sequence tag
ESTscan, protein prediction from consensus sequences, 121
Expressed sequence tag (EST),
  applications, 93
  chimeric tags, 94
  definition, 76, 93
  differential representation of genes, 94
  end preferences, 77
  functional annotation, 94
  genome survey sequence strategy, 76, 77
  leishmaniasis DNA vaccine development, 424
  medium-throughput analysis,
    clone picking and archiving,
      lambda phage libraries, 81, 82, 88
      plasmid libraries, 82, 88
    materials, 79–81, 87, 88
    polymerase chain reaction and gel electrophoresis of products, 82–84, 88, 89
    sequencing, 85–87, 89
    throughput capacity, 77, 79, 87
  software analysis,
    automated sequence calling and trimming, 100, 102, 103, 122, 123
    BLAST-based sequence similarity analysis,
      commands, 111, 112, 124
      local database building, 109–111, 123, 124
      multiple searches, 112, 113
      overview, 108, 109
      parsing of outputs, 113, 116, 118
    clustering into putative genes,
      CLOBB, 103, 105, 106, 123
      StackPACK, 106
    consensus sequence prediction for each cluster,
      CAP3, 107
      phrap, 107, 108
    database submission, 97, 98
    hardware, 95, 121
    protein prediction from consensus sequences,
      Decoder, 121
      DIANA-EST, 121
      ESTscan, 121
    relational databases, 118, 119
    sequence extraction from sequencer chromatograms and quality assurance, 96, 97, 122
    SimiTri for similarity analysis, 119–121
    software sources, 95, 96, 121, 122

F

FASTA, gene function prediction, 27–29, 33
Fiber-FISH, *see* Fluorescence *in situ* hybridization
FISH, *see* Fluorescence *in situ* hybridization
Fluorescence *in situ* hybridization (FISH),
  applications, 379, 380
  artificial chromosome clones, 379
  chromosome preparation,

filarial parasites and
nematodes, 383, 385, 392
room conditions, 392
schistosomes and trematodes,
381–383, 392
fiber-FISH,
applications, 396
chromatin extraction
approaches,
overview, 396, 397
protocol 1, 399–401
protocol 2, 400, 401
DNA preparation, 395
materials, 397–400
resolution, 396
hybridization, 386
labeling for probes,
bacterial artificial chromosome
clones, 385, 386, 392
small fragment DNA clones,
386, 392
yeast artificial chromosome
clones, 385, 386, 392
materials, 380, 381
microscopy and computation,
388, 389
signal location description, 389,
392
washing and detection of signals,
387, 393

G

GACK, gene abundance analysis,
244, 245, 247
GeneDB, features, 62, 63
Gene duplication, *see* DNA
microarray
Gene Ontology (GO),
annotation, 38, 39
assigning terms, 39
complementary ontology efforts,
39, 40
querying, 39
rationale, 37, 38
vocabulary, 38
Genome annotation, *see* Annotation
Genome sequencing strategies,
clone-by-clone strategy,
bacterial artificial chromosome
continuous sequence
generation,
gap filling, 8, 9
shotgun library
construction, 7
shotgun sequencing and
assembly, 7, 8
bacterial artificial chromosome
walking, 9, 10
map-as-you-go approach, 3
minimal tiling path, 3
paired end-sequence marker
generation, 3–6
physical map construction, 2, 3
validation, 10
selection of strategy, 14, 15
shotgun sequencing overview, 57
whole-chromosome shotgun, 11,
12
whole-genome shotgun, 12–14
GiardiaDB, features, 60
GO, *see* Gene Ontology

H–I

HMMer, gene function prediction,
32
In situ polymerase chain reaction on
chromosomes (PRINS),
applications, 379, 380, 387, 388
materials, 380, 381
reaction, 388, 393

Index

L

Leishmania,
  amastigotes, 424
  chromosome fragmentation, *see* Chromosome fragmentation
  diseases, 299
  DNA vaccines,
    administration, 435, 436
    expressed sequence tag analysis, 424
    materials, 426, 427, 436
    murine model studies, 425, 426
    plasmid preparation,
      maxipreparation, 434, 435
      minipreparation, 433, 434
    polymerase chain reaction of complementary DNA,
      amplification, 427, 428, 436
      expression vector preparation, 431–433, 436
      positive TA colony analysis, 430, 431, 436
      product purification, 428, 436
      TOPO TA cloning, 428–430
  promastigotes, 424
  shuttle mutagenesis,
    in vitro transposition, 312–315
    mariner toolkit, 308, 310, 311
    materials, 306, 307
    Mos1 transposase,
      expression, 311–313
      purification, 312–314
    principles, 300–302, 304, 305
    rationale, 300

M

mariner, *see Mos1*
Messenger RNA (mRNA),
  reverse transcription for gene identification, 76
  spliced leader sequences, 423, 424
Microarray, *see* DNA microarray
Minisatellite variant repeat-polymerase chain reaction (MVR-PCR),
  amplification of *MS42* locus,
    crude lysates, 196, 197, 200
    entire minisatellite repeat array, 197, 200
    primers, 199
    repeat types within minisatellite, 197, 198, 201
  genomic DNA and crude lysate preparation,
    bloodstream trypanosomes, 196
    procyclic cultures, 195, 196
  interpretation, 198, 199, 201
  materials, 193, 195, 199, 200
  minisatellites amenable to analysis, 190, 191
  principles, 188–190
  Trypanosoma brucei applications, 188–191, 193
Mos1,
  donor plasmids, 302, 304
  mariner toolkit, 308, 310, 311
  transposase,
    expression, 311–313
    purification, 312–314
  transposition mechanism, 302
mRNA, *see* Messenger RNA
MVR-PCR, *see* Minisatellite variant repeat-polymerase chain reaction

N

NCBI, databases, 50
Northern blot, *Trypanosoma brucei* RNA interference analysis, 294–296
NREMBASE, features, 61, 62

O

Open reading frames (ORFs), annotation, 19, 20
ORFs, *see* Open reading frames
Orthologs,
 evolution, 128, 144
 gene function prediction, 33, 34
 identification with BLAST,
  BLAST searches, 134, 135
  materials, 131, 132
  overview, 128, 129, 131
  reciprocal best match retrieval, 135, 144, 145
  sequence retrieval and processing, 134

P

PCR, *see* Polymerase chain reaction
PFGE, *see* Pulsed-field gel electrophoresis
phrap, consensus sequence prediction for expressed sequence tag clusters, 107, 108
PlasmoDB, features, 68
Plasmodium falciparum,
 dihydrofolate reductase gene random mutagenesis,
  degenerate oligonucleotides-*Pfu* mutagenesis,
   materials, 323, 324
   primer design, 328–330
   principles, 328
   reaction conditions, 330
  error-prone polymerase chain reaction,
   amplification reactions, 325, 331
   DNA extraction from gels, 326, 331
   DNA product desalting and preparation for cloning, 325, 326, 331
   electroporation and library harvesting, 327, 328, 332
   ligation reaction, 326, 327, 331, 332
   materials, 322, 323
   principles, 321, 322
  mutator *Escherichia coli*,
   DNA library construction, 324, 325, 331
   materials, 322
   principles, 320, 321
 transfection,
  analysis,
   polymerase chain reaction, 270, 275
   pulsed-field gel electrophoresis, 272
   Southern blot, 272, 275
  chloramphenicol acetyltransferase assay for transient transfectants, 273
  DNA preparation, 268, 269
  efficiency, 263, 264, 273
  electroporation and plating, 269
  materials, 264, 265
  parasite preparation, 268
  selection, 269, 270, 274

Index 447

vectors,
  pARL-1a, 268
  pHH1, 266, 274
  pHH2, 268, 274
  pHTK, 266, 274
  Rep20, 268
Polymerase chain reaction (PCR), *see also* Amplified restriction fragment length polymorphism; RACE cloning; RAGE cloning,
  complementary DNA probe amplification, 224, 232, 233
  epitope tagging, *see Trypanosoma brucei*
  error-prone polymerase chain reaction, see Random mutagenesis,
  expressed sequence tag analysis, 82–84, 88, 89
  gene disruption, *see Trypanosoma brucei*
  minisatellites, see Minisatellite variant repeat-polymerase chain reaction,
  Plasmodium falciparum transfectant analysis, 270, 275
  PRINS, *see In situ* polymerase chain reaction on chromosomes
  random mutagenesis, *see* Random mutagenesis,
Positive selection scanning
  applications, 128
  CLUSTAL W alignment, 137, 138, 147, 148
  interpretation, 143, 144, 148
  materials, 132, 133
  nonsynonymous-to-synonymous substitution ratio, 128
  nucleotide substitution rate calculation,
    file format conversion, 139, 140
    nonsynonymous substitution calculation, 141–143
    nonsynonymous-to-synonymous substitution ratio calculation, 143
    synonymous substitution calculation, 141–143
    transition-to-transversion ratio calculation, 140, 141
  sequence data retrieval and processing, 136, 137, 146, 147
PRINS, *see In situ* polymerase chain reaction on chromosomes
Prosite, gene function prediction, 32
Protein–protein interactions, *see* Yeast two-hybrid system
Proteomics, *see* Two-dimensional gel electrophoresis
Pseudogenes, identification, 24
PSORT, annotating without sequence similarity, 35
Pulsed-field gel electrophoresis (PFGE),
  chromosome fragmentation analysis, 369, 379
  chromosome resolution,
    advantages and limitations, 335, 336
    casting and running, 341, 349
  mapping,
    applications, 336
    digestion in gel and electrophoresis of fragments, 344, 349

excision from gels, 343, 344, 349
Southern blotting and hybridization of single-locus markers, 344–346
materials, 337–339, 348
whole-chromosome shotgun cloning, 337, 346–348
Plasmodium falciparum transfectant analysis, 272

R

RACE cloning,
gene family member cloning, 166, 168
materials, 155, 156, 168, 169
nucleic acid extraction, 157, 158, 160, 169
principles, 152, 153
Trachipleistophora hominis mitochondrial heat shock protein 70 gene, 160, 162, 168, 169
RAGE cloning,
materials, 155, 156, 168, 169
nucleic acid extraction, 157, 158, 160, 169
principles, 152, 153
Trachipleistophora hominis mitochondrial heat shock protein 70 gene,
polymerase chain reaction, 165, 166
vector preparation, 163, 165
Random mutagenesis,
mutagens, 320
Plasmodium falciparum dihydrofolate reductase gene,

degenerate oligonucleotides-*Pfu* mutagenesis,
materials, 323, 324
primer design, 328–330
principles, 328
reaction conditions, 330
error-prone polymerase chain reaction,
amplification reactions, 325, 331
DNA extraction from gels, 326, 331
DNA product desalting and preparation for cloning, 325, 326, 331
electroporation and library harvesting, 327, 328, 332
ligation reaction, 326, 327, 331, 332
materials, 322, 323
principles, 321, 322
mutator XL1-red *Escherichia coli*,
DNA library construction, 324, 325, 331
materials, 322
principles, 320, 321
RNA genes, identification, 24, 25
RNAi, *see* RNA interference
RNA interference (RNAi),
principles, 287
research applications, 287, 288
Trypanosoma brucei,
cloning of procyclic cells, 292, 296
materials, 288–290, 295
Northern blot analysis, 294–296
RNA construct generation, 290, 291, 295, 296

Index

RNA dot blot analysis, 293, 294
RNA expression induction, 292, 293
RNA isolation, 293
transfection by electroporation, 291, 292, 296

S

Sequencing, *see* Genome sequencing strategies
Shuttle mutagenesis,
  Leishmania,
    in vitro transposition, 312–315
    mariner toolkit, 308, 310, 311
    materials, 306, 307
    Mos1 transposase,
      expression, 311–313
      purification, 312–314
    principles, 300–302, 304, 305
    rationale, 300
  principles, 300–302, 304, 305
SimiTr, expressed sequence tag similarity analysis, 119–121
Single nucleotide polymorphisms (SNPs),
  Toxoplasma gondii genotyping,
    allele-specific primer extension,
      amino-allyl dye coupling, 258, 261
      controls, 257
      extension reactions, 257, 258
      primers, 256, 257
    generic oligo-dT tag array,
      controls, 255
      slide preparation, 255, 260
      tags, 254, 255

    markers, 253, 254
    materials, 251, 260
    microarray,
      hybridization, 258, 259
      interpretation, 259–261
    polymerase chain reaction, 255, 256
    principles, 250–252
    sample preparation, 253
    typing techniques, 250
SL exon, *see* Spliced leader exon
SMART, gene function prediction, 33
SNPs, *see* Single nucleotide polymorphisms
Southern blot,
  chromosome fragmentation analysis, 369, 375
  chromosome mapping, 344–346
  Plasmodium falciparum transfectant analysis, 272, 275
Spliced leader (SL) exon, messenger RNA identification, 76
StackPACK, clustering expressed sequence tags into putative genes, 106
Subtracted complementary DNA library,
  advantages over DNA microarrays, 204
  differential screening, 209, 216
  materials, 204
  messenger RNA isolation, 207, 208, 215
  parasite material preparation, 206, 215
  suppressive subtractive hybridization principles, 206–209, 215, 216

two-dimensional gel
electrophoresis combination,
213–215
Suppressive subtractive
hybridization, *see*
Subtracted complementary
DNA library
SWISS-PROT, content, 50

T

TIGR Gene Index, features, 60, 61
tigrdb, features, 62
TIGRFAMs, gene function
prediction, 33
TMHMM, annotating without
sequence similarity, 35
Toxoplasma gondii,
life cycle, 249
single nucleotide polymorphism
genotyping,
allele-specific primer
extension,
amino-allyl dye coupling,
258, 261
controls, 257
extension reactions, 257,
258
primers, 256, 257
frequency of polymorphisms,
250
generic oligo-dT tag array,
controls, 255
slide preparation, 255, 260
tags, 254, 255
markers, 253, 254
materials, 251, 260
microarray,
hybridization, 258, 259
interpretation, 259–261

polymerase chain reaction,
255, 256
principles, 250–252
sample preparation, 253
types, 249
Trachipleistophora hominis,
features, 154
mitochondrial heat shock protein
70 gene,
RACE, 160, 162, 168, 169
RAGE,
polymerase chain reaction,
165, 166
vector preparation, 163, 165
Transcriptomics, *see* DNA
microarray; Subtracted
complementary DNA library
Transfection, *see* Chromosome
fragmentation; *Plasmodium
falciparum*; RNA
interference
Transposon mutagenesis, *see* Shuttle
mutagenesis
TrEMBL, content, 46
Trichomonas vaginalis,
features, 154
RACE for gene family member
cloning, 166, 168
Trypanosoma brucei,
epitope tagging,
DNA modules, 279, 285
materials, 278, 285
polymerase chain reaction of
tagging cassettes, 279–282,
285
transformant identification
with Western blot, 282, 283,
285
fiber-FISH,
applications, 396

chromatin extraction
approaches,
overview, 396, 397
protocol 1, 399–401
protocol 2, 400, 401
DNA preparation, 395
materials, 397–400
resolution, 396
gene disruption,
materials, 278, 285
polymerase chain reaction of drug-resistance genes, 281, 282, 285
transformant identification, 283, 284, 286
minisatellite variant repeat-polymerase chain reaction, *see* Minisatellite variant repeat-polymerase chain reaction
RNA interference analysis of gene function, *see* RNA interference
variant identification techniques, 188
Two-dimensional gel electrophoresis,
denaturing gel electrophoresis, 211–213, 216
interpretation, 213, 216
isoelectric focusing, 210, 211
materials, 204, 205
parasite material preparation, 206, 215
principles, 210
sample preparation, 210, 216
subtracted complementary DNA library combination, 213–215
Two-hybrid system, *see* Yeast two-hybrid system

U–W

UniProt, content, 50

WCS, *see* Whole-chromosome shotgun
Western blot, *Trypanosoma brucei* transformant identification, 282, 283, 285
WGS, *see* Whole-genome shotgun
Whole-chromosome shotgun (WCS),
genome sequencing, 11, 12
library preparation,
blunt-end fragment preparation, 346–348, 350
chromosomal DNA isolation from LMP agarose, 346, 349, 350
ligation reactions, 348
Whole-genome shotgun (WGS), genome sequencing, 12–14

X

XL1-red, *see* Random mutagenesis

Y

Yeast two-hybrid system,
auxotrophic markers, 404
host strain selection, 410
LacZ reporter assays,
filter-lift colony assay, 414, 415, 419, 420
liquid assay, 415, 420
materials, 404–400
plasmid preparation, 416, 417, 420
principles, 404

transformation of yeast, 412–414
validation of protein-protein
    interactions, 417, 418, 420
vectors,
    cloning, 412, 418, 419
    selection, 409, 410, 418